Progress in Botany 63

Springer

Berlin
Heidelberg
New York
Barcelona
Hong Kong
London
Milan
Paris
Tokyo

63 PROGRESS IN BOTANY

Genetics
Physiology
Ecology

Edited by

K. Esser, Bochum
U. Lüttge, Darmstadt
W. Beyschlag, Bielefeld
F. Hellwig, Jena

With 43 Figures

ISSN 0340-4773
ISBN 3-540-42304-4 Springer-Verlag Berlin Heidelberg New York

The Library of Congress Card Number 33-15850

Springer-Verlag Berlin Heidelberg New York
a member of BertelsmannSpringer Science+Business Media GmbH
http://www.springer.de

Cover design: Design & Production, Heidelberg
Typesetting: M. Masson-Scheurer, Neckargemünd
SPIN 10799203 31/3130 - 5 4 3 2 1 0 - Printed on acid-free paper

Contents

Review

Genetics

Physiology

List of Editors

Professor Dr. Dr. h. c. mult. K. Esser
Lehrstuhl für Allgemeine Botanik, Ruhr Universität
Postfach 10 21 48
44780 Bochum, Germany

Phone: +49-234-32-22211; Fax: +49-234-32-14211
e-mail: karl.esser@ruhr-uni-bochum.de

Professor Dr. U. Lüttge
TU Darmstadt, Institut für Botanik, FB Biologie (10)
Schnittspahnstraße 3-5
64287 Darmstadt, Germany

Phone: +49-6151-163200; Fax: +49-6151-164630
e-mail: luettge@bio.tu-darmstadt.de

Professor Dr. W. Beyschlag
Fakultät für Biologie, Lehrstuhl für Experimentelle
Ökologie und Ökosystembiologie
Universität Bielefeld, Universitätsstraße 25
33615 Bielefeld, Germany

Phone: +49-521-106-5573; Fax: +49-521-106-6038
e-mail: w.beyschlag@biologie.uni-bielefeld.de

Professor Dr. F. Hellwig
Friedrich-Schiller-Universität Jena
Biologisch-Pharmazeutische Fakultät
Institut für Spezielle Botanik
Philosophenweg 16
07743 Jena, Germany

Phone +49-3641-949250; Fax +49-3641-949252
e-mail: hellwig@otto.biologie.uni-jena.de

Curriculum vitae Rudolf Hagemann

Rudolf Hagemann was born on October 21, 1931 in Aue/Erzgebirge, Germany, and grew up in Raschau/Erzgebirge.

School attendance: Primary school at Raschau; secondary school at Schwarzenberg; certificate Abitur 1950.

Scientific career:

1950–1952	Study of biology, Universität Leipzig
1952–1955	Study of biology and genetics, Martin-Luther-Universität Halle-Wittenberg
1955	Diploma in biology
1955–1958	Research work for Ph.D. thesis in the Institut für Kulturpflanzenforschung Gatersleben der Deutschen Akademie der Wissenschaften zu Berlin (=DAW), supervised by Hans Stubbe
1958	Doctor's degree (Dr.rer.nat.) of the Martin-Luther-Universität

1958–1967	Postdoctoral research work in the Institut für Kulturpflanzenforschung Gatersleben der DAW, Abteilung Genetik und Cytologie (Direktor: Hans Stubbe)
1966	Habilitation für Genetik, Martin-Luther-Universität
1967	Professor of Genetics and Director of the newly founded Institut für Genetik, Naturwissenschaftliche Fakultät, Martin-Luther-Universität Halle-Wittenberg
1969–1994	Full Professor of Genetics, Martin-Luther-Universität in Halle
1981–1984	Director of the Section of Biological Sciences of the Martin-Luther-Universität
1990, 1995, 1997	Guest Professorship for Molecular Genetics at the Universität Salzburg, Austria, during the respective summer semesters
1994–1996	Professor, Max-Planck-Gesellschaft (MPI für Züchtungsforschung Köln; MPI für Experimentelle Medizin Göttingen)

Research activities: Extranuclear inheritance in higher plants, especially genetics and molecular biology of plastids. Genetic instabiliy: paramutation in tomato. History of genetics.

Scientific publications: 201 publications in scientific journals and symposium proceedings.

Books:
Plasmatische Vererbung (Ziemsen Verlag, Lutherstadt Wittenberg) 1958
Russian translation: Plasmatitscheskaja Nasledstwennost, Moskwa 1962
Plasmatische Vererbung (Gustav Fischer Verlag, Jena) 1964
Allgemeine Genetik (unter Mitarbeit von T. Börner, R. Piechocki, F. Siegemund), 1st edition 1984, 2nd edition 1986, 3rd edition 1991 (Gustav Fischer Verlag, Jena und Stuttgart), 4th edition (Spektrum Verlag, Heidelberg)

Editions:
Beiträge zur Genetik und Abstammungslehre (with H. Böhme and R. Löther), Volk und Wissen Verlag, Berlin, 1st edition 1976, 2nd edition 1978
Gentechnologische Arbeitsmethoden - Ein Handbuch experimenteller Techniken und Verfahren. Akademie Verlag, Berlin, und Gustav Fischer Verlag, Stuttgart, 1990
Ergebnisse und Trends der Gentechnologie. Akademie Verlag, Berlin 1991

Member of the Editorial Board of the Journals:

Molecular and General Genetics (MGG, since 1988; now Molecular Genetics and Genomics)

Theoretical and Applied Genetics (TAG, since 1970)

Biologische Rundschau (1976–1990)

Biologisches Zentralblatt (1975–1994)

Biochemie und Physiologie der Pflanzen (1971–1987)

Honors:

1969	Election as Member of the "Deutsche Akademie der Naturforscher Leopoldina"; member of its senate between 1974 and 1993
1973, 1978, 1988	Research Awards of the Martin-Luther-Universität Halle-Wittenberg
1980	N.I. Vavilov-Medal of the All-Union Society of Geneticists and Breeders of USSR
1987	Appointment as a foreign member of the Fachbeirat des Max-Planck-Institutes für Züchtungsforschung Köln
1989	Thomasius-Medal of the Martin-Luther-Universität (for the promotion of rising generations of scientists)
1990	Election as Corresponding Member of the "Akademie der Wissenschaften der DDR" (an election several years before had been prevented for political reasons by the ruling party - SED - of GDR)

Milestones in Plastid Genetics of Higher Plants

By Rudolf Hagemann

1 The Discovery of Non-Mendelian Inheritance

Many biological disciplines have developed gradually from speculations and accidental findings, over systematic observations and investigations to well-aimed experiments. In retrospect, it often seems impossible to define exactly the historical starting point of a scientific discipline. Consequently, there are different opinions as to the date when a new field was opened up.

Such a problem does not exist for the origin of extranuclear genetics and plastid genetics. Its date of birth can be defined exactly: The publication of the third issue of the world-wide first genetics journal, *Zeitschrift für induktive Abstammungs- und Vererbungslehre*, in spring 1909.

In this issue the German geneticists and botanists, Carl Correns and Erwin Baur, published back-to-back two articles on non-Mendelian inheritance of plant variegations. Correns (1909a) wrote in the footnote of his paper: "The simultaneous publication of the paper with the following article of E. Baur is based on mutual agreement; however, each of us had no knowledge of the contents of the other ones paper."

These articles are:

Correns, Carl: Vererbungsversuche mit blass(gelb)grünen und buntblättrigen Sippen bei *Mirabilis jalapa, Urtica pilulifera* und *Lunaria annua*. Z Indukt Abstamm- Vererbungsl 1:291–329, 1909 (Inheritance experiments with pale(yellow)green and variegated varieties of *Mirabilis jalapa, Urtica pilulifera* and *Lunaria annua.)*

Baur, Erwin: Das Wesen und die Erblichkeitsverhältnisse der "Varietates albomarginatae hort." von *Pelargonium zonale*. Z Indukt Abstamm – Vererbungsl 1:330–351, 1909 (The nature and the inheritance properties of the "Varietates albomarginatae hort." of *Pelargonium zonale*.)

Correns and Baur reported for different plant species the non-Mendelian (nowadays: extranuclear) mode of inheritance of green-white or green-yellow leaf variegations.

Progress in Botany, Vol. 63

After crosses between variegated, yellow and green plants (or branches) of *Mirabilis, Urtica* and *Lunaria*, Correns observed a purely maternal inheritance of the trait green versus yellow: green branches always gave rise to green seedlings, yellow branches only yielded yellow offspring, while the variegated branches produced green, yellow and green-yellow variegated seedlings in widely varying ratios. The pollen parent had no influence on the character of the progeny.

Baur found another type of non-Mendelian inheritance in his crossing experiments with *Pelargonium zonale*. He crossed green and periclinal chimeric white-margined plants (or white shoots from otherwise white-margined plants). Reciprocal crosses revealed a biparental, though non-Mendelian inheritance of the leaf color trait green versus white. The F1 progeny consisted of green, green-white variegated and white seedlings. In many crosses, there was a bias in the F1 phenotypes towards that of the maternal parent.

Thus, in 1909, Baur and Correns simultaneously discovered and described the occurrence of non-Mendelian inheritance in higher plants. Their conclusion that in addition to the Mendelian inheritance of genes in the cell nucleus, there are other hereditary factors outside the nucleus (i. e. in the protoplasm) which exhibit a non-Mendelian mode of inheritance, marks a milestone in genetics and the date of birth of a new field of research.

2 The Foundation of the Theory of Plastid Inheritance by Erwin Baur

Correns and Baur had a general consensus of opinion about the existence of non-Mendelian inheritance of variegations in higher plants. However, they did not agree on the localization of these non-Mendelian factors within the cell.

Baur clearly expressed the view (already in 1909) that the **plastids** themselves are the **carriers of the hereditary factors** which determine that the plastids are green (= normal, non-mutated) or white or yellow, respectively (= mutated, incapable of becoming green). With his classic paper on *Pelargonium zonale* in 1909, he laid the foundation for the theory of plastid inheritance and thus the basis of the genetic discipline "plastid genetics".

Simultaneously with the *Pelargonium* studies, Baur performed experiments with green-white variegated plants of *Antirrhinum majus*. In this species, Baur (1910a,b) found after reciprocal crosses a purely maternal inheritance of the trait green versus white (the same mode of inheritance as had been described by Correns 1909a, for *Mirabilis, Urtica* and *Lunaria*).

Winge (1919) came up with a straightforward explanation for the difference between the modes of inheritance between *Pelargonium* (biparental inheritance) and *Antirrhinum, Mirabilis, Urtica* and *Lunaria* (uniparental maternal inheritance): The only difference is that, in the case of maternal inheritance, the plastids are transmitted by the egg cells only, whereas in the case of biparental inheritance (*Pelargonium*), the plastids are transmitted to the next generation by both the egg cells and the sperm cells of the pollen.

Baur (1919) fully accepted and supported this hypothesis, and included this explanation in his widely distributed textbook of genetics and its later editions; he upheld this view in all his later publications on this issue.

Later on, Otto Renner supported this view in many papers since 1922.

Even Thomas Hunt Morgan, who was very skeptical regarding many reports on the phenomena of "cytoplasmic inheritance", expressed in his book, *The Physical Basis of Heredity* (1919, 1921), his consent with Baur's theory that plastids are carriers of hereditary factors.

By contrast, Carl Correns developed a fundamentally different hypothesis. He expressed the opinion, that the difference between 'green' and 'white' (or 'yellow') resides in the **cytoplasm.** His line of argumentation is based on the terms **"healthy"** and **"ill" ("diseased").** The cytoplasm is – according to Correns (1909a) – either healthy or diseased; when the indifferent plastids come to lie in the healthy cytoplasm, they develop into green chloroplasts, but if they come to lie in diseased cytoplasm, they are (or become) white or yellow.

In his later papers on non-Mendelian inheritance, Correns (1922, 1928) added a new idea to his hypothesis: He assumed that, in plants that will become green-white variegated later, the cytoplasm of the embryonic cells (i.e. early meristematic cells) is in a **"labile cytoplasmic state".** During early development of the seedlings, this "labile state" *switches* either to a normal, permanently "healthy state" (allowing the formation of green chloroplasts) or to a permanently "diseased state" (causing white or yellow plastids and cells).

Correns was well aware of Baur's contrasting argumentation

1. that the plastids themselves are the carriers of the genetic differences between 'green' and 'white', and
2. that, in *Pelargonium*, pollen and egg cells transmit plastids to the next generation, while in *Antirrhinum* and *Mirabilis*, only the egg cells transmit plastids.

However, Correns was reluctant to accept Baur's view.

One of Correns' main arguments against Baur's theory was the presumed lack of so-called "mixed cells" within variegated leaves.

3 The Controversy About "Mixed Cells" and the Proof of Their Existence

According to Baur's theory, the hereditary constitution of the plastids themselves determines whether they are green (normal) or white (mutated), and during ontogenetic development a random sorting-out of plastids is taking place.

If somatic segregation of green and white plastids takes place at random, one has to expect not only cells with green plastids and cells with white plastids, but also cells which contain both types of plastids. Unless the two plastid types are mutually exclusive, "mixed cells" (Mischzellen) should be found containing both green chloroplasts and white plastids side by side within one and the same cell.

Correns repeatedly stated that such "mixed cells" have not been found in a sufficient quantity or not at all. However, astonishingly, Correns cited three authors who had in fact reported the finding of "mixed cells"; amongst them was his coworker Funaoka (1924) who observed "mixed cells" in variegated leaves of *Stellaria media* "relatively frequently" (Correns and F. von Wettstein 1937, pp.17, 22; cf. also Dahlgren 1925).

The controversy about the presence or absence of "mixed cells" in variegated leaves lasted for many years. For me personally, this controversy marked the first step into the field of plastid inheritance.

In 1958, I had received my Ph.D. on the basis of a genetic study on green-yellow variegated tomato plants; this variegation is the result of an interesting genetic phenomenon which –in accordance with Otto Renner – I at first termed 'somatic conversion', but later on – in agreement with the maize geneticists Alexander Brink and Edward Coe – "paramutation". At the start of my postdoctoral work, my academic teacher Hans Stubbe, Gatersleben, asked me whether I would be interested in studying his green-white variegated plants of *Antirrhinum majus*: "So far we have nobody in the Gatersleben Institute who is working on problems of "plasmatic inheritance" (extranuclear inheritance). I would be glad if you became interested in this field". I accepted his proposal, and thus, I have been working for many decades – in parallel – on paramutation in *Lycopersicon esculentum* (review: Hagemann 1993b) and on extranuclear inheritance, mainly plastid inheritance, in higher plants.

(It is certainly purely by chance, but it has always been intriguing to me that one of my scientific examples, Otto Renner, was throughout his scientific career actively working on just these two genetic phenomena in the model plant *Oenothera*: paramutation ('somatic conversion') and plastid inheritance.)

Inspired by the publications of Maly and Wild (1956) and Wild (1958) on the presence of mixed cells in variegated plants of *Antirrhinum majus,* I began cytological investigations in green-white variegated *An-*

Fig. 1. A green-white variegated plant of *Antirrhinum majus* L. The white areas and shoots contain the white plastome mutant *en:alba-1*. (Hagemann 1964)

tirrhinum plants (Fig. 1) of Stubbe's "Gatersleben line" (which contained the plastome mutant *en:alba-1*; cf. Sect. 8.a).

I was able to prove the regular presence of mixed cells in cotyledons, foliage leaves, bracts and sepals in leaf areas which were finely variegated ("checkered variegation pattern"). In mixed cells green and white plastids were found to be present in all possible ratios (Hagemann 1960, 1961).

Mixed cells were initially found by light microscopy (Fig. 2), and later also with the electron microscope (Döbel and Hagemann 1963).

During the following years I studied variegated plants from other species, and demonstrated the presence of mixed cells in mutants of *Lycopersicon pimpinellifolium* and *Pelargonium zonale*. With these studies - and by evaluating publications on other species - I could characterize complications which can prevent the identification of mixed cells:

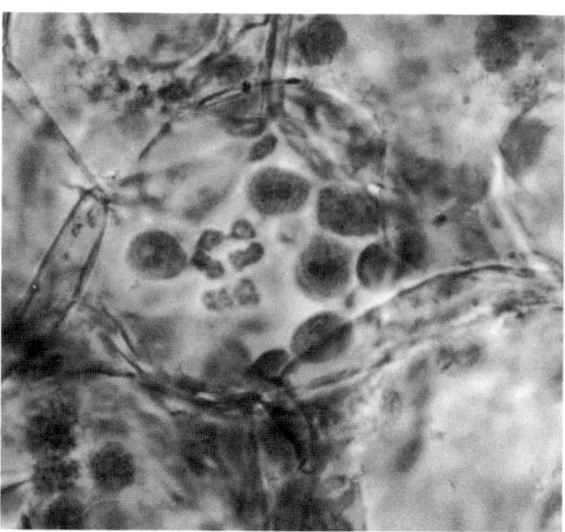

Fig. 2. A mixed cell in a foliage leaf of a variegated *Antirrhinum* plant. The green chloroplasts are larger, have a regular grana structure and contain starch grains. The mutant plastids (in the middle) are smaller and are white (appear darker in this picture due to the phase contrast used). (Hagemann 1964)

- In the variegated snapdragon line "Gatersleben" the deficient white plastids regularly degenerate, vacuolate and finally decay. Therefore the number of visible mixed cells is reduced with increasing age of the leaves. Consequently in those mixed cells, which can still be found in old leaves, the green plastids prevail (Hagemann 1961).
- In green-white and green-yellow variegated plants of *Pelargonium zonale* (produced by crosses of green plants with the varieties "Mrs. Parker" and "Mrs. Pollock"), I could not find any mixed cells at first glance. However, later on I regularly found mixed cells in variegated plants growing in late winter/early spring in the greenhouse under dim light. It turned out that in summer the mutant plastids had been vacuolated and destroyed by the bright sunlight; whereas under dim light they survived longer. Such a situation might have been the reason for Correns being unable to find mixed cells (in sufficient quantity).
- Wild (1959) has found that in a particular line of *Antirrhinum majus* (line F) the mutant plastids are yellow in young leaves, but in older leaves they become more and more green. Therefore mixed cells could be demonstrated in younger leaves only.
- For decades, numerous cases of variegated or striped plants have been known, in which the color intensity at the boundary between the green and the white leaf areas changes gradually. Such cases have been described in *Epilobium, Pelargonium* and *Zea*. This is obviously due to the action of metabolites that are transported from cell to cell.

> Even if different cells with different plastids influence each other, it is quite clear that one can expect genetically different plastids within one and the same cell to act upon each other. In cases like these, it may be impossible to demonstrate the presence of mixed cells (references in Hagemann 1964, 1965).

Today, as the result of many investigations conducted by numerous research workers in many species, we can clearly state that the presence of mixed cells harboring genetically different plastids is an established, proven fact. This result is in full agreement with Baur's theory of plastid inheritance.

4 Hybrid Plastid Deficiency in *Oenothera* and Other Genera

Baur's crosses with *Pelargonium* and *Antirrhinum* as well as Correns' experiments with variegated plants of several genera were based on so-called **loss mutations** ("Defekt-Mutationen"), plastid mutations causing *inability* of the plastids to become green in any nuclear background that had been tested.

Starting in 1922, Otto Renner published numerous papers on plastid inheritance (in which he fully accepted and supported Baur's theory). He performed crosses between different species of the genus *Oenothera* (subgenus *Euoenothera*). These analyses led him to the discovery of a new phenomenon in plastid genetics: **hybrid (plastid) deficiency ("Bastardbleichheit")**. Different species of the genus *Oenothera* differ not only in their genotypes, but also in the genetic constitution of their plastids (the plastome). Evolution did not only involve mutations in the genome complexes, but also a genetic diversification of the plastomes. The differences in the plastomes manifest in a way that specific plastid types are unable to become green when interacting with certain genome complexes; however, when interacting with their innate genome complexes, these plastids develop into normal green chloroplasts.

Here, the occurrence of a 'chlorophyll deficiency' is clearly not due to a plastid "loss mutation". Instead, during evolution of the genus *Oenothera,* **"differentiation mutations"** ("Differenzierungs-Mutationen") occurred that led – paralleled by nuclear mutations during species evolution – to genetic differences in the interactions between the nucleus and the plastids.

Plastids which have been interacting with an unsuitable genotype and have therefore been chlorophyll-deficient for many generations, at once become fully green when united again with their innate genotype (Renner 1924, 1929, 1934, 1936).

Plants containing green plastids (e.g. from the father) and deficient plastids (e.g. from the mother) are called **hybrid-variegated.**

The first examples of hybrid-variegation have been worked out by Renner. One case should be described: After reciprocal crosses between *Oenothera lamarckiana* (velans.gaudens) and *Oe. hookeri* (hookeri.hookeri), distinct reciprocal differences occur: The cross *Oe. hookeri* × *Oe. lamarckiana* leads to velans.hookeri hybrids, the great majority of which are entirely green (a few plants have some yellow speckles on the first leaves.). In contrast the cross *Oe. lamarckiana* × *Oe. hookeri* yields pale yellow seedlings, most of which die; but some of the seedlings have green speckles and therefore can survive. From these seedlings, some develop into yellow-green variegated plants. Renner's explanation is as follows: The hookeri-plastids are able to interact correctly with the hybrid genome complexes velans.hookeri and develop into green chloroplasts. However, the lamarckiana-plastids are not able to cooperate with the hybrid genome velans.hookeri; they become "hybrid deficient". In *Oenothera* the egg cells contribute the majority of plastids to the zygote; the sperm cells contribute much less. The small speckles on the F1 plants represent the paternal plastids. The green-yellow variegated plants (from the cross *Oe. lamarckiana* × *Oe. hookeri*) contain *Oe. hookeri* plastids in their green sectors, which can interact well with the hybrid genotype velans.hookeri, whereas the pale yellow branches contain hybrid-deficient lamarckiana-plastids.

Subsequent backcrosses with *Oe. hookeri* and with *Oe. lamarckiana* as well as test crosses with other species, e.g. *Oe. syrticola*, demonstrated the correctness of Renner's interpretation. Many more impressive examples of this sort have been described by Renner (1924, 1929, 1936; cf. Hagemann 1964).

During the following years, the genetic and physiological details of the interactions between different nuclear genomes and divergent wild-type plastids within the subgenus *Euoenothera* were worked out by Renner's former Ph.D. students and coworkers Wilfried Stubbe (1959, 1960, 1964) and Franz Schötz (1958).

In 14 wild species of the subgenus *Euoenothera*, Stubbe (1959, 1960, 1964) demonstrated the existence of 5 different types of normal plastids. These plastome wild types give full green pigmentation with the innate genotype of their own species. However, they react differently when combined with different genome complexes from the 14 species, resulting in green pigmentation or various types of chlorophyll deficiencies, e.g. light green or yellow or white pigmentation.

Herrmann et al. (1980) and Gordon et al. (1982) have characterized all five plastome wild types by specific restriction patterns produced by five individual restriction enzymes and their combinations. It is now easily possible to identify these five plastomes. Recently, even the complete nucleotide sequence of the *Oenothera elata* plastid DNA representing plastome I (of the five distinguishable *Euoenothera* plastomes) has been determined (Hupfer et al. 2000).

The genome complexes of the subgenus *Euoenothera* have been classified into three main genome groups (A, B, C) according to their interactions with the five plastome wild-types. The complex interaction patterns between different plastome wild-types and genome complexes have been summarized by Stubbe (1959, 1960) in a combination rectangle (Fig. 3).

The phenomenon of hybrid plastid deficiency (="hybrid bleaching") is even more complicated, because in a particular genome-plastome

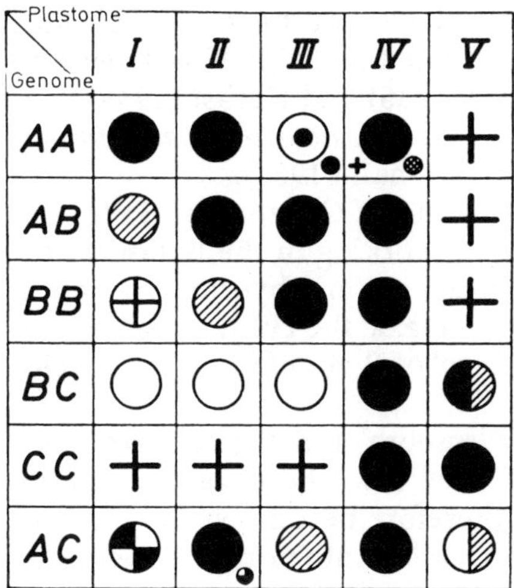

● normal green
green to grayish green
yellow green (lutescent)
periodically lutescent
yellow green to yellow
○ white or yellow
⊕ white and with inhibition of growth and germination
+ lethal; but white if occuring as an exception
slightly yellowing
periodically pale (diversivirescent
⊙ periodically pale (virescent)

Fig. 3. Combination rectangle for the relations between different genomes (*A*, *B*, *C*) and plastomes (*I–V*) in *Euoenothera*. (The use of more than one sign in some squares depends on slight differences between the A complexes). (Stubbe 1964)

combination, the extent of disharmony between the nucleus and the plastids and thus the intensity of bleaching changes during ontogenetic development of the *Oenothera* plants. Schötz (1958) has characterized a succession of stages with weak or strong disturbances during ontogeny of plants and their leaves, and has been able to characterize six different types of disturbed plastid-nucleus interactions.

Similar results, i.e. hybrid deficiency and hybrid variegation, have also been found in the subgenus *Munzia* (former: *Raimannia*) of *Oenothera* by Schwemmle et al. (1938) and Schwemmle (1943).

Without any doubt, the most thorough and intense investigations on hybrid plastid deficiency have been performed in the genus *Oenothera*. However, these phenomena are not confined to this genus. They have been described in three other genera of angiosperms: *Geranium*, *Hypericum* and *Pelargonium* (and without doubt they may exist in many more genera).

Dahlgren (1925) reported the occurrence of green-white variegated F1 plants after reciprocal crosses between *Geranium bohemicum* and *G. deprehensum*. He explained this variegation, in full agreement with Renner, as an example of hybrid variegation: the *bohemicum* plastids are normally able to become green in cooperation with the hybrid nucleus, whereas the *deprehensum* plastids cannot interact sufficiently with it and therefore become white.

After crosses between *Hypericum acutum* and *H. montanum* K.L. Noack (1931, 1934) and Herbst (1935) observed hybrid variegation in the F1 generation. In cooperation with the hybrid nucleus, the *montanum* plastids develop into normal green chloroplasts, but the *acutum* plastids show hybrid bleaching. It should be mentioned that K.L. Noack gave an alternative explanation, but Renner (1934, 1936) and Herbst (1935) convincingly showed that the variegation of the *H. acutum* × *montanum* hybrids is due to hybrid variegation (cf. Hagemann 1964).

Our plastid research group at the Institute of Genetics in Halle – in cooperation with F. Pohlheim, at that time working in the Botanical Institute Potsdam – has analyzed hybrid bleaching and hybrid variegation after crosses between different *Pelargonium* taxa: the cultivar 'roseum' of *Pelargonium zonale*, the cultivar 'Stadt Bern' of *P. zonale*, and the species *P. inquinans*.

The F1 hybrids between the cultivar 'roseum' and the cultivar 'Stadt Bern' are variegated, because the plastids of 'Stadt Bern' cannot readily interact with the hybrid genotype and therefore become yellow (hybrid bleached), whereas the 'roseum' plastids develop into normal green chloroplasts in the presence of the same genotype. Both plastome types have been characterized by their specific plastid DNA restriction patterns (Metzlaff et al. 1982).

After crossing *P. zonale* with *P. inquinans*, the plastids of *P. inquinans* become hybrid bleached in the F1 plants; the *P. zonale* plastids are green (Pohlheim 1986).

Research workers interested in this phenomenon, discuss several possibilities for the molecular basis of "hybrid plastid deficiency": (1) The synthesis of specific thylakoid proteins is partially (or entirely) blocked. (2) Different (thylakoid) proteins, partly synthesized in the cytoplasm and partly in the plastids, do not match correctly during the formation of a complex membrane structure. (3) Both mechanisms may lead to a premature and/or partial degradation of these proteins. However, so far no direct experimental approach has been found to prove or disprove these ideas.

5 The Plastids in Male Gametophytes of Angiosperms (Electron Microscopy)

a) Different Modes of Plastid Inheritance

In 1909/1910, Correns (1909a,b) and Baur (1909, 1910a,b) reported the existence of two modes of plastid inheritance in angiosperms (and numerous successors – above all Otto Renner – confirmed this fact):

(a) *Uniparentally maternal plastid inheritance*: In the majority of angiosperms, there is a uniparental, purely maternal, inheritance of plastids, for example in *Mirabilis, Antirrhinum, Beta, Hordeum, Zea, Lycopersicon* and many others (Hagemann 1964, 1992; Hagemann and Schroeder 1989).

(b) *Biparental plastid inheritance*: In a minority of species, a clear biparental plastid inheritance is found. The best-studied genera of this type are *Pelargonium, Oenothera, Hypericum,* and *Medicago* (Hagemann 1992).

Within this group, one has to distinguish three subtypes. (1) In the genera *Oenothera* and *Hypericum,* a distinct bias of the maternal plastids is observed. The cross green x white yields many green, a number of variegated seedlings, and (almost) no white seedlings, whereas the cross white × green yields variegated, many white and (almost) no green seedlings. (2) In *Pelargonium,* there is often a rather equal contribution of plastids from the mother and the father, although the results of several reciprocal crosses vary widely in that respect. (3) Reciprocal crosses in *Medicago sativa* proved a biparental plastid inheritance with a predominantly paternal transmission, i.e. a distinct bias towards the paternal plastids. This very seldom case in angiosperms is comparable with the situation in the gymnosperm species *Cryptomeria japonica,* where a very strong bias in favor of the paternal plastids has been observed in reciprocal crosses (Ohba et al. 1971; Masoud et al. 1990).

As already mentioned in Section 2, Winge (1919) expressed the idea that the difference between these two modes of plastid inheritance is solely based on the fact that in most species, only the egg cells transmit plastids to the next generation (leading to uniparentally maternal plastid inheritance), whereas in some species (e.g. *Pelargonium*) both the egg cells and the sperm cells of the pollen transmit plastids into the zygote leading to biparental plastid inheritance.

Baur (1909) expressed the view in his classical paper on *Pelargonium zonale* that the best and most straightforward explanation of his genetic results with *Pelargonium* is the assumption that paternal plastids are also transmitted into the zygote:

"But should it – in contrast to the hitherto ruling doctrine – turn out that the male sperm cells are also able to transmit plastids into the egg cells, then the hereditary processes of white-margined plants (of *Pelargonium*) would be fully understandable. *It is definitely necessary* that the *developmental processes regarding the plastids of higher plants* have to be carefully studied *with newer methods* in a continuous series from the sexual cell (through the developing organism) to the sexual cell of the next generation." (original German text in Baur 1909, p. 350; and Hagemann 2000, p. 103).

The light microscope methods, available at that time and during the following three decades, were insufficient to fulfill this demand. The demonstration of chlorophyll-fluorescing plastids in the generative cells of *Lupinus luteus* by Ruhland and Wetzel (1924) and the light microscope observations of Wylie (1941) pointed into the right direction (cf. Figs. 32 and 33 in Hagemann 1964). However, only later could clear observations be made by electron microscopy.

The investigations during the following decades led to the result that the differences between the divergent modes of plastid inheritance are obviously due to differences in the mode of distribution and transmission of plastids during gamete formation or fertilization. In angiosperms (and gymnosperms), there is no hint of the existence of restriction-modification processes acting on plastid DNA in zygotes or embryos (as have been found in *Chlamydomonas*).Therefore we came to the conclusion that cytological mechanisms acting during the development of male and female gametes are the basis for the different modes of plastid inheritance in angiosperms.

The ultimate proof was expected to be provided by a *combination of genetic studies and electron microscope observations using the same plant species*. Our plastid research group at the Institute of Genetics decided to pursue this problem which was not done by many other researchers in the field.

When our research group began to deal with this problem in the seventies, a purely paternal plastid transmission was not yet known. The first reports on exclusively paternal plastid transmission in gymnosperms were published in 1986, and the *first report on* **purely paternal plastid inheritance in an angiosperm** – *in the kiwi plant Actinidia deliciosa* – only appeared **in 1995** (Cypriani et al. 1995).

Thus we focused on the distribution of plastids during microsporogenesis, pollen development and fertilization in angiosperms with electron microscopy.

Our own investigations and the results of several other laboratories allowed the characterization of **four groups of angiosperm species**, in which different cytological and physiological mechanisms are acting in plastid inheritance (Hagemann and Schröder 1989; Hagemann 1992).

We defined the following four plant types: *Lycopersicon* type, *Solanum* type, *Triticum* type and *Pelargonium* type. The first three types all lead to uniparentally maternal plastid transmission, although the underlying mechanisms are rather different.

Lycopersicon type: During the first pollen mitosis, the generative cell does not receive any plastids. There is an extremely unequal distribution of plastids into the vegetative cell only. The generative cell does not contain plastids ab initio. (This means that, during microsporogenesis, cells without plastids are regularly formed in this large group of angiospermous species.) Thus, the sperm cells are free of plastids and cannot transfer plastids into the egg cell. A long list of species reflecting this type is given in Hagemann and Schröder (1989, p. 59).

Solanum type: In species of the *Solanum* type, the vegetative and the generative cells receive plastids (the vegetative cell receives many more than the generative cell). However, the plastids in the generative cells disappear during maturation of these cells; in consequence, the sperm cells do not contain plastids.

Triticum type: Both the generative cell and the sperm cell contain plastids. However, during the process of fertilization the plastids are *not* transmitted into the egg cell. Obviously the plastids are stripped off the sperm nucleus during fertilization. The same seems to be true for the mitochondria. Consequently, in species of this type, e.g. *Triticum* and *Triticale*, there is a uniparentally maternal plastid (and mitochondrial) inheritance.

Pelargonium type: In contrast to the previous types, in the genera and species of the *Pelargonium* type, the generative and the sperm cells (fertilizing the egg cell) do contain plastids which are regularly transmitted into the zygote and thus into the next generation. This results in biparental or uniparentally paternal inheritance of plastids. This mode of plastid transmission has been characterized in intense studies in the genera *Pelargonium, Oenothera, Hypericum*, and in the species *Plumbago zeylanica, Medicago sativa* and *Actinidia deliciosa*.

b) Cytological Mechanisms Underlying Plastid Distribution and Transmission

Five cytological mechanisms are acting within different angiospermous species, which determine the mode of plastid distribution and transmission during microsporogenesis and fertilization. They are the basis of the four types characterized above:

1. Equal plastid distribution during male sperm development (no plastid degeneration and no exclusion)
   During pollen development in plants of the genera *Pelargonium, Oenothera, Hypericum, Medicago* and *Actinidia*, there is an equal distribution of plastids during the first and the second pollen mitosis. This results in the presence of plastids in the sperm cells which are transmitted into the egg cell with great regularity (Fig. 4).
2. Plastid degeneration
   In species of the *Solanum* type, a specific degeneration of plastids takes place in the generative cells, but not in the vegetative cells. These degeneration processes have been described by Clauhs and Grun (1977) and Schroeder (1986).
3. Plastid exclusion during the first pollen mitosis
   In species of the *Lycopersicon* type (i.e. in the majority of angiosperm species) generative cells without plastids are formed. In *Chlorophytum* and *Gasteria*, it was demonstrated that during the prophase of the first pollen mitosis, the plastids become polarized and are clustered in the center or at the proximal pole of the microspore, whereas the dividing nucleus is located at the distal pole. Therefore all plastids of the microspore are exclusively transmitted into the vegetative cell (cf. Figs. 1 and 3 in Hagemann 1992).
   Extensive electron microscope investigations on this subject have been performed in the plastid research group in Halle between 1983

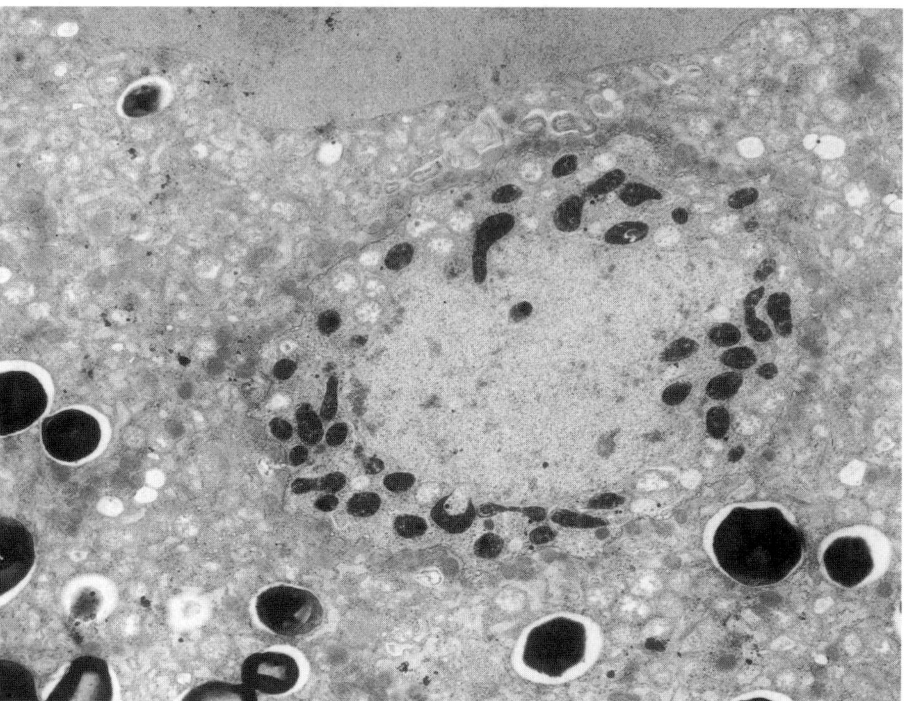

Fig. 4. Pollen grain of *Pelargonium zonale*. The generative cell (in the middle) contains many plastids. In the surrounding cytoplasm of the vegetative cell there are many plastids with big starch grains. (Henrike Stein, Institute of Genetics, Halle)

and 1989 by my former coworker M.-B. Schroeder, who is now professor of botany at the Geisenheim Research Center (all references in Hagemann and Schroeder 1989).

4. Plastid exclusion during sperm cell formation or development:
   In *Plumbago zeylanica*, Russel (reviews 1987, 1992) has described the formation of a striking sperm cell heteromorphism. In the course of second pollen mitosis, two dimorphic sperm cells are formed which differ in size, morphology and organelle content (Fig. 5). The larger sperm cell contains many mitochondria and no (or hardly any) plastids; this mitochondria-rich sperm cell fuses with the central cell. The smaller sperm cell contains numerous (up to 46) plastids and relatively few mitochondria; this sperm cell with many plastids fuses (in more than 94% of the fertilization events) with the egg cell. Thus, the male plastids are excluded from transmission into the central cell, but are transmitted into the egg cell.
   It is easy to imagine that in other species the opposite situation could also occur: the exclusion of plastids from the sperm cell which fuses

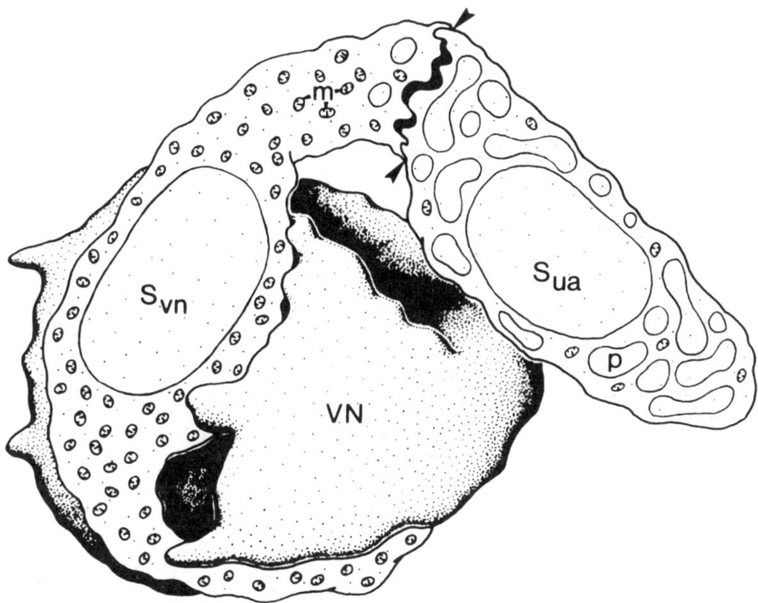

Fig. 5. Reconstruction of the two sperm cells (S_{vn}, S_{ua}) of *Plumbago zeylanica* and the associated vegetative nucleus (*VN*) with superimposed profiles of mitochondria and plastids. The S_{vn} contains a majority of the mitochondria and two plastids near the sperm cross-wall (←). The S_{ua} contains most (and usually all) of the plastids and significantly fewer mitochondria. (Russell 1984)

with the egg cell. (However, such a case has not been described so far.)

5. Plastid exclusion during fertilization:
   In several cereals, especially in wheat and triticale, both generative and sperm cells regularly contain plastids. Nevertheless, genetic differences in plastids (white mutant plastids) and mitochondria (cytoplasmic male sterility) are inherited in a uniparentally maternal mode (Hagemann and Schroeder 1985). This situation can only be explained by assuming that the plastids and the mitochondria are stripped off the sperm nucleus during the process of fertilization; therefore they are not transmitted into the zygote.
   Mogensen (1988) observed remnants of the sperm cell cytoplasm containing plastids and mitochondria within the degenerated synergid in barley. This seems to be a mechanism for preventing plastids to be transmitted into the egg cell.
   Van Went and Willemse (1984) have given a survey about the fertilization processes in angiosperms; they have clearly outlined different possibilities, as observed in different species, regarding the transmission of the nucleus, the plastids and the mitochondria of the sperm cells into the egg cell and the central cell.

Corriveau and Coleman (1988) reported a rapid screening method – using the fluorescent dye DAPI – for the detection of plastid DNA in generative and sperm cells of the pollen. They could find positive results for many (43) species, and they concluded that these species have the potential of biparental plastid inheritance. *Other* authors oversimplified this observation and stated: 'presence of plastids in generative and sperm cells means biparental plastid inheritance'.

This conclusion is clearly wrong. As explained above in point (5) a number of plants, e.g. those of the *Triticum* type, contain plastids in the sperm cells; nevertheless they do not transmit these plastids into the egg cell. Only plants of the *Pelargonium* type transmit their male plastids into the next generation. Any conclusion about the transmission of male plastids into the zygote can only be drawn on the basis of reciprocal crosses, RFLP analyses or detailed electron microscopic studies of the zygotes and early embryos of hybrid plants.

With these investigations, paralleled by physiological, biochemical and molecular biological studies on plastid genetics, we were able to elaborate – on the basis of electron microscope studies combined with the analysis of reciprocal crosses – a clear picture about the different modes of transmission of plastids by sperm cells.

Sometimes we are a little bit sad to read in some *recently published German and English/American textbooks of genetics and botany* the simple and general statement that the plastids (in higher plants) are transmitted by the mother only, as if the genetic studies by Baur, Renner and their followers since 1909 and the cytological investigations in several laboratories in different countries did not exist.

6 Plastome Mutations

Plastome mutations have been found and used in genetic experiments as well as in cytological and in physiological/biochemical investigations for many decades. There are four sources of plastome mutations for the experimenter: spontaneous mutations; mutations experimentally induced by mutagens, nuclear-gene-induced plastome mutations and mutations introduced by genetic transformation of the plastid genome.

On the whole, three phenotypes have been found which are caused by the effects of plastid mutations:

1. deficiencies in the light or dark reactions of photosynthesis, mostly connected with changes in the leaf color (light green, yellow, creme or white leaves),
2. herbicide resistance, and
3. antibiotic resistance (mostly in algae).

The great majority of these mutations are spontaneous mutations.

a) Spontaneous Mutations

Most plastome mutants belong to group (1); the molecular basis of some of them will be dealt with in detail in Section 8.

Here, the group of **herbicide-resistant mutants** will be discussed briefly. The extensive use of herbicides of the s-triazine type, including the herbicide atrazine, in many countries with well-developed agriculture in North America, Europe and Israel led to a long lasting and very strong selection pressure in favor of spontaneous herbicide resistant weed plants, e.g. of the species *Amaranthus hybridus, Brassica campestris, Chenopodium album, Poa annua* and *Solanum nigrum*. It was soon found out that the target of the herbicide is the chloroplast protein D1 of the photosystem II complex, encoded by the plastid gene psbA. Spontaneous missense mutations leading to a substitution of the amino acid serine in position 264 by another amino acid (glycine, threonine, or alanine) causes a herbicide resistance against triazine herbicides. Because the target protein of the herbicide action was known, it was possible to directly identify the molecular change in the D1 protein (and in the psbA gene). Additional studies with atrazines and other similarly acting compounds in *Chlamydomonas reinhardtii* and *Euglena gracilis* revealed additional mutational sites and changes within the D1 protein (detailed references in: Hagemann 1990, 1992; Trebst et al. 1990).

In *Chlamydomonas reinhardtii* a great number of **antibiotic resistance mutations** have been isolated and studied which are inherited in a non-Mendelian uniparental mode and represent plastome mutations. They have been intensely studied by the research groups of Sager, Boynton, Gillham and colleagues (references in Boynton et al. 1992; Gillham 1994; Harris et al. 1989; Rochaix 1992).

b) Experimental Induction of Plastome Mutations

In 1927/28, H.J. Muller and L.J. Stadler proved that X-rays are a powerful means for inducing mutations in *Drosophila* and barley, respectively. In 1929/30, H. Stubbe and N.W. Timofeef-Ressovsky confirmed these results for *Antirrhinum* and *Drosophila* (H. Stubbe 1938). Very soon afterwards the experimental induction of mutations by ionizing radiation, by UV, and later by a great variety of chemical compounds became a widely used method in many fields of genetics of eukaryotic and prokaryotic organisms.

Researchers working on plastid genetics in higher plants, however, faced tremendous problems over many years. For a long time they had tried to experimentally induce plastome mutations in higher plants by physical and chemical agents in the same way mutations were successfully induced in practically all genetic objects, including the induction of mitochondrial mutations in yeast and of plastid mutations in *Chlamydomonas* and *Euglena*. However, for many years all experimental attempts to mutagenize higher plant plastid genomes failed entirely or gave unsatisfactory or inconclusive results.

Therefore, spontaneous plastome mutants had been the only source of new genetic material for about six decades.

In 1969, the group of Beletski in Rostov/Russia reported the occurrence of variegated plants of sunflower, *Helianthus annuus*, after treatment with nitroso-methyl-urea (NMU). The same compound was successfully used by Hentrich and Beger (1974), Pohlheim and Beger (1974) and Pohlheim (1974) with the small ornamental plant *Saintpaulia ionantha*. These two species heretofore had played no role in plastid genetics. Moreover, the *Saintpaulia* variety used was only vegetatively propagated.

This led our plastid research group in the Institute of Genetics in Halle to perform mutation experiments with standard objects of plant genetics: snapdragon (*Antirrhinum majus*), tomato (*Lycopersicon esculentum*) and evening primrose (*Oenothera hookeri*). In addition we not only used N-nitroso-N-methyl-urea (NMU), but also tested the related compound N-nitroso-N-ethyl-urea (NEU) in order to compare the action of these compounds, both of which are strong mutagens and effective carcinogens. These experiments in our lab were conducted in cooperation with Ulrike Grimmer, Franziska Lieberwirth, Monika Lindenhahn-Hagemann and Magdalena Scholze.

The technical aspects of these studies and their main results have been summarized in an article for the handbook "Methods in Chloroplast Molecular Biology" (Hagemann 1982).

In summary, the following results were obtained.

1. Successful mutagenesis experiments:
   We were able to induce plastome mutations using solutions of either NMU or NEU in *Antirrhinum majus, Lycopersicon esculentum* and *Oenothera hookeri* (Hagemann 1976, 1982).
2. Susceptibility of different plants to the chemical mutagens:
   NMU and NEU are highly mutagenic, cancerogenic, toxic compounds. In order to effectively induce plastome mutations we had to use solutions of NMU and NEU, in concentrations which were significantly higher than those typically used for the induction of nuclear gene mutations. For the induction of plastome mutations, solutions of NMU and NEU were tested in concentrations between 1 and 30 millimol/l. The most efficient concentrations for *Antirrhinum* and *Lycopersicon* were between 7 and 15 mM (for 2–3 h); they led to a high percentage of variegated plants. In *Oenothera*, the effective concentrations were found to be lower: 4–8 mM (Hagemann and Lindenhahn 1983).

   Similar mutagenesis experiments in the pea, *Pisum sativum*, were unsuccessful because NMU concentrations required for the induction of plastome mutations turned out to be lethal for *Pisum* (Grimmer, pers. comm.).
3. Differences between species:
   Antirrhinum majus proved to be the most suitable object for the experimental induction of plastome mutations. The mutagenized seeds and the plants obtained gave

highly reproducible results without any dependence upon environmental factors or endogenous influences. Therefore we used *Antirrhinum* as a positive control in all subsequent experiments aiming at the induction of plastome mutations in other species (and also as a quality control for the chemical compounds used).

The three species – snapdragon, tomato and evening primrose – showed remarkable differences regarding the speed of sorting-out of genetically different plastids. Sorting-out appears to occur relatively quickly in *Oenothera* and *Lycopersicon*; therefore many of the finely checkered green-white variegations disappeared rather quickly. By contrast sorting-out in *Antirrhinum* proceeds much more slowly; in consequence, many more variegated plants are recovered and can be propagated vegetatively and even sexually.

In *Antirrhinum*, the number of surviving seedlings greatly decreases with increasing concentrations of the mutagenic compound. On the other hand, with increasing concentrations of the mutagen an increasing proportion of variegated seedlings occurs among the survivors (cf. Hagemann 1982).

In a typical experiment, 200 seeds were treated for 3 h with NMU. For three different NMU concentrations the following numbers of survivors and variegated plants were obtained:

3 mM: 137 surviving seedlings, 8.0% variegated plants
9 mM: 120 surviving seedlings, 58.3% variegated plants
13 mM: 92 surviving seedlings, 92.4% variegated plants

The experiments with *Oenothera* gave comparable results, but only about half as many variegated plants were obtained as for *Antirrhinum* under similar conditions.

4. Comparison of the mutagenicity of NMU and NEU:
   For *Antirrhinum majus* plastids, we systematically compared the action of NMU with that of NEU. We found an approximately two-fold dose of NEU to be as efficient as NMU.
5. Proportion of plastome mutants among the induced mutations:
   We were fully aware of the fact that NMU and NEU not only induce plastome mutations, but can also cause nuclear gene and chromosome mutations. Recessive gene mutations are detected only in the M2 generation. Therefore, we focused our attention on variegated M1 plants. Whereas part of them certainly carried dominant nuclear mutations or chromosome aberrations, others were clearly plastome mutations. They were identified by applying three criteria:
   a) the finely checkered pattern of variegation in leaf color which is typical for sorting-out of genetically different types of plastids,
   b) the cytological demonstration (by light and electron microscopy) of the presence of mixed cells in these plants, and
   c) the proof of the non-Mendelian, uniparental inheritance of these traits in both *Antirrhinum* and *Lycopersicon*.

These investigations gave us chemical compounds to hand, which enabled researchers in the field of plastid inheritance to experimentally induce plastome mutations in any given higher plant species. Since the publication of our results (1982), several research groups have used this method to obtain plastome mutations (and nuclear mutations, too, e.g. Hosticka and Hanson 1984; Sears and Sokalski 1991).

Other research groups also tried other compounds, EMS (ethyl-methane-sulfonate), MNNG (methyl-nitro-nitroso-guanidine), and 5-bromodeoxyuridine; but these compounds seem to be effective only for a limited number of species, whereas NMU and NEU are successfully applicable in a wide range of species (Hagemann 1982).

c) Nuclear Gene-Induced Plastome Mutations

This type of mutation represents a very special mode of interaction between the nucleus and plastids. The basic genetic phenomenon seems to be rather simple: A recessive mutant allele of a *nuclear* gene induces, in homozygous condition, the rather frequent occurrence of *plastid (=plastome) mutations*. These plastome mutations lead to cells and plants containing genetically different types of plastids: wild-type plastids which develop into normal green chloroplasts, and mutant plastids which become white or yellow or pale/light green. The sorting-out of wild-type and mutant plastids results in the occurrence of variegated (dicots) or striped (monocots) plants according to the different mode of leaf growth (in dicots versus monocots). Once the plastome mutations are induced by the mutant nuclear gene, they are subsequently inherited in a typical non-Mendelian, extranuclear mode: They show uniparentally maternal inheritance (in most species) or biparental inheritance (in a minority of species, e.g. *Oenothera hookeri*).

The phenomenon of nuclear gene-induced plastome mutations was described in a number of species: *Hordeum vulgare* L., *Oryza sativa* L., *Zea mays* L., *Arabidopsis thaliana* (L.) Heynh., *Epilobium hirsutum* L., *Nepeta cataria* L., *Petunia hybrida* Hort. and *Oenothera hookeri* L. (detailed references in Hagemann 1986; Epp and Parthasarthy 1987; Sears and Sokalski 1991).

In the last decades, mainly three of these mutants have been used in intense studies:

1. the mutant *albostrians (as)* of *Hordeum vulgare* (Börner et al. 1976 and Hagemann 1986 in Halle, Börner, Hess and colleagues in Berlin,
2. the mutant *plastome mutator (pm)* of *Oenothera elata*, var. *hookeri* (Epp 1973; Epp and Parthasarthy 1987; Sears and Sokalski 1991), and
3. the mutants *chloroplast mutator 1 and 2 (chm1 and chm2)* of *Arabidopsis thaliana* (Redei 1973; Redei and Plurad 1973; Martinez-Zapater et al. 1992).

The mutant line "*albostrians*" of *Hordeum vulgare* consists of white, green-white striped and green seedlings. It has been intensely studied, because the cells of the white seedlings and the white parts of the striped seedlings have a distinct **plastid ribosome deficiency**: Biochemical and electron microscope studies revealed a specific absence of plastid ribosomes: In the white mutant cells, the cytoplasmic 80 S ribosomes and their RNAs (25 S and 18 S rRNAs) are present in normal amounts; in contrast, plastid ribosomes and plastidal ribosomal RNAs (23 S and 16 S) could not be found in plastids of the white cells. This complete plastid ribosome deficiency was the starting point for several research projects which will be dealt with in Section 8.

The line "*plastome mutator*" of *Oenothera elata*, var. *hookeri*, has been investigated with the aim of characterizing the molecular mechanism(s) of the nuclear gene-induced plastome mutagenesis (Epp and Parthasarthy 1987; Sears and Sokalski 1991).

The mutants "*chloroplast mutator 1 and 2*" of *Arabidopsis thaliana* were found and described by George Redei (1973). These mutants stirred some interest, because later a new variegated mutant of *Arabidopsis thaliana* was found, which is allelic to the *chm* locus. This mutant, according to Martinez-Zapater et al. (1992), does not primarily affect the plastome, but causes rearrangements of the mitochondrial genome.

Therefore the possibility was discussed that this primary action on mitochondria may also be true for other genes causing "gene-induced plastome mutations". At present, however, this remains an open question.

7 The Path into Molecular Genetics of Plastids

A very important new era of plastid genetics began in the years 1962–1964. This time period marks the opening of the field of molecular genetics of plastids. Three decisive steps characterize the path into this new field which, during the following decades, exerted a profound influence on our knowledge of the genetics, biology and evolution of plastids.

a) Discovery and Characterization of Plastid DNA

During the period of 1962–1964, it became clear that plastids contain their own specific DNA, and that this plastid DNA (= chloroplast DNA) has specific characteristics different from nuclear DNA.

(In parallel, the same development took place regarding mitochondrial DNA.)

The proof for the existence of plastid DNA was based on several lines of evidence:

1. Electron microscopic pictures of (*Chlamydomonas*) chloroplasts provided evidence for the presence of DNA strands within the chloroplasts (Ris and Plaut 1962).
2. Demonstration of the specific incorporation of ^{3}H-thymidine into young chloroplasts (Stocking and Gifford 1959; Wollgiehn and Mothes 1963).
3. CsCl buoyant density centrifugation allowed the separation of major (nuclear) DNA bands and minor DNA bands, one of which was plastid DNA. These investigations were complemented by the determination of the GC content of the DNA from purified chloroplasts (Sager and Ishida 1963; Kirk 1963).

4. Analysis of normal versus mutant, chloroplast-free cells of *Euglena gracilis* and characterization of the minor DNA band present in normal and absent in chloroplast-free cells (Edelman, Schiff and Epstein 1965).
5. These findings were later supported by the characterization of the physical and chemical differences between plastid DNA and nuclear DNA: lack of 5-methyl-cytosine in plastid DNA (of most species), but presence in nuclear DNA; rapid renaturation of plastid DNA (Ray and Hanawalt 1964; Wells and Birnstiel 1967).

The identification and characterization of the specific plastid DNA was a complicated process with several side paths, errors, mistaken identities and disagreements, which caused John Kirk (1971) during a symposium to make the (despairing or amused?) exclamation: "Will the real chloroplast DNA please stand up !"

Around 1971, the scientific community of plastid researchers came to the generally accepted opinion that in all higher plants and many algae the buoyant density of plastid DNA was in the range of 1,696–1,698 g/cm^3, corresponding to a GC content of 36–38%. A distinct exception was *Euglena gracilis* with a density of 1,685 (detailed references in Gillham 1978; Kirk 1986; Kirk and Tilney-Bassett 1967).

During the seventies, a wealth of data was elaborated by many laboratories about the genome size, GC content, conformation and other basic physicochemical properties of plastid DNA (cf. Hagemann and Börner 1981; Palmer 1985).

b) Application of Restriction Enzymes and the Construction of Restriction Maps and Physical Maps

A new and very strong impetus for the progress of plastid genetics came from the application of restriction enzymes. It was shown that type II restriction enzymes recognize particular sequences and cut within the recognition sequences at specific target sites.

The use of many different type II restriction enzymes allowed the construction of restriction maps for the plastid DNAs of many angiospermous species.

The first **restriction maps** of plastid DNA (= ptDNA) were published for *Zea mays* and *Spinacia oleracea* in 1976, quickly followed during the next years by reports for the ptDNA of many more species (cf. Table 1 in Hagemann and Metzlaff 1983, Progress in Botany 45).

The great majority of angiospermous species analyzed, have an "inverted repeat" (IR) in their plastid chromosome, which separates the "small single copy region" (SSC) from the "large single copy region" (LSC). The inverted repeat contains the operons for the ribosomal RNAs

5 S, 4.5 S, 23 S and 16 S. A few species, however, have apparently lost one inverted repeat (pea, broad bean).

By contrast, *Euglena gracilis* turned out to have quite a different overall organization of its ptDNA; it does not contain inverted repeats, instead there are tandem repeats of the rDNA operon in the ptDNA molecule (e.g. 3 or 3.5).

In general, *Euglena gracilis* exhibits a number of exceptional characteristics: a GC content different from most other species, a divergent organization of the plastid genome; the existence of cells without chloroplasts and others. That is why people often made the joke: "*Euglena* is not a plant at all, it is a green animal!"

The use of restriction fragments of ptDNA for DNA-RNA hybridization and the use of linked transcription-translation systems allowed the localization of specific genes for rRNAs, tRNAs and mRNAs for particular polypeptides. In this way, an increasing number of plastid genes could be localized, and thus the restriction maps were gradually transformed into **physical maps of the ptDNA.**

c) DNA Sequencing of Plastid Genes and Genomes

The refinement of the physical maps has benefited greatly from the development of efficient methods for DNA sequencing (Sanger et al. 1977; Maxam and Gilbert 1977), and their application for the total sequencing of the human mitochondrial DNA by Sanger's group (Anderson et al. 1981). Already in 1980, the first two plastid genes were completely sequenced: the gene for the 16 S rRNA of maize (Schwarz and Kössel 1980) and the gene for the large subunit of Rubisco of maize (McIntosh et al. 1980). This was the starting signal for many research groups. In the following years, the DNA sequences of many plastid genes have been determined (cf. Table 1 in Hagemann et al. 1985, Progress in Botany 47). In this way, physical maps of the plastid genomes of many higher and lower plants were constructed and led to a detailed knowledge of the molecular structure, organization and coding capacity of the ptDNAs of many plant species.

In 1986 an exciting breakthrough was achieved: The plastid DNA molecules of two different plant species were completely sequenced by two Japanese research groups. Ohyama and coworkers reported the total nucleotide sequence of the plastid DNA molecule of the liverwort *Marchantia polymorpha* (Ohyama et al. 1986) and the research group of Sugiura sequenced the plastid DNA molecule of tobacco, *Nicotiana tabacum* (Shinozaki et al. 1986).

The plastid DNA of *Marchantia* was reported to consist of 121,024 base pairs, and that of *Nicotiana* of 155,844 base pairs.

The chloroplast genomes of both the liverwort (bryophyta) and tobacco (angiospermae) were remarkably similar with regard to their general structure, gene content and genome organization. The circular molecules consist of a "large single copy region" (LSC), a small single copy region" (SSC) and two "inverted repeats" (IRs) separating LSC and SSC. A significant difference between the two species was only found regarding the size of the IR: in *Marchantia* it is about 15,000 base pairs smaller than that of *Nicotiana* (*Marchantia*: 10,058 bp versus *Nicotiana*: 25,339 bp).

The complete sequencing of these two plastid DNAs marks an exciting event in the field of molecular biology and genetics of plastids. With these results, a reference system was established which all research groups dealing with plastid DNA sequences can refer to. During the following years, the plastid genomes of several other species of higher plants were completely sequenced: *Oryza sativa* (1989), *Epifagus virginiana* (1992), *Pinus thunbergii* (1994) and *Zea mays* (1995); in addition, the plastid DNA of the green alga *Euglena gracilis* was fully sequenced (1993) (cf. Hagemann and Hagemann 1994; Hagemann et al. 1996, in Progress in Botany 55 and 57).

Thus, the picture emerged in the first half of the nineties that the plastid DNAs of the land plants have a rather conservative structure and a characteristic coding capacity (with the exception of the parasite *Epifagus* which has lost a considerable amount of plastid photosynthesis-related genes).

Moreover, a number of characteristic features were found:

a) the presence of operons in plastid DNA, i.e. many plastid genes are organized in a polycistronic arrangement (comparable to the situation in eubacteria); the polycistronic precursor RNAs are processed into smaller RNA species.
b) many plastid genes contain introns representing three different groups; the splicing processes have been extensively characterized (both *cis*- and *trans*-splicing).

(comprehensive references are given in Herrmann 1992; Hagemann 1993; and by Hagemann et al. 1983–1996, in Progress in Botany).

However, this uniform picture changed and was supplemented by new results obtained from the sequencing of the **plastid DNAs of different algae.** In 1995, the complete sequences of the plastid genomes of three non-green algae were published: *Porphyra purpurea* (Rhodophyta: Rhodophyceae) by Reith and Munholland, *Cyanophora paradoxa* (Rhodophyta: Glaucocystophyceae; cyanelles) by Stirewalt et al., and *Odontella sinensis* (Chromophyta: Bacillariophyceae, Diatomeae) by Kowallik et al. Two years earlier, the total sequence of the plastid genome of *Euglena gracilis* (Euglenophyta) had been reported (Hallick et al. 1993). Regarding its size and coding capacity, the *Euglena* plastid genome did not differ greatly from those of the land plants.

However, the plastid genomes of *Porphyra*, *Cyanophora* and *Odontella* were significantly *larger* and had a *distinctly higher coding capacity* (the details are given in Table 1 of Hagemann et al. 1998 in Progress in Botany 59).

The circular plastid genome of *Porphyra purpurea* is 191,028 bp in length and thus the largest plastid genome, sequenced to date. The plastid genome contains 251 genes and open reading frames; this is more than twice the number of plastid genes in land plants.

Two main reasons account for the increased number of plastid genes:

1. Groups of plastid genes which are also present in land plants comprise significantly more genes in *Porphyra* (for photosystem I, for ribosomal proteins and even for Rubisco: not only the rbcL gene for the large subunit, but also rbcS for the small subunit, is plastid-encoded in red and brown algae).
2. More importantly, the algal plastid DNA contains entire groups of genes which are not present in plastid DNAs of land plants (but present in their nucleus), such as genes for the control of DNA replication, for gene expression and for several biosynthetic functions.

The plastid DNA of *Odontella sinensis* comprises 119,704 bp and that of *Cyanophora paradoxa* 135,599 bp. Both *Odontella* and *Cyanophora* also have many more plastid genes than the land plants.

The results elaborated with these three algae indicate that they may represent primitive types of plastids where less genes were evolutionarily transferred to the nucleus. The plastid genomes of land plants may be the most derived type of present-day plastids and thus the temporary end of plastid evolution. The study of algal plastids led Kowallik (1992, 1994), Gray (1993) and Reith (1995) to the conclusion that – in accordance with the endosymbiont theory – there was only one process of endosymbiotic incorporation of an ancestral cyanobacterium, which then led to the plastids in all plant lineages: *a monophyletic origin of all present-day plastids.*

8 Molecular Analysis of Specific Plastome Mutants

a) From the Mutant Phenotype to the Gene

In many species of higher plants, numerous variegated plants occurred spontaneously or were experimentally induced by chemical mutagens. A considerable percentage of them showed a non-Mendelian mode of inheritance indicating a plastid mutation. Such plants have deficiencies in the light or dark reactions of photosynthesis, usually connected with changes in the leaf color (white, yellow, cream or light green leaves). The white sectors of these plants harbor only mutant plastids and thus are incapable of autotrophic growth. Such a plastome mutant would be lethal in a homoplasmic mutant plant (at least when grown in soil). In a chimeric plant, however, the survival of the white tissue is facilitated by the adjacent green tissue which supplies all essential metabolites not synthesized by the mutant plastids.

The aim of the researchers has always been to find the mutated gene. This task was relatively easy, when the changed protein was known. As outlined above (6.1), many herbicide resistances were connected with the D1 protein. Thus the analysis could be concentrated on this protein and its encoding gene. Comparable was the finding of a drastic change in the function of Rubisco in *Oenothera hookeri*, which could be attributed to its large subunit. The molecular analysis revealed a 5-bp deficiency in the rbcL gene of *Oenothera* (Winter and Herrmann 1988).

However, when many candidate genes exist, in one of which the mutational change may have occurred, then the procedure to identify the mutated gene is (or has been) much more difficult.

So far, at least six plastid genes have been identified in land plants which encode polypeptides of the photosystem I complex, and altogether 13 genes encoding polypeptides of the photosystem II complex. In addition, there are at least 15 plastid genes for polypeptides of the cytochrome b/f complex, the ATP synthase complex, and for stromal polypeptides. Hence, intense efforts are necessary to identify the affected gene in a plastome mutant.

The first step of such an analysis is the *physiological and biochemical characterization* of a particular plastome mutant, in order to narrow down the site of its photosynthetic deficiency. Having obtained this information (deficiency in photosystem I or II, in the cytochrom b/f complex, in the ATP synthase or in the dark reactions of photosynthesis) the second step aims at the *identification of the specific gene* changed by the plastome mutation. The third step is *sequencing* the mutant gene (or at least part of it) which contains the molecular change.

The analysis of some mutants which have been studied in detail by our plastid research group in Halle and colleagues from other institutes will be described here.

b) The Mutant *en:alba-1* of *Antirrhinum majus*

My work with the plastome mutant *en:alba-1* began with the demonstration of true "mixed cells" in variegated leaves, which contained normal wild-type plastids and mutant plastids side by side within the same cell (cf. Sect. 3). We soon decided to make an attempt to analyze its physiological and biochemical deficiencies in order to gain insights into the underlying genetic defect. (In the seventies, we used the neutral designation *en:* as an abbreviation for an"extranuclear" mutant; nowadays we would prefer "*pt*" for plastid mutant. However, we refrained from renaming the mutant.)

The plastome mutant *en:alba-1* has been kept and propagated over many years through variegated plants. Branches containing only mutant plastids are yellowish-white under normal light conditions. Under dim light (shaded conditions) they become distinctly more green. They can accumulate up to 38% of the chlorophyll content of the wild-type. Nevertheless, *en:alba-1* plastids exhibit a total block in photosynthesis. Biochemical analyses demonstrated a deficiency in photosystem I, whereas photosystem II functions normally. These findings were supported by biophysical methods, such as measurement of electron spin resonance, delayed light emission and intense red fluorescence (F. Herrmann et al. 1974; Hagemann and Börner 1979).

Our initial results suggested we should proceed with an analysis of the SDS-solubilised pigment-protein-complexes and lamellar proteins of mutant and wild-type plastids of *Antirrhinum majus.* A clear-cut result was obtained: In the mutant *en:alba-1,* the major pigment-protein complex I and the minor complex Ia are absent, and in agreement with this finding, the lamellar protein bands 1 and 2 are missing. These bands represent key subunits of photosystem I.

These investigations were performed by my first Ph.D. student, Falko Herrmann (1971a,b).

We concluded from our findings: The deficiency in the mutant *en:alba-1* is the result of a specific mutation in its plastid DNA. Obviously, the plastid DNA encodes specific thylakoid proteins. As a result of the plastome mutation one (or a few) lamellar proteins are changed (or absent). As a secondary consequence, pigments, especially chlorophylls, cannot be correctly attached to these proteins, and hence photodestruction occurs. These effects ultimately cause the complete lack of photosystem I and hence of all photosynthetic activity (F. Herrmann 1971a, b; Hagemann et al. 1974).

Our conclusions appeared straightforward at this time (1971/72); however, warnings of caution came from some colleagues (Could these be indirect effects?).

However, direct analyses of the plastid DNA were not possible at that time.

In the following two decades, although our work focused on other aspects of plastid biology, the *en:alba-1* mutant was not forgotten.

In the early nineties, my coworker Claudia Schaffner studied the mutant with newly available physiological and molecular genetical methods. She was able to confirm the results from our previous investigations and in addition gained interesting new insights: The analysis of primary photosynthetic reactions in mutant branches indicated a dysfunction of photosystem I (PSI). The peak wavelength of PSI-dependent chlorophyll fluorescence emission at 77 K was shifted by 4 nm to 730 nm, as compared to fluorescence from the wild type. There were no redox transients of the reaction center Chl P 700 upon illumination of leaves with continuous far-red light or with rate-saturating flashes of white light. Also, the PSI reaction center proteins PsaA and PsaB were not detectable by SDS-PAGE in mutant plastids.

The question of greatest interest to us was , how to identify the primary defect of this plastome mutation in the plastid DNA. C. Schaffner ultimately succeeded by using the PCR-SSCP method, a combination of SSCP (analysis of single-strand conformation polymorphism) with PCR.

The SSCP analysis is based on the fact that single-stranded DNA molecules take on specific sequence-based secondary structures (conformers) under non-denaturing conditions. DNA single strands differing by as little as one single base substitution may form different conformers and thus migrate differently in a nondenaturing polyacrylamide gel.

In order to identify the mutation in the plastid DNA Claudia Schaffner employed a modified SSCP procedure using large PCR fragments that were cleaved with various restriction enzymes. When DNA from the

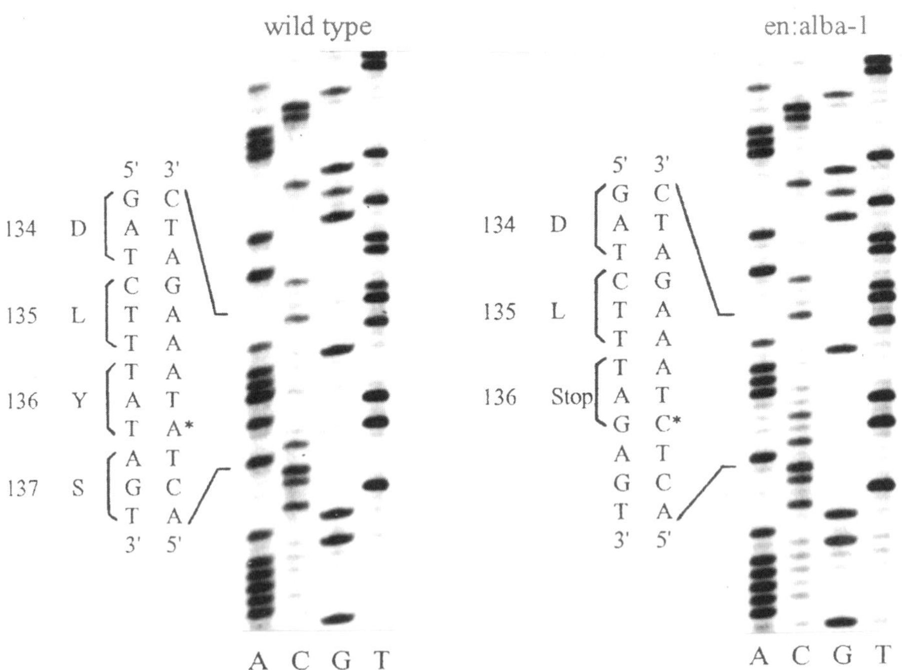

Fig. 6. Direct sequencing of the PCR-amplified *psaB* gene from wild-type and *en:alba-1* DNA of *Antirrhinum majus* using oligonucleotide primer PSAB5. *Asterisks* show the position of the mutation. (Schaffner et al. 1995)

wild-type and the mutant *en:alba-1* was submitted to SSCP analyses, a single-stranded HinfI fragment of a PCR product from the psaB gene showed differences in electrophoretic mobility. Sequence analysis revealed that the observed SSCP was caused by **a single base substitution (transversion)** at codon 136 (TAT→TAG) within the **plastid psaB gene** (Fig. 6). The point mutation produces a new stop codon that leads to a truncated PsaB protein. This result indicates that the mutation prevents the assembly of a functional photosystem I complex (Schaffner et al. 1995).

The successful molecular analysis of the plastome mutant *en:alba-1* had a special meaning for my scientific career as plastid geneticist. The physiological and biochemical analysis of the mutant was initiated in the late sixties by my *first* diploma student, who later became my *first* Ph.D. student, Falko H. Herrmann (who is now Professor of Human Genetics at the Ernst-Moritz-Arndt-Universität Greifswald). The final investigations, elucidating that the underlying molecular change in this mutant is a single base pair substitution in the plastid gene psaB, were performed in the nineties by my *last* Ph.D. student, Claudia Schaffner (who is now at the Deutsches Krebsforschungszentrum, Heidelberg).

C. Schaffner also performed parallel investigations with another plastome mutant of *Antirrhinum majus* which was experimentally induced by treating seeds with nitroso-methyl-urea (NMU, cf. Sect. 6.b), *en:alba-4*. In this case, the PCR-SSCP analyses of the psbD-psbC operon revealed mobility shifts of single-stranded DNA fragments (after digestion with the restriction enzymes AluI or Sau3AI). By DNA sequencing, this plastome mutant was shown to carry **a transition of a GC to an AT base pair in psbD**. This transition in position 1027 of the psbD gene changes the codon 343 for proline into a codon for serine (CCT→TCT), and thus leads to an amino acid exchange near the C terminus of the D2 protein.

Together with D1, the D2 protein forms a heterodimer which binds most of the cofactors for charge separation in photosystem II, and thus represents the main part of the reaction center of photosystem II. The C-terminus of D2 is highly conserved and obviously plays an important functional role. Its mutational change in the plastome mutant *en:alba-4* severely impairs the function of photosystem II (Schaffner 1995; Hagemann et al. 1996).

Studies with mutants of *Pelargonium zonale* were performed in parallel and in comparison to the *Antirrhinum* mutants. The periclinal chimera "Mrs. Pollock" is yellow-margined. The yellow parts of these plants contain the plastome mutation *en:gilva-1*. Investigation of the mutant plastids by F. Herrmann demonstrated a deficiency in photosystem I - comparable to the deficiency in the mutant *en:alba-1* of *Antirrhinum majus*. Several biophysical methods were used to characterize this mutant. Analysis of electron spin resonance signals and measurement of the intense red fluorescence revealed the deficiency of photosystem I. The analyses of enzymatic reactions in *Pelargonium* turned out to be very difficult, because homogenization of leaf material sets free many inhibiting (especially phenolic) compounds interfering with enzymatic reactions.

Subsequent analyses of the thylakoid proteins demonstrated the absence of the major chlorophyll-protein complex I and the absence of the respective lamellar proteins in PAA gels (Herrmann and Hagemann 1971; Herrmann et al. 1974, 1976).

Thus the plastome mutation in the *Pelargonium* variety "Mrs. Pollock" (mutant *en:gilva-1*) turned out to cause a deficiency in photosystem I, very similar to that in the *en:alba-1* mutant of *Antirrhinum* (Hagemann 1979).

c) The Mutant albostrians of *Hordeum vulgare*: Plastid Ribosome Deficiency

In this section, the mutant *albostrians* will be dealt with in some detail. This has two reasons:

Firstly, as a result of extensive investigations on plastid ribosome deficiency, the barley mutant *albostrians* seems to be one of the most intensely studied mutants of higher plants.

Secondly, the study of this mutant has been initially performed by myself and my colleagues in Gatersleben and Halle; later on, when my former coworker Thomas Börner received the Professorship of Genetics at the Humboldt-University in Berlin, the investigations were continued and greatly expanded by Börner and many of his colleagues in Berlin.

In 1960, a colleague at the Gatersleben Institute, Friedrich Scholz, proposed to me a collaboration about a mutant line of *Hordeum vulgare*, which had originated after X-ray treatment of the spring barley variety 'Haisa' and which showed a conspicuous green-white striping. The mutant line itself segregated in green (5%), green-white striped (60%) and pure white seedlings (35%). Reciprocal crosses with green plants of 'Haisa' revealed a very interesting genetic character: All plants of the mutant line are homozygous for the nuclear mutant gene "*albostrians*" (*as as*). This recessive allele *as* induces in most of its homozygotes (but not in heterozygotes) plastid mutations, which – after induction – are inherited purely maternally. Cells containing only mutant plastids are white. This plastid mutation is also stable, after replacing the *albostrians* allele *as* with its wild-type allele *As*. Normal and mutant plastids are sorted out at random during vegetative divisions. This sorting-out leads in monocots to a green-white striping (comparable to the green-white variegations in dicots). Whether white stripes on leaves, white sectors, white heads or entirely white seedlings occur, depends upon the developmental stage, in which sorting-out of genetically different plastids takes place, and in which cells with only mutant plastids occur. Reciprocal crosses show a uniparentally maternal inheritance of the plastid difference. Thus, the *albostrians* line represents an example of "nuclear-gene-induced plastid mutations" or "plastome mutators" (Hagemann and Scholz 1962; Hagemann 1986; cf. Sect. 6.c).

Biochemical studies revealed a specific **absence of the plastid ribosomes** from the white cells and tissues of *albostrians*: In the white mutant cells, the *cytoplasmic ribosomes* and their RNAs (25 S and 18 S rRNAs) *are present* in normal amounts and show a normal structure. In contrast, the plastids of the white cells do not contain plastid ribosomes and ribosomal (23 S and 16 S) RNAs (Börner et al. 1976; Hagemann and Börner 1979). This fact has been proven by both biochemical and electron microscope studies.

This complete plastid ribosome deficiency in the white *albostrians* seedlings and cells was the starting point for several research projects, the results of which will be outlined in the following brief survey.

- The white mutant plastids are smaller than the green chloroplasts, but they are present in approximately normal numbers per cell. They contain plastid DNA and their plastid DNA is replicated, as shown by autoradiography. Thus the DNA polymerase which performs this replication is encoded in the nucleus, is synthesized on cytoplasmic ribosomes and then transported into the ribosome-free plastid.
- As a consequence of the absence of plastid ribosomes the enzyme ribulose 1,5 bisphosphate carboxylase (= Rubisco) and the coupling factor CF1 could not be found. This finding was based on immunological, gel electrophoretic and electron microscope studies (Börner et al. 1976; Hagemann and Börner 1979).
- Soon it became obvious that there is a strong relationship between the protein synthesizing systems in the plastids and in the cytoplasm: The small subunit of Rubisco is encoded by the nucleus, synthesized in the cytoplasm and then transported into the plastid; the same is true for two subunits of CF1. Nevertheless the whole complexes of Rubisco and CF1 are absent. Moreover, it was found that in the white *albostrians* cells and plants, two cytoplasmically synthesized plastid enzymes, phosphoribulokinase and D-glyceraldehyde-3-phosphate:NADP$^+$ oxidoreductase are found only in considerably reduced amounts. Also the activity and the apoprotein of nitrate reductase is lacking in this mutant. This led to the hypothesis that *a plastid factor* or plastid signal, which indicates the developmental state of the plastid, is involved in the expression of certain nuclear genes (Bradbeer, Atkinson, Börner and Hagemann 1979; Börner et al. 1986).
- A very interesting finding was made by Siemenroth et al. (1981). They observed – although plastid ribosomes are not present in *albostrians* plastids – an RNA polymerase activity and found that the plastid

genes for 23 S and 16 S rRNA are transcribed in mutant plastids and that these RNAs are processed. They concluded that the enzymes involved in the transcription and processing of these rRNAs are synthesized on cytoplasmic ribosomes and then transported into the mutant plastids. These findings gave an indication for the existence of an RNA polymerase which is synthesized in the cytoplasm and afterwards transported into the organelle (in addition to the plastid-DNA encoded RNA polymerase).

- This finding was extended to the general question, which mRNAs and tRNAs are transcribed in the ribosome-deficient *albostrians* plastids. Börner and Hess (1993), Hess et al. (1993), and Hess and Börner (1999) observed the transcription of several chloroplast gene clusters in the ribosome-less plastids (rps15-ndhF, rpoB/C1/C2, rpl23/trnI/trnL and several others). Hübschmann and Börner (1998) could identify promoters recognized by the nuclear-encoded RNA polymerase. Recently Börner's group could identify two *Arabidopsis* nuclear genes coding for chloroplast RNA polymerases (Hedtke et al. 1997, 2000).
- A further interesting finding is the observation that there are intense interactions between the physiological activities of plastids and mitochondria: In *albostrians* cells, the synthesis of transcripts of genes encoding components of the respiratory chain (coxII and III, atp 6 and A, cob) accumulate to a distinctly higher rate (about four-fold) than in green leaves (Hedtke et al. 1999).
- The ribosome deficiency in *albostrians* also has effects on the splicing processes of plastid RNAs. The barley mutant *albostrians* is defective in the splicing of all subgroup IIA introns in the plastid (acting e.g. on the pre-RNAs of the plastid genes trnA, I, K and G). The plastid gene matK encodes the only currently known chloroplast intron-specific maturase (MatK). This polypeptide is lacking in *albostrians*; its absence in the ribosome-deficient mutant seems to be connected with the observed splicing defects (Vogel et al. 1999).
- Unfortunately, detailed knowledge about the molecular structure and sequence of the nuclear, *albostrians* gene is still lacking. Hopefully, knowing its molecular structure will allow us to unravel the way from the mutated gene in the nucleus to the change of the genetic information in the plastids.

9 RNA Editing in Plastids

About ten years ago, a new genetic phenomenon in plastid genetics was observed: RNA editing (Hoch et al. 1991; Kudla et al. 1992).

Editing of mRNAs as a posttranscriptional process was first discovered in *kinetoplasts of trypanosomes* (Benne et al. 1986) and soon afterwards in the nuclear encoded mRNA of the *human apolipoprotein B* (Powell et al. 1987). In 1989, three groups described the occurrence of RNA editing in many mRNAs encoded by the *mitochondrial* genome of higher plants (Covello and Gray 1989; Gualberto et al. 1989; Hiesel et al. 1989).

The main characteristic of RNA editing is: The nucleotide sequence of the preRNA, as synthesized by transcription from the genomic DNA, is secondarily changed at specific sites. The biological significance is based on the fact that the pre-mRNA with its primary nucleotide sequence would encode a non-functional polypeptide; only after editing specific sites, the secondary RNA sequence is able to encode a fully functional polypeptide. In this sense, RNA editing has been viewed as a mechanism 'regenerating the right information'.

My plastid research group at the Institute of Genetics in Halle came into contact with the research on RNA editing in 1990/1991. After the breakdown of the 'Berlin wall' and the border between East and West Germany, my coworker Jörg Kudla took the initiative and started a scientific cooperation of our plastid group with the research team of Hans Kössel at the University of Freiburg. As a consequence of the close scientific and personal contacts between our groups, two of my former students and coworkers, Jörg Kudla and Ralph Bock, moved from the University of Halle to Hans Kössel's team in Freiburg just at the time when the first case of RNA editing in plastids was found and published by this group (Hoch et al. 1991). They were soon fully integrated in the research work on RNA editing in plastids in the following years (Kudla et al. 1992; Bock et al. 1993 and many subsequent papers, cf. Bock 1998). Hans Kössel not only integrated these two young scientists into his group, he also opened up the way for them to perform experimental work during long-term stays in leading US universities.

Since the initial discovery of RNA editing in trypanosomes, a variety of obviously different types and systems of editing in different cellular compartments and taxa have been characterized (see Table 1).

Obviously, very different types and systems of RNA editing exit. This also suggests the existence of different editing mechanisms.

In this article, only **RNA editing in plastids** will be dealt with in some detail.

The RNA editing processes in plastids exhibit the following features (references in Maier et al. 1996; Bock et al. 1997; Bock 2000, 2001):

- RNA editing has been observed in the plastids of seed plant and fern species analyzed so far.
- It has not been found in plastids of certain bryophyte lineages, e.g. in the intensely studied liverwort *Marchantia polymorpha*, and also not in algae and cyanobacteria, the presumptive ancestor of the present-day plastids.

Table 1. Systems of editing

Compartment	Organism (genus)	Type of editing
Kinetoplast (= Mitochondrion)	*Trypanosoma*	U insertion/deletion
Mitochondrion	*Physarum*	Nucleotide (C,U,G,A) insertions C to U conversion
Mitochondrion	Higher plants	Mostly C to U conversion, seldom U to C conversion
Plastids	Higher plants	C to U conversion
Nucleus	Mammals (including man)	C to U conversion A to I conversion
Nucleus	*Drosophila*	A to I conversion
Mitochondrion	Snails, *Monotremata, Marsupialia*	Various nucleotide conversions
Host cytoplasm	Paramyxoviruses	G insertions

(References in: Maier et al. 1996; Bock et al. 1997; Bock 2000; Maas and Rich 2000).

- In plastids of higher plants RNA editing always proceeds by conversion of single cytosine residues to uracil (C to U conversion). However, U to C conversions - "reverse editing" - have been reported in the hornwort *Anthoceros formosa.*
- RNA editing is an early posttranscriptional process which obviously can precede splicing and cleavage of polycistronic into di- or monocistronic mRNAs.
- It predominantly occurs within coding regions; but editing in a noncoding intergenic region has also been described.
- Editing frequently affects internal codons within a reading frame. It results in the restoration of codons for conserved and functionally important amino acid residues (as revealed by comparisons with other species). The codons - as encoded in the plastid DNA and transcribed into pre-mRNA - would lead to a non-functional polypeptide; only after editing, the synthesis of a functional polypeptide becomes possible.
- In some cases, RNA editing affects the initiation codon and creates a functional AUG start codon (by changing an initially present ACG triplet which if not changed would initiate translation with a very low efficiency only, if at all).
- Editing in plastids mostly changes the second nucleotide of a codon, less frequently the first nucleotide. Examples of silent editing at the third codon position are very seldom.

- Transcripts which are edited in one species (e.g. in spinach) may already contain the "correct" nucleotide sequence at the plastid DNA level in another species (e.g. in *Antirrhinum majus*), as found for example for the psbF/L transcript.
- There are plastid genes, the mRNAs of which are edited, but there are also other genes whose transcripts are not edited, because they already have the "correct" sequence.
- In general, editing of plastid RNAs is not as frequent as editing in mitochondrial transcripts.
- Organ- and tissue-specific differences of RNA editing have been found in special cases: Two sites in the psbF/L transcript in spinach are only partially edited in roots and seeds; but editing is complete in plastids of all other tissues and developmental stages.
- The most important aim of present-day research on RNA editing in plastids is the elucidation of the underlying molecular mechanism. Several approaches have been used to tackle this problem:
- In higher plants editing in plastids involves a cytosine to uridine conversion which is - in chemical terms - a deamination. The C to U editing in the human mRNA for apolipoprotein B is carried out by an "editosome" containing a cytidine deaminase activity. Such an enzyme activity may well be involved in plastid editing also.
- Another problem concerning the editing process is its extraordinarily high specificity. How are the specific editing sites selected by the "editing apparatus"? Are there *cis*-acting "editing motifs" and are there also *trans*-acting factors?
- A great experimental help in connection with this has been the elaboration and use of chloroplast transformation technologies: the 'biolistic system' (the particle gun) as well as the polyethylene glycol (PEG) based chemical transformation of plastids. (These technologies have been summarized by Hagemann et al. 1996, 1998; and Bock and Hagemann 2000, in Progress in Botany, Vols. 57, 59, 61, and in several reviews: Bock 1998, 2000, Bock et al. 1996, 1997; Chaudhuri and Maliga 1996; Goulds et al. 1993; Koop et al. 1996).
- The use of this transformation technology allowed the identification and characterization of mRNA sequences which flank the editing sites and are involved in editing site recognition: the major *cis*-acting recognition elements were found to reside in the 5′ upstream region (Bock et al. 1996).
- In addition *trans*-acting factors play an important role in the editing process (Bock and Koop 1987; Chaudhuri and Maliga 1996). Presumably, the essential upstream *cis*-sequence element serves as binding site for a *trans*-acting factor.

These studies on RNA editing in plastids are also relevant to RNA editing in plant mitochondria, in which many similar investigations are performed; both systems seem to have many molecular features in common. Detailed studies on RNA editing processes in mammals have already led to stimulating insights into the participating enzymatic activities and the molecular structure of the editing complex termed "editosome" (Maas and Rich 2000).

The elaboration of faithful in vitro editing systems, the development of genetic screens for RNA editing mutants and the further characterization of *cis*-acting elements as well as *trans*-acting factors will lead to further progress in our understanding of this curious genetic phenomenon which is of equally great interest for plastid genetics and general genetics.

10 Experimental Gene Transfer into Plastids

Experimental gene transfer is one of the most important methods in modern genetics. A variety of techniques has been elaborated for the transfer of genes into the nucleus of eukaryotes. Gene transfer into chloroplasts turned out to be particularly difficult, because such a method requires the transfer of foreign DNA through: (1) the cell wall, (2) the plasma membrane and (3) the double membrane of the plastid. Moreover – after the successful DNA transfer into the plastids – marker systems must be available which facilitate sorting of transformed plastid DNA molecules from wild-type copies.

A milestone for organelle genetics was the construction of a device capable of accelerating DNA-coated tungsten (or gold) particles to penetrate cells in an evacuated chamber (Sanford 1988). The use of this "biolistic" (= biological + ballistic) system, often termed "particle gun", allowed a successful gene transfer into, and thus, genetic transformation of the single large chloroplast of the green alga *Chlamydomonas reinhardtii* (Boynton et al. 1988). (In parallel experiments Johnston et al. 1988, transformed yeast mitochondria.)

The successful gene transfer experiments in *Chlamydomonas* also allowed the detailed study of the recombination processes within the organelle following the delivery of foreign DNA into the chloroplast (Boynton et al. 1992).

Only 2 years later, the research group of Pal Maliga accomplished the genetic transformation of plastids of the higher plant *Nicotiana tabacum*, also by particle bombardment (Svab et al. 1990; Maliga 1993). Although the transformation frequency was low in the first experiments, the methods for the selection of plastid transformants were considerably improved during the following years, so that already in 1993 plastid transformation in tobacco became an efficient and reproducible genetic technique (Carrer et al. 1993; Maliga 1993; Svab and Maliga 1993).

In parallel investigations, an alternative procedure for delivering foreign DNA into chloroplasts was developed: polyethylene glycol (PEG) treatment of tobacco protoplasts; these experiments also resulted in stable plastid transformation (Goulds et al. 1993; O'Neil et al. 1993; Koop et al. 1996).

Thus, nowadays two alternative methods are available for efficient gene transfer into higher plant chloroplasts.

Current efforts in the field of gene transfer into higher plants plastids represent three main directions:

1. Improvement of the methodology of gene transfer procedures and the selection of homoplasmic cells including e.g.
   - Analysis of the processes of homologous recombination between transforming DNA and the endogenous plastid DNA.
   - Development of efficient procedures for the selection of homoplasmic cells.
   - Elaboration of procedures to remove – after the successful incorporation of the desired foreign genes – the unnecessary vector parts from the transformed plastid DNA (Fischer et al. 1996; Zubko et al. 2000)
2. Use of gene transfer into chloroplasts and of transgenic chloroplasts for studies on gene expression including e.g.
   - Studies on the regulation of transgene expression at the transcriptional, posttranscriptional and translational levels.
   - Elucidation of plastid genome-encoded gene functions by reverse genetics (e.g. Ruf et al. 1997; Hager et al. 1999; Ruf et al. 2000)
3. Use of transgenic chloroplasts in plant biotechnology:
   The great perspectives of the use of plants with transgenic chloroplasts in biotechnology and the first promising results obtained have been reviewed in detail in the previous volume of Progress in Botany (Bock and Hagemann 2000) and will not be reviewed here again.

   I shall emphasize just one aspect which is of particular importance for the public discussion about the possible risks of transgenic plants. Most crop plants transmit their plastids purely maternally (cf. Sect. 5), i.e. the plastids are not transmitted via the pollen. When the transgenes, e.g. for herbicide resistance, for resistance against pathogenic insects or other traits, are integrated into the plastid genome, then there is no longer any danger of transgene transmission via the pollen to neighboring fields or weedy relatives of the crop plants.

   The combination of this advantage with the elimination of selectable marker genes after the stable incorporation of the desired transgenes, as mentioned under (1), is a further step towards the reduction of possible ecological risks connected with the use of transgenic crop plants.

An important task for future investigations on gene transfer in higher plants is the application and adaptation of the methodology, which is now so successful in tobacco (and recently has become possible in *Arabidopsis*; Sikdar et al. 1998) to important agricultural crop plants.

Acknowledgements. The author wishes to sincerely thank Prof. Dr. Ralph Bock, University of Münster, for helpful discussions of the manuscript and his valuable advice on the improvement of the text.

I wrote my first contributions for *Fortschritte der Botanik*, Vols. 28 and 30 on "Extrachromosomal Inheritance" (1966, 1968). Two years later, at the height of the Cold War, the Ministry of Universities in East-Berlin prohibited my further publication in this "West German publication series". Only in 1981, could I start writing again for . Since then, every second year, I have written , together with one or two coworkers the review "Extranuclear Inheritance: Plastid Genetics". I wish to emphasize and I gratefully appreciate that every year (between 1969 and 1980) Springer Publishers kept sending me the new volumes of *Fortschritte der Botanik* as a present (although I could not contribute to the series during these years).

References

Anderson S, Bankier AT, Barrell BG, DeBruijn MHL, Coulson AR, Drouin J, Eperon IC, Nierlich DP, Roa BA, Sanger F, Schreier PH, Smith AJH, Sraden R, Young IG (1981) Sequence and organisation of the human mitochrondrial genome. Nature 290:457–465

Baur E (1909) Das Wesen und die Erblichkeitsverhältnisse der "Varietates albomarginatae hort." von *Pelargonium zonale.* Z Indukt Abstammungs- Vererbungsl 1:330–351

Baur E (1910a) Vererbungs- und Bastardierungsversuche mit *Antirrhinum.* Z Indukt Abstammungs- Vererbungsl 3:34–98

Baur E (1910b) Untersuchungen über die Vererbung von Chromatophorenmerkmalen bei *Melandrium*, *Antirrhinum* und *Aquilegia.* Z Indukt Abstammungs- Vererbungsl 4:81–102

Baur E (1911) Einführung in die experimentelle Vererbungslehre, 1. Aufl. Gebrüder Borntraeger, Berlin

Baur E (1919) Einführung in die experimentelle Vererbungslehre, 3. Aufl. Gebrüder Borntraeger, Berlin

Beletski YD, Razoriteleva EK, Zhdanov YA (1969) Cytoplasmic mutations in sunflower, induced by N-Nitrosomethylurea. Dokl Acad Nauk SSSR 186:1425–1426

Benne R, Van den Burg J, Brakenhoff J, Sloof P, Van Boom JH, Tromp MC (1986) Major transcript of the frameshifted coxII from trypanosome mitochondria contains four nucleotides that are not encoded in the DNA. Cell 46:819–826

Bock R (1998) Analysis of RNA editing in plastids. Methods: A Companion to Methods in Enzymology 15:75–83

Bock R (2000) Sense from nonsense: how the genetic information of chloroplasts is altered by RNA editing. Biochimie (Paris) 82:549–557

Bock R (2001) RNA editing in plant mitochondria and chloroplasts. In: Bass BL (ed) RNA editing. Frontiers in molecular biology. Oxford University Press, Oxford, pp 38–60

Bock R, Hagemann R (2000) Extranuclear inheritance: plastid genetics: manipulation of plastid genomes and biotechnological applications. Prog Bot 61:76–90

Bock R, Koop H-U (1997) Extraplastidic site-specific factors mediate RNA editing in chloroplasts. EMBO J 16:3282–3288

Bock R, Hagemann R, Kössel H, Kudla J (1993) Tissue-specific and stage-specific modulation of RNA editing of the psbF and psbL transcript from spinach plastids - a new regulatory mechanism? Mol Gen Genet 247:439–443

Bock R, Hermann M, Kössel H (1996) In vivo dissection of *cis*-acting determinants for plastid RNA editing. EMBO J 15:5052–5059

Bock R, Kössel H, Maliga P (1994) Introduction of a heterologous editing site into the tobacco plastid genome: The lack of RNA editing leads to a mutant phenotype. EMBO J 13:4623–4628

Bock R, Albertazzi F, Freyer R, Fuchs M, Ruf S, Zeltz P, Maier RM (1997) Transcript editing in chloroplasts of higher plants. In: Schenk HEA, Herrmann R, Jeon KW, Müller NE, Schwemmler W (eds) Eukaryotism and symbiosis. Springer, Berlin Heidelberg New York, pp 123–137

Börner T, Hess W (1993) Altered nuclear, mitochondrial and plastid gene expression in white barley cells containing ribosome-deficient plastids. In: Kück U, Brennicke A (eds) Plant Mitochondria. Verlag Chemie, Weinheim, pp 207–220

Börner T, Schumann B, Hagemann R (1976) Biochemical studies on a plastid ribosome-deficient mutant of *Hordeum vulgare*. In: Bücher T, Neupert W, Sebald W, Werner S (eds) Genetics and biogenesis of chloroplasts and mitochondria. Elsevier/North-Holland, Amsterdam, pp 331–338

Börner T, Mendel RR, Schiemann J (1986) Nitrate reductase is not accumulated in chloroplast-ribosome deficient mutants of higher plants. Planta 169:202–207

Boynton J, Gillham NW, Harris EH, Hosler JP, Johnson AM, Jones AR, Randolph-Anderson BL, Robertson D, Klein TM, Shark KB, Sanford JC (1988) Chloroplast transformation in *Chlamydomonas* with high velocity microprojectiles. Science 240:1534–1538

Boynton J, Gillham NW, Newman SM, Harris EH (1992) Organelle genetics and transformation of *Chlamydomonas*. In: Herrmann RG (ed) Cell organelles. Springer, Wien New York, pp 3–64

Bradbeer JW, Atkinson YE, Börner T, Hagemann R (1979) Cytoplasmic synthesis of plastid polypeptides may be controlled by plastid-synthesized RNA. Nature 279:816–817

Carrer H, Hockenberry TN, Svab Z, Maliga P (1993) Kanamycin resistance as a selectable marker for plastid transformation in tobacco. Mol Gen Genet 241:49–56

Chaudhuri S, Maliga P (1996) Sequences directing C to U editing of the psbL mRNA are located within a 22 nucleotide segment spanning the editing site. EMBO J 15:5956–5964

Chaudhuri S, Maliga P (1997) New insights into plastid RNA editing. Trends Plant Sci 2:5–6

Clauhs RP, Grun P 1977) Changes in plastid and mitochondrion content during maturation of generative cells of *Solanum* (Solanaceae). Am J Bot 64:377–383

Correns C (1909a) Vererbungsversuche mit blaß(gelb)grünen und buntblättrigen Sippen bei *Mirabilis jalapa*, *Urtica pilulifera* und *Lunaria annua*. Z Indukt Abstammungs-Vererbungsl 1:291–329

Correns C (1909b) Zur Kenntnis der Rolle von Kern und Plasma bei der Vererbung. Z Indukt Abstammungs- Vererbungsl 2:331–340

Correns C (1922) Vererbungsversuche mit buntblättrigen Sippen. VI. Einige neue Fälle von Albomaculatio. Sitzungsber Preuß Akad Wiss 33:460–471

Correns C (1928) Über nichtmendelnde Vererbung. Z Indukt Abstammungs Vererbungsl Suppl I:131–168

Correns C, von Wettstein F (ed) (1937) Nicht mendelnde Vererbung. Handbuch der Vererbungswissenschaft, Bd. II H, Gebrüder Borntraeger, Berlin

Corriveau JS, Coleman AW (1988) Rapid screening method to detect potential biparental inheritance of plastid DNA and results for over 200 angiosperm species. Am J Bot 75:1443–1458

Covello PS, Gray MW (1989) RNA editing in plant mitochondria. Nature 341:662–666

Cypriani G, Testolin R, Morgante M (1995) Paternal inheritance of plastids in interspecific hybrids of the genus *Actinidia* revealed by PCR-amplification of chloroplast DNA fragments. Mol Gen Genet 247:693–697
Dahlgren KVO (1925) Die reziproken Bastarde zwischen *Geranium bohemicum* L. und seiner Unterart *deprehensum* Erik Almq. Hereditas 6:237–256
Döbel P, Hagemann R (1963) Elektronenmikroskopischer Nachweis echter Mischzellen bei *Antirrhinum majus* status albomaculatus. Biol Zentralbl 82:749–751
Edelman M, Schiff JA, Epstein HT (1965) Studies of chloroplast development in *Euglena.* XII. Two types of satellite DNA. J Mol Biol 11:769–774
Epp M (1973) Nuclear gene-induced plastome mutations in *Oenothera hookeri.* I. Genetic analysis. Genetics 75:465–483
Epp M, Parthasarthy MV (1987) Nuclear gene-induced plastome mutations in *Oenothera hookeri.* II. Phenotype description with electron microscopy. Am J Bot 74:143–151
Fischer N, Stampacchia O, Redding K, Rochaix J-D (1996) Selectable marker recycling in the chloroplast. Mol Gen Genet 251:373–380
Funaoka S (1924) Beiträge zur Kenntnis der Anatomie panaschierter Blätter. Biol Zentralbl 44:343–384
Gillham NW (1978) Organelle heredity. Raven Press, New York
Gillham NW (1994) Organelle genes and genomes. Oxford Univ Press, New York
Goulds T, Maliga P, Koop H-U (1993) Stable plastid transformation in PEG-treated protoplasts of *Nicotiana tabacum* Bio/Technology 11:95–97
Gordon KHJ, Crouse EJ, Bohnert HJ, Herrmann RG (1982) Physical mapping of differences in chloroplast DNA of the five wild-type plastomes in *Oenothera* subsection *Euoenothera.* Theoret Appl Genet 61:373–384
Gray MW (1993) Origin and evolution of organelle genomes. Curr Opin Genet Dev 3:884–890
Gualberto JM, Lamattina L, Bonnard G, Weil JH, Grienenberger JM (1989) RNA editing in wheat mitochondria results in the conservation of protein sequences. Nature 341:660–662
Hagemann R (1958) Somatische Konversion bei *Lycopersicon esculentum* Mill. Z Vererbungsl 89:587–613
Hagemann R (1960) Das Vorkommen von Mischzellen bei einer Gaterslebener Herkunft des Status albomaculatus von *Antirrhinum majus* L. Kulturpflanze 8:168–184
Hagemann R (1961) Über die Mischungsverhältnisse in Mischzellen des Status albomaculatus von *Antirrhinum majus* L. Kulturpflanze 9:163–170
Hagemann R (1964) Plasmatische Vererbung. Gustav Fischer Verlag, Jena
Hagemann R (1965) Advances in the field of plastid inheritance in higher plants. In:Geerts SJ (ed) Genetics today. Proc XI Int Congr Genetics, The Hague, vol 3. Pergamon Press, London, pp 613–625
Hagemann R (1966) Extrachromosomale Vererbung. Fortschr Bot 28, Springer, Berlin Heidelberg New York, pp 202–216
Hagemann R (1968) Extrachromosomale Vererbung. Fortschr Bot 30, Springer, Berlin Heidelberg New York, pp 225–241
Hagemann R (1976) Plastid distribution and plastid competition in higher plants and the induction of plastom mutations by Nitroso-urea-compounds. In: Bücher T, Neupert W, Sebald W, Werner S (eds) Genetics and biogenesis of chloroplasts and mitochondria. Elsevier/North-Holland, Amsterdam, pp 331–338
Hagemann R (1979) Genetics and molecular biology of plastids of higher plants. Stadler Genetics Symp (Univ Missouri, Columbia) 11:91–116
Hagemann R (1982) Induction of plastome mutations by nitroso-urea-compounds. In: Edelman M, Hallick RB, Chua N-H (eds) Methods in chloroplast molecular biology. Elsevier Biomedical Press, Amsterdam, pp 119–127

Hagemann R (1986) A special type of nucleus-plastid-interactions: nuclear gene induced plastome mutations. In: Akoyunoglou G, Senger H (eds) Regulation of chloroplast differentiation. AR Liss, New York, pp 455–466
Hagemann R (1990) Molecular basis of plastome mutations and their effects on photosynthesis. In: Baltscheffsky M (ed) Current research in photosynthesis, vol 3. Kluwer, Dordrecht, pp 429–436
Hagemann R (1992) Plastid genetics in higher plants. In: Herrmann RG (ed) Cell organelles. Springer, Wien, New York, pp 65–96
Hagemann R (1993a) Neuere molekulare und cytologische Aspekte der Plastiden-Genetik. Biol Zentralbl 112:244–287
Hagemann R (1993b) Studies towards a genetic and molecular analysis of paramutation at the sulfurea locus of *Lycopersicon esculentum* Mill. In: Yoder J (ed) Molecular biology of tomato. Technomic Publ Co Inc, Lancaster, PA, pp 75–82
Hagemann R (2000) Erwin Baur (1875–1933). Pionier der Genetik und Züchtungsforschung. Seine wissenschaftlichen Leistungen und ihre Ausstrahlung auf Genetik, Biologie und Züchtungsforschung von heute. Verlag R Kovar, Eichenau
Hagemann R, Börner T (1979) Genetische, cytologische und molekularbiologische Analyse der Plastiden höherer Pflanzen. Wiss Z Univ Halle 28:49–60
Hagemann R, Börner T (1981) Extranuclear inheritance: plastid genetics. Prog Bot 43:159–173
Hagemann R, Hagemann MM (1994) Extranuclear inheritance: plastid genetics. Prog Bot 55:260–275
Hagemann R, Lindenhahn M (1983) Induction of plastome mutants in *Oenothera* by nitroso-methyl-urea (NMU). Proc XV Int Congr Genetics, New Delhi, Part I Abstr 506. Oxford and IBH Publ Co, New Delhi, p 293
Hagemann R, Metzlaff M (1983) Extranuclear inheritance: plastid genetics. Prog Bot 45:212–227
Hagemann R, Scholz F (1962) Ein Fall geninduzierter Mutationen des Plasmotypus bei Gerste. Züchter 32:50–59
Hagemann R, Schroeder M-B (1985) New results about the presence of plastids in generative and sperm cells of Gramineae. In: Sexual reproduction in seed plants, ferns and mosses. PUDOC, Wageningen, pp 53–55
Hagemann R, Schroeder M-B (1989) The cytological basis of the plastid inheritance in angiosperms. Protoplasma 153:57–64
Hagemann R, Börner T, Herrmann F, Knoth R (1974) The use of plastome and gene mutants of higher plants in studying the genetic control of plastid development and function. Ontogenes (Moskwa) 5:273–283 (in Russian with English Summary)
Hagemann R, Lindenhahn M, Metzlaff M (1985) Extranuclear inheritance: plastid genetics. Prog Bot 47:208–227
Hagemann R, Hagemann MM, Metzlaff M (1987) Extranuclear inheritance: plastid genetics. Prog Bot 49:245–262
Hagemann R, Hagemann MM, Metzlaff M (1989) Extranuclear inheritance: plastid genetics. Prog Bot 51:238–250
Hagemann R, Bock R, Hagemann MM (1996) Extranuclear inheritance: plastid genetics. Prog Bot 57:197–217
Hagemann R, Hagemann MM, Bock R (1998) Extranuclear inheritance: plastid genetics. Prog Bot 59:108–129
Hager M, Biehler K, Illerhaus J, Ruff S, Bock R (1999) Targeted inactivation of the smallest plastid genome-encoded open reading frame reveals a novel and essential subunit of the cytochrome b_6f complex. EMBO J 18:5834–5842
Hallick RB, Hong L, Drager RG, Favreau MR, Montfort A, Orsat B, Spielmann A, Stutz E (1993) Complete sequence of *Euglena gracilis* chloroplast DNA. Nucleic Acids Res 21:3537–3544

Harris EH, Burkhart GD, Gillham NW, Boynton JE (1989) Antibiotic resistance mutations in the chloroplast 16 S and 23 S rRNA genes of *Chlamydomonas reinhardtii*: correlation of genetic and physical maps of the chloroplast genome. Genetics 123:281–292

Hedtke B, Börner T, Weihe A (1997) Mitochondrial and chloroplast phage-type RNA polymerase in *Arabidopsis*. Science 277:809–811

Hedtke B, Wagner I, Börner T, Hess WR (1999) Inter-organellar crosstalk in higher plants: impaired chloroplast development affects mitochondrial gene and transcript levels. Plant J 19:635–643

Hedtke B, Börner T, Weihe A (2000) One RNA polymerase serving two genomes. EMBO Rep 1:435–440

Hentrich W, Beger B (1974) Untersuchungen über die mutagene Effizienz von N-Nitroso-N-Methylharnstoff bei *Saintpaulia ionantha* H. Wendl. Arch Züchtungsf 4:29–43

Herbst W (1935) Über Kreuzungen in der Gattung *Hypericum* mit besonderer Berücksichtigung der Buntblättrigkeit. Flora 129:235–259

Hermann M, Bock R (1999) Transfer of plastid RNA-editing activity to novel sites suggests a critical role for spacing in editing-site recognition. Proc Natl Acad Sci USA 96:4856–4861

Herrmann F (1971a) Struktur und Funktion der genetischen Information in den Plastiden. II. Untersuchung der photosynthesedefekten Plastommutante (en:)alba-1 von *Antirrhinum majus* L. Photosynthetica 5:258–266

Herrmann F (1971b) Genetic control of pigment-protein complexes I and Ia of the plastid mutant en:alba-1 of *Antirrhinum majus*. FEBS Lett 19:267–269

Herrmann F, Hagemann R (1971) Struktur und Funktion der genetischen Information in den Plastiden. III. Genetik, Chlorophylle und Photosyntheseverhalten der Plastommutante "Mrs. Pollock" und der Genmutante "Cloth of Gold" von *Pelargonium zonale*. Biochem Physiol Pflanz (BPP) 162:390–409

Herrmann F, Matorin D, Timofeev K, Börner T, Rubin AB, Hagemann R (1974) Structure and function of the genetic information in plastids. IX. Studies on primary reactions of photosynthesis in plastome mutants of *Antirrhinum majus* and *Pelargonium zonale* having impaired photosynthesis. Biochem Physiol Pflanz (BPP) 165:393–400

Herrmann F, Schumann B, Börner T, Knoth R (1976) Struktur und Funktion der genetischen Information in den Plastiden. XI. Die plastidalen Lamellarproteine der photosynthesedefekten Plastommutanten en:gil-1 ("Mrs. Pollock") und der Genmutante "Cloth of Gold" von *Pelargonium zonale* Ait. Photosynthetica 10:164–171

Herrmann FH, Börner T, Hagemann R (1980) Biosynthesis of thylakoids and the membrane-bound enzyme systems of photosynthesis. In: Reinert J (ed) Chloroplasts. Results and problems in cell differentiation, vol 10. Springer, Berlin Heidelberg New York, pp 147–177

Herrmann RG (ed)(1992) Cell Organelles. Springer Verlag Wien New York

Herrmann RG, Maier RM (2000) Chloroplast thylakoid membranes. A paradigm for biogenetic and evolutionary complexity. In: Yunus M, Pathre U, Mohanty P (eds) Probing Photosynthesis. Mechanisms, regulation and adaptation. Taylor & Francis, London, pp 127–147

Herrmann RG, Possingham JV (1980) Plastid DNA - The Plastome. In: Reinert J (ed) Chloroplasts. Results and problems in cell differentiation, vol 10. Springer, Berlin Heidelberg New York, pp 45–96

Herrmann RG, Seyer P, Schedel R, Gordon K, Bisanz C, Winter P, Hildebrandt JW, Wlaschek M, Alt J, Driesel AJ, Sears BB (1980) The plastid chromosomes of several dicotyledons. In: Bücher T, Sebald W, Weiss (eds) Biological chemistry of organelle formation. Springer, Berlin Heidelberg New York, pp 97–112

Hess W, Börner T (1999) Organellar RNA polymerases of higher plants. Int Rev Cytol 190:1–59

Hess W, Prombona A, Fieder B, Subramanian AR, Börner T (1993) Chloroplast rps15 and the rpoB/C1/C2 gene cluster are strongly transcribed in ribosome-deficient plastids: evidence for a functioning non-chloroplast-encoded RNA polymerase. EMBO J 12:563–571

Hiesel R, Wissinger B, Schuster W, Brennicke A (1989) RNA editing in plant mitochondria. Science 246:1632–1634

Hoch B, Maier RM, Appel K, Igloi GL, Kössel H (1991) Editing of a chloroplast mRNA by creation of an initiation codon. Nature 353:178–180

Hosticka LP, Hanson MR (1984) Induction of plastid mutations in tomatoes by nitrosomethylurea. J Hered 75:242–246

Hübschmann T, Börner T (1998) Characterization of transcript initiation sites in ribosome-deficient plastids. Plant Mol Biol 36:493–496

Hupfer H, Swiatek M, Hornung S, Herrmann RG, Maier RM, Chiu W-L, Sears B (2000) Complete nucleotide sequence of the *Oenothera elata* plastid chromosome, representing plastome I of the five distinguishable *Euoenothera* plastomes. Mol Gen Genet 263:581–585

Johnston SA, Anziano PQ, Shark K, Sanford JC, Butow RA (1988) Mitochondrial transformation in yeast by bombardment with microprojectiles. Science 240:1538–1541

Kirk JTO (1963) The deoxyribonucleic acid of broad bean chloroplasts. Biochem Biophys Acta 76:417–424

Kirk JTO (1971) Will the real chloroplast DNA please stand up? In: Boardman NK, Linnane W, Smillie RM (eds) Autonomy and biogenesis of mitochondria and chloroplasts. Elsevier/North Holland, Amsterdam, pp 267–276

Kirk JTO (1986) The discovery of chloroplast DNA. BioEssays 4:36–38

Kirk JTO, Tilney-Bassett RAE (1967) The Plastids. Their chemistry, structure, growth and inheritance. WH Freeman, San Francisco

Kirk JTO, Tilney-Bassett RAE (1978) The Plastids. Their chemistry, structure, growth and inheritance, 2nd edn. Elsevier/North-Holland, Amsterdam

Koop H-U, Steinmüller K, Wagner H, Rößler C, Eibl C, Sacher L (1996) Integration of foreign sequences into tobacco plastome via polyethylene glycol-mediated protoplast transformation. Planta 199:193–201

Kowallik KV (1992) Origin and evolution of plastids from chlorophyll a+c containing algae: suggested ancestral relationships to red and green algal plastids. In:Lewin RA(ed) Origins of plastids. Chapman and Hall, New York, pp 223–263

Kowallik KV (1994) From endosymbionts to chloroplasts: evidence for a single prokaryotic/eukaryotic endocytobiosis. Endocytobiosis Cell Res 10:137–149

Kowallik KV, Stroebe B, Schaffran I, Kroth-Pancic P, Freier U (1995) The chloroplast genome of a chlorophyll a+c-containing alga, *Odontella sinensis*. Plant Mol Biol Rep 13:336–342

Kudla J, Igloi G, Metzlaff M, Hagemann R, Kössel H (1992) RNA editing in tobacco chloroplasts leads to the formation of a translatable psbL messenger RNA by a C to U substitution within the initiation codon. EMBO J 11:1099–1103

Leff J, Mandel M, Epstein HT, Schiff JA (1963) DNA satellites from cells of green and aplastidic algae. Biochem Biophys Res Comm 13:126–130

Maas S, Rich A (2000) Changing genetic information through RNA editing. BioEssays 22:790–802

Maier RM, Zeltz P, Kössel H, Bonnard G, Gualberto JM, Grienenberger JM (1996) RNA editing in plant mitochondria and chloroplast. Plant Mol Biol 32:343–365

Maly R, Wild A (1956) Ein cytologischer Beitrag zur "Entmischungstheorie" verschiedener Plastidensorten. Z Indukt Abstammungs- Vererbungsl 87:493–496

Maliga P (1993) Towards plastid transformation in flowering plants. Trends Biotechnol 11:101–107

Maliga P, Carrer H, Kanevski I, Staub JM, Svab Z (1993) Plastid engineering in land plants: a conservative genome is open to change. Philos Trans R Soc Lond B 342:203–208

Martinez-Zapater JM, Gil P, Capel J, Somerville CR (1992) Mutations at the *Arabidopsis* CHM locus promote rearrangements of the mitochondrial genome. Plant Cell 4:889–899

Masoud SA, Johnson LB, Sorensen EL (1990) High transmission of paternal plastid DNA in alfalfa plants demonstrated by restriction fragment polymorphic analysis. Theor Appl Genet 79:49–55

Maxam AM, Gilbert W (1977) A new method for sequencing DNA. Proc Natl Acad Sci USA 74:560–564

McIntosh L, Poulsen C, Bogorad L (1980) Chloroplast gene sequence for the large subunit of ribulose bisphosphate carboxylase of maize. Nature 288:556–560

Metzlaff M, Pohlheim F, Börner T, Hagemann R (1982) Hybrid variegation in the genus *Pelargonium*. Curr Genet 5:245–249

Mogensen HL (1988) Exclusion of male mitochondria and plastids during syngamy in barley as a basis for maternal inheritance. Proc Natl Acad Sci USA 85:2594–2597

Morgan TH (1919) The physical basis of heredity. JB Lippincott, Philadelphia

Morgan TH (1921). Die stoffliche Grundlage der Vererbung. (Übersetz. H Nachtsheim). Gebrüder Borntraeger, Berlin

Noack KL (1931) Über *Hypericum*-Kreuzungen. I. Die Panaschüre der Bastarde zwischen *Hypericum acutum* Moench und *Hypericum montanum* L. Z Indukt Abstamm- Vererbungsl 59:77–101

Noack KL (1934) Über *Hypericum*-Kreuzungen. IV. Die Bastarde zwischen *Hypericum acutum* Moench, *montanum* L, *quadrangulum* L, *hirsutum* L und *pulchrum* L. Z Bot 28:1–71

Ohba K, Iwakawa M, Ohada Y, Murai M (1971) Paternal transmission of a plastid anomaly in some reciprocal crosses of Suzi, *Cryptomeria japonica* D. Don. Silvae Genet 210:101–107

Ohyama K, Fukuzawa H, Kohchi T, Shirai H, Sano T, Sano S, Umesono K, Shiki Y, Takeuchi M, Chang Z, Aota S-I, Inokuchi H, Ozeki H (1986) Chloroplast gene organization deduced from complete sequence of liverwort *Marchantia polymorpha* chloroplast DNA. Nature 322:572–574

O'Neill C, Horvath GV, Horvath E, Dix PJ, Medgyesy P (1993) Chloroplast transformation in plants: polyethylene glycol (PEG) treatment of protoplasts is an alternative to biolistic delivery systems. Plant J 3:729–738

Palmer JD (1985) Comparative organization of chloroplast genomes. Annu Rev Genet 19:325–354

Pohlheim F (1974) Nachweis von Mischzellen in variegaten Adventivsprossen von *Saintpaulia*, entstanden nach Behandlung isolierter Blätter mit N-Nitroso-N-Methylharnstoff. Biol Zentralbl 93:141–148

Pohlheim F (1986) Hybrid variegation in crosses between *Pelargonium zonale* (L.) l'Herit.ex Ait. and *Pelargonium inquinans* (L.) l'Herit.ex Ait. Plant Breed 97:93–96

Pohlheim F, Beger B (1974) Erhöhung der Mutationsrate im Plastom bei *Saintpaulia ionantha* H. Wendl. Biochem Physiol Pflanz 169:377–383

Powell LM, Wallis SC, Pease RJ, Edwards YH, Knott TJ, Scott J (1987) A novel form of tissue-specific RNA processing produces apolipoprotein-B48 in intestine. Cell 50:831–840

Ray DS, Hanawalt PC ((1964) Properties of the satellite DNA associated with the chloroplasts of *Euglena gracilis*. J Mol Biol 9:812–824

Redei GP (1973) Extrachromosomal mutability determined by a nuclear gene locus in *Arabidopsis*. Mutat Res 18:149–162

Redei G, Plurad SB (1973) Hereditary structural alterations of plastids induced by a nuclear gene in *Arabidopsis*. Protoplasma 77:361–380

Reith M (1995) Molecular biology of rhodophyte and chromophyte plastids Annu Rev Plant Physiol Plant Mol Biol 46:549–575
Reith M, Munholland J (1995) Complete nucleotide sequence of the *Porphyra purpurea* chloroplast genome. Plant Mol Biol Rep 13:333–335
Renner O (1922) Eiplasma und Pollenschlauchplasma bei den Oenotheren. Z Indukt Abstammungs- Vererbungsl 27:235–237
Renner O (1924) Die Scheckung der Oenotherenbastarde. Biol Zentralbl 27:309–336
Renner O (1929) Artbastarde bei Pflanzen. Handbuch der Vererbungswissenschaft, Bd. IIA. Gebrüder Borntraeger, Berlin
Renner O (1934) Die pflanzlichen Plastiden als selbständige Elemente der genetischen Konstitution. Ber Verhandl Sächs Akad Wiss Leipzig Math-phys Kl 86:241–266.
Renner O (1936) Zur Kenntnis der nichtmendelnden Buntheit der Laubblätter. Flora 130:218–290.
Renner O (1937) Zur Kenntnis der Plastiden- und Plasmavererbung. Cytologia (Tokyo) Fuji Jub vol, Pars II, pp 644–653
Ris H, Plaut W (1962) Ultrastructure of DNA-containing areas in the chloroplast of *Chlamydomonas.* J Cell Biol. 13:383–391
Rochaix J-D (1992) Post-transcriptional steps in the expression of chloroplast genes. Annu Rev Cell Biol 8:1–28
Ruhland W, Wetzel K (1924) Der Nachweis von Chloroplasten in generativen Zellen von Pollenschläuchen. Ber Dtsch Bot Ges 42:3–14
Ruf S, Kössel H, Bock R (1997) Targeted inactivation of a tobacco intron-containing open reading frame reveals a novel chloroplast-encoded photosystem I related gene. J Cell Biol 139:95–102
Ruf S, Biehler K, Bock R (2000) A small chloroplast-encoded protein as a novel architectural component of the light-harvesting antenna. J Cell Biol 149:369–377
Russel SD (1984) Ultrastructure of the sperm cell of *Plumbago zeylanica.* II. Quantitative cytology and three-dimensional organization. Planta 162:385–391
Russel SD (1987) Quantitative cytology of the egg and central cell of *Plumbago zeylanica* and its impact on cytoplasmic inheritance patterns. Theor Appl Genet 74:693–699
Russel SD (1992) Double fertilization. Int Rev Cytol 140:357–388
Sanger F, Micklen S, Coulson AR (1977) DNA sequencing with chain-terminating inhibitors. Proc Natl Acad Sci USA 74:5463–5467
Sager R, Ishida MR (1963) Chloroplast DNA in *Chlamydomonas.* Proc Natl Acad Sci USA 50:725–730
Schaffner C (1995) Molekulargenetische und physiologische Untersuchungen an photosynthesedefizienten Plastommutanten von *Antirrhinum majus* L. Diss, Math-Nat Tech Fak, Martin-Luther-Univ Halle
Schaffner C, Laasch H, Hagemann R (1995) Detection of point mutations in chloroplast genes of *Antirrhinum majus* L. I. Identification of a point mutation in the psaB gene of a photosystem I plastome mutant. Mol Gen Genet 249:533–544
Schötz F (1958) Periodische Ausbleichungserscheinungen des Laubes bei *Oenothera.* Planta 43:182–240
Schuster G, Bock R (2001) Editing, polyadenylation and degradation of mRNA in the chloroplast. In: Andersson B, Aro E-M (eds) Advances in Photosynthesis. Kluwer, Dordrecht, pp 121–136
Schroeder M-B (1986) Ultrastructural studies on plastids of generative and vegetative cells in Liliaceae. 4. Plastid distribution during generative cell maturation in *Convallaria majalis* L. Biol Zentralbl 105:427–433
Schwarz Z, Kössel H (1980) The primary structure of the 16 S rDNA from *Zea mays* chloroplast is homologous to E. coli 16 S rRNA. Nature 283:739–742
Schwemmle J (1940) Plastidenmutationen bei Eu-Oenotheren. Z Indukt Abstammungs-Vererbungsl 75:358–800
Schwemmle J (1943) Plastiden und Genmanifestation. Flora 137:81–72

Schwemmle J, Haustein E, Sturm A, Binder M (1938) Genetische und zytologische Untersuchungen an Eu-Oenotheren, Teil I – VI. Z Indukt Abstamm- Vererbungsl 73:358–800

Sears BB, Sokalski MB (1991) The *Oenothera* plastome mutator: effect of UV irradiation and nitroso-methyl-urea on mutation frequencies. Mol Gen Genet 229:245–252

Shinozaki K, Ohme M, Tanaka M, Wakasugi T, Hayashida N, Matsubayashi T, Zaita N, Chunwongse J, Obokata J, Yamaguchi-Shinozaki K, Ohto,C, Torazawa K, Meng B-Y, Sugita M, Deno H, Kamogashira T, Yamada K, Kusuda J, Takaiwa F, Kato A, Tohdoh N, Shimada H, Sugiura M (1986) The complete nucleotide sequence of the tobacco chloroplast genome: its gene organization and expression. EMBO J 5:2043–2049

Siemenroth A, Wollgiehn R, Neumann D, Börner T (1981) Synthesis of ribosomal RNA in ribosome-deficient plastids of the mutant "albostrians" of *Hordeum vulgare* L. Planta 153:547–555

Sikdar SR, Serino G, Chaudhuri S, Maliga P (1998) Plastid transformation in *Arabidopsis thaliana.* Plant Cell Rep 18:20–24

Staub JM, Maliga P (1995) Expression of a chimeric uidA gene indicates that polycistronic mRNAs are efficiently translated in tobacco plastids. Plant J 7:845–848

Stirewalt VL, Michalowski CB, Löffelhardt W, Bohnert H, Bryant DA (1995) Nucleotide sequence of the cyanelle genome from *Cyanophora paradoxa.* Plant Mol Biol Rep 13:327–332

Stocking CR, Gifford EM (1959) Incorporation of thymidine into chloroplasts of *Spirogyra.* Biochem Biophys Res Comm 1:159–164

Stubbe H (1938) Genmutation. 1. Allgemeiner Teil. Handbuch der Vererbungswissenschaft, Bd IIF. Gebrüder Borntraeger, Berlin

Stubbe W (1959) Genetische Analyse des Zusammenwirkens von Genom und Plastom bei *Oenothera.* Z Vererbungsl 90:288–298

Stubbe W (1960) Untersuchungen zur genetischen Analyse des Plastoms von *Oenothera.* Z Bot 48:191–218.

Stubbe W (1964) The role of the plastome in the evolution of the genus *Oenothera.* Genetica 35:28–33

Svab Z, Maliga P (1993) High-frequency plastid transformation in tobacco by selection for a chimeric aadA gene. Proc Natl Acad Sci USA 90:913–917

Svab Z, Hajdukiewicz P Maliga P (1990) Stable transformation of plastids in higher plants. Proc Natl Acad Sci USA 87:913–917

Trebst A, Depka B, Kipper M (1990) The topology of the reaction center polypeptides of photosystem II. In: Baltscheffsky M (ed) Current research in photosynthesis, vol 1. Kluwer, Dordrecht, pp 217–222

Van Went, JL, Willemse MTM (1984) Fertilization. In: Johri BM (ed) Embryology of angiosperms. Springer, Berlin Heidelberg New York, pp 273–317

Vogel J, Börner T, Hess WR (1999) Comparative analysis of splicing of the complete set of chloroplast group II introns in three higher plant mutants. Nucl Acids Res 27:3866–3874

Wells R, Birnstiel M (1967) A rapidly renaturing deoxyribonucleic component associated with chloroplast preparations. Biochem J 105:53P-54P

Wild A (1958) Experimentelle Beeinflussung des Granamusters einer abweichenden Plastidensorte von *Antirrhinum majus.* Planta 50:379–387

Wild A (1959) Untersuchung zweier albomakulater Linien von *Antirrhinum majus* auf ihr Verhalten in Teilreaktionen der Photosynthese. Beitr Biol Pflanzen 35:137–175

Winge Ö (1919) On the non-mendelian inheritance in variegated plants. C R Trav Labor Carlsberg 14:1–20.

Winter P, Herrmann RG (1988) A five-base-pair-deletion in the gene for the large subunit causes the lesion in the ribulose bisphosphate carboxylase/oxygenase-deficient plastome mutant sigma of *Oenothera hookeri.* Bot Acta 101:68–75

Wollgiehn R, Mothes K (1963) Über DNS in den Chloroplasten von *Nicotiana rustica*. Naturwissenschaften 50:95–96

Wylie RB (1941) Some aspects of fertilization in *Vallisneria*. Am J Bot 38:419–434

Zubko E, Scutt C, Meyer P (2000) Intrachromosomal recombination between attP regions as a tool to remove selectable marker genes from transgenes. Nat Biotechnol 18:442–445

Note added in proof:
Further new aspects of plastid genetics and genomics are dealt with in the article
Bock R, Hippler M.: Extranuclear Inheritance: Functional genomics in chloroplasts
in *this* volume of Progress in Botany 63:106–131, 2002

Prof. Dr. Rudolf Hagemann
Jägerplatz 3
06108 Halle (Saale), Germany

Genetics

Structural Genome Analysis Using Molecular Cytogenetic Techniques

By Rod Snowdon, Barbara Kusterer, and Renate Horn

1 Introduction

Knowledge of the physical organisation of DNA sequences within the genome is critical for the understanding of genome structure and function. Classical cytogenetic methods rely on karyological landmarks that may be more or less informative depending on the chromosomal distribution of banding patterns. Because the frequency of such bands generally depends on the distribution of repetitive sequences in the genome, chromosome identification is often extremely difficult in plant species with genomes containing little repetitive DNA. The prime example in this respect is *Arabidopsis thaliana*, where surprisingly little is known about the physical chromosome structure (Fransz et al. 1998; Heslop-Harrison 1998) in comparison to the enormous amount of information now available from *Arabidopsis* genome analyses using molecular markers and sequences (Meinke et al. 1998; Kaul et al. 2000).

The introduction of **fluorescence in situ hybridisation (FISH)** for plants by Schwarzacher et al. (1989) made a new tool for cytological studies available that proved to be extremely efficient in analysing introgressions, intergenomic translocations, relationships between genomes and species as well as the evolution of genomes and repetitive sequences. In combination with molecular analyses (e.g. RFLP) or conventional staining methods (e.g. C-banding) fluorescence in situ hybridisation, especially using tandem repeats, allows chromosomes in many species and hybrids to be identified unambiguously.

2 Molecular Cytogenetic Tools

a) Classical Versus Molecular Cytological Analysis

Molecular cytogenetic methods utilising fluorescence in situ hybridisation (Schwarzacher and Heslop-Harrison 2000) allow the precise physical localisation of genes or DNA sequences directly on cytological preparations. In cases where classical karyotype analysis is insufficient, FISH

Progress in Botany, Vol. 63

can give important physical information on the location and distribution of labelled DNA probes, regardless of genome or chromosome size. Various classes of repetitive DNA sequences – particularly 45 S/5 S ribosomal DNA and centromeric repeat sequences – have now been localised on the chromosomes of most important plant genomes (reviewed by Heslop-Harrison 2000), including those with cytogenetically less amenable chromosomes like *Arabidopsis* (Maluszynska and Heslop-Harrison 1993a; Fransz et al. 1998, 2000; Heslop-Harrison et al. 1999), tobacco (Lim et al. 2000), rice (Kamisugi et al. 1994; Ohmido and Fukui 1995) and *Brassica* (Maluszynska and Heslop-Harrison 1993b; Harrison and Heslop-Harrison 1995; Snowdon et al. 1997a, 2000a). Molecular cytogenetic techniques also provide a valuable complement to systems like the cereals where extensive karyotype information is already available. The primary example of this is wheat, where FISH has been particularly helpful for the characterisation of agronomically important wheat-alien translocation lines (e.g. Friebe et al. 1996; Schubert et al. 1998).

As ongoing technological development has constantly improved the resolution and detection of fluorescence signals over the past two decades, FISH has been applied increasingly for studies of the physical genome organisation of various DNA sequence classes in plant genomes. The diverse applications of FISH in plants for physical genome analysis are covered in excellent reviews by Jiang and Gill (1994, 1996), Gill and Friebe (1998) and de Jong et al. (1999).

b) Genome Analysis Using FISH and GISH

FISH enables the physical localisation of labelled DNA probes to discrete chromosomes or chromosomal regions (de Jong et al. 1999). Depending on the type of DNA sequence used as a probe it is possible with FISH to localise all categories of DNA from large, single-copy probes through to various classes of repetitive sequences (Linares et al. 2000; Pickering et al. 2000; Taketa et al. 2000; Schubert et al. 2001).

In plants, the highly condensed chromatin and the difficulty of obtaining preparations free of cell-wall debris and cytoplasm means that in most cases single-copy sequences less than about 10 kb in length cannot be reliably detected on metaphase chromosomes by FISH. For investigations of the genomic organisation of repetitive sequences, however, in situ hybridisation is the method of choice. With FISH it is possible to obtain direct physical information on the spatial distribution of repetitive DNA at the chromosomal level (e.g. Brandes et al. 1997; Schmidt and Heslop-Harrison 1998). Hence a more complete overview can be obtained of the function of various chromosomal structures (Houben et al. 1996; Fuchs et al. 1996; Heslop-Harrison et al. 1999), particularly when the physical organisation of the chromosome can be compared to the detailed information that is now available from the complete sequencing of *Arabidopsis thaliana* chromosomes (Lin and Kaul 1999; Mayer and Schuller 1999; Salanoubat et al. 2000; Theologis et al. 2000) and of internal chromosome features (McCombie and de la Bastide 2000).

Genomic in situ hybridisation (GISH; Heslop-Harrison and Schwarzacher 1996) enables the visualisation of different genome components in polyploids or interspecific hybrids. Where the parental or ancestral genomes are sufficiently heterogeneous, GISH is an extremely effective method for analysis of chromosome additions, introgressions, and genome homology which has been applied to plants as diverse as rice (Fukui et al. 1997), cereals (Chen et al. 1995; Pickering et al. 1997; Zhou et al. 1998), grasses (Cao et al. 2000), potato, tomato (Garriga-Calderé et al. 1997; Escalante et al. 1998), tobacco (Kitamura et al. 1997), onions (Hou and Peffley 2000; Khrustaleva and Kik 2000), peanut (Raina and Mukai 1999), *Brassica* (Fahleson et al. 1997; Skarzhinskaya et al. 1998; Snowdon et al. 1997a, 2000b), banana (D'Hont et al. 2000) and citrus (Pedrosa et al. 2000), along with many others.

Because GISH signals are conferred by genome-specific tandem and dispersed repeat DNA, it is possible with GISH to visualise the distribution of such sequences within the genome. However, apart from the distribution of repetitive sequence over the chromosomes the degree of condensation of the chromosomes may play a role in "painting" chromosomes (Fahleson et al. 1997; Schubert et al. 2001). The small chromosomes of species like *Brassica*, rice and *Arabidopsis* show an unusually low proportion of tandem and dispersed repeat sequences in their chromosome arms, with GISH signals almost exclusively at the pericentromeric heterochromatin (Skarzhinskaya et al. 1998; Snowdon et al. 2000a). Hence it is difficult or impossible in these species to detect small chromosome introgressions on chromosome arms by GISH.

The extent to which karyotype similarity and meiotic pairing reflects interspecies genome relationships may often be difficult to determine using conventional methods. GISH can offer answers to questions like (1) how do identical karyotypes in related species indicate close molecular homology, and (2) to what extent are homoeologous meiotic pairing partners homologous in molecular terms (Parokonny et al. 1997).

c) Development of Addition, Deletion and Substitution Lines

Chromosome addition, deletion and substitution lines represent an extremely valuable tool for genome analysis, because they allow direct localisation of genes to specific chromosomes based on their expression or non-expression in lines carrying the respective added, deleted or substituted chromosome or chromosome segment. For example, characterisation of the now vast collection of wheat deletion stocks (reviewed by Endo and Gill 1996) has played a crucial role not only in the physical genome mapping of wheat itself but also in studies of the biology of plant chromosomes in general. Improvements in embryo rescue and somatic hybridisation methodologies have greatly increased the range of possibilities for development of chromosome addition lines by interspecific and intergeneric hybridisation. Two leading examples of this are the recovery of maize addition chromosomes in oat (Riera-Lizarazu et al. 1996) and the development of tomato addition lines in potato (Garriga-Calderé et al. 1997). In *Brassica* and other species with a high degree of genome duplication, cytological stocks are a valuable asset for gene localisation and studies of genome organisation. Addition lines of *Brassica nigra* in *B. napus* and *B. oleracea*, respectively, were used to localise blackleg resistance to chromosome 4 of *B. nigra* (Chévre et al. 1996) and to investigate the genetics of *B. nigra* self-incompatibility (Chévre et al. 1997b). Chen et al. (1997a) developed *B. campestris-B. alboglabra* addition lines which helped to identify homologous regions in the A and C genomes of the amphidiploid *B. napus* and to assign seed-colour genes to a C-genome chromosome (Chen et al. 1997b). For breeding purposes, addition or substitution lines containing genes of economic interest are an ideal starting point for chromosomal introgression of novel traits into important agricultural crops.

In addition, alien chromosome additions are also becoming increasingly useful in molecular cytogenetic studies (Garriga-Calderé 1999a). Examples include: (1) determination of synteny between species and increased accuracy of genetic maps (Chen et al. 1997a; Suen et al. 1997), (2) molecular localisation and cloning of alien genes (Ishii et al. 1994; Potz et al. 1996; Iwano et al. 1998), (3) physical mapping of chromosomes (Gill et al. 1996), (4) generation of chromosome-specific DNA libraries (Riera-Lizarazu et al. 1996; Ananiev et al. 1997; Cheng et al. 2001), (5) introgression mapping (King et al. 1997) and (6) unravelling of the molecular organisation of individual chromosomes or chromosome segments (Fransz et al. 1996b; Zhong et al. 1996, 1998).

3 Chromosome Introgressions from Foreign Genomes

Interspecific or intergeneric hybrids have been used in many crops to significantly broaden the genetic variability available to the modern plant breeder (e.g. Friebe et al. 1996; Korell et al. 1996a, 1996b; Wolters et al. 1994). If such hybrids – produced by sexual and somatic hybridisation or microprotoplast fusion – are to be further used in breeding, the following aspects have to be considered (Jacobsen et al. 1995; Garriga-Calderé et al. 1997): (1) the possibility of backcrossing the hybrids to the parents, (2) the transmission of the individual genomes, or individual chromosomes, from the hybrids to the progeny in subsequent generations, (3) the occurrence and extent of any genetic recombination between the homoeologous chromosomes, and (4) the structural integrity of individual chromosomes in an alien genomic background and the possibility of producing lines with alien chromosome additions and/or substitutions. The combination of GISH and RFLP allows analysis of the chromosome constitution and enables the fate and inheritance of alien chromosomes and chromosome translocations to be followed. RFLP analysis can identify the presence of particular individual chromosomes using chromosome specific DNA probes, but it cannot determine whether a chromosome is present in disomic or monosomic form, nor is it possible to estimate whether each chromosome is totally or partially represented (Garriga-Calderé et al. 1997). GISH of mitotic (root tips) and meiotic (pollen mother cells) preparations can be exceptionally helpful to elucidate the chromosome constitution (Jacobsen et al. 1995). Larger alien chromosome segments might contain detrimental genes linked to the desired trait, resulting in linkage drag (Fahleson et al. 1997). Therefore, intergenomic translocation of a small alien chromosome segment is preferable in most cases.

The introgression of resistance to blackleg disease to *Brassica napus* (AACC, 2n=38) can help to identify the mechanism of chromosome introgression. In existing rapeseed cultivars, resistance to the blackleg complex has its origin almost exclusively in introgressions from the B genome of either diploid *B. nigra* (BB, 2n=16) or *B. juncea* (AABB, 2n=36). From B. *juncea* crosses, the resistance genes were found to have been transferred at the homoeologous position in the *B. napus* genome; the introgressions replaced the corresponding *B. napus* fragment (Chévre et al. 1997a; Barret et al. 1998). Similar studies are in progress from other introgressed lines (Anne-Marie Chévre, personal communication) but it seems that the existence of homology between the region of the donor genome containing the gene or genes of interest, and a corresponding region in the recipient genome, is prerequisite for the natural chromosomal introgression of alien germplasm into a crop genome by homologous recombination. Fortunately, comparative genome studies (reviewed in Heslop-Harrison 2000) have shown that genome homoeology is widespread not only between modern crop plants and their close relatives but also across genera and even between different plant families. This has great implications for the expansion of the available genetic diversity in crop species and the potential introduction (or reintroduction) of valuable germplasm into plant breeding programmes.

a) Structural Genome Differentiation in Solanaceae

A large number of both symmetric and asymmetric hybrids and cybrids between tomato and other *Lycopersicon*, *Solanum* and even *Nicotiana* species have been produced by somatic hybridisation and microprotoplast fusions that could not be obtained via sexual hybridisation due to incompatibilities and loss of fertility (reviewed in Wolters et al. 1994).

The value of genomic in situ hybridisation (GISH) in combination with RFLP analysis using chromosome-specific markers has been especially demonstrated in intergeneric somatic hybrids between *Lycopersicon* and *Solanum* (Jacobsen et al. 1995; Garriga-Calderé et al. 1997, 1998) but also in interspecific hybrids within the genus *Lycopersicon* (Parokonny et al. 1997).

In modern tomato cultivars most of the disease resistances, as well as other useful traits such as a high level of soluble solids, fruit characteristics and stress tolerances, have been introduced from several different wild *Lycopersicon* species into *Lycopersicon esculentum* (Rick et al. 1987). Genomic in situ hybridisation to investigate genome interactions in allohexaploid (2n=6x=72) *Lycopersicon esculentum* (+) *L. peruvianum* somatic hybrids and their backcross progenies with *L. esculentum* required an increase in hybridisation stringency to enable distinction of sequences with 90–95% homology (Parokonny et al. 1997), as standard GISH procedures allow sequences sharing only 80–85% homology to remain hybridised.

Somatic karyotypes of the two closely related species *L. esculentum* and *L. peruvianum* are morphologically almost identical and most of the somatic chromosomes cannot be distinguished using conventional cytological procedures such as Feulgen, aceto-orcein staining or even C-banding. GISH revealed that both of the investigated somatic hybrids (2n=6x=72) comprised a diploid chromosome set (2n=2x=24) from *L. esculentum* and a tetraploid chromosome set (2n=4x=48) from *L. peruvianum* (Parokonny et al. 1997). In BC_1, the somatic chromosome complement of allodiploid plants (2n=2x=24) consisted of a haploid chromosome set from both *L. esculentum* and *L. peruvianum*. The allodiploid nature of the BC_1 generation allowed the entire haploid sets of each species to engage in homoeologous crossing-over and recombination. The BC_2 and BC_3 generations were characterised by segmental allodiploids (2n=24) showing a gradual reduction in the number of *L. peruvianum* chromosomes and between zero and five recombinant chromosomes (Parokonny et al. 1997).

Somatic hybridisation provides a way of transferring genes, via homoeologous crossing-over and recombination, across the incompatibility barriers between species such as *L. esculentum* and *L. peruvianum* (both 2n=2x=24) which – although closely related – are sexually incompatible and difficult to cross (Parokonny et al. 1997).

Somatic hybrids between *L. esculentum* and *S. lycopersicoides* (Hossain et al. 1994) were analysed by GISH revealing that from eight investigated plants four were tetraploids (2n=48) with an equal number of chromosomes derived from each parent, and the other four were hexaploid containing an average of two sets of tomato chromosomes and one set from the wild parent (Escalante et al. 1998). RFLP analysis of the progenies showed the presence of both parent-specific alleles, the loss of some as well as the presence of a few non-parental alleles, indicating rearrangements and/or recombinations of the nuclear DNA.

Dong et al. (1999) performed GISH on somatic hybrids of *Solanum etuberosum* and *S. tuberosum*, together with five BC_1 and three BC_2 plants, to elucidate their chromosome constitution. Pollination of tetraploid potatoes by S. *phureja*, known as a "dihaploid inducer", allows production of dihaploids. In potato, dihaploids (2n=24) play a critical role in the genetic improvement of *Solanum tuberosum* (2n=48) by simplifying the complex genetics of the cultivated crop. Genomic in situ hybridisation of the dihaploid PDH55 (*Solanum tuberosum*) demonstrated that DNA from the dihaploid inducer is stably incorporated by somatic translocation (Wilkinson et al. 1995).

Sexual hybridisation of potato and tomato is not possible, however somatic fusion hybrids have been successfully produced (Melchers et al. 1978; Shepard et al 1983; Jacobsen et al. 1992; Schoenmakers et al. 1992). Although they are intergeneric hybrids the parental species, potato and tomato, possess some common cytogenetic features: (1) their chromosomes are morphologically similar (Yeh and Peloquin 1965; Ramanna and Wagenvoort 1976), (2) their molecular linkage maps are nearly homosequential (Tanksley et al. 1992), and (3) homoeologous chromosome pairing and recombination might occur in fusion hybrids (de Jong et al. 1993). Compared to somatic tomato hybrids (Parokonny et al. 1997), tomato and potato genomes can be easily distinguished by GISH (Garriga-Calderé et al. 1997, 1998; Jacobsen et al. 1995), which is consistent with the results of Ganal et al. (1988) indicating that highly repetitive DNA sequences found in tomato have undergone rapid divergence since the separation of *Lycopersicon* from *Solanum*. In addition, the presence or absence of individual chromosomes can be determined through RFLP analysis using chromosome-specific DNA probes (Jacobsen et al. 1995; Garriga-Calderé et al. 1997, 1999a).

Investigating BC_1 and BC_2 backcross progenies derived from a hexaploid potato (+) tomato fusion hybrid (2n=6x=72), Jacobsen et al. (1995) detected more alien tomato chromosomes with GISH than by RFLP. This indicated the presence of some tomato chromosomes in duplicate and others in haploid condition (Jacobsen et al. 1995). In BC_1 tomato chromosome 1, 3 and 6 were present in duplicate, and chromosome 8, 9 and 10 in haploid condition. In BC_2 plants the number of tomato chromosomes varied from one to six.

For the development of a complete series of tomato-chromosome addition-substitution lines in a potato background Garriga-Calderé et al. (1997) used hexaploid potato (+) tomato fusion hybrids backcrossed to the tetraploid (2n=2x=48) potato. Three BC_1 parents were further backcrossed to different tetraploid potato pollinators to produce 97 BC_2 plants (Garriga-Calderé et al. 1998). The number of alien tomato chromosomes transmitted through the female BC_1 ranged from zero to six depending on the chromosome constitution of the BC_1. By a combination of GISH and RFLP analyses the genome composition of the BC_1 progenies was established (Garriga-Calderé et al. 1997). Among the BC_2 plants generated, a total of 27 single additions were detected for as many as seven different chromosomes (1, 2, 4, 6, 8, 10, and 12) out of 12 possible types (Garriga-Calderé et al. 1998, 1999b). One BC_2 derivate 2101-1 consisted of four genomes of potato with two chromosomes 10 of tomato, which however had a different size due to a deletion in the paracentromeric heterochromatic region of the long arm (Garriga-Calderé et al. 1999a). FISH using the telomeric repeat pAtT4 from *Arabidopsis thaliana* and the sub-telomeric repeat TGR1 of tomato, showed intact telomeres and subtelomeres for both alien chromosomes. The molecular and cytological organisation of the telomeric repeat (TR) and the sub-telomeric repeat (TRG1) of tomato had been studied in detail by FISH on extended DNA fibres by Zhong et al. (1998).

b) Genome Analysis in Brassicaceae

Molecular cytogenetic studies in Brassicaceae have been considerably hampered by the small genome and chromosome sizes of *Brassica* species and particularly of *Arabidopsis thaliana*. However, a detailed karyotype of *Arabidopsis* can be achieved using meiotic pachytene cells in combination with fluorescence in situ hybridisation (Fransz et al. 1998). Analysed pachytene bivalents proved to be 20–25 times longer than mitotic metaphase chromosomes.

The A and C genome components of *B. napus* (AACC, 2n=38) cannot be clearly distinguished from one another using GISH (Snowdon et al. 1997c, 1999), confirming the extremely high homoeology between these genomes that has been well documented using molecular markers (Parkin et al. 1995; Cheung et al. 1997). In *B. juncea* (AABB, 2n=36) and *B. carinata* (BBCC, 2n=34), on the other hand, the 16 B genome chromosomes can be readily distinguished from the A and C genomes by GISH (Snowdon et al. 1997c), indicating that little recombination has occurred among the diploid genomes in these amphidiploids.

Comparative genome mapping in *Brassica* has revealed that the three genomes A, B and C have distinct chromosomal structures differentiated by a large number of rearrangements, but collinear regions involving virtually the whole of each of the three genomes have been identified (Lagercrantz and Lydiate 1996). However, the cytological difficulties associated with *Brassica* hinder a cytogenetic analysis of these chromosome rearrangements. Interstitial telomere sequences present in the distal chromosome arms of *B. oleracea* (Snowdon et al. 1999) support the view that *Brassica* genomes have originated from extensive genome replication, chromosome rearrangements and fusions (Lagercrantz 1998). Analysis of FISH patterns of rDNA probes can also give information about structural similarities between the chromosomes of *Brassica* amphidiploid species and those of their respective diploid progenitors (e.g. Maluszynska and Heslop-Harrison 1993b; Snowdon et al. 1997b).

Genomic in situ hybridisation has been effectively used in *Brassica* for cytological analysis of numerous interspecific and intergeneric hybrids. Analysis of such hybrids was previously undertaken using classical cytogenetics and molecular markers, for example the mapping in *B. napus* introgression lines of B genome resistance genes against blackleg caused by *Leptosphaeria maculans* (Chévre et al. 1996, 1997b) and the localisation of *B. juncea* introgressions in the rapeseed genome (Plieske et al. 1998). Using GISH, Nielen et al. (1997) investigated the chromosomal composition of plants derived from asymmetric somatic hybridisation between *B. nigra* and *B. napus*. In the first backcross generation a remarkable reduction in chromosome number was detectable and additionally the number of 'mixed' chromosomes was reduced. Plants with resistance to *Leptosphaeria maculans* or *Plasmodiophora brassicae* were selfed or further backcrossed and the genome composition quantified by Southern hybridisation and GISH (Nielen et al. 1997). Winter et al. (1999) generated *B. napus-B. juncea* lines and backcross progenies from the hybrids *B. napus* × *S. arvensis* and *B. napus* × *C. monensis*, respectively, and used GISH to characterise the genome composition of backcross progenies resistant to aggressive *L. maculans* isolates. Dixelius and Wahlberg (1999) demonstrated by RFLP markers that resistance to *Leptosphaeria maculans* is conserved in a specific region of the *Brassica* B genome.

The introduction of resistance against *Heterodera schachtii* from *Raphanus sativus* (RR, 2n=18) into *Brassica napus* (AACC) was described by Voss et al. (1999a). Backcrosses to the oilseed rape variety 'Lisandra' were characterised cytologically using GISH (Snowdon et al.1997a, 1997c). The BC_1 showed 47 chromosomes corresponding to an aneuploid AACCR genome consisting of 38 *B. napus* chromosomes plus a full haploid set of nine *R. sativus* chromosomes. In the BC_2 generation the number of R-chromosomes varied from 3 to 5, whereas in the BC_3 only 1 to 3 R-chromosomes were detected. Highly resistant BC_4 offspring were subsequently generated that exhibited only a monosomic *R. sativus* addition chromosome, suggesting that the resistance is carried on a single chromosome (Snowdon et al. 1999; Voss et al. 1999b, 2000).

Lesquerella fendleri (2n=12) is considered to be an important gene donor to rapeseed, since its oil contains large amounts of lesquerolic acid, an economically important hydroxy fatty acid, and several valuable amino acids are present in the seed meal (Munse et al. 1992). Skarzhinskaya et al. (1998) studied the chromosome complements of somatic hybrids produced between *Brassica napus* (+) *Lesquerella fendleri* by karyotype analysis and GISH. Hybrids from symmetric fusions contained *L. fendleri* chromosome additions varying between two chromosomes and two chromosome complements. In asymmetric fusion experiments, plants with 38 to 76 chromosomes were observed. Intra- and intergenomic recombinations were observed in hybrids from symmetric and asymmetric fusions (Skarzhinskaya et al. 1998). In a similar manner, somatic hybrids between *Brassica napus* and *Eruca sativa* (2n=22) were characterised using FISH with species-specific repetitive sequences and GISH (Fahleson et al. 1997). The repetitive DNA sequences isolated from *E. sativa* correspond to the rDNA intergenic spacer and a telomere-associated repeat. GISH revealed that the somatic hybrid progeny contained one or two complete sets of *Eruca sativa* chromosomes, but no intergenomic translocations were detected. Two plants displaying one *E. sativa* chromosome had 38 non-labelled *B. napus* chromosomes, indicating that these two plants represent addition lines.

In summary, GISH methods have proven to be extremely beneficial for monitoring chromatin transfer and introgression in interspecific and intergeneric *Brassica* hybrids, particularly for selection of backcross offspring containing genes of interest from the donor species within a minimal donor genome component.

c) Chromosome Constitution and Recombination in Poaceae

Chromosome engineering methodologies, based on the manipulation of pairing control mechanisms and induced translocations, have been employed to transfer specific disease and pest resistance genes from annual (e.g. rye) or perennial (e.g. *Lophopyrum* spp., *Agropyron* spp.) members of the wheat tribe, Triticeae, into wheat (Jauhar and Chibbar 1999).

In *Triticum aestivum* (2n=6×=42, AABBDD) but also in wheat × alien hybrids, the dominant *Ph1* allele suppresses pairing between homoeologues in favour of that between homologues. *Ph1* (homoeologous pairing) locus was localised on the long arm of chromosome 5B. Nullisomy for chromosome 5B and deletion mutants have been applied to promote homoeologous pairing in wheat × alien hybrids.

Mikhailova et al. (1998) analysed effects of different alleles of the *Ph1* locus on the behaviour and morphology of two 5RL rye telosomes in a wheat background by genomic in situ hybridisation (GISH), using rye genomic DNA as a probe. The results indicate that *Ph1* is involved in chromosome condensation and/or scaffold organisation. This would account for the various effects of this locus on both premeiotic associations of homologues, regulation of meiotic homo(eo)logous chromosome pairing and synapsis, the resolution of bivalent interlockings and centromere behaviour (Mikhailova et al. 1998). Benavente et al. (1996, 1998) analysed wheat-wheat and wheat-rye homoeologous pairing in metaphase I and wheat-rye recombination at anaphase I by GISH in wild-type (*Ph1Ph2*) and mutant *ph1b* and *ph2b* wheat × rye hybrids. Three types of wheat-rye metaphase I association could be visualised by GISH: (1) end-to-end extremely distal association, (2) end-to-end distal association and (3) interstitial association (Benavente et al. 1996). The frequency of wheat-rye metaphase I association exceeded the frequency of wheat-rye recombination in both *ph* mutant hybrids. The promoting effect of the *ph2b* mutations seemed to be evenly distributed among all possible homoeologous association whereas the effect of *ph1b* mutations was greater between distant homoeologous partners (Benavente et al. 1998).

Rye chromosome distribution in backcross progenies and doubled haploid lines of hybrids between octoploid triticale and wheat was followed by GISH (Wang et al. 1995, 1996). Lines varied in their wheat and rye genome composition depending on the two introgressions systems (anther culture or conventional backcrossing), and were either wheat-rye chromosome multiple-addition lines or had spontaneous substitutions and/or wheat-rye translocations. Rye telocentric chromosomes were observed in the backcross progenies. Most plants of the DH lines contained a pair of 4R chromosomes, whereas 1R or 7R were present in others (Wang et al. 1996). Non-Robertsonian wheat-rye translocations have been reported by Wang et al. (1998).

Thinopyrum intermedium (2n=6×=42, Syn. *Agropyron intermedium*) and *Thinopyrum ponticum* (2n=10×=70, Syn. *Agropyron elongatum*) have been two of the most important perennial Triticeae species for wheat improvement because of their resistance to a number of diseases and pests like the wheat streak mosaic virus (WSMV) vectored by wheat curl mite (WCM), as well as stress tolerance, and high cross ability with various *Triticum* species (Dewey 1984). Based on GISH results, the genomic constitutions of *Th. intermedium* and *Th. ponticum* were redesignated JJ^SS and JJJ^SJ^S, respectively (Chen et al. 1998c). The S genome was homologous to the S genome of *Pseudoroegneria strigosa* while the J^S referred to modified J or E genomes distinguished by the presence of S genome specific sequences close to the centromere. The higher frequencies of autosyndetic pairing among *Thinopyrum* chromosomes than among wheat chromosomes in wheat × *Thinopyrum* spp. hybrids indicated that the relationships among the three genomes of *Th. intermedium* and among the five genomes of *Th. ponticum* are closer than those among the three genomes of *T. aestivum* (Cai and Jones 1997).

Characterising five different partial amphiploids by GISH, it could be shown that the alien genomes from Agrotana (resistant to WSMV and WCM), OK7211542 and ORRPX were derived from *Th. ponticum* and not from *Th. intermedium* (Chen et al. 1998a, b). In backcrosses of a Agrotana × *Triticum turgidum* cross Thomas et al. (1998) could recover two independent Robertsonian translocations that probably involved the reunion of the short arm of a group-6 chromosome of *Thinopyrum ponticum* and the long arm of 6D of wheat. These lines showed resistance to wheat curl mite, a vector of wheat streak mosaic virus. TAF46 and Zhong 5 proved to be wheat × *Th. intermedium* partial hybrids derived from different combinations of the J, J^S and S chromosomes of *Th. intermedium*. However, different interpretations of the genome composition for Zhong 5 were also reported (Tang et al. 2000). Discrepancies are likely to be due to the highly polyploid nature of the *Thinopyrum* species and the close relationship among E, J, J^S and S genomes within the Triticeae (Chen et al. 1998b).

Three wheat germplasm lines possessing resistance to WSMV, derived from *T. aestivum* × *Th. intermedium* crosses, were analysed by C-banding and GISH to determine the amount and location of alien chromatin in the transfer lines (Chen et al. 1998d). One line was confirmed as a disomic substitution line in which wheat chromosome 4 A was replaced by *Th. intermedium* chromosome 4Ai#2, the other two lines carried an identical Robertsonian translocation chromosome in which the complete short arm of chromosome Ai4#2 was transferred to the long arm of wheat chromosome 4 A.

Hohmann et al. (1996) described the detailed physical location and size of the transferred BYDV resistant *A. intermedium* chromosome segments in nine independently produced families of bread wheat (Banks et al. 1995) by means of C-banding, *Agropyron*-amplified repetitive DNA sequences, GISH, RFLP, and comparative physical deletion mapping. GISH was used to investigate allo- and autosyndetic chromosome pairing in the first self-fertile allotetraploid *Triticum-Agropyron*, obtained by crossing the autotetraploid forms of *T. tauschii* and *A. cristatum* (Martín et al. 1999).

The genomic constitution of *Aegilops cylindrica* Host (2n=4×=28, $D^cD^cC^cC^c$) was analysed by C-banding and fluorescence in situ hybridisation (FISH) using the DNA clones pSc119, pAs1, pTa71, and pTA794 and genomic DNA. The C-banding patterns of the D^c and C^c genome chromosomes of *Ae. cylindrica* are similar to those of D and C genome chromosomes of the diploid progenitor species *Ae. tauschii* Coss and *Ae. caudata* L., respectively (Linc et al. 1999). Simultaneous genomic in situ hybridisation with probe preannealing (SP-GISH) was used for discriminating *Aegilops speltoides* chromosome regions by their relatedness to DNA of other species e.g. *Ae. bicornis*, *Ae tauschii* and *Hordeum spontaneum* (Belyayev and Raskina 1998). This approach could elucidate the function and dynamics of the non-coding DNA fraction in the evolutionary process. A complete set of *Triticum aestivum-Aegilops speltoides* chromosome addition lines was developed and cytologically characterised in terms of chromosome length, arm ratio, distribution of marker C-bands, and FISH sites using a *Ae. speltoides*-specific repetitive element, GclR-1, as probe (Friebe et al. 2000).

In the genus *Oryza*, more than 20 wild species are known, designated in genomic constitution as AA, BB, BBCC, CC, CCDD, EE, FF, GG, and HHJJ, respectively (Aggarwal et al. 1997; Fukui et al. 1997; Shishido et al. 1998). Rice B and D genome could be unequivocally discriminated from the C genome in two amphidiploid wild rice species, *O. minuta* (2n=4×=48, BBCC) and *O. latifolia* (2n=4×=48, CCDD) by using *O. officinalis* (2n=2×=24, CC) as C genome specific DNA for genomic in situ hybridisation (Fukui et al. 1997). Using the C genome as a pivotal genome the genetic distance from the B genome is larger than that from the D genome. In somatic hybrids between *O. sativa* (2n=24, AA) and *O. punctata* (2n=48, BBCC) the three different genomes A, B and C were distinguishable by GISH applying genomic DNA of the diploid rice species *O. sativa* (AA), *O. punctata* (BB) and *O. officinalis* (CC) (Shishido et al. 1998). In the somatic hybrids, specific chromosome reduction was only observed in the B and C genomes but not in the A genome.

BPH-resistant gene(s) could be successfully transferred from *O. eichingeri* (2n=24, CC) to *O. sativa* (2n=24, AA) cv. 02428 by Yan et al. (1997, 1999). GISH allowed to detect 12 chromosomes of *O. eichingeri* in F_1, F_2, BC_1, and 24 chromosomes in plant E_{24} derived from anther culture, thus confirming that both BC_1 and F_2 were allotriploids (2n=36, AAC) while plant E_{24} was an amphiploid (2n=48, AACC). Rice telotrisomics were developed from an indica rice variety "Zhongxian 3037", characterised by FISH using a rice centromeric BAC clone. Application of the telotrisomics in microdissection and development of chromosome-specific DNA markers were demonstrated by Cheng et al. (2001).

Somatic hybridisation of *O. sativa* and *Proteresia coarctata* (2n=4×=48), a saline-tolerant wild species, allowed to obtain a line comprising an allohexaploid chromosome complement (2n=6×=72) which represents full chromosome sets of both species (Jelodar et al. 1999).

Development of diploid *Lolium*-tribes possessing high forage quality of ryegrasses combined with persistence, drought tolerance and frost resistance from the *Festuca*-species can be improved by applying genomic in situ hybridisation (Zeller 1999). Allohexaploid *Festuca arundinacea* (2n=6×=42) was derived from a hybrid between *F. pratensis* (2n=2×=28) and *F. glaucescens* (2n=4×=28) (Humphreys et al. 1995). The genomic constitution of progenies obtained by backcross programmes (Humphreys and Paskinskiene 1996) or anther culture (Zwierzykowski et al. 1998) of pentaploid hybrids between *Festuca arundinacea* and *Lolium multiflorum* were characterised by GISH.

All plants regenerated from anther culture contained a complete set of chromosomes of both *Lolium* and *Festuca*. In addition, these plants except one had at least one translocated *Lolium-Festuca* chromosome. *Lolium multiflorum*-like drought-resistant plants obtained by backcrosses all represented introgression lines carrying a single *Festuca* recombinant chromosome (Humphreys and Paskinskiene 1996) which hybridised to a DNA probe from *F. pratensis* indicating that the *F. pratensis* chromosome in *F. arundinacea* which is homoeologous to chromosome 2 in *L. multiflorum* carries genes for drought resistance. Chromosome substitutions and recombination in the amphiploid *Lolium perenne* × *Festuca pratensis* cv Prior (2n=4×=28) could be revealed by GISH (Canter et al. 1999). There was a substitution of *Festuca*-origin chromosomes by those of *Lolium*-origin, resulting in a mean of 17.9 (15–21) *Lolium* and 9.7 (7–13) *Festuca* chromosomes per genotype.

Repetitive sequences – DNA motifs that are repeated hundreds or thousands of times in the genome – make up the majority of most plant genomes and examples both with widespread distribution and with high species specificity have been reported (Schmidt and Heslop-Harrison 1998). Inter- and intraspecific variation of different cloned repetitive DNA sequences were used to obtain information about intergenomic translocations, interrelationship of species and evolution of repetitive sequences in *Hordeum* (Pickering et al. 2000; Taketa et al. 2000), *Avena* (Linares et al. 2000), and maize (Chen et al. 2000).

Fluorescence in situ hybridisation using two abundant tandemly repeated DNA sequences, dpTA1 and pSc119.2, in six wild *Hordeum* taxa, representing the four basic genomes of the genus, revealed the presence of pSc119.2 in tetraploid *Hordeum murinum*, but absence in the diploid form. This suggests that the tetraploid is not likely to be a simple autotetraploid of the diploid (Taketa et al. 2000).

4 Molecular Cytogenetic Approaches for Physical Mapping

a) Integrating Genetic Maps with Karyotype Information

Most molecular markers used to generate genetic maps cannot be directly physically localised by FISH. Larger genomic clones like BACs, YACs and cosmids, on the other hand, are suitable for FISH and enable marker sequences to be localised indirectly. Localisation of single-copy or low-copy BACs to metaphase chromosomes has been accomplished in numerous plant genomes including barley (Pedersen et al. 1995), rice (Jiang et al. 1995), cotton (Hansen et al. 1995), potato and tomato (Fuchs et al. 1996), sorghum (Gómez et al. 1997) and *Arabidopsis* (Fransz et al. 2000). In this way it is now possible to compare genetic and cytogenetic maps and obtain information on the physical distribution of molecular markers. The use of FISH to relate genetic markers to karyotype information will be particularly useful for important crop plants with small chromosomes, where the ability to localise BAC clones containing mapped molecular markers will enable the integration of genetic maps with karyotype information. In many cases, for example in *Brassica* (Armstrong et al. 2000), this represents the first opportunity to align molecular marker linkage groups with karyotype information.

b) High Resolution FISH Techniques

In mitotic metaphase, however, chromosomes are normally too highly condensed to allow signals of closely linked markers to be resolved. FISH to interphase nuclei can significantly increase resolution, but little spatial information is available in interphase and thus it is impossible to assign markers to specific chromosomes. Hybridisation to meiotic pachytene chromosomes, on the other hand, can give an up to 40-fold increase in resolution compared with metaphase, depending on genome size and heterochromatin organisation (de Jong et al. 1999). Using pachytene FISH, Song et al. (2000) were able to localise BACs containing RFLP markers representing the outermost extremes of potato linkage group I to the distal chromosome ends, indicating that this linkage group covers the entire potato chromosome 1. Peterson et al. (1999) localised genomic clones containing mapped RFLP markers to barley chromosome 11, and surprisingly discovered that the arrangement of two markers was reversed in comparison to the genetic map. Zhong et al. (1999) used FISH to tomato pachytene chromosomes to locate the root-knot nematode-resistance gene Mi-1 and the acid phosphatase gene Aps-1 with respect to the junction of euchromatin and pericentromeric heterochromatin. Such information cannot be obtained using genetic mapping techniques.

The optimal resolution currently available for in situ hybridisation to plant chromosomes is provided by FISH to extended chromatin fibres, also known as fibre-FISH (de Jong et al. 1999). Because extended chromatin – in contrast to condensed chromosomes – is comparatively linear, physical measurements of the lengths of or between fibre-FISH signals can give a surprisingly accurate estimation of physical distances associated with labelled DNA sequences. In this respect FISH to extended chromatin fibres is potentially a powerful tool for physical mapping and positional cloning. Moreover, fibre-FISH makes it possible to visualise the linear positions of DNA sequences in mapping contigs and thus to exactly localise and order clones on their chromosome targets (de Jong et al. 1999).

The first application of fibre-FISH in plants was described by Fransz et al. (1996b), whose comparison of physical signal lengths with the molecular size of the probes used confirmed the accuracy and reproducibility of calibrated fibre-FISH experiments. By comparing signal lengths with size standards, the length of the 5 S rDNA cluster in tomato was estimated to be around 600 kb. Subsequently, Jackson et al. (1998) used fibre-FISH with labelled BAC clones to estimate the sizes of gaps in the physical contig map of *A. thaliana* chromosome II. The size of gap 2 was measured at 31 kb, considerably less than the 340 kb estimated previously from the genetic distance between markers flanking this gap.

Detailed examination by fibre-FISH of the physical distribution of repetitive sequence elements can also provide much-needed information for the understanding of chromosome structure and function. For example, the centromeric repeats Sau3A10 in rice (Dong et al. 1998) and pSau3A9 in sorghum (Miller et al. 1998) were shown to be present in long uninterrupted arrays resembling previously reported tandem repeats located in the centromeres of human and *A. thaliana* chromosomes. Such sequences are thought to be central to centromere function.

High-resolution FISH can also be useful to obtain physical information about the integration of T-DNA constructs in transgenic plants. Regarding factors that might influence transgene expression, chromosomal location and organisation of the integration site may play an important role. The composition of complex loci and the copy number and arrangement of transgene inserts cannot be examined so accurately by other methods as by FISH.

Moscone et al. (1996) demonstrated that combined FISH/GISH along with DAPI counterstaining allows distinction of 20–21 out of 24 chromosome pairs of *Nicotiana tabacum* and simultaneous detection of the chromosome integration site and sub-genomic allocation of the transgene locus 271. The FISH signal of this transgene locus was detected on the long arm telomeres of a nucleolus organising region (NOR)-bearing chromosome pair (Moscone et al. 1996). This transgene locus probably consists of 6–7 copies of the 8.3 kb pRiN construct (Park et al. 1996), comprising approximately 50–60 kb. In an aneuploid tobacco line Papp et al. (1996) were able to observe a 30 kb transgene locus in the S subgenome. Transgene loci ranging in size from 2.7 kb (Fransz et al. 1996a) to 17 kb (Ambros et al. 1986a, 1986b; Wang et al. 1995) have been localised on *Petunia* chromosomes by FISH. Wolters et al. (1998) were able with fibre-FISH to determine the copy number and arrangement of T-DNA and vector DNA sequences at transgene loci comprising multiple T-DNA inserts in potato. Transgene integration sites into hexaploid oat following DNA delivery by microprojectile bombardment was investigated by FISH (Svitashev et al. 2000). The structural complexity of the transgene integration sites ranged from simple integration structures of apparently contiguous transgene copies to tightly linked clusters of multiple copies of transgenes interspersed with oat DNA.

Jackson et al. (2000) also demonstrated the utility of fibre-FISH for studies of genome evolution in polyploids. By FISH to metaphase chromosomes they were able to demonstrate that a 431 kb contig consisting of six BACs from *A. thaliana* was, as expected, duplicated four to six times in *B. rapa*. FISH to extended chromatin fibres showed, furthermore, that signal lengths in *B. rapa* were not longer than those in *A. thaliana*, supporting the theory that chromosomal duplications rather than amplification of repetitive sequences in intergenic regions have played the main role in the evolution of *Brassica* genomes.

5 Future Trends and Perspectives

The examples described here demonstrate that molecular cytogenetic techniques have become an important tool in the structural analysis of plant genomes. Alongside a more detailed description of polyploid and hybrid genomes and chromosome introgression, we are now able to visualise the distribution of various classes of DNA sequences at the chromosomal level and to integrate karyotype information with molecular marker maps. Schmidt and Heslop-Harrison (1998) used FISH to suggest a universal model for plant chromosome structure, and since then the development of molecular cytogenetic techniques has continued to a point where important new information can now be obtained with respect to chromosome function and genome structure. Moreover, the application of high-resolution FISH methods to plant genomes now gives molecular cytogenetic techniques, in parallel with molecular marker methods, the potential to play a major role in physical mapping efforts.

References

Aggarwal RK, Brar DS, Khush GS (1997) Two new genomes in the *Oryza* complex identified on the basis of molecular divergence analysis using total genomic DNA hybridization. Mol Gen Genet 254:1–12

Ambros PF, Matzke AJM, Matzke MA (1986a) Localization of *Agrobacterium rhizogenes* T-DNA in plant chromosomes by in situ hybridisation. EMBO J 5:2073–2077

Ambros PF, Matzke AJM, Matzke MA (1986b) Detection of a 17 kb unique sequence (T-DNA) in plant chromosomes by in situ hybridisation. Chromosoma 94:11–18

Ananiev EV, Riera-Lizarazu O, Raines HW, Philips RL (1997) Oat-maize chromosome addition lines: a new system for mapping the maize genome. Proc Natl Acad Sci USA 94:3524–3529

Armstrong SJ, Howell EC, Franz P, Jones GH, Kearsey MJ, King GJ, Kop E, Ryder CD, Teakle GR, Vicente JG (2000) Integrating the genetic and physical maps of *Brassica oleracea* var. *alboglabra*. Acta Hort 539:77–82

Banks PM, Larkin PJ, Bariana HS, Lagudah ES, Appels R, Waterhouse PM, Brettell RIS, Chen X, Xu HJ, Xin ZY, Qian YT, Zhou XM, Cheng ZM, Zhou GH (1995) The use of cell culture for subchromosomal introgressions of barley yellow dwarf virus resistance from *Thinopyrum intermedium* to wheat. Genome 38:395–405

Barret P, Guerif J, Reynoird JP, Delourme R, Eber F, Renard M, Chevre AM (1998) Selection of stable *Brassica napus-B. juncea* recombinant lines resistant to blackleg (*Leptosphaeria maculans*) 2: a 'to and fro' strategy to localise and characterise interspecific introgressions on *B. napus* genome. Theor Appl Genet 96:1097–1103

Belyayev A, Raskina O (1998) Heterochromatin discrimination in *Aegilops speltoides* by simultaneous genomic in situ hybridization. Chromosome Res 6:559–565

Benavente E, Fernández-Calvín B, Orellana J (1996) Relationship between the levels of wheat-rye metaphase I chromosomal pairing and recombination revealed by GISH. Chromosoma 105:92–96

Benavente E, Orellana J, Fernández-Calvín B (1998) Comparative analysis of the meiotic effects of wheat *ph1b* and *ph2b* mutations in wheat × rye hybrids. Theor Appl Genet 96:1200–1204

Brandes A, Heslop-Harrison JS, Kamm A, Kubis SE, Doudrick RL, Schmidt T (1997) Comparative analysis of the chromosomal and genomic organization of Ty1-*copia*-like retrotransposons in pteridophytes, gymnosperms and angiosperms. Plant Mol Biol 33:11–21

Cai X, Jones S (1997) Direct evidence for high level of autosyndetic pairing in hybrids of *Thinopyrum intermedium* and *Th. ponticum* with *Triticum aestivum*. Theor Appl Genet 95:568–572

Canter PH, Pasakinskienė I, Jones RN, Humphreys MW (1999) Chromosome substitutions and recombination in the amphiploid *Lolium perenne* × *Festuca pratensis* cv. Prior (2n=4×=28). Theor Appl Genet 98:809–814

Cao MS, Sleper DA, Dong FG, Jiang JM (2000) Genomic in situ hybridization (GISH) reveals high chromosome pairing affinity between *Lolium perenne* and *Festuca mairei*. Genome 43:398–403

Chen BY, Cheng BF, Jorgensen RB, Heneen WK (1997a) Production and cytogenetics of *Brassica campestris-alboglabra* chromosome addition lines. Theor Appl Genet 94:633–640

Chen BY, Jorgensen RB, Cheng BF, Heneen WK (1997b) Identification and chromosomal assignment of RAPD markers linked with a gene for seed colour in a *Brassica campestris-alboglabra* addition line. Hereditas 126:133–138

Chen CC, Chen CM, Hsu FC, Wang CJ, Yang JT, Kao YY (2000) The pachytene chromosomes of maize as revealed by fluorescence in situ hybridization with repetitive DNA sequences. Theor Appl Genet 101:30–36

Chen Q, Conner RL, Laroche A (1995) Identification of the parental chromosomes of the wheat-alien amphiploid *Agrotana* by genomic in situ hybridization. Genome 38:1163–1169

Chen Q, Ahmad F, Collin J, Comeau A, Fedak G, St-Pierre CA (1998a) Genomic constitution of partial amphiploid OK7211542 used as a source immunity to barley yellow dwarf virus for bread wheat. Plant Breeding 117:1–6

Chen Q, Conner RL, Ahmad F, Laroche A, Fedak G, Thomas JB (1998b) Molecular characterization of the genome composition of partial amphiploids derived from *Triticum aestivum* × *Thinopyrum ponticum* and *T. aestivum* × *Th. intermedium* as sources of resistance to wheat streak mosaic virus and its vector, *Aceria tosichella*. Theor Appl Genet 97:1–8

Chen Q, Conner RL, Laroche A, Thomas JB (1998c) Genome analysis of *Thinopyrum intermedium* and *Thinopyrum ponticum* using genomic in situ hybridization. Genome 41:580–586

Chen Q, Friebe B, Conner RL, Laroche A, Thomas JB, Gill BS (1998d) Molecular cytogenetic characterization of *Thinopyrum intermedium*-derived wheat germplasm specifying resistance to wheat streak mosaic virus. Theor Appl Genet 96:1–7

Cheng Z, Yan H, Yu H, Tang S, Jiang J, Gu M, Zhu L (2001) Development and applications of a complete set of rice telotrisomics. Genetics 157:361–368

Cheung WY, Champagne G, Hubert N, Landry BS (1997) Comparison of the genetic maps of *Brassica napus* and *Brassica oleracea*. Theor Appl Genet 94:569–582

Chévre AM, Eber F, This P, Barret P, Tanguy X, Brun H, Delseny M, Renard M (1996) Characterization of *Brassica nigra* chromosomes and of blackleg resistance in *B. napus-B. nigra* addition lines. Plant Breed 115:113–118

Chévre AM, Barret P, Eber F, Dupuy P, Brun H, Tanguy X, Renard M (1997a) Selection of stable *Brassica napus-B. juncea* recombinant lines resistant to blackleg (*Leptosphaeria maculans*). Theor Appl Genet 95:1104–1111

Chévre AM, Eber F, Barret P, Dupuy P, Brace J (1997b) Identification of the different *Brassica nigra* chromosomes from both sets of *B. oleracea- B. nigra* and *B. napus-B. nigra* addition lines with a special emphasis on chromosome transmission and self-incompatibility. Theor Appl Genet 94:603–611

De Jong JH, Wolters AMA, Kok JM, Verhaar H, Van Eden J (1993) Chromosome pairing and potential for intergeneric recombination in some hypotetraploid somatic hybrids of *Lycopersicon esculentum* (+) *Solanum tuberosum*. Genome 36:1032–1041

De Jong JH, Fransz P, Zabel P (1999) High resolution FISH in plants – techniques and applications. Trends Plant Sci 4:258–263

Dewey DR (1984) The genomic system of classification as a guide to intergeneric hybridisation with perennial Triticeae. In: Gustafson JP (ed) Gene manipulation in plant improvement, vol 16. Plenum Press, New York, pp 209–279

D' Hont A, Paget-Goy A, Escoute J, Carreel F (2000) The interspecific genome structure of cultivated banana, *Musa* spp. revealed by genomic DNA in situ hybridization. Theor Appl Genet 100:177–183

Dixelius C, Wahlberg S (1999) Resistance to *Leptosphaeria maculans* is conserved in a specific region of the *Brassica* B genome. Theor Appl Genet 99:368–372

Dong FG, Miller JT, Jackson SA, Wang GL, Ronald PC, Jiang JM (1998) Rice (*Oryza sativa*) centromeric regions consist of complex DNA. Proc Natl Acad Sci USA 95:8135–8140

Dong F, Novy RG, Helgeson JP, Jiang J (1999) Cytological characterization of potato – *Solanum etuberosum* somatic hybrids and their backcross progenies by genomic in situ hybridization. Genome 42:987–992

Endo TR, Gill BS (1996) The deletion stocks of common wheat. J Hered 87:295–307

Escalante A, Imanishi S, Hossain M, Ohmido N, Fukui K (1998) RFLP analysis and genomic in situ hybridization (GISH) in somatic hybrids and their progeny between *Lycopersicon esculentum* and *Solanum lycopersicoides*. Theor Appl Genet 96:719–726

Fahleson J, Lagercrantz U, Mouras A, Glimelius K (1997) Characterization of somatic hybrids between *Brassica napus* and *Eruca sativa* using species-specific repetitive sequences and genomic in situ hybridization. Plant Sci 123:133–142

Fransz PF, Stam M, Montijn B, Ten Hoopen R, Wiegant J, Kooter JM, Oud O, Nanninga N (1996a) Detection of single-copy genes and chromosome rearrangements in *Petunia hybrida* by fluorescence in situ hybridisation. Plant J 9:767–774

Fransz PF, Alonso-Blanco C, Liharska TB, Peeters AJM, Zabel P, de Jong JH (1996b) High-resolution physical mapping in *Arabidopsis thaliana* and tomato by fluorescence in situ hybridization to extended DNA fibres. Plant J 9:421–430

Fransz PF, Armstrong S, Alonso-Blanco C, Fischer TC, Torres-Ruiz RA, Jones G (1998) Cytogenetics for the model system *Arabidopsis thaliana*. Plant J 13:867–876

Fransz PF, Armstrong S, de Jong JH, Parnell L, van Drunen C, Dean C, Zabel P, Bisseling T, Jones GH (2000) Integrated cytogenetic map of chromosome arm 4 s of *A. thaliana*: structural organization of heterochrochromatic knob and centromere region. Cell 100:367–376

Friebe B, Jiang J, Raupp WJ, McIntosh RA, Gill BS (1996) Characterization of wheat-alien translocations carrying resistance to diseases and pests: current status. Euphytica 91:59–87

Friebe B, Qi LL, Nasuda S, Zhang P, Tuleen NA, Gill BS (2000) Development of a complete set of *Triticum aestivum-Aegilops speltoides* chromosome addition lines. Theor Appl Genet 101:51–58

Fuchs J, Kloos DU, Ganal MW, Schubert I (1996) In situ localization of yeast artificial chromosome sequences on tomato and potato metaphase chromosomes. Chromosome Res 4:277–281

Fukui K, Shishido R, Kinoshita T (1997) Identification of the rice D-genome chromosomes by genomic in situ hybridisation. Theor Appl Genet 95:1239–1245

Ganal MW, Lapitan NLV, Tanksley SD (1988) A molecular and cytogenetic survey of major repeated DNA sequences in tomato (*Lycopersicum esculentum*). Mol Gen Genet 213:262–268

Garriga-Calderé F, Huigen DJ, Filotico F, Jacobsen E, Ramanna MS (1997) Identification of alien chromosomes through GISH and RFLP analysis and the potential for establishing potato lines with monosomic additions of tomato chromosomes. Genome 40:666–673

Garriga-Calderé F, Huigen DJ, Angrisano A, Jacobsen E, Ramanna MS (1998) Transmission of alien tomato chromosomes from BC_1 to BC_2 progenies derived from backcrossing potato (+) tomato fusion hybrids to potato: the selection of single additions for seven different tomato chromosomes. Theor Appl Genet 96:155–163

Garriga-Calderé F, Huigen DJ, Jacobsen E, Ramanna MS (1999a) Origin of an alien disomic addition with an aberrant homologue of chromosome-10 to tomato and its meiotic behaviour in a potato background revealed through GISH. Theor Appl Genet 98:1263–1271

Garriga-Calderé F, Huigen DJ, Jacobsen E, Ramanna MS (1999b) Prospects for introgressing tomato chromosomes into the potato genome: an assessment through GISH analysis. Genome 42:282–288

Gill BS, Friebe B (1998) Plant cytogenetics at the dawn of the 21st century. Curr Opin Plant Biol 1:109–115

Gill KS, Gill BS, Endo TR, Boyko EV (1996) Identification and high-density mapping of gene-rich regions in chromosome group 5 of wheat. Genetics 143:1001–1012

Gómez MI, Islam-Faridi MN, Woo SS, Schertz KF, Czeschin D, Zwick MS, Sing RA, Stelly DM, Price HJ (1997) FISH of a maize sh2-selected sorghum BAC to chromosomes of *Sorghum bicolor*. Genome 40:475–478

Hanson RE, Zwick MS, Choi SD, Islam-Faridi MN, McKnight TD, Wing RA, Price HJ, Stelly DM (1995) Fluorescent in situ hybridization of a bacterial artificial chromosome. Genome 38:646–651

Harrison GE, Heslop-Harrison JS (1995) Centromeric repetitive DNA sequences in the genus *Brassica*. Theor Appl Genet 90:157–165

Heslop-Harrison JS (1998) Cytogenetic analysis of *Arabidopsis*. In: Martinez-Zapater J, Salinas J (eds) Methods in molecular biology, vol 82: *Arabidopsis* protocols. Humana Press, Totowa, NJ, pp 119–127

Heslop-Harrison JS (2000) Comparative genome organization in plants: from sequence and markers to chromatin and chromosomes. Plant Cell 12:617–635

Heslop-Harrison JS, Schwarzacher T (1996) Genomic Southern and in situ hybridization for plant genome analysis. In: Jauhar PP (ed) Methods of genome analysis in plants. CRC Press, Boca Raton, pp 163–179

Heslop-Harrison JS, Murata M, Ogura Y, Schwarzacher T, Motoyoshi F (1999) Polymorphisms and genomic organization of repetitive DNA from centromeric regions of *Arabidopsis* chromosomes. Plant Cell 11:31–42

Hohmann U, Badaeva K, Busch W, Friebe B, Gill BS (1996) Molecular cytogenetic analysis of *Agropyron* chromatin specifying resistance to barley yellow dwarf virus in wheat. Genome 39:336–347

Hossain M, Imanishi S, Matsumoto A (1994) Production of somatic hybrids between tomato (*Lycopersicon esculentum*) and night shade (*Solanum lycopersicosdes*) by electrofusion. Breed Sci 44:405–412

Hou A, Peffley EB (2000) Recombinant chromosomes of advanced backcross plants between *Allium cepa* L. and *A. fistulosum* L. revealed by in situ hybridization. Theor Appl Genet 100:1190–1196

Houben A, Brandes A, Pich U, Manteuffel R, Schubert I (1996) Molecular cytogenetic characterization of a higher plant centromere/kinetochore complex. Theor Appl Genet 93:477–484

Humphreys MW, Pasakinskiene I (1996) Chromosome painting to locate genes for drought resistance transferred from *Festuca arundinacea* into *Lolium multiflorum*. Heredity 77:530–534

Humphreys MW, Thomas HM, Morgan WG, Meredith MR, Harper JA, Thomas H, Zwierzykowski Z, Ghesquiere M (1995) Discriminating the ancestral progenitors of hexaploid *Festuca arundinacea* using genomic in situ hybridisation. Heredity 75:171–174

Ishii T, Brar DS, Multani DS, Khush GS (1994) Molecular tagging of genes for plant hopper resistance and earliness introgressed from *Oryza australiensis* into cultivated rice, *O. sativa*. Genome 37:217–221

Iwano M, Sakamoto K, Suzuki G, Watanabe M, Takayama S, Fukui K, Hinata K, Isogai A (1998) Visualization of a self-incompatibility gene in *Brassica campestris* L. by multicolour FISH. Theor Appl Genet 96:751–757

Jackson SA, Wang ML, Goodman HM, Jiang JM (1998) Application of fiber-FISH in physical mapping of *Arabidopsis thaliana*. Genome 41:566–572

Jackson SA, Cheng Z, Wang ML, Goodman HM, Jiang J (2000) Comparative fluorescence in situ hybridization mapping of a 431 kb *Arabidopsis thaliana* bacterial artificial chromosome contig reveals the role of chromosomal duplications in the expansion of the *Brassica rapa* genome. Genetics 156:833–838

Jacobsen E, Reinhout P, Bergervoet JEM, De Looff J, Abidin PE, Huigen DJ, Ramanna MS (1992) Isolation and characterization of potato-tomato somatic hybrids using an amylose-free potato mutant as parental genotype. Theor Appl Genet 85:159–164

Jacobsen E, De Jong JH, Kamstra SA, Van den Berg PM, Ramanna MS (1995) Genomic in situ hybridisation (GISH) and RFLP analysis for the identification of alien chromosomes in the backcross progeny of potato (+) tomato fusion hybrids. Heredity 74:250–257

Jauhar PP, Chibbar RN (1999) Chromosome-mediated and direct gene transfer in wheat. Genome 42:570–583

Jelodar NB, Blackhall NW, Hartmann TPV, Brar DS, Kush G, Davey MR, Cocking EC, Power JB (1999) Intergeneric somatic hybrids of rice [*Oryza sativa* L (+) *Porteresia coarctata* (Roxb.) Tateoka]. Theor Appl Genet 99:570–577

Jiang JM, Gill BS (1994) Nonisotopic in situ hybridization and plant genome mapping: the first 10 years. Genome 37:717–725

Jiang JM, Gill BS (1996) Current status and potential of fluorescence in situ hybridization in plant genome mapping. In: Patterson AH (ed) Genome mapping in plants. Academic Press, San Diego, pp 127–135

Jiang J, Gill BS, Wang GL, Ronald PC, Ward DC (1995) Metaphase and interphase fluorescence in situ hybridization mapping of the rice genome with bacterial artificial chromosomes. Proc Natl Acad Sci USA 92:4487–4491

Kamisugi Y, Nakayama S, Nakajima R, Ohtsubo H, Ohtsubo E, Fukui K (1994) Physical mapping of the 5 S ribosomal RNA genes on rice chromosome 11. Mol Gen Genet 245:133–138

Kaul S, Koo HL, Jenkins J, Rizzo M, Rooney T, Tallon LJ, Feldblyum T, Nierman W, Benito MI, Lin XY, Town CD, Venter JC, Fraser CM, Tabata S, Nakamura Y, Kaneko T, Sato S, Asamizu E, Kato T, Town CD, Sasamoto S, Ecker JR, Theologis A, Federspiel NA, Palm CJ (2000) Analysis of the genome sequence of the flowering plant *Arabidopsis thaliana*. Nature 408:796–815

Khrustaleva LI, Kik C (2000) Introgression of *Allium fistulosum* into *A. cepa* mediated by *A. roylei*. Theor Appl Genet 100:17–26

King IP, Morgan WG, Harper JA, Meredith MR, Jones RN, Thomas HM (1997) Proc Aberystwyth Cell Genet Group, 7th Annu Meeting. Aberystwyth, 7–10 January 1997. Exp Biol (on-line)

Kitamura S, Inoue M, Ohmido N, Fukui K (1997) Identification of parental chromosomes in the interspecific hybrids of *Nicotiana rustica* L × *N. tabacum* L. and *N. gossei domin* × *N. tabacum* L., using genomic in situ hybridization. Breeding Sci 47:67–70

Korell ML, Brahm L, Horn R, Friedt W (1996a) Interspecific and intergeneric hybridization in sunflower, I. General breeding aspects. Plant Breeding Abstr 66:925–931

Korell ML, Brahm L, Friedt W, Horn R (1996b) Interspecific and intergeneric hybridization in sunflower breeding, II. Specific use of wild germplasms. Plant Breeding Abstr 66:1081–1091

Lagercrantz U (1998) Comparative mapping between *Arabidopsis thaliana* and *Brassica nigra* indicates that *Brassica* genomes have evolved through extensive genome replication accompanied by chromosome fusions and frequent rearrangements. Genetics 150:1217–1228

Lagercrantz U, Lydiate DJ (1996) Comparative genome mapping in *Brassica*. Genetics 144:1903–1910

Lim KY, Matyásek R, Lichtenstein CP, Leitch AR (2000) Molecular cytogenetic analyses and phylogenetic studies in the *Nicotiana* section Tomentosae. Chromosome Res 109:245–258

Lin XY, Kaul SS (1999) Sequence and analysis of chromosome 2 of the plant *Arabidopsis thaliana*. Nature 402:761–768

Linares C, Irigoyen ML, Fominaya A (2000) Identification of C-genome chromosomes involved in intergenomic translocations in *Avena sativa* L., using cloned repetitive DNA sequences. Theor Appl Genet 100(3–4):353–360

Linc G, Friebe BR, Kynast RG, Molnar-Lang M, Köszegi B, Sutka J, Gill BS (1999) Molecular cytogenetic analysis of *Aegilops cylindrica* Host. Genome 42:497–503

Maluszynska J, Heslop-Harrison JS (1993a) Molecular cytogenetics of the genus *Arabidopsis*: in situ localization of rDNA sites, chromosome numbers and diversity in centromeric heterochromatin. Ann Bot 71:479–484

Maluszynska J, Heslop-Harrison JS (1993b) Physical mapping of rDNA loci in *Brassica* species. Genome 36:774–781

Martín A, Cabrera A, Esteban E, Hernández P, Ramírez MC, Rubiales D (1999) A fertile amphiploid between diploid wheat (*Triticum tauschii*) and crested wheat grass (*Agropyron cristatum*). Genome 42:519–524

Mayer K, Schuller C (1999) Sequence and analysis of chromosome 4 of the plant *Arabidopsis thaliana*. Nature 402:769–777

McCombie WR, de la Bastide M (2000) The complete sequence of a heterochromatic island from a higher eukaryote. Cell 100:377–386

Meinke DW, Cherry JM, Dean C, Rounsley SD, Koornneef M (1998) *Arabidopsis thaliana*: a model plant for genome analysis. Science 282:662–682

Melchers G, Sachristan MD, Holder AA (1978) Somatic hybrid plants from potato and tomato regenerated from fused protoplasts. Calsberg Res Commun 43:203–218

Mikhailova EI, Naranjo T, Shepherd K, Wennekes-van Eden J, Heyting C, De Jong JH (1998) The effect of the wheat *Ph1* locus on chromatin organisation and meiotic chromosome pairing analysed by genome painting. Chromosoma 107:339–350

Miller JT, Jackson SA, Nasuda S, Gill BS, Wing RA, Jiang J (1998) Cloning and characterization of a centromere-specific repetitive DNA element from *Sorghum bicolor*. Theor Appl Genet 96:832–839

Moscone EA, Matzke MA, Matzke AJM (1996) The use of combined FISH/GISH in conjunction with DAPI counterstaining to identify chromosomes containing transgene inserts in amphidiploid tobacco. Chromosoma 105:231–236

Munse BG, Cuperus FP, Derksen JTP (1992) Composition and physical properties of oils from new oilseed crops. Ind Crops Prod 1:57–65

Nielen S, Schneider O, Gerdemann-Knörck M (1997) Analysis of somatic *Brassica* hybrids by genomic in situ hybridisation (GISH). In: Lelley T (ed) Current topics in plant cytogenetics related to plant improvement. WUV-Universitätsverlag, Vienna, pp 206–213

Ohmido N, Fukui K (1995) Physical mapping of rice DNAs by an improved FISH method. Jpn Agric Res Q 29:83–88

Papp I, Iglesias VA, Moscone EA, Michalowski S, Spiker S, Park YD, Matzke MA, Matzke AJM (1996) Structural instability of a transgene locus in tobacco associated with aneuploidy. Plant J 10:469–478

Park YD, Papp I, Moscone EA, Iglesia VA, Vaucheret H, Matzke AJM, Matzke MA (1996) Gene silencing mediated by promotor homology occurs at the level of transcription and results in meiotically heritable alterations in methylation and gene activity. Plant J 9:183–194

Parkin IAP, Sharpe AG, Keith DJ, Lydiate DJ (1995) Identification of the A and C genomes of amphidiploid *Brassica napus* (oilseed rape). Genome 38:1122–1131

Parokonny AS, Marshall JA, Bennett MD, Cocking EC, Davey MR, Power JB (1997) Homoeologous pairing and recombination in backcross derivatives of tomato somatic hybrids [*Lycopersicon esculentum* (+) *L. peruvianum*]. Theor Appl Genet 94:713–723

Pedersen C, Giese H, Linde-Laursen I (1995) Towards an integration of the physical and the genetic chromosome maps of barley by in situ hybridization. Hereditas 123:77–88

Pedrosa A, Schweizer D, Guerra M (2000) Cytological heterozygosity and the hybrid origin of sweet orange [*Citrus sinensis* (L.) Osbeck]. Theor Appl Genet 100:361–367

Peterson DG, Lapitan NLV, Stack SM (1999) Localization of single- and low-copy sequences on tomato synaptonemal complex spreads using fluorescence in situ hybridization (FISH). Genetics 152:427–439

Pickering RA, Hill AM, Kynast RG (1997) Characterization by RFLP analysis and genomic in situ hybridization of a recombinant and a monosomic substitution plant derived from *Hordeum vulgare* L. × *Hordeum bulbosum* L. crosses. Genome 40:195–200

Pickering RA, Malyshev S, Künzel G, Johnston PA, Korzun V, Menke M, Schubert I (2000) Locating introgressions of *Hordeum bulbsoum* chromatin within the *H. vulgare* genome. Theor Appl Genet 100:27–31

Plieske J, Struss D, Röbbelen G (1998) Inheritance of resistance derived from the B-genome of *Brassica* against *Phoma lingam* in rapeseed and the development of molecular markers. Theor Appl Genet 97:929–936

Potz H, Schubert V, Houben A, Schubert I, Weber WF (1996) Aneuploids as a key for new molecular cloning strategies: development of DNA markers by microdissection using *Triticum aestivum* – *Aegilops markgrafii* chromosomes addition line B. Euphytica 89:41–47

Raina SN, Mukai Y (1999) Genomic in situ hybridization in *Arachis* (Fabaceae) identifies the diploid wild progenitors of cultivated (*A. hypogaea*) and related wild (*A. monticola*) peanut species. Plant Syst Evol 214:251–262

Ramanna MS, Wagenvoort M (1976) Identification of the trisomic series in diploid *Solanum tuberosum* L., group *tuberosum*. I. Chromosome identification. Euphytica 30:15–31

Rick CM, DeVerna JW, Chetelat RT, Stevens MA (1987) Potential contributions of wide crosses to improvement of processing tomatoes. Acta Hort 200:45–55

Riera-Lizarazu O, Rines HW, Phillips RL (1996) Cytological and molecular characterization of oat × maize partial hybrids. Theor Appl Genet 93:123–135

Salanoubat M, Lemcke K, Rieger M, Ansorge W, Unseld M, Fartmann B, Valle G, Blocker H, Perez-Alonso M, Obermeier B, Delseny M, Boutry M, Grivell LA, Mache R, Puigdomenech P, De Simone V, Choisne N, Artiguenave F, Robert C, Brottier P, Wincker P, Cattolico L, Weissenbach J, Saurin W, Quetier F (2000) Sequence and analysis of chromosome 3 of the plant *Arabidopsis thaliana*. Nature 408:820–822

Schmidt T, Heslop-Harrison JS (1998) Genomes, genes and junk: the large-scale organization of plant chromosomes. Trends Plant Sci 3:195–199

Schoenmakers HCH, Nobel EM, Koornneef M (1992) Use of leaky nitrate reductase-deficient mutant of tomato (*Lycopersicon esculentum* Mill.) for selection of somatic hybrids cell lines with wild type potato (*Solanum tuberosum* L.). Plant Cell Tissue Organ Cult 31:151–154

Schubert I, Shi F, Fuchs J, Endo TR (1998) An efficient screening for terminal deletions and translocations of barley chromosomes added to common wheat. Plant J 14:489–495

Schubert I, Fransz PF, Fuchs J, De Jong JH (2001) Chromosome painting in plants. Cell Science (in press)

Schwarzacher T, Heslop-Harrison JS (2000) Practical in situ hybridisation. BIOS Scientific, Oxford, 203 pp

Schwarzacher T, Leitch AR, Bennet MD, Heslop-Harrison JS (1989) In situ localization of parental genomes in a wide hybrid. Ann Bot 64:315–324

Shepard JF, Bidney D, Barsby T, Kemble R (1983) Genetic transfer in plant through interspecific protoplast fusion. Science 219:683–688

Shishido R, Apisitwanich S, Ohmido N, Okinaka Y, Mori K, Fukui K (1998) Detection of specific chromosome reduction in rice somatic hybrids with the A, B, and C genomes by multi-color genomic in situ hybridization. Theor Appl Genet 97:1013–1018

Skarzhinskaya M, Fahleson J, Glimelius K, Mouras A (1998) Genome organization of *Brassica napus* and *Lesquerella fendleri* and analysis of their somatic hybrids using genomic in situ hybridization. Genome 41:691–701

Snowdon R, Köhler W, Köhler A (1997a) Identification and characterization of rDNA loci in the *Brassica* A and C genomes. Genome 40:582–587

Snowdon R, Köhler W, Friedt W, Lühs W, Köhler A (1997b) Detecting R-genome chromatin in nematode-resistant *Brassica napus* × *Raphanus sativus* hybrids. In: Lelley T (ed) Current topics in plant cytogenetics related to plant improvement. WUV-Universitätsverlag, Vienna, pp 220–226

Snowdon RJ, Köhler W, Friedt W, Köhler A (1997c) Genomic in situ hybridization in *Brassica* amphidiploids and interspecific hybrids. Theor Appl Genet 95:1320–1324

Snowdon RJ, Köhler A, Köhler W, Friedt W (1999) FISH-ing for new rapeseed lines: the application of molecular cytogenetic techniques to *Brassica* breeding. Proc 10th Int Rapeseed Congr, Canberra, Australia. Groupe Consultatif International de Recherche sur le Colza (http://www.regional.org.au/papers/rapeseed/Breeding/196.htm)

Snowdon RJ, Friedt W, Köhler A, Köhler W (2000a) Molecular cytogenetic localisation and characterisation of 5 S and 25 S rDNA loci for chromosome identification in oilseed rape (*Brassica napus* L.). Ann Bot 86:201–204

Snowdon RJ, Winter H, Diestel A, Sacristan MD (2000b) Development and characterisation of *Brassica napus-Sinapis arvensis* addition lines exhibiting resistance to *Leptosphaeria maculans*. Theor Appl Genet 101:1008–1014

Song JQ, Dong FG, Jiang JM (2000) Construction of a bacterial artificial chromosome (BAC) library for potato molecular cytogenetics research. Genome 43:199–204

Suen DF, Wang CK, Lin RF, Kao YY, Lec FM, Chen CC (1997) Assignment of DNA markers to *Nicotiana sylvestris* chromosomes using monosomic alien addition lines. Theor Appl Genet 94:331–337

Svitashev S, Ananiev E, Pawlowski WP, Somers DA (2000) Association of transgene integration sites with chromosome rearrangements in hexaploid oat. Theor Appl Genet 100:872–880

Taketa S, Ando H, Takeda K, Harrison GE, Heslop-Harrison JS (2000) The distribution, organization and evolution of two abundant and widespread repetitive DNA sequences in the genus *Hordeum*. Theor Appl Genet 100:169–176

Tang S, Li Z, Jia X, Larkin PJ (2000) Genomic in situ hybridisation (GISH) analyses of *Thinopyrum intermedium*, its partial amphiploid Zhong 5, and disease-resistant derivates in wheat. Theor Appl Genet 100:344–352

Tanksley SD, Ganal MW, Prince JP, De Vicente MC, Bonierbale MW, Broun P, Fulton TM, Giovanonni JJ, Grandillo S, Martin GB, Messeguer R, Miller JC, Miller L, Paterson AH, Pineda O, Roger MS, Wing RA, Wu W, Young ND (1992) High density molecular linkage maps of tomato and potato genomes. Genetics 132:1141–1160

Theologis A, Ecker JR, Palm CJ, Federspiel NA, Kaul S, White O, Alonso J, Altafi H, Araujo R, Bowman CL, Brooks SY, Buehler E, Chan A, Chao QM, Chen HM, Cheuk RF, Chin CW, Chung MK, Conn L, Conway AB, Conway AR, Creasy TH, Dewar K, Dunn P, Etgu P (2000) Sequence and analysis of chromosome 1 of the plant *Arabidopsis thaliana*. Nature 408:816–820

Thomas J, Chen Q, Tablert L (1998) Genetic segregation and the detection of spontaneous wheat-alien translocations. Euphytica 100:261–267

Voss A, Lühs WW, Snowdon RJ, Friedt W (1999a) Development and molecular characterisation of nematode-resistant rapeseed (*Brassica napus* L.). In: Scarascia GT, E Porceddu and MA Pagnotta (eds) Genetics and breeding for crop quality and resistance. Kluwer, Dordrecht, pp 195–202

Voss A, Lühs WW, Snowdon RJ, Friedt W (1999b) Development and molecular characterization of rapeseed (*Brassica napus* L.) resistant against beet cyst nematodes. Proc 10th Int Rapeseed Congr, Canberra, Australia. Groupe Consultatif International de Recherche sur le Colza (http://www.regional. org.au/papers/rapeseed/Breeding/443.htm)

Voss A, Snowdon RJ, Lühs WW, Friedt W (2000) Intergeneric transfer and introgression of nematode resistance from *Raphanus sativus* into the *Brassica napus* genome. Acta Hort 539:129–134

Wang E, Xing H, Wen Y, Zhou W, Wie R, Han H (1998) Molecular and biochemical characterization of a non-Robertsonian wheat-rye chromosome translocation line. Crop Sci 38:1076–1080

Wang J, Lewis ME, Whallon JH, Sink KC (1995) Chromosome mapping of T-DNA inserts in transgene *Petunia* by in situ hybridisation. Transgene Res 4:241–246

Wang YB, Hu H, Snape JW (1995) Spontaneous wheat/rye translocations from female meiotic products of hybrids between octoploid triticale and wheat. Euphytica 81:265–270

Wang YB, Hu H, Snape JW (1996) The genetic and molecular characterization of pollen-derived plant lines from octoploid triticale × wheat hybrids. Theor Appl Genet 92:811–816

Wilkinson MJ, Bennett ST, Clulow SA, Allainguillaume J, Harding K, Bennett MD (1995) Evidence for somatic translocation during potato dihaploid production. Heredity 74:146–151

Winter H, Gaertig S, Diestel A, Sacristán MD (1999) Blackleg resistance of different origin transferred into *Brassica napus*. Proc 10th Int Rapeseed Congr, Canberra, Australia. Groupe Consultatif International de Recherche sur le Colza (http://www.regional. org.au/papers/rapeseed/Breeding/593.htm)

Wolters AMA, Jacobsen E, O'Connell M, Bonnema G, Ramulu KS, De Jong H, Schoenmakers H, Wijbrandi J, Koornneef M (1994) Somatic hybridization as a tool for tomato breeding. Euphytica 79:265–277

Wolters AMA, Trindade LM, Jacobsen E, Visser RGF (1998) Fluorescence in situ hybridisation on extended DNA fibres as a tool to analyse complex T-DNA loci in potato. Plant J 13:837–847

Yan H, Min S, Zhu L (1999) Visualization of *Oryza eichingeri* chromosomes in intergenomic hybrid plants from *O. sativa* × *O. eichingeri* via fluorescent in situ hybridization. Genome 42:48–51

Yan HH, Xiong ZM, Min SK, Hu HY, Zhang ZT, Tian SL, Fu Q (1997) The production and cytogenetical studies of *Oryza sativa–Oryza eichingeri* amphiploid. Acta Genet Sin 24:23–29

Yeh BP, Peloquin SJ (1965) Pachytene chromosomes of the potato (*Solanum tuberosum* Group Andigena). Am J Bot 52:1014–1020

Zeller FJ (1999) Gentransfer mittels Genom- und Chromosomen-Manipulationen zwischen *Festuca*- und *Lolium*-Arten. J Appl Bot 73:43–49

Zhong XB, Fransz PF, Wennekers-Van Eden J, Zabel P, Van Kammen AB, De Jong JH (1996) High-resolution mapping on pachytene chromosome and extended DNA fibers by fluorescence in situ hybridisation. Plant Mol Bio Rep 14:232–242

Zhong XB, Fransz PF, Wennekes-Van Eden J, Ramanna MS, Van Kammen A, Zabel P, De Jong JH (1998) FISH studies reveal the molecular and chromosomal organization of individual telomere domains in tomato. Plant J 13:507–517

Zhong XB, Bodeau J, Fransz PF, Williamson VM, van Kammen A, De Jong JH, Zabel P (1999) FISH to meiotic pachytene chromosomes of tomato locates the root-knot nematode resistance gene *Mi*-1 and the acid phosphatase gene *Aps*-1 near the junction of euchromatin and pericentromeric heterochromatin of chromosome arms 6 S and 6L, respectively. Theor Appl Genet 98:365–370

Zhou R, Jia J, Dong Y, Schwarzacher T, Reader SM, Wu S, Gale MD, Miller TE (1998) Characterization of *Triticum aestivum/Psathyrostachys juncea* derivatives by genomic in situ hybridization. Euphytica 99:85–88

Zwierzykowski Z, Lukaszewski AJ, Lesniewska A, Naganowska B (1998) Genomic structure of androgenic progeny of pentaploid hybrids, *Festuca arundinacea* × *Lolium multiflorum*. Plant Breeding 117:457–462

PD Dr. Renate Horn
Dr. Rod Snowdon
Dipl.-Ing. agr. Barbara Kusterer
Institut für Pflanzenbau und Pflanzenzüchtung, IFZ
Heinrich-Buff-Ring 26–32
35392 Giessen, Germany
Tel.: +49 641 99 37423
Fax : +49 641 99 37429
e-mail: renate.horn@agrar.uni-giessen.de

Function of Genetic Material: Genes Involved in Quantitative and Qualitative Resistance

By Thomas Lübberstedt, Volker Mohler, and Gerhard Wenzel

1 Introduction

A higher plant contains a minimum of 20,000 genes (Kaul et al. 2000). A successful new variety – the end product of the function of genetic material – is never the result of the addition of just one gene but rather a better combination of several genes. Thus, **the challenge in plant breeding is the optimal combination of many genes.** Good luck and the "green thumb" of the breeder, are still important prerequisites for successful plant breeding. The question is whether increased knowledge of the function of genetic material will help in offering reliable tools for optimal combinations of the $[(n+1) \cdot n]^k$ alleles (n=number of alleles per k loci) in a better genome (Rommens and Kishore 2000).

During the period of progress which is reported in this Volume, a huge amount of structural data of the DNA composition and the localization of genes on genetic maps has been collected for many plant species and for many characters (Oberhagemann et al. 1999; Ma et al. 2000; Chen et al. 2001). No longer are all data published on hard copy but also in the Internet as, e.g., a catalogue of AFLP markers covering the potato genome by Van der Voort et al. (1998) under www.spg.wau.nl/pv/aflp/ catalog.htm. Since it may not be so interesting to just sum up these new data, we would rather focus on a few questions concerning the nature of important characters of plants, which might be answered using this new information.

Characters expressed by a plant have different genetic bases. They might be inherited very simply due to the presence or absence of one specific allele, or the expression relies on a more complex genetic configuration. In the simpler situation a **qualitative inheritance** of one gene (monogenic) is normal while the latter more complex one results in a continuous trait expression due to several or many genes involved (polygenic inheritance). This difference is of particular interest for disease resistance genes (R genes) since it has some additional consequences. A monogenic qualitatively expressed trait can be added to a genome via a backcross approach within ten generations. Its expression is, however, rather specific and can be overcome by a simple mutation in the pathogen population thus; it is not very stable and is called vertical resistance. In contrast, **quantitative trait loci (QTL)** affect complex in-

Progress in Botany, Vol. 63

herited resistances more nonspecifically, and give a partial, incomplete or horizontal resistance which is harder to overcome by mutations in the pathogen and thus more stable or durable (Van der Plank 1978; Wenzel et al. 1985). QTL for resistance are called **quantitative resistance loci (QRL)** (Young 1996). Breeding programs aiming at disease resistance favor the incorporation of QRLs. Unfortunately, increasing the level of a quantitative trait needs many more complex breeding strategies (Geiger and Heun 1989). To pyramid as many QRL as possible, success depends presently on the existence of a distinct quantitative test system.

By marker-assisted selection (MAS) (e.g., in barley, Dehmer et al. 1991) an intermediate process for building up a broader resistance may be performed: several monogenic traits are combined in one breeding line through backcrossing. An example of the successful use of MAS on the basis of localized powdery mildew resistance genes is given by Wenzel et al. (2000). Since monogenic mildew resistance is not very durable, different resistances were pyramided. The combination of the three powdery mildew genes *Pm1c* (Hartl et al. 1999), *Pm24* (Huang et al. 2000) and *Pm29* (Zeller et al. 2001) in one wheat line could be achieved.

The molecular structure of simple monogenic characters is increasingly being understood (Young 2000). For disease resistances (fungi, bacteria and viruses) most genes code for a limited number of similar proteins involved in the signal transduction chain with transmembrane activity (leucine-rich repeats, e.g., Jones et al 1996; Meyers et al. 1998), another hint for a functional and local similarity; in addition these genes are arranged in gene clusters (Meyers et al. 1999). (Insect resistance probably uses another path.)

Many new data for quantitatively inherited resistance became available since the review of Young (1996) on QRL appeared, who concluded that with the help of DNA markers and QTL mapping complex forms of disease resistance will become accessible. He states that the distinction between qualitative and quantitative disease resistance may disappear. At the end of the premolecular time Robertson (1985) postulated that qualitative and quantitative traits may be the result of different types of variation of DNA. Variations of a minor nature may result in different (quantitative) efficiencies of alleles responsible for gene products while major gene rearrangements or changes in the region of the gene essential for a normal functioning gene product, result in qualitative expression differences. Quantitative genes may be alleles of qualitative ones involving the same trait locus.

For several species, qualitative R genes are arranged in gene clusters (McMullen and Simcox 1995; Kaul et al. 2000). The available genome data on QRL will be analyzed here to find out whether they are **randomly distributed over the genome, or whether QRL clusters exist.** It might even be possible that in QRL clusters, at least several quantitative loci are close or very close to qualitative ones and form functional units, comparable with the operon structure in microorganisms. Such information would simplify the task of finding qualitative and quantitative loci responsible for disease resistance and give additional hints about the basis of complex gene functions. Further, it would have consequences for plant breeding: To recombine tightly linked alleles demands large populations and very tightly linked markers. A successful combination will result, however, in a very powerful complex QRL. Otherwise - randomly distributed genes - may be combined easily but their stability is lower.

2 Localization of Genes for Disease Resistance (R Genes)

a) Classical Approaches

Methods and objectives of conventional genetic analysis have been extensively reviewed by Koorneef and Stam (1992). Besides the segregation analysis of crosses with multiple marker stocks, aneuploids are applied to assign new genes to distinct chromosomes. In diploid plant species, trisomic analysis is the classical method to allocate genes on specific chromosomes, while in polyploids monosomic analysis is the method of choice. In maize, moreover, B-A translocations represent the most effective stocks to be used for mapping purposes. The classical chromosomal assignment of loci controlling quantitative disease resistance is mainly done by the introduction of single chromosomes or chromosome pairs from the same or a related species into a susceptible genetic background (thereby replacing the homologues) and the subsequent measurement of their effects on the phenotype. A disomic substitution line will identify all additive gene actions on the respective chromosome plus all epistatic interactions among these genes as well as those with genes from other chromosomes, while a monosomic substitution line is used to estimate effects of dominance and their interactions. Despite steady refinements in marker-based QTL analyses, the use of precise genetic stocks for estimating the effects of individual chromosomes on the trait is still popular, especially in the cytogenetically well-studied common wheat.

Examination of a series of 21 chromosome substitution lines of *Fusarium* head blight resistant *T. macha* into the susceptible *T. aestivum* 'Hobbit sib' showed that *T. macha* chromosomes 1B, 4A and 7A carry genes that have a major influence on resistance to *Fusarium* (Mentewab et al. 2000). Using the same substitution lines, Grausgruber et al. (1998) located loci for resistance to initial infection and invasion of the host on chromosomes 3A, 4A, 5A and 6B of *T. macha*. Another study from Buerstmayr et al. (1999) dealt with a backcross reciprocal monosomic analysis of *Fusarium* head blight resistance using the highly resistant Hungarian winter wheat line 'U-136.1', harboring Chinese and Japanese resistance donors in its pedigree, and the highly susceptible 'Hobbit sib'. Five hemizygous families containing 'U-136.1' chromosomes 6B, 5 A, 6D, 1B, and 4B had a visually reduced spread of infection compared to lines having the 'Hobbit-sib' chromosomes while chromosome 2B from 'U-136.1' had an increased spread of infection. In the report of Ellerbrook et al. (1999) a substitution series of 'Synthetic 6x' into 'Chinese Spring' has been studied to determine which of the 21 chromosomes of 'Synthetic 6x' conferred resistance to *Stagonospora nodorum*. The consecutive development of single chromosome recombinant lines not only enabled the identification of chromosome 5D of 'Synthetic 6x' to be most effective against that disease but is also a starting point for map-based QTL analysis.

In conclusion, the information obtained from the analysis of individualized chromosomes is highly useful when attempting to genetically map disease resistance QTL in large-genome species because **molecular marker work can be focussed on significantly acting chromosomes** only.

b) Quantitative Trait Loci for Disease Resistance (QRL)

The major goals of QTL analyses are:

1. to locate QTL in the genome,
2. to determine gene effects,
3. to measure the degree of dominance, and
4. to estimate the presence of interactions with environmental factors or other loci for these QTL.

Although the term QTL was coined by Geldermann in 1975, the basic concept was first employed as early as 1923 in a study on the association of seed-coat patterns and pigmentation in *Phaseolus vulgaris* (Sax 1923). The major limitation for the application of systematic (genome-wide) QTL analyses until 1980 was the lack of a sufficient number of genetic markers.

With the advent of restriction fragment length polymorphism (RFLP) markers (Botstein et al. 1980) and subsequently a number of polymerase chain reaction (PCR)-based molecular markers such as SSR (simple sequence repeat), RAPD (random amplified polymorphic DNA), and AFLP (amplified length polymorphism) markers (Lübberstedt et al. 2000), this limitation has been overcome. In major crop species such as maize and barley thousands of molecular markers are now available (http:// www.agron.missouri.edu/; http://wheat.pw.usda.gov/). Consequently, the number of experimental QTL studies in plants has dramatically increased during the last decade also including studies on resistance traits (Young 1996).

In parallel to the developments in molecular biology, **statistical tools for QTL analyses** improved greatly during the last two decades (Jansen 1996; Melchinger 1998). Originally, regression analyses were conducted individually for each marker to detect significant differences between marker genotype classes, e.g., between the homozygote and heterozygote class at a given marker locus in a backcross population. However, single-marker analyses have the major disadvantage of confounding estimates for QTL position and gene effect. In cases where there are significant differences between marker genotype classes, closely linked QTL with small gene effects can not be differentiated from distant QTL with large gene effects.

α) Simple Interval Mapping (SIM)

This problem has been overcome by the more powerful interval mapping approach (Lander and Botstein 1989), using intervals of linked markers as a unit of analysis rather than single markers. By a maximum likelihood (Lander and Botstein 1989) or a regression approach (Haley and Knott 1992), the most likely position of a QTL as well as the respective gene effect can be estimated. The "simple" interval mapping (SIM) procedure has been further improved by combining it with a multiple regression analysis on markers associated with other QTL.

β) Composite Interval Mapping (CIM)

The composite interval mapping (CIM) approach (Jansen and Stam 1994; Zeng 1994) is more powerful compared to SIM with respect to (1) QTL detection, (2) more accurate estimation of gene effects, and (3) separation of linked QTL. Further improvements based on the CIM approach relate to, e.g., the joint analysis of multiple traits (Jiang and Zeng 1995) and populations (Haley 1999), as well as the evaluation of the accuracy of the obtained estimates (Melchinger et al. 1998; Utz et al. 2000). A number of software packages have been developed implementing the CIM approach such as PLABQTL (Utz and Melchinger 1996), MapQTL, QTL cartographer (http://linkage.rockefeller.edu/soft/list.html), and MultiQTL (Korol 2001).

γ) Consequences

Since this review summarizes the results of a large number of QTL studies, the difference in data quality among the studies has to be taken into consideration. Besides the statistical method employed in QTL analysis, data quality depends on

- the trait heritability,
- the population size and type, and
- the number of environments and markers employed.

According to Utz and Melchinger (1994), the major consequences of sub-optimal experimental conditions such as small QTL mapping populations and low inheritance are (1) **non-detection of QTL** and (2) **overestimation of gene effects.**

In contrast, **estimation of QTL positions remains largely unaffected,** which is of relevance for this review when comparing positions of

- QTL-affecting resistance traits (QRL) and
- major resistance genes or resistance gene analogues (RGAs).

Most QTL studies included in this review employed CIM. Hence, the risk of inaccurately estimated QTL positions due to inappropriate statistical methods or experimental conditions, should be low.

c) Resistance Gene Analogs (RGAs)

The cloning of many different R genes from plants in recent years and the assembly of their sequence data led to the identification of conserved domains that can account for many predicted functions of R genes. The largest group of known R gene products are from the NBS-LRR type (Meyers et al. 1999), others carry a protein kinase domain (Martin et al. 1993) or are composed of all three autonomously folding protein structures (Song et al. 1995).

The use in PCR assays, of degenerate oligonucleotide primers designed from conserved domains between R proteins of *N* (tobacco), *RPS2* (*Arabidopsis*) and *L6* (flax) has been referred to as a promising approach to obtain disease resistance candidate genes.

In that way, resistance gene analogs (RGAs) have been isolated from dicot (Kanazin et al. 1996; Leister et al. 1996; Yu et al. 1996; Gentzbittel et al. 1998; Shen et al. 1998; Pflieger et al. 2000) and monocot species (Chen et al. 1998; Collins et al. 1998; Leister et al. 1998). Specific PCR primers derived from conserved regions of an NBS-LRR sequence at the *Cre3* cereal cyst nematode resistance locus in *Triticum tauschii* L. and other known R genes (Grant et al. 1995) have been applied to isolate resistance gene-like sequences in wheat and barley (Seah et al. 1998). Feuillet et al. (1997) designed oligonucleotides corresponding to the conserved subdomains II to VIII in Ser/Thr protein kinases, to successfully amplify a receptor-like kinase gene encoded at the *Lr10* disease resistance locus of wheat, while Chen et al. (1998) reported the scoring of RGAs with specific primers based on LRR regions of genes *RPS2*, *Xa21* (rice), *N* and *Cf-9* (tomato) in the small-grain cereals wheat, barley and rice.

Many of the retrieved amplification products from the above mentioned works cosegregate or map in close vicinity to known major resistance genes or QTL. Furthermore, a resistance gene-analogous sequence was used as a probe to identify transposon-induced *Rp1-D* (gene for resistance to maize common rust) mutations in maize and thus helped to isolate the gene.

However, since **RGAs were shown to be members of multigene families** (e.g., about 200–300 NBS-LRR sequences accounting for 1–2% of coding sequences were estimated in *Arabidopsis*), it is difficult to identify RGAs as the active gene copy within a cluster of paralogs (Meyers et al. 1998; Wei et al. 1999). Therefore, Graham et al. (2000), in their investigation on the expression of NBS-LRR resistance gene analogues in soybean, proposed examining other R gene motifs rather than the highly conserved signatures, to differentiate a functional gene from its non-functional paralog. The significantly increasing knowledge of the structural organization and sequence evolution of NBS-LRR sequences, the majority of RGAs known so far, combined with a rigorous analysis of expressed copies only, will make it feasible in the future to develop primers for the amplification of NBS-LRR groups most likely to be true R gene candidates.

3 Genomic Distribution of QRL

With the many data available today an attempt will be made, using the examples of maize and barley, to answer the following questions:

1. Are the QRL randomly distributed across the chromosomes?
2. Are the QRL randomly distributed across the 100 BINs (see Sects. 3.a, 3.b)?
3. Do the QRL map preferentially to clusters of major resistance genes and RGAs?

To clarify these questions on the basis of the available literature, chi-square tests have been applied for each question.

a) Maize

Important maize diseases are southern corn leaf blight, different rusts and smuts, as well as the sugarcane and maize dwarf mosaic caused by virus (Agrios 1997). The respective pathosystems have been major targets in QTL mapping studies (Table 1). The same applies to the European corn borer which causes significant economic losses in maize production in several countries (Bohn et al. 2000). However, since the LRR-type of resistance genes have not been identified for insect resistance so far, the respective QTL studies were excluded from this review. Most likely different biochemical pathways are involved in insect resistance compared to disease (fungus, bacteria, virus, nematode) resistance (McMullen et al. 1998; Dangl and Holub 1997).

Clustering of major resistance genes in the maize genome has been reported by McMullen and Simcox (1995; Table 2). These genes include different viral and fungal resistance genes, the latter mainly acting against *Puccinia sorghi* (rp genes) and *Setosphaeria turcia* (ht genes).

Table 1. Maize resistance quantitative trait loci (QTL) studies evaluated in this review

Reference	Pathogen	Abbreviation (Fig. 1)
Fungal		
Lübberstedt et al. (1998a)	*Puccinia sorghi*	Ps
Lübberstedt et al. (1998b)	*Ustilago maydis*	Um
Welz et al. (1999)	*Setospheria turcica*	St
Lübberstedt et al. (1999)	*Sporisorium reiliana*	Sr
	Ustilago maydis	Um
	Puccinia sorghi	Ps
Agrama et al. (1999)	*Peronoscrospora sorghi*	Pe
Freymark et al. (1999)	*Setospheria turcica*	St
Dingerdissen et al. (1996)	*Setospheria turcica*	St
Schechert et al. (1999)	*Setospheria turcica*	St
Saghai Maroof et al. (1996)	*Cercospora zeae*	Cz
Kerns et al. (1999)	*Puccinia sorghi*	Ps
	Ustilago maydis	Um
Pe et al. (1993)	*Gibberella zeae*	Gz
Viral		
Xia et al. (1999)	SCMV	Scm
Marcon et a. (1999)	HPV	Hpv
Pernet et al. (1999a, b)	MSV	Msv
Lu et al. (1999)	MSV	Msv

Two of these resistance genes are not known to be members of the LRR-resistance genes, hm1 and bx1 (McMullen and Simcox 1995; Frey et al. 1997), in contrast to rp1 (Hulbert et al. 1999). Collins et al. (1998) mapped 20 RGA loci to the maize genome. These **RGA loci are more frequently located in the same chromosomal BINs** (Neuffer et al. 1997) as major resistance genes than would be expected by chance (chi-square test).

In total, 16 peer-reviewed QTL studies on fungal (14) or viral (5) maize diseases were selected for this consideration (Table 1), whereas respective QTL studies on bacterial or nematode resistance are lacking so far. Most frequently, resistances to *S. turcica* (4), *P. sorghi* (3), *U. maydis* (3), and MSV (3) were investigated (Table 1). Major maize resistance genes have been described for only some of the pathogens listed in Table 1 (see Table 2): ht genes (*S. turcica*), rp genes (*P. sorghi*), scmv1 and scmv2 (SCMV), msv1 (MSV) (McMullen and Simcox 1995; Xu et al. 1999). In the case of *Sporisorium reiliana*, a major resistance gene has been identified in *Sorghum* but not in maize so far (Lübberstedt et al. 1999).

In this review, when possible, QRL have been included from analyses across environments, and otherwise from single-environment experiments. If the same pathosystem was investigated in different studies, QRL from each individual study were included for further calculations, even if located on the same chromosomal BIN. Since these BINs of about

Table 2. Major resistance genes and resistance gene analogs (RGAs) in maize[a]

Chromosome/BIN[a]	Major gene	RGA
1.04–1.05	*msv1*	Pic12, Pic13a
1.06	*hm1*	Pic15
2.03		Pic17
2.07–2.08	*ht1*	Pic18a
3.04–05	*rp3, mv1, wsm2, Scmv2*	Pic13b, Pic21, ssCS4a
4.01–02	*bx1, rp4*	Pic18b, ssCS4b
4.08		Pic14
5.03		Pic11
6.00–6.01	*mdm1, wsm1, Scmv1, rhm1*	Pic19, ssCS4c
6.05–6.06		Pic13c, Pic16
7.02		Pic13d
8.05–8.06	*ht2, htn1*	
9.00–9.01		SsCS4d
9.04–9.05	*hm2*	
10.01	*rp1, rp5, rp1-G, rpp9*	Pic20
10.05	*wsm3*	

[a]According to McMullen and Simcox (1995), Neuffer et al. (1997), Collins et al. (1998), Ming et al. (1999), Xu et al. (1999).

20 cM contain physically large regions (about 4×10^7 bp on average) there is little evidence for identical genes behind co-segregating QRL. However, there might be a tendency of "double counting" the same QTL. In total, 205 QRL including 180 "fungal" and 25 "viral" QRL were employed in this review (Table 3).

Regarding question (1), the observed number of QTL was compared with the expected number of QTL for each chromosome based on its genetic length according to Neuffer et al. (1997). To address question (2), a Poisson distribution was compared with the observed distribution of all QRL to the 100 maize BINs. As a first step, the number of BINs was reduced to 80, since the most proximal and distal BINs contained far fewer QTL than expected at random (Fig. 1). This is most likely due to the smaller size of these "extreme" BINs compared to the intermediate BINs (average size 20 cM). Therefore, the "extreme" BINs were fused with the neighboring BINs (e.g.: 1.00+1.01 or 1.11+1.12).

For question (3), the maize genome was divided into three fractions, containing (I) no R-gene or RGAs, (II) exactly one R-gene and/or RGA, and (III) two or more R-genes and/or RGAs (clusters) (Table 3). A random distribution of QRL was compared with a preferential mapping to genome fraction (III) or fraction (II+III).

Table 3. Summary maize resistance quantitative trait loci (QTL) mapping

	QTL for fungal resistance (No. / interval length, cM)			QTL for viral resistance (No. / interval length, cM)			
Chromo-some	Wthin R-gene/ RGA cluster	Close to single R-gene/RGA	Unlinked to unknown R-genes/RGAs	Wthin R-gene/ RGA cluster	Close to single R-gene/RGA	Unlinked to unknown R-genes/RGAs	Sum
1	4 / 47	4 / 31	20 / 182	3 / 47	0 / 31	1 / 182	32 / 229
2	-	9 / 60	13 / 156	-	1 / 60	2 / 156	25 / 216
3	5 / 50	-	17 / 165	3 / 50	-	3 / 165	28 / 215
4	0 / 26	2 / 23	14 / 138	0 / 26	0 / 23	1 / 138	17 / 187
5	-	6 / 28	22 / 155	-	1 / 28	2 / 155	31 / 183
6	5 / 64	-	6 / 87	3 / 64	-	0 / 87	14 / 151
7	-	3 / 25	11 / 94	-	0 / 25	0 / 94	14 / 119
8	8 / 44	-	11 / 127	0 / 44	-	1 / 127	20 / 171
9	-	8 / 83	6 / 91	-	0 / 83	1 / 91	15 / 174
10	0 / 21	1 / 19	5 / 137	0 / 21	2 / 19	1 / 137	9 / 177
Sum	22 / 252	33 / 269	125 / 1332	9 / 252	4 / 269	12 / 1332	205/1853

Chrom.	BIN 00	01	02	03	04	05	06	07	08	09	10	11	12
1		Ps	Um St	Ps Sr Pc Ps	Um Um	Um Ps msv msv msv	[illegible]	St Pe Gz	scm Ps Ps Um	Um Ps	Ps Um	Ps St	Um
2			St	[illegible]	Um msv St	Um Ps Um	msv Ps St Um Um Um	[illegible]	[illegible]	Ps Ps			
3		Ps St	msv	St	hpv msv scm Ps Ps	Sr Um Um	Ps Ps St Um Um	Gz Ps St St	Ps St Um Um	msv msv	St		
4				Ps St Um	msv Gz Ps St Um		Sr St Um	Um Um	[illegible]	Um	Ps		
5		scm Ps Ps Ps St Um Um	msv Gz St Um Um	[illegible]	Gz Ps Ps Ps Sr St	Cz Ps St	St Um	St					
6	hpv scm Sr	msv Um			Ps Ps		Ps Sr St	Ps Ps Sr Um					
7		Ps Ps Um	[illegible]	Ps St St Um	Ps Ps Um Um								
8	Um	Ps	Ps Sr St	Ps St St	Ps	Ps Ps St	Cz Ps Sr St Um	msv	Ps Um				
9		[illegible]	msv St		[illegible]	[illegible]	Ps Um Um	Ps Ps					
10	Um		Ps		Um Um	[illegible]	msv Gz Ps						

Fig. 1. Distribution of disease resistance QTL across the maize genome subdivided into BINs (Neuffer et al. 1997). *White*, *black*, and *gray* BINs reflect genome fractions I, II, and III, respectively (Sect. 3)

Answer to question (1) for maize

The 205 QRL were almost distributed at random among the 10 maize chromosomes. Only chromosome 5 contained more QRL than expected by chance. This finding is interesting, since chromosome 5 as well as chromosome 7 were the only chromosomes without an R-gene/RGA cluster. Possible explanations are (1) an underestimation of the physical size of chromosome 5 because of either suppressed recombination or a lack of proximal or distal markers, (2) a higher gene density in general or specific for QRL, and (3) a higher number of QRL by chance. Another explanation might be "double counting" of the same QTL, since three BINs contained three QRL for *P. sorghi* (5.01, 5.04) or *U. maydis* (5.03) (Fig. 1).

Answer to question (2) for maize

The distribution of QRL across all 80 (100) chromosomal BINs is in agreement with a Poisson distribution, indicating **no significant clustering of QRL.** However, there is a tendency of overrepresentation of BINs containing 0 or 5–7 QTL, which might be due to QRL clusters or "double counting" (see above).

Answer to question (3) for maize

Across all QRL there was **no significant clustering with R-genes and RGAs.** The same applied to "fungal" QRL, whereas viral QRL displayed significant clustering with genome fraction (III). Therefore, the majority of QRL identified with regard to fungal diseases seem to be randomly distributed across the genome. In contrast, several "viral" QRL identified so far might be related to LRR-type R-genes.

b) Barley

The same questions asked for maize were analyzed in barley. The barley genome has been estimated to contain around 5.5 pg of DNA per haploid nucleus, equivalent to approximately 5.3×10^9 bp (Bennett and Smith 1976). The genome consists of a complex mixture of unique and repeated nucleotide sequences (Flavell 1980) and a gene density of one gene per 123–212 kb can be expected if genes are distributed equidistantly (Panstruga et al. 1998). However, grass genomes seem to contain regions that are highly enriched in genes with very little or no repetitive

DNA. Feuillet and Keller (1999) found five genes on a 23 kb DNA around the receptor-like kinase gene *Lrk10*.

Disease and pest resistance genes in barley can be viewed in the yearly updated 'Coordinators Report' section in the 'Barley Genetics Newsletter' that can be accessed in the GrainGenes database (http://wheat.pw.usda.gov). There does not seem to be a prevalence for the occurrence of either qualitative or quantitative disease reactions in barley. QTL have been identified in recent years for the most intensively investigated disease reactions of the host, for which major R genes were described in the past. The only exception seems to be *Fusarium* head blight (FHB) against which no major gene action has been reported so far and, therefore, resistance is considered to be quantitatively inherited.

To get an idea about genomic organization of QRL, the distribution of mapped QRL in the barley genome was examined and compared with the positions of major disease resistance genes and RGAs. This process of data merging has become feasible since the barley genome has been divided into approximately 10-cM intervals designated as BINs (Kleinhofs et al. 1998). **The majority of major R genes, QRL and RGAs from the various studies had been mapped with original BIN markers**

Table 4. List of diseases for which quantitative trait loci (QTL) have been integrated in the 'Steptoe × Morex' BIN map

Disease/Pathogen	References
Powdery mildew (*Ml*) (*Erysiphe graminis* f.sp. *hordei*)	Heun (1992); Saghai Maroof et al. (1994); Backes et al. (1995, 1996); Spaner et al. (1998)
Leaf rust (*Lr*) (*Puccinia hordei*)	Spaner et al. (1998); Kicherer et al. (2000)
Stem rust (*Sr*) (*Puccinia graminis* f.sp. *hordei*)	Spaner et al. (1998)
Stripe rust (*Yr*) (*Puccinia striformis* f.sp. *hordei*)	Chen et al. (1994); Hayes et al. (1996); Toojinda et al. (2000)
Scald (*Rh*) (*Rhynchosporium secalis*)	Backes et al. (1995); Spaner et al. (1998)
Net blotch (*Nb*) (*Pyrenophora teres* f. sp. *teres*)	Steffenson et al. (1996); Richter et al. (1998); Spaner et al. (1998)
Spot blotch (*Sb*) (*Cochliobolus sativus*)	Steffenson et al. (1996)
Leaf stripe (*Ls*) (*Pyrenophora graminea*)	Pecchioni et al. 1996)
Fusarium head blight (*Fs*) (*Fusarium* sp.)	de la Pena et al. (1999); Zhu et al. (1999); Ma et al. (2000)
Barley yellow dwarf virus (BYDV)	Toojinda et al. (2000)
Bacterial leaf streak (*Bls*) (*Xanthomonas campestris* pv. *hordei*)	El Attari et al. (1998)

from the 'Steptoe × Morex' map which allowed us to present thorough allocation results. A total of 38 major R genes and 21 RGAs were assigned to 21 and 13 BINs, respectively (Table 5). For the analysis of QRL distribution, the barley genome was fractionated, consisting of (I) empty BINs, (II) BINs which are populated by two or more R genes and RGAs, respectively or are mixtures of both with at least three members, (termed as clusters and furnished with a black label in the skeletal BIN map) and (III) BINs with remaining configurations (classified as random and highlighted in gray). This resulted in the identification of 14 clusters as well as 14 randomly occupied BINs (Table 5, Fig. 2). Besides the pure clusters of major resistance genes (such as BIN 3(3H)-016 harboring four divergent genes against barley yellow mosaic virus disease, or BIN 3(3H)-006 representing a heterogeneous R gene cluster), coincidences of

Table 5. Major resistance genes and resistance gene analogs (RGAs) in barley

Chromosome/BIN[a]	Major gene	RGA
1(7H)-001	*mlt, Rpg1, Rrs2*	pic20, pic15, Hv-b9, ABG331, ABG333, ssCH4
1(7H)-003	*Rcs5*	
1(7H)-007	*rms1*	RSB001, ssCH4
1(7H)-011.012	*Mlf, Rph3*	ssCH4, ssCH5
1(7H)-013		ssCH4
2(2H)-001	*Mlhb1.a, Rph17*	
2(2H)-006	*Rph16*	
2(2H)-011.012	*Ha2, Rph15*	ssCH4
2(2H)-015	*MlLa, Rph18*	ssCH4
3(3H)-001	*Rph7.g*	
3(3H)-006	*Rrs1, Rpt.a, Ryd2*	
3(3H)-008		pic20
3(3H)-016	*rym4, rym5, rym6, rym10*	
4(4H)-006	*Mlg, rym11*	Hv-b3
4(4H)-010	*mlo*	
4(4H)-013	*rym8, rym9*	
5(1H)-002	*Rti, Mla, Rph4*	Hv-b6
5(1H)-003	*Yr4, Mlk*	
5(1H)-013		pic20
6(6H)-003	*Rrs13*	
7(5H)-004	*rym3, Rph2*	ssCH4, ssCH5
7(5H)-007	*Mlj*	RSB001
7(5H)-010		ssCH5
7(5H)-011.012	*Rph9*	
7(5H)-013	*rpg4*	

[a]According to the website http://barleygenomics.wsu.edu
RGA sources: Ayliffe et al. (2000), Leister et al. (1998), Seah et al. (1998), barleygenomics website.

Chr.	BINs[a]															
	01	02	03	04	05	06	07	08	09	10	11	12	13	14	15	16
1(7H)	*Ml(2) Sr, Nb*	*Fs*	*Sb Fs(2)*	*Rh, Sb*	*Ml*		*Ml, Nb Ls*			*Ml*	*BYDV*					
2(2H)			*Lr(2) Ls*	*Lr, Nb Ls*	*Nb, Lr Fs*		*Fs*	*Ml*	*Fs*	*Lr*	*Fs*		*Ml*		*Ml, Rh Fs*	
3(3H)		*Bls, Nb Fs(2)*				*Nb, Fs*		*Bls*	*Nb*			*Nb*	*Nb, Fs Yr*		*Rh*	*Fs, Rh*
4(4H)	*Ls*	*BYDV*		*Fs(2)*	*Ml, Nb*	*Ml(2) Nb(2)*	*Lr*	*Ml, Sr Rh, Nb Fs*		*Ml, Yr(2)*	*BYDV*					
5(1H)		*Ml Yr(2)*	*Ml*	*Ml*		*Sb, Fs*	*Ml, Sb*	*Nb*	*BYDV*			*Fs(2)*				
6(6H)		*Ml*		*Yr*		*Ml, Lr Rh Nb(3)*	*Nb*	*Fs(2)*								
7(5H)			*Ml, Fs*	*Ml(2) Fs, Nb*	*Ml*		*Yr*		*Yr*	*Lr, Nb*	*Ml*	*Ml*	*Ml*		*Ml(2)*	

[a] BIN size: 10 cM

Fig. 2. Distribution of disease resistance QTL in the barley genome subdivided by the BIN method (Kleinhofs et al. 1998). *White*, *black*, and *gray* BINs reflect genome fractions I, II, and III, respectively (Sect. 3)

Table 6. Organization of disease resistance quantitative trait loci (QTL) in the barley genome

Chr.	QTL location	QTL for disease resistance (nos.)											Σ	
		Ml	*Sr*	*Nb*	*Fs*	*Sb*	*Rh*	*Ls*	*Yr*	*Lr*	*BYDV*	*Bls*	A[a]/BIN	B[b]/BIN
1(7H)	A[a]	3	1	2				1			1		8/3	
	B[b]	2			3	2	1							8/5
2(2H)	A	1			2		1						4/2	
	B	2		2	3			2		5				14/8
3(3H)	A			1	2		1						4/2	
	B			4	3		1		1			2		11/6
4(4H)	A	2		2									4/1	
	B	3	1	2	3		1	1	2	1	2			16/8
5(1H)	A	1							2				3/1	
	B	3		1	3	2					1			10/7
6(6H)	A													
	B	2		4	2		1		1	1				11/5
7(5H)	A	2		1	1								4/1	
	B	7		1	1				2	1				12/9
Σ		28	2	20	23	4	6	4	8	8	4	2	27/10	82/48

[a] A, Inside R gene/RGA cluster
[b] B, Outside R gene/RGA cluster.

map positions between major R genes and RGAs were observed in eight BINs (Table 5). The most noticeable cluster describes BIN 1(7H)-001 with major genes conferring resistance against powdery mildew (*mlt*), stem rust (*Rpg1*) and scald (*Rrs2*) which have been co-localized with six RGAs.

A total of 109 QTL from 23 investigations on fungal (9), viral (1) and bacterial (1) barley diseases (Table 4, Table 6) were added to the BIN map (Fig. 2). Of the 100 barley BINs, 58 were found to harbor QRL. As in maize, the same questions were addressed, and assessed with the same statistics.

Answer to question (1) for barley

The **109 QRL were randomly distributed among the seven barley** chromosomes rather than confined to particular chromosomes regardless of 'double counting' the same QTL.

Answer to question (2) for barley

The comparison of QRL across the 100 BINs based on Poisson distribution revealed the observation that **particular BINs are preferentially occupied by QRL** (number of BINs consisting of 0 and 3–6 QRL were significantly higher than expected), but only when the number of QRL was determined by 'double counting'.

Answer to question (3) for barley

QRL in barley showed a preferred mapping to BINs belonging to genome subclass (II) regardless of the way the number of QRL had been specified. This behavior disappeared when genome portion (III) was added for analysis, hence revealing an even distribution of QRL throughout the genome.

With respect to the above-made definition, 27 BINs representing QRL cluster were identified. Of them, 16 were attributed to the (I) and (III) fractions and eight were allocated to the (II) fraction of the barley genome. Linkages of QRL with RGAs suggest that components of quantitative resistance may be controlled by factors which are similar to genes encoding qualitative resistance or that RGAs may directly display quantitative disease resistance phenotypes. To what extent QRL might be associated with RGAs in barley remains unknown due to the limited number of RGAs mapped in the barley genome. Coincidences of BIN positions between major R genes and QRL for the same pathosystem have been observed for powdery mildew (*mlt*, *MlLa*, *Mlg*, *Mla*; Saghai Maroof et al. 1994; Backes et al. 1995), stem rust (*Rpg1*; Spaner et al. 1998) and spot blotch (*Rcs5*; Steffenson et al. 1996). If these major genes were not tagged in the respective QTL analyses, these loci may possibly represent defeated major genes which act as QTL (loss of dominance) (Li et al. 1999).

4 Comparison Maize – Barley

The major difference between maize and barley was a significant clustering of QRL as well as an association of QRL with R-gene/RGA clusters in barley in contrast to maize. In addition, one chromosome of maize displayed significantly more QRL than expected by chance.

For both species the same number of BINs (100) was defined (http://www.agron.missouri.edu/, http://barleygenomics.wsu.edu) and about the same number of BINs containing R-genes and RGAs have been identified so far (Figs. 1, 2), organized in 10 (maize) or 7 (barley) chromosomes. On average, barley BINs are smaller (~14 cM) than maize BINs (~20 cM). However, due to the physically larger genome of barley (~5×10^9 bp) compared to maize (~3×10^9 bp), barley BINs contain more DNA than maize BINs. Thus, the number of crossovers per meiosis is much lower in barley, if compared to the physical genome size, but about equal if related to the number of chromosomes. Furthermore, recombination tends to be suppressed in chromosome regions containing repetitive DNA, resulting in BINs of equal genetic but rather different physical size.

With respect to the distribution of QRL, the distribution of functional genes across BINs rather than the (average) genetic or physical size of BINs would be of interest. So far, only a few regions of "large genome cereals" such as maize, barley, and wheat (not rice) have been sequenced. Some of these studies indicate the organization of gene-rich regions in islands interspersed by large segments of repetitive DNA (Panstruga et al. 1998; Feuillet and Keller 1999). In consequence, the gene number per BIN would be expected to be rather variable. In this situation, an excess of BINs with 0 or with several (>3 in this review) QRL would be expected, agreeing with the observed QRL distributions in both maize and barley. Since the QRL distribution to BINs deviated significantly from a Poisson distribution in barley but not maize, the tendency of genes to be organized in islands might increase with physical genome size. An above-average gene density might also be an explanation for the significantly higher number of QRL on maize chromosome 5. However, this finding (1 out of 17 chromosomes) might also be due to chance.

The significant association of QRL with R-gene/RGA clusters in barley but not maize might be explained by the different pathosystems investigated in QTL studies of both species. Major R-genes are known for almost all diseases investigated in QTL studies of barley but not maize.

However, QRL for resistance against *P. sorghi* and *S. turcica* in maize, both resistances include major genes, are not associated with R-gene/RGA clusters in maize. Therefore, **the presence of major resistance genes for a given pathosystem does not seem to be sufficient for a preferential localization of respective QRL to R-gene/RGA clusters.** Nevertheless, since maize virus QRL were significantly associated with R-gene/RGA clusters, the pathosystems included in such comparative studies need to be taken into account. Another explanation for differences in QRL distribution among maize and barley might be different resistance strategies developed during evolution. A major difference between maize and barley is the fertilization system. Maize is an allogamous, barley an autogamous species. Consequently, maize plants are usually heterozygous in contrast to barley, allowing exploitation of dominance effects and heterosis. In combination with a higher absolute level of recombination, genes with small favourable gene effects should be pyramided in a shorter time in maize.

Preferential localization into R-gene/RGA BINs is not proof of an RGA-like sequence of QRL. These BINs might contain a large number of additional genes. In addition, single RGA-like sequences have also been identified outside R-gene/RGA clusters. Hence, QRL mapping outside R-gene/RGA clusters might code for NBS-LRR proteins. Nevertheless, analysis of the complete sequence of *Arabidopsis thaliana* revealed the organization of >90% of the approximately 250 RGA-like sequences in gene clusters. In addition, these gene clusters might cover several Megabasepairs of DNA (Meyers et al. 1999), representing a significant fraction of a genetic BIN.

If more studies on the genomic organization of R-genes/RGAs as well as QTL studies on resistance traits become available, the results obtained in this review might change. So far, syntenic relationships between the distantly related grass species maize and barley do not seem to be useful for QRL identification.

5 Consequences and Perspectives for Application

Clustering of QRL as indicated for barley (Sect. 3.b) would have consequences (1) for the isolation of QRL or closely linked markers and (2) with regard to plant breeding. For gene or marker isolation of QRL, the initial genetic or physical characterization could be focused on a limited fraction of the genome (corresponding to fraction III in Sect. 3.a). In a long-term perspective, sequencing of such clusters would be of interest, especially if the genes are organized in islands. However, sequencing of gene clusters might result in a large number of candidate genes of similar sequence impairing identification of the gene of interest. This is true for clusters of RGAs, which might in some cases act as QRL. The ultimate goal would be the development of allele-specific markers for genes of interest for large scale genotyping, e.g., in plant breeding programs.

Preferential clustering of QRL would affect plant breeding in different ways. If more than one QRL involved in resistance to a given disease is located within a gene cluster, it will be difficult to estimate the effects of individual genes by QTL analysis. Instead a net gene effect will be obtained which is the result of different individual QRL and their interaction. Consequently, **a cluster of QRL with several positive alleles might decay in the process of selection.** Furthermore, a positive allele at one QRL might be masked by a negative allele at another closely linked QRL. Resolution of closely linked QRL might be rather difficult due to the need to identify rare recombinants but could be facilitated by marker-assisted selection using closely linked markers. If a rare recombinant combining positive acting QRL alleles can be identified, the respective haplotype of QRL alleles would be comparatively stable and behave almost like a single locus.

Clustering of QRL would also affect the use of wild species as donors of resistance as well as the information transfer by synteny. If QRL of interest would be introduced from wild species, it would be rather likely that additional QRL with untested resistance properties would be simultaneously introduced by linkage drag. These might turn out to be deleterious after infection with other pests. If QRL would cluster, the localization of gene clusters might be conserved across species and allow transfer of information by use of syntenic relationships. However, these syntenic relationships have not been found between maize and barley.

Random distribution of QRL as indicated for maize (Sect. 3.a) would require genome-wide approaches such as QTL mapping or expression profiling for QRL identification. Markers need to be developed for a larger number of genome regions compared to clustered QRL. Combination of favourable QRL alleles should, however, be comparatively easier.

A detailed understanding of the relationship between genomic organization and function of QRL requires complete sequence information as well as detailed functional studies using new approaches like genomics, proteomics and metanomics. For large-genome grasses, accumulation of these data can be expected in the next 1–2 decades. This general question on the relationship of genome organization and the function of genetic material will become even more complex for traits such as yield or in the case of polyploid species such as potato (Ross 1986). This time interval can be bridged by intermediate meta-analyses of genomes.

Another recent approach to engineer disease resistance is the transfer of genes coding for antimicrobial proteins (AMPs) or antifungal proteins (AFPs) by gene technology (Woytowich and Khachatourians 2001). Such peptides while being only recent discoveries, evolved to a limited extent in order to conserve their defense role for various species against pathogen attack.

A clever, economical and ecological combination of conventional and molecular approaches will result in crop plants which are genetically protected against a pathogen attack. Perhaps man will then for the first time have the chance to be faster in breeding higher plants for disease resistance than evolution in developing new virulences of the pathogen.

References

Agrama HA, Moussa ME, Naser ME, Tarek MA, Ibrahim AH (1999) Mapping of QTL for downy mildew resistance in maize. Theor Appl Genet 99:519–523

Agrios GN (1997) Plant pathology, 4th edn. Academic Press, San Diego

Ayliffe MA, Collins NC, Ellis JG, Pryor A (2000) The maize *rp1* rust resistance gene identifies homologues in barley that have been subjected to diversifying selection. Theor Appl Genet 100:1144–1154

Backes G, Graner A, Foroughi-Wehr B, Fischbeck G, Wenzel G, Jahoor A (1995) Localization of quantitative trait loci (QTL) for agronomic important characters by the use of a RFLP map in barley (*Hordeum vulgare* L.). Theor Appl Genet 90:294–302

Backes G, Schwarz G, Wenzel G, Jahoor A (1996) Comparison between QTL analysis of powdery mildew resistance in barley based on detached primary leaves and on field data. Plant Breed 115:419–421

Bennett MD, Smith LB (1976) Nuclear DNA amounts in angiosperms. Philos Trans R Soc Lond B Biol Sci 274:227–274

Bohn MO, Schulz B, Kreps R, Klein D, Melchinger AE (2000) QTL mapping for resistance against the European corn borer (*Ostrinia nubilalis* H.) in early maturing European dent germplasm. Theor Appl Genet 101:907–917

Botstein D, White RL, Skolnick M, Davis RW (1980) Construction of a genetic linkage map in man using restriction fragment length polymorphisms. Am J Hum Genet 32:314–331

Buerstmayr H, Lemmens M, Fedak G, Ruckenbauer P (1999) Back-cross reciprocal monosomic analysis of Fusarium head blight resistance in wheat (*Triticum aestivum* L.). Theor Appl Genet 98:76–85

Chen FQ, Prehn D, Hayes PM, Mulrooney D, Corey A, Vivar H (1994) Mapping genes for resistance to barley stripe rust (*Puccinia striiformis* f. sp. *hordei*). Theor Appl Genet 88:215–219

Chen X, Salamini F, Gebhardt C (2001) A potato molecular-function map for carbohydrate metabolism and transport. Theor Appl Genet 102:284–295

Chen XM, Line RF, Leung H (1998) Genome scanning for resistance-gene analogs in rice, barley, and wheat by high-resolution electrophoresis. Theor Appl Genet 97:345–355

Collins N, Drake J, Ayliffe M, Sun Q, Ellis JG, Hulbert SH, Pryor AJ (1999) Molecular characterization of the maize Rp1-D rust resistance haplotype and its mutants. Plant Cell 11:1365–1376

Collins NC, Webb CA, Seah S, Ellis JG, Hulbert SH, Pryor T (1998) The isolation and mapping of disease resistance gene analogs in maize. Mol Plant-Microbe Interact 11:968–978

Dangl JL, Holub EB (1997) La Dolce Vita: a molecular feast in plant-pathogen interactions. Cell 91:17–24

Dehmer K, Graner A, Wenzel G (1991) Screening for defined DNA sequences in minimal amounts of barley tissue by PCR. Plant Breed 107:70–72

De la Pena RC, Smith KP, Capettini F, Muehlbauer GJ, Gallo-Meagher M, Dill-Macky R, Somers DA, Rasmusson DC (1999) Quantitative trait loci associated with resistance to

Fusarium head blight and kernel discoloration in barley. Theor Appl Genet 99:561–569
Dingerdissen AL, Geiger HH, Lee M, Schechert A, Welz HG (1996) Interval mapping of genes for quantitative resistance of maize to *Setosphaeria turcica*, cause of Northern leaf blight, in a tropical environment. Mol Breed 2:143–156
El Attari H, Rebai A, Hayes PM, Barrault G, Dechamp-Guillaume G, Sarrafi A (1998) Potential of doubled-haploid lines and localization of quantitative trait loci (QTL) for partial resistance to bacterial leaf streak (*Xanthomonas campestris* pv. *hordei*) in barley. Theor Appl Genet 96:95–100
Ellerbrook CM, Korzun V, Worland AJ (1999) Using precise genetic stocks to investigate the control of *Stagonospora nodorum* resistance in wheat. In: van Ginkel M, McNab A, Krupinsky J (eds) Septoria and Stagonospora diseases of cereals: a compilation of global research. Procs 5th Int Septoria Worksh Mexico, DF, Mexico, pp 150–153
Feuillet C, Keller B (1999) High gene density is conserved at syntenic loci of small and large grass genomes. Proc Natl Acad Sci USA 96:8265–8270
Feuillet C, Schachermayr G, Keller B (1997) Molecular cloning of a new receptor-like kinase gene encoded at the *Lr10* disease resistance locus of wheat. Plant J 11:45–52
Flavell R. (1980) The molecular characterization and organization of plant chromosomal DNA sequences. Annu Rev Plant Physiol 31:569 –596
Frey M, Chomet P, Glawischnig E, Stettner C, Grün S, Winklmair A, Eisenreich W, Bachner A, Meeley RB, Briggs SP, Simcox K, Gierl A (1997) Analysis of chemical defense mechanism in grasses. Science 277:696–699
Freymark PJ, Lee M, Woodman WL, Martinson CA (1993) Quantitative and qualitative trait loci affecting host-plant response to *Exserohilum turcicum* in maize (*Zea mays* L.). Theor Appl Genet 87:537–544
Geiger HH, Heun M (1989) Genetics of quantitative resistance to fungal diseases. Annu Rev Phytopathol 27:317–341
Geldermann H (1975) Investigations on inheritance of quantitative characters in animals by gene markers. I. Methods. Theor Appl Genet 46:319–330
Gentzbittel L, Mouzeyar S, Badaoui s, Mestries E, Vear F, Tourvieille D, Nicolas P (1998) Cloning of molecular markers for disease resistance in sunflower, *Helianthus annuus* L. Theor Appl Genet 96:519–525
Graham MA, Marek LF, Lohnes D, Cregan P, Shoemaker RC (2000) Expression and genome organization of resistance gene analogs in soybean. Genome 43:86–93
Grant MR, Godiard L, Straube E, Ashfield T, Lewald J, Sattler A, Innes RW, Dangl JL (1995) Structure of the *Arabidopsis RPM1* gene enabling dual specificity disease resistance. Science 269:843–846
Grausgruber H, Buerstmayr H, Lemmens M, Ruckenbauer P (1998) Chromosomal location of Fusarium head blight resistance and *in vitro* toxin tolerance in wheat using the Hobbit "sib" (*Triticum macha*) chromosome substitution lines [*Triticum aestivum* L.]. J Genet Breed 52:173–180
Haley C (1999) Advances in Quantitative trait locus mapping. http://agbio.cabweb.org
Haley CS, Knott SA (1992) A simple regression method for mapping quantitative trait loci in line crosses using flanking markers. Heredity 69:315–324
Hartl L, Mohler V, Zeller FJ, Hsam SLK, Schweizer G (1999) Identification of AFLP markers closely linked in the powdery mildew resistance gene *Pm1c* and *Pm4a* in common wheat (*Triticum aestivum* L.) Genome 42:322–329
Hayes P, Prehn D, Vivar H, Blake T, Comeau A, Henry I, Johnston M, Jones B, Steffenson B, St Pierre CA, Chen F (1996) Multiple disease resistance loci and their relationship to agronomic and quality loci in a spring barley population. J Agric Genomics (http://www.ncgr.org/research/jag/papers96/paper296/jqtl22.html)
Heun M (1992) Mapping quantitative powdery mildew resistance of barley using a restriction fragment length polymorphism map. Genome 35:1019–1025

Huang XQ, Hsam SLK, Zeller FJ, Wenzel G, Mohler V (2000) Molecular mapping of the wheat powdery mildew resistance gene *Pm24* and marker validation for molecular breeding. Theor Appl Genet 101:407–414

Jansen RC (1996) Complex plant traits: time for polygenic analysis. Trends Plant Sci 1:89–94

Jansen RC, Stam P (1994) High resolution mapping of quantitative traits into multiple loci via interval mapping. Genetics 136:1447–1455

Jiang C, Zeng ZB (1995) Multiple trait analysis of genetic mapping for quantitative trait loci. Genetics 140:1111–1117

Jones JDG (1996) Plant disease resistance genes: structure function and evolution. Curr Opin Biotechnol 7:155–160

Kanazin V, Marek LF, Shoemaker RC (1996) Resistance gene analogs are conserved and clustered in soybean. Proc Natl Acad Sci USA 93:11746–11750

Kaul S, Koo HL, Jenkins J et al. (2000) Analysis of the genome sequence of the flowering plant *Arabidopsis thaliana*. Nature 408:796–815

Kerns MR, Dudley JW, Rufener III GK (1999) QTL for resistance to common rust and smut in maize. Maydica 44:37–45

Kicherer S, Backes G, Walther U, Jahoor A (2000) Localising QTLs for leaf rust resistance and agronomic traits in barley (*Hordeum vulgare* L.). Theor Appl Genet 100:881–888

Kleinhofs A, Kudrna D, Matthews D (1998) Integrating barley molecular and morphological/physiological marker maps. Barley Genet Newsl 28:89–91

Koorneef M, Stam P (1992) Genetic analysis. In: Koncz C, Chua N-H, Schell J (eds) Methods in *Arabidopsis* research. World Scientific Publ, River Edge, MN, pp 83–99

Korol A (2001) MultiQTL. http://www.incubators.org.il/26020.htm

Lander ES, Botstein D (1989) Mapping Mendelian factors underlying quantitative traits using RFLP linkage maps. Genetics 121:185–199

Leister D, Ballvora A, Salamini F, Gebhardt CA (1996) PCR-based approach for isolating pathogen resistance genes from potato with potential for wide application in plants. Nat Genet 14:421–429

Leister D, Kurth J, Laurie DA, Yano M, Sasaki T, Devos K, Graner A, Schulze-Lefert P (1998) Rapid reorganization of resistance gene homologues in cereal genomes. Proc Natl Acad Sci USA 95:370–375

Li ZK, Luo LJ, Mei HW, Paterson AH, Zhao XH, Zhong DB, Wang YP, Yu XQ, Zhu L, Tabien R, Stansel JW, Ying CS (1999) A "defeated" rice resistance gene acts as a QTL against a virulent strain of *Xanthomonas oryzae* pv. *oryzae*. Mol Gen Genet 261:58–63

Lu XW, Brewbaker JL, Nourse SM, Moon HG, Kim SK, Khairallah M (1999) Mapping of quantitative trait loci conferring resistance to maize streak virus. Maydica 44:313–318

Lübberstedt T, Klein D, Melchinger AE (1998a) Comparative quantitative trait loci mapping of partial resistance to *Puccinia sorghi* across four populations of European flint maize. Phytopathology 88:1324–1329

Lübberstedt T, Klein D, Melchinger AE (1998b) Comparative QTL mapping of resistance to *Ustilago maydis* across four populations of European flint maize. Theor Appl Genet 97:1321–1330

Lübberstedt T, Xia XC, Tan G, Liu X, Melchinger AE (1999) QTL mapping of resistance to *Sporisorium reiliana* in maize. Theor Appl Genet 99:593–598

Lübberstedt T, Melchinger AE, Dußle C, Vuylsteke M, Kuiper M (2000) Relationships among early European maize inbreds, IV. Genetic diversity revealed with AFLP and comparison with RFLP, RAPD, and pedigree data. Crop Sci 40:783–791

Ma Z, Steffenson BJ, Prom LK, Lapitan NLV (2000) Mapping of quantitative trait loci for Fusarium head blight resistance in barley. Phytopathology 90:1079–1088

Marcon A, Kaeppler SM, Jensen SG, Senior L, Stuber C (1999) Loci controlling resistance to high plains virus and wheat streak mosaic virus in a B73 x Mo17 population of maize. Crop Sci 39:1171–1177

Martin GB, Brommonschenkel SH, Chungwongse J, Frary A, Ganal MW, Spivey R, Wu T, Earle ED, Tanksley SD (1993) Map-based cloning of a protein kinase gene conferring disease resistance in tomato. Science 262:1432–1436

McMullen MD, Simcox KD (1995) Genomic organization of disease and insect resistance genes in maize. Mol Plant-Microbe Interact 8:811–815

McMullen MD, Byrne P, Snook ME, Wiseman BR, Lee EA, Widstrom NW, Coe EH (1998) Quantitative trait loci and metabolic pathways. Proc Natl Acad Sci USA 95:1996–2000

Melchinger AE (1998) Advances in the analysis of data on quantitative trait loci. In: Chopra VL, Singh RB, Varma A (eds). Crop productivity and sustainability - shaping the future. Oxford and IBH, New Delhi, pp 773–791

Melchinger AE, Utz HF, Schön CC (1998) Quantitative trait locus (QTL) mapping using different testers and independent population samples in maize reveals low power of QTL detection and large bias in estimates of QTL effects. Genetics 149:383–403

Mentewab A, Rezanoor HN, Gosman N, Worland AJ, Nicholson P (2000) Chromosomal location of Fusarium head blight resistance genes and analysis of the relationship between resistance to head blight and brown foot rot. Plant Breed 119:15–20

Meyers BC, Shen KA, Rohani P, Gaut BS, Michelmore RW (1998) Receptor-like genes in the major resistance locus in lettuce are subject to divergent selection. Plant Cell 11:1833–1846

Meyers BC, Dickermann AW, Michelmore RW, Sivaramakrishnan S, Sobral BW, Young ND (1999) Plant disease resistance genes encode members of an ancient and diverse protein family within the nucleotide binding superfamily. Plant J 20:312–322

Ming R, Brewbaker JL, Moon HG, Musket TA, Holley R, Pataky JK, McMullen MD (1999) Identification of a major gene, sw1, conferring resistance to Stewart's wilt in maize. Maydica 44:519–523

Neuffer MG, Coe EH, Wessler SR (1997) Mutants of maize. Cold Spring Harbor Laboratory Press, New York

Oberhagemann P, Chatot-Bandras C, Schäfer-Pregel R, Wegener D, Palomino C, Salamini F, Bonnel E, Gebhardt C (1999) A genetic analysis of quantitative resistance to late blight in potato: towards marker-assisted selection. Mol Breed 5:399–415

Panstruga R, Büschges R, Piffanelli P, Schulze-Lefert P (1998) A contiguous 60 kb genomic stretch from barley reveals molecular evidence for gene islands in a monocot genome. Nucleic Acids Res 26:1056–1062

Pe ME, Gianfranceschi L, Taramino G, Tarchini R, Angelini P, Dani M, Binelli G (1993) Mapping quantitative trait loci (QTLs) for resistance to *Gibberella zeae* infection in maize. Mol Gen Genet 241:11–16

Pecchioni N, Faccioli P, Toubia-Rahme H, Vale G, Terzi V, Giese H (1996) Quantitative resistance to barley leaf stripe (*Pyrenophora graminea*) is dominated by one major locus. Theor Appl Genet 93:97–101

Pernet A, Hoisington D, Franco J, Isnard M, Jewell D, Jiang C, Marchand JI, Reynaud B, Glaszmann JC, Gonzalez de Leon D (1999a) Genetic mapping of maize streak virus resistance from the Mascarene source. I. Resistance in line D211 and stability against different virus clones. Theor Appl Genet 99:524–539

Pernet A, Hoisington D, Dintinger J, Jewell D, Jiang C, Khairallah M, Letourmy P, Marchand JL, Glaszmann JC, Gonzalez de Leon (1999b) Genetic mapping of maize streak virus resistance from the Mascarene source. II. Resistance in line CIRAD390 and stability across germplasm. Theor Appl Genet 99:524–539

Pflieger S, Lefebvre V, Caranta C, Blattes A, Goffinet B, Palloix A (2000) Disease resistance gene analogs as candidates for QTLs involved in pepper-pathogen interactions. Genome 42:1100–1110

Richter K, Schondelmaier J, Jung C (1998) Mapping of quantitative trait loci affecting *Drechslera teres* resistance in barley with molecular markers. Theor Appl Genet 97:1225–1234

Robertson DS (1985) A possible technique for isolating genic DNA for quantitative traits in plants. J Theor Biol 117:1–10
Rommens CM, Kishore GM (2000) Exploiting the full potential of disease-resistance genes for agricultural use. Curr Opin Biotechnol 11:120–125
Ross H (1986) Potato breeding: problems and perspectives. Adv Plant Breed13:1–78
Saghai Maroof MA, Zhang Q, Biyashev RM (1994) Molecular marker analyses of powdery mildew resistance in barley. Theor Appl Genet 88:733–740
Saghai Maroof MA, Yue YG, Xiang ZX, Stromberg EL, Rufener GK (1996) Identification of quantitative trait loci controlling gray leaf spot disease in maize. Theor Appl Genet 93:539–546
Sax K (1923) Association of size differences with seed-coat pattern and pigmentation in *Phaseolus vulgaris*. Genetics 8:552–560
Schechert AW, Welz HG, Geiger HH (1999) QTL for resistance to *Setosphaeria turcica* in tropical African maize. Crop Sci 39:514–523
Seah S, Sivasithamparam K, Karakousis A, Lagudah ES (1998) Cloning and characterisation of a family of disease resistance gene analogs from wheat and barley. Theor Appl Genet 97:937–945
Shen KA, Meyers BC, Islam-Faridi MN, Chin DB, Stelly DM, Michelmore RW (1998) Resistance gene candidates identified by PCR with degenerate oligonucleotide primers map to clusters of resistance genes in lettuce. Mol Plant-Microbe Interact 11:815–823
Song W-Y, Wang G-L, Chen L-L, Kim H-S, Holsten T, Wang T, Zhai W-X, Zhu L-H, Franquet C, Ronald P (1995) A receptor kinase-like protein encoded by the rice disease resistance gene, *Xa21*. Science 270:1804–1806
Spaner D, Shugar LP, Choo TM, Falak I, Briggs KG, Legge WG, Falk DE, Ulrich SE, Tinker NA, Steffenson BJ, Mather DE (1998) Mapping of disease resistance loci in barley on the basis of visual assessment of naturally occurring symptoms. Crop Sci 38:843–850
Steffenson BJ, Hayes PM, Kleinhofs A (1996) Genetics of seedling and adult plant resistance to net blotch (*Pyrenophora teres* f. *teres*) and spot blotch (*Cochliobolus sativus*) in barley. Theor Appl Genet 92:552–558
Thomas WTB, Powell W, Waugh R, Chalmers KJ, Barua UM, Jack P, Lea V, Forster BP, Swanston JS, Ellis RP, Hanson PR, Lance RCM (1995) Detection of quantitative trait loci for agronomic, yield, grain and disease characters in spring barley (*Hordeum vulgare* L.). Theor Appl Genet 91:1037–1047
Thümmler F, Wenzel G (2000) Function of genetic material: From gene structure to gene function – approaches to understanding the action of genes in higher plants. Prog Bot 61:54–75
Toojinda T, Broers LH, Chen XM, Hayes PM, Kleinhofs A, Korte J, Kudrna D, Leung H, Line RF, Powell W, Ramsay L, Vivar H, Waugh R (2000) Mapping quantitative and qualitative disease resistance genes in a doubled haploid population of barley (*Hordeum vulgare*). Theor Appl Genet 101:580–589
Utz HF, Melchinger AE (1994) Comparison of different approaches to interval mapping of quantitative trait loci. In: Van Oijen JW, Jansen J (eds). Proc 9th meeting of the EUCARPIA section biometrics in plant breeding. Meeting reports, University of Wageningen, pp 195–204
Utz HF, Melchinger AE (1996) PLABQTL: a program for composite interval mapping of QTLs. J Quant Trait Loci 2, article 1, on-line (http://probe.nalusda.gov:8000/otherdocs/jqtl/)
Utz HF, Melchinger AE, Schön CC (2000) Bias and sampling error of the estimated proportion of genotypic variance explained by quantitative trait loci determined from experimental data in maize using cross validation and validation with independent samples. Genetics 154:1839–1849
Van der Plank JE (1978) Genetic and molecular basis of plant pathogenesis. Springer, Berlin, Heidelberg, New York

Van der Voort JNAM, Van Eck HJ, Draaistra J, Vanzandervoort PM, Jacobsen E, bBakker J (1998) An oline catalogue of AFLP markers covering the potato genome. Mol Breed 4:73–77
Wei F, Gobelman-Werner K, Morroll SM, Kurth J, Mao L, Wing R, Leister D, Schulze-Lefert P, Wise RP (1999) The *Mla* (powdery mildew) resistance cluster is associated with three NBS-LRR gene families and suppressed recombination within a 240-kb interval on chromosome 5 S (1HS) of barley. Genetics 153:1929–1948
Welz HG, Schechert AW, Geiger HH (1999) Dynamic gene action at QTLs for resistance to *Setosphaeria turcica* in maize. Theor Appl Genet 98:1036–1045
Wenzel G (1997) Function of genetic material responsible for disease resistance in plants. Prog Bot 59:80–107
Wenzel G, Lind V, Walther H (1985) Resistenzzüchtung – der genetische Beitrag zum Pflanzenschutz. Naturwissenschaften 72:25–31
Wenzel G, Lössl A, Frei U, Mohler V, Hsam SLK, Huang XQ, Thümmler F, Zeller FJ (2000) Genomics as a tool for an efficient utilisation of genetic resources using potato and wheat as examples. In: Oono K, KomatsudaT, Vaughan T (eds) Integration of biodiversity and genome technology for crop improvement. NIAR, Tsukuba, pp 7–10
Woytowich AE, Khachatourian GG (2001) Plant fungal peptides and their use in transgenic crop plants. Appl Mycol Biotechnol 1:145–164
Xia XC, Melchinger AE, Kuntze L, Lübberstedt T (1999) QTL mapping of resistance to sugarcane mosaic virus in maize. Phytopathology 34:479–501
Xu ML, Melchinger AE, Xia XC, Lübberstedt T (1999) High-resolution mapping of loci conferring resistance to sugarcane mosaic virus in maize using RFLP, SSR, and AFLP markers. Mol Gen Genet 261:574–581
Young ND (1996) QTL mapping and quantitative disease resistance in plants. Annu Rev Phytopathol 34:479–501
Young ND (2000) The genetic architecture of resistance. Curr Opin Plant Biol 3:285–290
Yu YG, Buss GR, Saghai-Maroof MA (1996) Isolation of a superfamily of candidate disease-resistance genes in soybean based on a conserved nucleotide-binding site. Proc Natl Acad Sci USA 93:11751–11756
Zeller FJ, Kong L, Hartl L, Mohler V, Hsam SLK (2001) Chromosomal location of genes for resistance to powdery mildew in common wheat (*Triticum aestivum* L em. Thell) 7. Gene *Pm29* in line Pova. Euhytica (in press)
Zeng ZB (1994) Precision mapping of quantitative trait loci. Genetics 136:1457–1468
Zhu H, Gilchrist L, Hayes P, Kleinhofs A, Kudrna D, Liu Z, Prom L, Steffenson B, Toojinda T, Vivar H (1999) Does function follow form? Principal QTLs for Fusarium head blight (FHB) resistance are coincident with QTL for inflorescence traits and plant height in a doubled-haploid population of barley. Theor Appl Genet 99:1221–1232

Professor Dr. G. Wenzel
Dr. Thomas Lübberstedt
Dr. Volker Mohler
Lehrstuhl für Pflanzenbau und -züchtung
TU München
85350 Freising-Weihenstephan, Germany

Extranuclear Inheritance: Functional Genomics in Chloroplasts

By Ralph Bock and Michael Hippler

1 Introduction: Structural versus Functional Genomics

Over the last two decades, our knowledge about essentially all aspects of modern biology has benefited immensely from the huge amount of data generated by rapidly progressing genome projects. As organellar DNAs can easily be purified and are of a relatively small size, they were among the first targets of genome projects. The complete sequences of two chloroplast genomes (of the liverwort *Marchantia polymorpha* and the angiosperm plant *Nicotiana tabacum*) were determined as early as in 1986 (Ohyama et al. 1986; Shinozaki et al. 1986). Up to now, more than a dozen chloroplast genome projects have been completed, covering a wide range of phylogenetically diverse taxa.

Thorough computer analyses of chloroplast sequence data soon revealed striking similarities with bacterial genes (Schwarz and Kössel 1979; Schwarz and Kössel 1980) confirming the prokaryotic ancestry and thus the endosymbiotic origin of chloroplasts. Homology with known eubacterial genes allowed many of the sequenced plastid genome-encoded reading frames to be assigned tentative functions. In this way, for example, a number of plastid ribosomal proteins, subunits of an *Escherichia coli*-like RNA polymerase and several photosynthesis-related proteins were identified. However, the functions of those potential plastid genes that lacked significant homology with known prokaryotic genes, remained elusive. During the 10 years following the completion of the first chloroplast genome projects, the functions of only relatively few of the remaining open reading frames could be elucidated. This illustrates a problem encountered by practically all genome projects: the difficult transition from structural genomics (i.e. DNA sequence determination) to functional genomics (i.e. determination of the biological information content of DNA sequences). Sequencing genes and generating kilobases or even megabases of DNA sequence data nowadays is a relatively simple exercise, whereas precisely defining the function(s) of a gene is usually a much more demanding and technically challenging task.

Progress in Botany, Vol. 63

Due to their compact (basically prokaryotic) genome organization and the large amount of available sequence data from a great number of phylogenetically diverse species, plastid genomes provide a useful and relatively simple model system for functional genomics. This review summarizes recent progress made in chloroplast functional genomics and focuses on reverse genetics as a particularly powerful tool for investigating functional aspects of plastid-encoded genes.

2 Approaches to Elucidate Plastid-Encoded Gene Functions

A number of different experimental strategies have been employed in order to complete our picture of the coding capacity and gene content of the plastid genome. These approaches comprise a variety of biochemical as well as genetic tools and, in many cases, it has been a combination of different approaches that eventually allowed a plastid open reading frame to be assigned a well-defined function. Below, successfully used methods are briefly summarized and illustrated by representative examples of identified novel chloroplast gene functions.

a) Structural Analysis of Plastid Protein Complexes

The functions of a number of chloroplast genome-encoded reading frames were discovered by careful biochemical analysis of known multiprotein complexes. Isolation of the complexes followed by purification of individual subunits and N-terminal amino acid sequencing produced in several cases sequences with clear homology to plastid reading frames with a heretofore unknown function.

For example, the 50S subunit of chloroplast ribosomes was purified from tobacco leaves, the ribosomal proteins were fractionated and the N-terminal amino acid sequence of a novel 14-kDa protein was determined. This sequence was found to match the N-terminal sequence deduced from the open reading frame ORF55 located between *ndhF* and *trnL* within the small single-copy region of the chloroplast genome (Yokoi et al. 1990). As the amino acid sequence deduced from ORF55 showed some homology to L32 proteins from *Escherichia coli* and *Bacillus stearothermophilus*, ORF55 was renamed *rpl32* (ribosomal protein of the large subunit no. 32).

b) Immunobiochemical Identification of Plastid-Encoded Gene Products

Another biochemical approach is the generation of specific antibodies directed against the putative protein product encoded by a given open reading frame. These antibodies are then used for localizing the gene product by (1) analysis of fractionated protein extracts and/or purified protein complexes or (2) immunolocalization using electron microscopy techniques.

In this way, the gene product of the chloroplast open reading frame *ycf10* was recognized as a protein component of the chloroplast envelope membrane (Katoh et al. 1996). The *ycf10* reading frame was overexpressed in *Escherichia coli*, and antibodies against the purified putative Ycf10 protein were raised. With these antisera, the Ycf10 protein was immunochemically detected in chloroplast and etioplast inner envelope membranes from pea plants (Sasaki et al. 1993a). The *ycf10* reading frame was therefore renamed *cemA* gene (chloroplast envelope membrane; Table 1). More recently, the biochemical function of the *ycf10/cemA* gene product was defined more precisely by forward genetics in cyanobacteria (Katoh et al. 1996) and by reverse genetics in chloroplasts (Rolland et al. 1997).

c) Analysis of Plastome Mutants: Forward Genetics

The starting point in classical ("forward") genetic analyses is the isolation of an interesting mutant phenotype not knowing anything about the underlying mutation and the affected gene(s). When present in the plant nuclear genome, such mutations can be physically mapped by classical linkage analyses and the mutated genes can ultimately be isolated using molecular methods (e.g. chromosome walking/map-based cloning, differential display, genomic subtraction). Occasionally, mutant phenotypes are found that exhibit non-Mendelian inheritance. This points to either organellar inheritance (i.e. through genes located on the plastid or mitochondrial genomes) or epigenetic mechanisms interfering with the normal mode of inheritance for nuclear genes (e.g. genomic imprinting, maternal predetermination etc.). When chloroplast localized, such mutations can be instrumental in identifying new gene functions in plastid genomes.

In an analysis of *Chlamydomonas reinhardtii* mutants with a cytochrome c-negative phenotype, one of the identified mutations exhibited uniparentally maternal inheritance as typical of chloroplast traits in *Chlamydomonas* and most higher plants (Howe and Merchant 1992). This mutant turned out to be unable to synthesize functional forms of cytochromes f and c6 due to a chloroplast genome mutation preventing heme attachment to c-type cytochromes (i.e. those cytochromes that carry a covalently attached heme prosthetic group). Chloroplast encoding of a cytochrome c assembly factor drew the attention to a plastid open reading frame (*ycf5*) that displayed limited sequence homology to bacterial genes (*ccl1/cycK*) required for the biogenesis of c-type cytochromes. Subsequent molecular analyses identified a frameshift mutation within the *ycf5* reading frame of the *Chlamydomonas* mutant. Moreover, the mutant cytochrome c-deficient phenotype could be complemented by introducing the wild-type *ycf5* sequence by genetic transformation (Xie and Merchant 1996; Xie et al. 1998). Also, targeted inactivation of *ycf5* was performed (by reverse genetics; see below) and, as expected, resulted in a cytochrome c6 and cytochrome f-deficient mutant phenotype. The *ycf5* reading frame was therefore renamed *ccsA* (for c-type cytochrome synthesis; Table 1).

Table 1. Conserved open reading frames (*ycf*s) in plastid genomes. The size of each reading frame is given for the completely sequenced plastid genomes from selected spermatophyte species (*Zea mays, Oryza sativa, Nicotiana tabacum, Epifagus virginiana, Pinus thunbergii*), the bryophyte *Marchantia polymorpha*, the green alga *Chlorella vulgaris*, the red alga *Porphyra purpurea* and the brown alga *Odontella sinensis*. – indicates absence of a homologous reading frame from the plastid genome of the respective species. Reading frames recently confirmed as genuine genes are indicated by their new gene names or by their tentatively assigned functions. ORF, Open reading frame; IRF, intron-containing open reading frame; *ycf*, hypothetical chloroplast reading frame; PSI, photosystem I-related gene; *ccsA*, c-type cytochrome biogenesis; *petN*, *petL*, subunits of the cytochrome b_6f complex; *psbT*, subunit of photosystem II (PSII); *lhbA*, structural subunit of the PSII light-harvesting complex; *cemA*, inner membrane protein (inorganic carbon uptake); *accD*, acetyl-CoA carboxylase subunit; *matK*, putative group II intron maturase (splicing factor)

Reading frame	Zea	Oryza	Nicotiana	Epifagus	Pinus	Marchantia	Chlorella	Porphyra	Odontella	References
ycf1	–	–	ORF1901	ORF1738	ORF1756	ORF464/1068	ORF819	–	–	Drescher et al. (2000)
ycf2	–	–	ORF2280	ORF2216	ORF2054	ORF2136	ORF1720	–	–	Drescher et al. (2000)
ycf3 / (PSI)	IRF170	IRF170	IRF168	–	IRF169	IRF168	ORF167	ORF173	ORF179	Ruf et al. (1997); Boudreau et al. (1997b)
ycf4 / (PSI)	ORF185	ORF184	ORF184	–	ORF184	ORF184	ORF183	ORF186	ORF181	Boudreau et al. (1997b)
ycf5 / *ccsA*	ORF321	ORF321	ORF313	–	ORF320	ORF320	ORF315	ORF319	ORF312	Xie and Merchant (1996)
ycf6 / *petN*	ORF29	ORF29	ORF29	–	ORF29	ORF29	–	ORF29	ORF29	Hager et al. (1999)
ycf7 / *petL*	ORF31	ORF31	ORF31	–	ORF33	ORF31	–	ORF31	ORF31	Takahashi et al. (1996)
ycf8 / *psbT*	ORF33	ORF35	ORF34	–	ORF35	ORF35	–	ORF31	ORF32	Monod et al. (1994)
ycf9 / *lhbA*	ORF62	ORF62	ORF62	–	ORF62	ORF62	ORF62	ORF62	ORF61	Ruf et al. (2000)
ycf10 / *cemA*	ORF230	ORF230	ORF229	–	ORF261	ORF434	ORF266	–	–	Rolland et al. (1997)
ycf11 / *accD*	–	–	ORF512	ORF493	ORF321	ORF316	–	ORF288	–	Sasaki et al. (1993b)
ycf12	–	–	–	–	–	ORF33	ORF37	ORF34	ORF34	
ycf14 / *matK*	ORF544	ORF542	ORF509	ORF439	ORF515	ORF370	–	–	–	
ycf15	ORF99	–	ORF87	–	–	–	–	–	–	

d) Targeted Generation of Plastome Mutants: Reverse Genetics

In contrast to forward genetics, reverse genetics starts from a known DNA sequence containing an open reading frame of unknown (or uncertain) function(s). A mutant for this reading frame is then specifically generated by suitable experimental procedures. For plant nuclear genes, inactivation by insertional mutagenesis (accomplished e.g. via transposon tagging or T-DNA tagging) has become a very powerful tool in functional genomics (for an overview see e.g. Bouchez and Höfte 1998; Azpiroz-Leehan and Feldman 1997). The development of these technologies has proven particularly valuable since most plants lack an efficient homologous recombination system in their nucleo-cytoplasmic compartment. Consequently, plant nuclear genes are usually not amenable to targeted inactivation or site-specific manipulation by homologous recombination. By contrast, chloroplasts have inherited an active homologous recombination system from their prokaryotic ancestors (Cerutti et al. 1992). This finding, together with the development of facile methods for introducing foreign DNA into chloroplasts (by particle bombardment or polyethyleneglycol-mediated transformation; reviewed e.g. in Bock and Hagemann 2000), has provided the basis for transgenic approaches in chloroplast functional genomics. During the past decade, reverse genetics analyses based on transformation technologies for plastids have undoubtedly become the most powerful tool for detailed functional analyses of plastid genes.

3 Reverse Genetics in Algal and Higher Plant Chloroplasts

The successful development of transformation technologies for chloroplasts (Boynton et al. 1988; Svab et al. 1990; Svab and Maliga 1993) has paved the way for addressing functional aspects of plastid genome-encoded genes and open reading frames by reverse genetics. Any chloroplast gene can be mutated by standard manipulation techniques in vitro or in *E. coli*. The mutant allele is then re-introduced into the plastid genome by chloroplast transformation where it replaces the endogenous intact wild-type allele by homologous recombination (Fig. 1).
Today, chloroplast transformation technologies are routinely available for two model plants, the unicellular green alga *Chlamydomonas reinhardtii* (Boynton et al. 1988) and the higher plant tobacco, *Nicotiana tabacum* (Svab et al. 1990; Svab and Maliga 1993). Both of these model plants have been extensively used for conducting functional genomics by reverse genetics. *Chlamydomonas reinhardtii* combines simple culture and selection procedures with an excellent, well-developed genetics. It is because of these advantages that Chlamydomonas is occasionally also referred to as "green yeast" or "photosynthetic yeast". *Nicotiana tabacum*

is a tetraploid species and therefore a less suitable object for classical genetics. However, in chloroplast functional genomics, it offers the possibility of studying the phenotypic effects of mutated plastid genes in the context of plant development and organelle differentiation.

Below, the two principle applications of reverse genetics approaches in chloroplasts are described: (1) the inactivation of plastid genome-encoded genes by insertional or deletional mutagenesis ("gene knockout") and (2) the introduction of point mutations by site-directed mutagenesis of plastid genes (Fig. 1).

a) Plastid Gene Knockouts

Construction of a null allele by deletional or insertional mutagenesis is the most appropriate strategy particularly in those cases where the function of a chloroplast gene/open reading frame (ORF) is entirely unknown. Theoretically, complete gene inactivation is also possible by site-directed mutagenesis, for example, through introducing nonsense mutations (stop codons) in the reading frame of the target gene. However, as compared to site-directed mutagenesis, targeted gene disruption ("gene knockout") has the advantage of ensuring 100% linkage between the insertion of the selectable marker gene (typically a chimeric *aadA* gene conferring resistance to aminoglycoside antibiotics, such as spectinomycin or streptomycin; Goldschmidt-Clermont 1991; for review see Rochaix 1997; Bock and Hagemann 2000) and the mutation in the chloroplast target gene (Fig. 1).

Interspecific conservation of an open reading frame is generally taken as a good indication for the reading frame being indeed a genuine gene. Such evolutionarily conserved open reading frames in chloroplasts are named *ycf* (hypothetical chloroplast reading frame; Table 1). Often, such *ycf*s are found conserved in the genomes of all photosynthetically active organisms from cyanobacterial genomes to higher plant chloroplast genomes, suggesting that they encode proteins involved in photosynthesis-related cellular functions.

Both *Chlamydomonas reinhardtii* and *Nicotiana tabacum* have been employed for "knockout studies" in order to elucidate the functions of conserved open reading frames encoded in plastid genomes (Table 1). As the chloroplast genome is present in multiple copies (in higher plants up to 10,000 per cell), stable targeted manipulation of plastid genes requires homoplasmy. A homoplasmic (or "homoplastomic") state is considered to be present when the population of chloroplast genomes in a cell is homogeneous, i.e. uniformly consists of either wild-type or mutant (transformed) genome copies. Homoplasmy can be achieved through repeated subcloning of chloroplast transformants under high

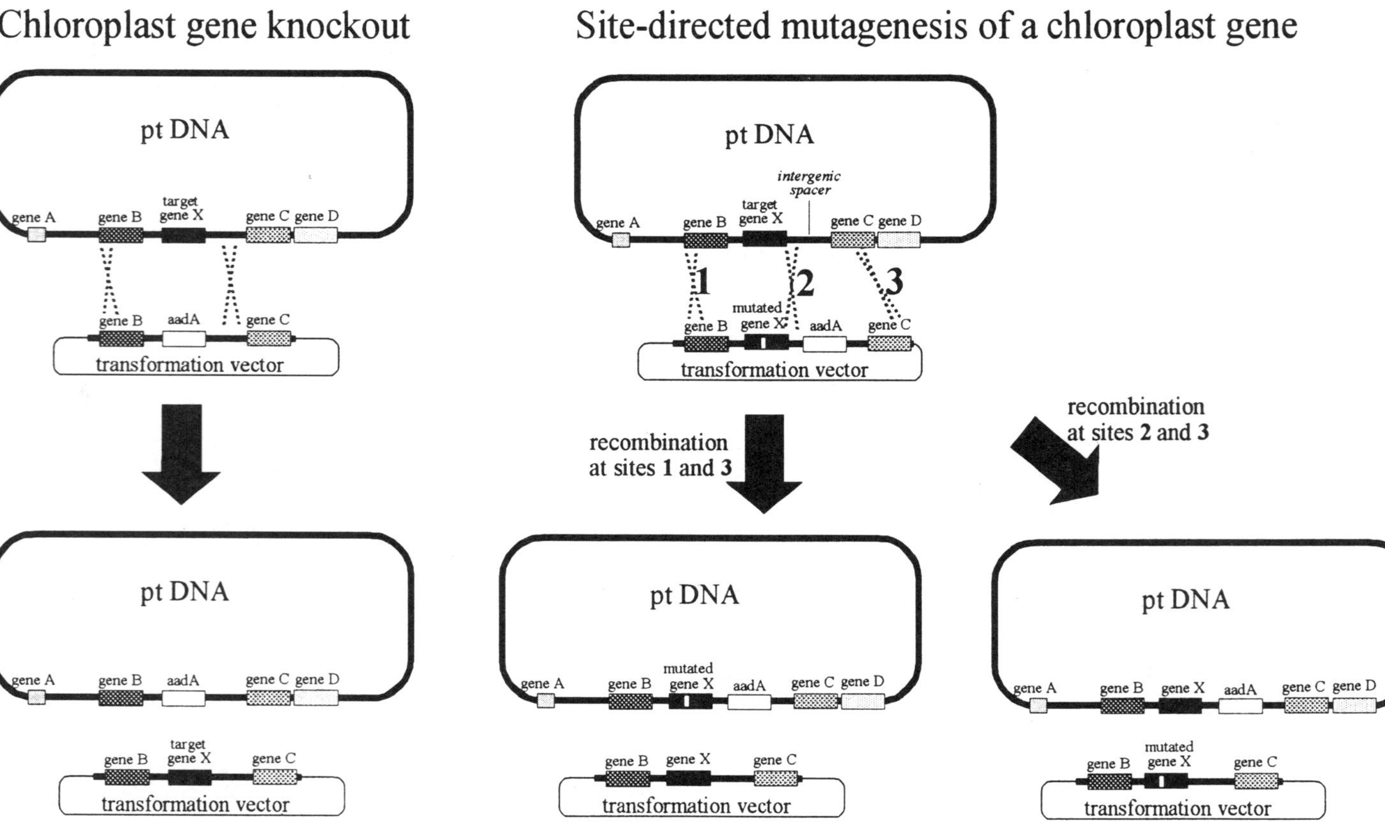
Chloroplast gene knockout
pt DNA
gene A
gene B
target gene X
gene C
gene D
aadA
transformation vector
Site-directed mutagenesis of a chloroplast gene
intergenic spacer
mutated gene X
1
2
3
recombination at sites 1 and 3
recombination at sites 2 and 3

selective pressure (i.e. high concentrations of the selecting antibiotic) resulting in a gradual sorting out of wild-type genome copies. Cells homoplasmic for a chloroplast gene "knockout" will entirely lack the wild-type allele and, thus, will reveal the phenotype associated with the lack of the respective gene product.

Homoplasmic targeted gene inactivations can easily be achieved for all photosynthesis-related reading frames since photosynthesis is optional under heterotrophic in vitro culture conditions: Non-photosynthetic *Chlamydomonas* cells can be grown on acetate-containing medium whereas tobacco plants without photosynthetic activity are grown on sucrose-containing synthetic media. Systematic reverse genetics analyses in both systems have revealed a number of theretofore unknown gene functions involved in photosynthesis, including small subunits of photosystem II (Monod et al. 1994), the cytochrome b_6f complex (Takahashi et al. 1996; Hager et al. 1999) and the light-harvesting antenna (Ruf et al. 2000; Table 1). Surprisingly, chloroplast gene disruptions also led to the discovery of novel photosynthesis-related proteins that are required for the stable accumulation of multi-protein complexes in the thylakoid membrane but are not integral components of these complexes. In this way, the gene products of the conserved chloroplast reading frames *ycf3* and *ycf4* were identified as novel proteins essential for the assembly of stable photosystem I units in the thylakoid membrane without actually being part of photosystem I (Ruf et al. 1997; Boudreau et al. 1997b; Table 1).

Reverse genetics by targeted gene inactivation has also begun to shed some light on the enigmatic role of the *ndh* genes in higher plant chlo-

Fig. 1. Reverse genetics in chloroplasts. *Left panel* Targeted inactivation of a chloroplast open reading frame (gene X; *filled box*) by deletional mutagenesis (= gene knockout). The gene X is excised from a cloned plastid DNA fragment and replaced with a chimeric spectinomycin resistance gene (*aadA*; *open box*) yielding the vector for biolistic chloroplast transformation. In successfully transformed chloroplasts, two homologous recombination events (*dotted lines*) in the flanking regions result in replacement of the target gene X with the selectable marker gene *aadA*. *Thick lines* Chloroplast sequences; *thin lines* sequences of a standard cloning vector (e.g. pUC, pBS). *Right panel* Site-directed mutagenesis of a chloroplast gene. The target gene X is linked to the selectable marker gene *aadA* (typically inserted into an adjacent non-coding region) and the desired mutation (*white dot*) is introduced by standard in vitro mutagenesis techniques. Two classes of chloroplast transformants will be obtained: homologous recombination at sites 2 and 3 leads to incorporation of only the spectinomycin resistance gene leaving the target gene X unchanged, whereas recombination at sites 1 and 3 results in plastid genomes carrying both the selectable marker gene and the mutation in gene X. (Recombination events at sites 1 and 2 do not introduce the *aadA* gene and ehence will not allow selection of plastid transformants.) Homoplasmic transplastomic lines harboring both the *aadA* and the mutated version of gene X are analyzed with respect to the phenotypic consequences of the introduced mutation whereas plants carrying the *aadA* gene but not the mutation in gene X can serve as control plants

roplast genomes. Because of their homology with mitochondrial complex I genes, these genes had long been suspected to encode subunits of a putative plastid NAD(P)H dehydrogenase possibly being involved in chlororespiration. Generation of *ndh* gene knockouts in tobacco has now made feasible the detailed functional characterization of this multiprotein complex employing molecular and physiological tools (Burrows et al. 1998; Kofer et al. 1998; Sazanov et al. 1998; Shikanai et al. 1998; Endo et al. 1999; Horváth et al. 2000). Remarkably, *ndh* gene knockout plants showed no phenotype under standard growth conditions. However, comparative analysis of chlorophyll fluorescence in the wild-type and the knockout plants has implicated the putative plastid Ndh complex in post-illumination reduction of the plastoquinone pool in vivo. In addition, recording of photosystem I re-reduction kinetics has suggested that the complex may also contribute to the re-reduction of the PSI reaction center chlorophyll $P700^+$ in the dark. Interestingly, the physiological effects observed in the *ndh* knockout plants appeared to be particularly pronounced when the plants were exposed to various stress conditions.

Recently, knockout experiments in chloroplasts have also led to the discovery of a number of plastid genome-encoded gene functions that appear to be essential for cell survival. The essential function of a chloroplast gene is evidenced by the impossibility to generate homoplasmic knockout cells: Under selective conditions, cells remain heteroplasmic with wild-type and transformed genomes coexisting in a fairly constant ratio (balanced selection). This stable heteroplasmy has been observed, for example, in knockout experiments for subunits of an *E. coli*-like plastid RNA polymerase (*rpo* genes; Rochaix 1997) in Chlamydomonas (the homologous genes in tobacco are, however, not essential and can be fully inactivated; Allison et al. 1996; Serino and Maliga 1998) and for some large open reading frames of unknown function in tobacco (Drescher et al. 2000) and *Chlamydomonas* (Boudreau et al. 1997a) chloroplasts.

b) Site-Directed Mutagenesis of Plastid Genes

The advent of dependable chloroplast transformation methods has opened the door for a thorough analysis of the structure-function relationship of photosynthetic complexes and has complemented the classical genetic analyses in a powerful way (for recent reviews see Hippler et al. 1998; Ruffle and Sayre 1999; Spreitzer 1999; Strotmann et al. 1999; Webber and Bingham 1999; Wollman 1999; Rochaix et al. 2000). Because of the efficient homologous recombination system in chloroplasts, specific deletions or site-directed mutations can be created relatively easily. In this article, we focus on site-directed mutagenesis work performed on chloroplast genes encoding subunits of photosystem I (PSI) and the cytochrome b_6/f complex. This work has contributed greatly to our understanding of the function of these protein complexes and their roles in oxygenic photosynthesis and thus may illustrate how this new technology has been applied to the study of bioenergetic processes within the chloroplast.

α) Generation of Chloroplast Site-Directed Mutants

Biolistic chloroplast transformation technology together with the availability of a chloroplast-specific selectable marker gene (*aadA*, encoding an aminoglycoside adenylyl transferase, see above) allows site-directed mutagenesis of any chloroplast gene facilitated by the efficient homologous recombination system. However, any directed chloroplast gene manipulation requires repeated time-consuming subcloning of the transformants because of the polyploidy of the chloroplast genome. Thus it is of advantage to first delete the gene to be mutagenized. Subsequent recycling of the *aadA* cassette (which is needed to select for the deletion) through excision by homologous recombination between repeats flanking the *aadA* gene (Fischer et al. 1996) permits rapid generation of mutant strains. Hence, it is no longer necessary to wait until all of the many copies of the chloroplast genome have been replaced by the newly introduced mutant allele of the gene (Newman et al. 1991; Redding et al. 1998).

β) Photosystem I

PSI is a large multimeric protein-pigment complex that contains at least eleven subunits, eight of which are inserted into the thylakoid membrane whereas the remaining three are extrinsic subunits located at the stromal side of the thylakoid membrane (see Golbeck and Bryant 1991). The PSI core consists of the two large integral membrane proteins PsaA and PsaB which bind most of the redox cofactors involved in electron transfer from the lumenal to the stromal side of the thlykoids: P_{700}, a pair of chlorophylls *a*; A_0, a chlorophyll *a*; A_1, a phylloquinone; the 4Fe-4S clusters F_X. The other two iron-sulfur clusters F_A and F_B are bound to the PsaC subunit and act as terminal electron acceptors. PSI functions as a light-driven plastocyanin/cytochrome c_6-ferredoxin oxidoreductase that is involved in linear electron transfer together with photosystem II and the cytochrome b_6f complex to generate NADPH and a proton gradient. PSI from *Synechococcus elongatus* was recently crystallized and the structure has been determined at a resolution of less than 4 Å which is sufficient for distinguishing 31 transmembrane and 14 parallel helices as well as the three iron-sulfur clusters (Schubert et al. 1997).

i) Reaction Center

Several site-directed mutations have been generated in the chloroplast-encoded genes for the large PSI reaction center subunits PsaA and PsaB to identify residues which act as ligands to the redox cofactors P700, A_O, A_1 and F_X. The 4Fe-4S cluster F_X is most likely coordinated by cysteines from both PsaA and PsaB which are present within the highly conserved motif PCDGPGRGGTC in the stromal loop between helices 8 and 9 (Fig. 2A; Smart et al. 1993). Site-directed changes of the first cysteine of this motif in either PsaB or PsaA in *C. reinhardtii* chloroplasts lead to a drastic destabilization of the PSI complex (Webber et al. 1993; Hallahan et al. 1995), whereas the same changes do not prevent PSI accumulation in cyanobacteria (Smart et al. 1993; Vassiliev et al. 1995). The change PsaB–P560A,L near the first cysteine does not significantly affect the stability and the electron transfer properties of PSI (Webber et al. 1993). A stronger effect is observed when the conserved PsaA–D576 is changed to L. This change results in a two-fold lower accumulation of PSI, prevents photoautotrophic growth and alters the electron paramagnetic resonance (EPR) spectrum of F_A and F_B suggesting a conformational change of PsaC (Hallahan et al. 1995). Changes of PsaB–P560L and

Fig. 2. A, B. Site-directed mutagenesis of chloroplast-encoded PSI subunits. A Topological model of PsaA and PsaB (Schubert et al. 1997; adapted from Rochaix et al. 2000) with the emphasis of the five C-terminal transmembrane helices which bind the prosthetic groups needed for electron transfer within the PSI complex. Transmembrane helices are represented by *vertical open boxes* and are numbered from *7* to *11*. The six N-terminal transmembrane helices are considered to function as light-harvesting antenna, since they bind most of the light-harvesting pigments of the complex (Schubert et al. 1997). The two *horizontal boxes* represent the parallel α-helices n and l as revealed by the crystal structure (Schubert et al. 1997). The conserved histidine residues which have been mutated in *C. reinhardtii* are marked by *dots* (Cui et al. 1995; Webber et al. 1996; Redding et al. 1998). The lumenal and stromal side of the thylakoid membrane are indicated as well as the stromal loop involved in coordination of the (4Fe–4S) F_X cluster. Amino acid residues within the stromal loop of PsaA or PsaB which were changed by site-directed mutagenesis are indicated in the sequence enlargement (Webber et al. 1993; Hallahan et al. 1995; Rodday et al. 1995). B Schematic representation of the PsaC amino acid sequence from *C. reinhardtii* (Takahashi et al. 1991; adapted from Rochaix et al. 2000). Critical cysteine residues involved in coordination of the (4Fe–4S) F_A and F_B clusters are indicated in *bold*. The C-terminal extension and the internal loop are boxed. Mutations discussed in the text are marked by *arrows* and the newly introduced amino acid(s) in the one letter code

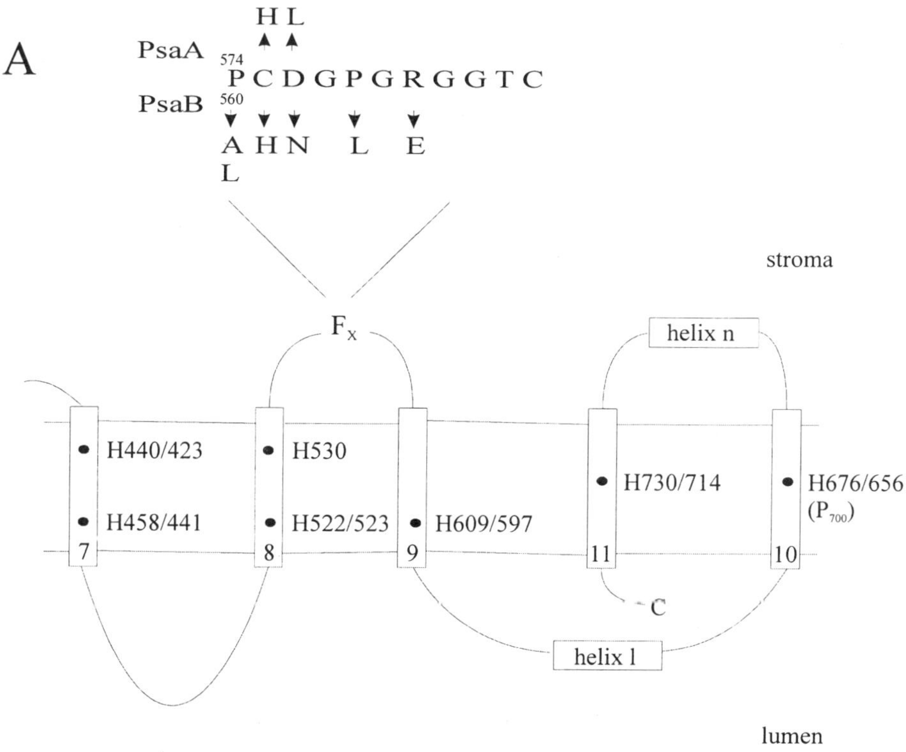
A
PsaA
574
H L
P C D G P G R G G T C
PsaB
560
A H N L E
L
stroma
F$_X$
helix n
H440/423
H458/441
H530
H522/523
H609/597
H730/714
H676/656
(P$_{700}$)
7
8
9
11
10
C
helix l
lumen
Reaction center

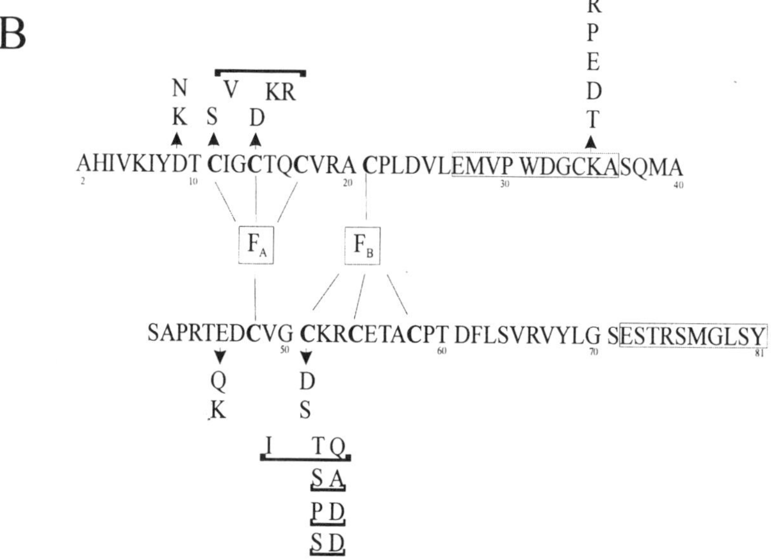
B
R
P
E
D
T
N V KR
K S D
AHIVKIYDT CIGCTQCVRA CPLDVLEMVP WDGCKASQMA
2 10 20 30 40
F$_A$
F$_B$
SAPRTEDCVG CKRCETACPT DFLSVRVYLG SESTRSMGLSY
50 60 70 81
Q D
K S
I TQ
SA
PD
SD

PsaB–564L appear to affect the interaction between the PSI core and PsaC (Rodday et al. 1995).

To identify possible ligands to chlorophyll and quinone cofactors, a systematic survey of conserved histidines in the last six transmembrane domains of PsaA and PsaB was performed by chloroplast site-directed mutagenesis (Redding et al. 1998). Nine sites in these helices are always occupied by histidines in both proteins. Seven histidines were altered in PsaA or PsaB or in both proteins together (two in helix 6, two in helix 7, and one each in helices 9, 10 and 11; Fig. 2A; Redding et al. 1998). In another study, histidine residues of the transmembrane PsaB helices 8 and 10 were changed (Cui et al. 1995; Webber et al. 1996).

The PsaB–H656N mutant in helix 10 had a noticeably changed visible difference spectrum ($P_{700}{}^{+}$–P_{700}). In addition, the redox potential of the $P_{700}/P_{700}{}^{+}$ couple was shifted up from 447 to 487 mV in the PsaB–H656N mutant. Finally, the $P_{700}{}^{+}$ electron nuclear double resonance (ENDOR) spectrum was changed. Mutants in which the analogous histidines in PsaA and PsaB had been substituted with glutamine were examined spectroscopically for changes in P_{700} (Redding et al. 1998). None of the double mutants presented any changes in the $P_{700}{}^{+}$–P_{700} visible difference spectrum, Fourier transform infrared difference spectrum (FTIR), or $P_{700}{}^{+}$ EPR spectrum, except for the double mutant in helix 10 (PsaA–H676Q/PsaB–H656Q) which showed changes in all three spectra.

From these results it appears that differences in the properties of P700 are observed only when histidines of helix 10 (PsaA–H676 and PsaB–H656) are altered, suggesting that these histidines are the likely ligands of P700. This is in line with structural information obtained from the PSI crystal structure that revealed two quasi symmetrical branches, implicating that the P700-coordinating residues must be located in symmetrical positions in the PsaA/PsaB heterodimer. This view is further supported by the fact that the localization of the histidine ligands of P700 within helix 10 is similar to that of the histidine ligands of P680 within helix D of the PSII reaction center polypeptides D1 and D2.

ii) The PsaC Subunit – the Acceptor Side of PSI

The chloroplast-encoded PsaC subunit, located at the stromal side of PSI, binds the terminal electron acceptors F_A and F_B that are two 4Fe–4S clusters coordinated by two sets of four cysteine residues (Hayashida et al. 1987; Hoj et al. 1987). Site-directed mutagenesis of the cysteine ligands in cyanobacteria demonstrated that C11, 14, 17 and 58 coordinate the iron atoms of cluster F_B whereas C21, 48, 51 and 58 coordinate cluster F_A (Fig. 2B; Zhao et al. 1992). Deletion of the *psaC* gene from cyanobacteria and *Chlamydomonas* results in an inability to grow photoautotrophically. Whereas non-functional PSI still accumulates in cyanobacteria lacking PsaC, PsaD or PsaE, the loss of PsaC in Chlamydomonas leads to complete destabilization of the PSI complex (Takahashi et al.

1991; Yu et al. 1995; Mannan et al. 1996). These differences also become apparent when cysteines coordinating the clusters are altered in *Chlamydomonas* or cyanobacteria. Site-directed mutagenesis of the algal cysteines results in complete destabilization of PSI, whereas cyanobacteria tolerate some changes and accumulate mixed ligand 4Fe–4S clusters (Takahashi et al. 1992; Yu et al. 1995).

Recent work in *C. reinhardtii* indicates that the PsaC subunit of PSI plays a crucial role in the binding of ferredoxin to the stromal side of PSI. PsaC resembles the 2(4Fe–4S) ferredoxin from the anaerobic non-photosynthetic bacterium *Peptococcus asaccharolyticus* with the exception of two domains that are absent from soluble 2(4Fe–4S) ferredoxins (Dunn and Gray 1988). These extra sequences, an internal loop between the two (4Fe–4S) cluster-binding motifs and a C-terminal extension of PsaC, have been proposed to interact with the PsaA/B heterodimer and the PsaD subunit (Naver et al. 1996, 1998). In vivo degenerated oligonucleotide-directed mutagenesis of the internal loop of PsaC identified K35 as a key interaction site between PSI and ferredoxin (Fig. 2B; Fischer et al. 1998). This is based on the observation that the site-directed mutations K35T, K35D and K35E drastically affect electron transfer from PSI to ferredoxin as measured by flash-absorption spectroscopy, whereas the K35R change has no effect on binding of ferredoxin. Chemical cross-linking experiments revealed that ferredoxin interacts not only with PsaD and PsaE, but also with PsaC at the level of K35. The exchange of K35 by T, D, E or R abolishes ferredoxin cross-linking to PsaC and cross-linking to PsaD and PsaE is reduced in the K35T, K35D and K35E mutants. These results demonstrate that PsaC provides an important residue for electrostatic interaction of PSI with ferredoxin to establish close protein contact needed for fast electron transfer between the terminal electron acceptors (F_A; F_B) and the iron-sulfur cluster of ferredoxin.

The positions of the three 4Fe–4S clusters are clearly defined in the 4 Å resolution crystal structure of PSI (Schubert et al. 1997). However, the order of the F_A and F_B iron-sulfur centers within PsaC has been determined only recently. Chemical inactivation of F_B (Diaz-Quintana et al. 1998; Vassiliev et al. 1998) as well as the analysis of six *psaC* site-directed mutants of *C. reinhardtii* in which residues located close to F_A and F_B were changed (Fig. 2B; Fischer et al. 1999) support the orientation with F_B as the distal cluster. The acidic residues D9 and E46, located one residue upstream of the first Cys coordinating F_B and F_A, respectively, were changed to a neutral or basic amino acid. When charged residues close to the F_B cluster were changed (to D9N or D9K; Fig. 2B), a 1.3–2-fold increase of ferredoxin affinity was observed, whereas the changes E46Q and E46K did not result in any difference. Introduction of further positively charged amino acids near the F_B cluster in the triple mutant I12V/T15K/Q16R resulted in a 60-fold increase in ferredoxin affinity as measured by flash absorption spectroscopy. In the triple mutant V49I/K52T/R53Q carrying alterations of three residues near the F_A cluster (Fig. 2B), PSI assembled to lower amounts as compared to the wild-type. The fact that the change of neutral to positively charged amino acids near the F_B cluster strongly facilitates ferredoxin binding to PSI supports the view that electron transfer in the terminal part of PSI follows the F_X–F_A–F_B–Fd pathway. This implicates that electron transfer within PsaC occurs against the redox potential gradient as F_A is more electropositive than F_B (Evans et al. 1974; Heathcote et al. 1978).

When two positively charged amino acids of PsaC from Chlamydomonas, K52 and R53 located between the cysteine ligands of F_A were changed to PsaC-K52S/R53A, low temperature electron paramagnetic resonance (EPR) revealed that F_B is preferentially photoreduced in this mutant (Fischer et al. 1997), whereas the F_A center is predominately reduced in wild-type *Chlamydomonas* as well as in vascular plants. Interestingly, the PsaC-like subunit in the green sulfur bacterium *Chlorobium limicola* has alanine and serine in the analogous positions (Büttner et al. 1992), also resulting in a preferential reduction of F_B as measured by low temperature EPR (Nitschke et al. 1990). Since the altered redox equilibrium in the Chlamydomonas PsaC double mutant K52S/R53A did not alter forward electron transfer to ferredoxin, it appears that electron transfer from one cluster to the other is not rate limiting (Fischer et al. 1997). However, this K52S/R53A double mutant as well as *Chlamydomonas* mutants having a less efficient ferredoxin reduction are strongly light sensitive (reviewed in Rochaix et al. 2000) under aerobic but not under anaerobic conditions. This supports the idea that the preferential reduction of the innermost cluster F_A is important under physiological conditions, since electron escape from the outermost cluster F_B to oxygen may result in the production of harmful reactive oxygen species. It is interesting to note that *C. limicola*, where F_B is the preferentially reduced iron-sulfur cluster, belongs to the obligate anaerobic bacteria. In this line, it has been shown that in the highly light-sensitive site-directed mutant PsaC–K35E, more lipid peroxidation occurs than in wild-type cells grown under identical light conditions (Hippler et al., submitted). As lipid peroxidation serves as a measure of oxygen radical production, this indicates that an efficient acceptor side of PSI is essential for cell survival in an aerobic environment by preventing the production of deleterious reactive oxygen species.

γ) The Cytochrome b_6/f Complex

The cytochrome b_6f complex is a large multimeric integral membrane protein complex that mediates electron transfer between the PSII and PSI reaction centers in the thylakoid membrane. This protein complex oxidizes plastoquinol and reduces plastocyanin (or cytochrome c_6 under copper deficiency in *Chlamydomonas*) and participates in the establishment of the proton-motive force used for the generation of ATP. In addition, the cytochrome b_6f complex appears to be involved in the sensing and transduction of a redox signal, thereby regulating a protein kinase, responsible for LHC phosphorylation and hence mediating state transitions (see Fig. 3 and below).

The cytochrome b_6f complex belongs, like the bc_1 complex of mitochondria and bacteria, to the family of quinol-oxidizing complexes. The molecular structure of the bc_1 complex from beef-heart or chicken mitochondria was determined at high resolution (Xia et al. 1997; Zhang et al. 1998) and revealed that the complex is organized as a dimer. Molecular weight determination (Breyton et al. 1997) and stoichiometry measurements (Pierre et al. 1995) of purified cyt b_6f complexes from *C. reinhardtii* have shown that this complex is also a dimer. The cyt b_6f complex comprises four major subunits, three of which are chloroplast-encoded: cytochrome f (*petA* gene product), cytochrome b_6 binding the

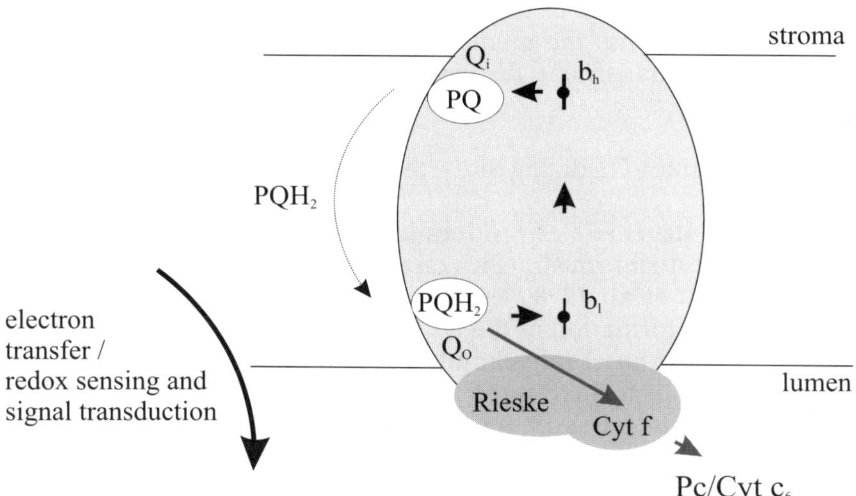

Fig. 3. Schematic representation of the structure and functions of the cytochrome b_6/f complex. Electron transfer via the high potential (b_h) and low potential (b_l) chains are indicated by *grey* and *black arrows*, respectively. Plastoquinols are oxidized at the Q_O site and plastoquinones are reduced at the Q_i site. The *broken arrow* indicates that the quinones transfer redox equivalents between the two sites. For the Rieske and cytochrome f subunits, which both possess one transmembrane domain, only the lumenal portions are depicted. Besides its role in photosynthetic electron transfer, the cytochrome b_6/f complex is also required for redox sensing and signal transduction. b_l and b_h represent the two hemes of cytochrome b_6; Pc, plastocyanin; Cyt c_6, cytochrome c_6

b_H and b_L hemes (*petB* gene product) and subunit IV (*petD* gene product) and the nuclear-encoded Rieske iron-sulfur protein (*PetC* gene product). In addition, four low molecular weight subunits, PetG, PetL PetM and PetN are part of the complex (Pierre et al. 1995; Takahashi et al. 1996; Hager et al. 1999). Recently, a fifth major subunit, the nuclear-encoded PetO gene product, has been identified and shown to be loosely associated with the complex in *Chlamydomonas* (Hamel et al. 2000)

Electron transfer within the bc-type protein complexes functions according to the Q-cycle model, a hypothesis initially proposed by Mitchell (1976) and later modified by Crofts et al. (1983). This model postulates the existence of two electron transfer chains operating in the cyt b_6f complex: the plastoquinol-Rieske-cytf high potential chain and the plastoquinol-cytb_L-cytb_H low potential chain (Fig. 3). In line with this, it is assumed that the cyt b_6f complex contains a site for plastoquinol oxidation, Q_O, near the lumenal side of the thylakoid membrane and a site for plastoquinone reduction, Q_i, near the stromal side of the membrane. The oxidation of plastoquinol at the Q_O site occurs through a concerted reduction of cytb_L and the Rieske Fe–S center and a subsequent rapid transfer of the electrons across the membrane to reduce cytb_H accompanied by the release of two protons on the lumenal side. Thus the physiological importance of the Q-cycle mechanism is the translocation of extra protons across the membrane thereby increasing the net ATP yield per electron transferred through the complex.

i) The Plastoquinol-Oxidizing Site – the Q_o Site of the b_6/f Complex

Site-directed mutagenesis of residues located in the loops between helices C and D of cytochrome b_6 (Finazzi et al. 1997) and helices E and F of subunit IV (Zito et al. 1998, 1999) have indicated that these residues contribute to the formation of the Q_O site. The crystal structure of the bc complex from beef-heart mitochondria indicates that the Rieske protein also participates in formation of the Q_o site. The lumenal EF loop in subunit IV possesses a short sequence of the four amino acids PEWY which is strictly conserved in all bc-type complexes (Degli-Esposti et al. 1993). When this sequence is mutated at position E78 to K or L, these changes resulted in a decrease of the concerted oxidation of plastoquinol at the Q_O site by a factor of 3 and 5, respectively. The fact that the leucine substitution has a more severe effect than the lysine one indicates most likely that steric hindrance may explain the reduced rate of plastoquinol oxidation. These results show that this glutamate residue plays a critical role in the turn-over of the complex, but is not essential for its function. However, when the PEWY sequence was altered by site-directed mutagenesis to PWYE, no concerted oxidation of plastoquinol at the Q_o site was measurable in the mutagenized *Chlamydomonas* strain. Although this strain assembled cyt b_6f complex, it was unable to grow under photoautotrophic conditions (Zito et al. 1999). As rationalized from the structure of the bc-complexes (Xia et al. 1997; Zhang et al. 1998), the proline, glutamate and tyrosine residues form the bottom of the Q_O pocket, whereas the tryptophane is pointing towards the exterior of the protein. Therefore the change of the PEWY sequence to PWYE must have resulted in a complete distortion of the binding pocket, thereby abolishing binding at the Q_O site.

A further exciting finding was that this mutant strain is locked in state I. State transitions are known as short-term chromatic adaptation processes (Bonaventura and Myers 1969; Murata 1969). Under high light conditions when the plastoquinol pool is reduced, light-harvesting complex proteins (LHCP), which serve as photon-capturing antenna for PSII, become phosphorylated and migrate towards PSI to participate in PSI light harvesting (state II; Delosme et al. 1994, 1996). This process is reversible when the plastoquinol pool is reoxidized (state I). A participation of the Q_O site in the activation of the LHCP kinase was suggested by Vener et al. (1995, 1997) since activation of the kinase turned out to require the presence of plastoquinol at the Q_O site. The finding that no phosphorylation of LHC proteins and no transitions of LHCP to PSI occurred in the PWYE mutant under state II conditions demonstrated that plastoquinol binding by the Q_O pocket is indeed required for LHCII kinase activation (Zito et al. 1999).

ii) The Plastocyanin/Cytochrome c_6-Reducing Site – Interaction of Cytochrome f with Plastocyanin or Cytochrome c_6

As part of the high potential chain, cytochrome f reduces plastocyanin or cytochrome c_6. A soluble redox-active 252 residue fragment of cytochrome f from turnip was crystallized and its structure was determined at a resolution of 2.3 Å (Martinez et al. 1994). The polypeptide is organized into a large and a small domain in which ß structures are dominant. The axial ligands of the heme are H25 and the N-terminal amino group of the polypeptide (Martinez et al. 1994). At the interface of the large and the small domain of cytochrome f, a positively charged motif is formed including K58, 65 and 66 within the large domain and K188 and 189 within the small domain (see Cramer et al. 1996). This interface is believed to interact with the two alternative electron acceptors (plastocyanin or cytochrome c_6). Site-directed mutagenesis of the three K of the large domain to Q, S and E, respectively and expression of the altered proteins in *C. reinhardtii* did not drastically change the amplitude and the kinetics of flash-induced photooxidation of cytochrome f in vivo compared to the wild-type (Soriano et al. 1996). In addition, a quintuple K mutant (K58Q–K65S–K66E–K188N–K189Q; Soriano et al. 1996) also did not significantly change cytochrome f photooxidation or dark reduction in vivo. However, when cytochrome f photooxidation was measured using the altered b_6/f complexes and plastocyanin as electron acceptor in vitro, it appeared that these basic residues are implicated in the electrostatic docking of plastocyanin at low ionic strength (Soriano et al. 1998). The alteration of the net charge of K58 and K65–66 from +3 to −1 resulted in at least a ten-fold decrease in the rate constant of cytochrome f oxidation by plastocyanin.

How can this discrepancy between the in vitro and the in vivo data be explained? One possibility is that the rather small effect on cytochrome f photooxidation in vivo is a consequence of the relatively high ionic strength in the thylakoid lumen. Alternatively, the release of oxidized plastocyanin from PSI, which is found to be rate-limiting in electron transfer from cytochrome f to P700 via plastocyanin (Drepper et al. 1996), is still slower in vivo than the rate of cytochrome f oxidation in both the wild-type and the different cytochrome f mutants.

iii) Analysis of an Internal Water Chain Within Cytochrome f

Another interesting feature of cytochrome f is the presence of a linear internal chain of five H_2O molecules and six hydrogen-bonding side chains that can be deduced from the 1.96 Å structure of turnip cytochrome f (Martinez et al. 1996). Four of the water molecules extend 11 Å from the heme towards K66 on the cytochrome surface. The function of the water chain was investigated by site-directed mutagenesis. Six mutants, N153Q, Q158N, Q158L, N168Q, N168F and N233L, were generated in *C. reinhardtii* to change the conserved amino acids that provide the hydrogen-bonding side chains (Ponamarev and Cramer 1998). Out of the six mutants, one mutant strain (N168F) did not grow phototrophically. A major effect of the mutations was a pronounced decrease in the rate of cytochrome f reduction as measured in vivo under phototrophic conditions. The largest effect was observed with mutation Q158L, where a six-fold slower rate was determined. To investigate the molecular structure of the "water chain" mutants, the 251-residue functional domain of cytochrome f from wild-type *Chlamydomonas* and the mutants N168F, Q158L and N153Q were over-expressed in *E. coli* and the structure of the crystallized proteins was determined (Sainz et al. 2000). The structural information revealed that mutant N168F had the most severe perturbation with two water molecules being displaced from the internal chain whereas mutant Q158L had a more weakly bound water at position 1. The structural data together with data obtained from the site-directed mutagenesis imply that the water molecules of the internal chain exert an important function in the concerted reduction of the high and low potential chains of the cytochrome b_6/f complex, which may include a proton transfer along a proton wire formed by the water chain.

4 Perspectives

Owing to technical limitations, chloroplast functional genomics by reverse genetics is currently feasible only in the unicellular green alga *Chlamydomonas reinhardtii* and in the higher plant tobacco. Tobacco may serve as a suitable model system for higher plants in general since, with only a few exceptions (Table 1), the gene content of the plastid DNA is well conserved among higher plant species. By contrast, the plastid genomes of algae exhibit greatly varying coding capacities: Whereas the gene number in green algae is similar to that in higher plants, non-green algae such as Chromophyta and Glaucocystophyta harbor significantly more genes in their plastid DNAs (Stirewalt et al. 1995; Kowallik et al. 1995). The plastid genome of the red alga *Porphyra purpurea* even contains twice as many genes as the plastid genomes of higher plants and green algae (Reith and Munholland 1995). Approximately half of the additional genes present in Porphyra are open reading frames of unknown function. The development of plastid transformation technology and hence efficient strategies for successful functional genomics in algal systems, represents one of the major challenges in current and future research.

Looking forward, we can anticipate the gradual revealing of plastid-encoded gene functions and a more and more detailed understanding of structure-function relationships in plastid multiprotein complexes. Our understanding of plastid genome evolution, gene expression and photosynthesis should increase concomitantly.

Acknowledgments. Critical reading of the manuscript by Dr. Stephanie Ruf is gratefully acknowledged. Work on chloroplast functional genomics in the authors' laboratories is supported by grants from the Deutsche Forschungsgemeinschaft and the State of Baden-Württemberg (Landesforschungsschwerpunkt "Evolutionäre Dynamik komplexer makromolekularer Interaktionen in pflanzlichen Zellorganellen").

References

Allison LA, Simon LD, Maliga P (1996) Deletion of rpoB reveals a second distinct transcription system in plastids of higher plants. EMBO J 15:2802–2809

Azpiroz-Leehan R, Feldman KA (1997) T-DNA insertion mutagenesis in Arabidopsis: going back and forth. Trends Genet 13:152–156

Bock R, Hagemann R (2000) Extranuclear inheritance: Plastid genetics: manipulation of plastid genomes and biotechnological applications. Prog Bot 61:76–90

Bonaventura C, Myers J (1969) Fluorescence and oxygen evolution *from Chlorella pyrenoisdosa.* Biochim Biophys Acta 189:366–383

Bouchez D, Höfte H (1998) Functional genomics in plants. Plant Physiol 118:725–732

Boudreau E, Turmel M, Goldschmidt-Clermont M, Rochaix JD, Sivan S, Michaels A, Leu S (1997a) A large open reading frame (orf1995) in the chloroplast DNA of Chlamydomonas reinhardtii encodes an essential protein. Mol Gen Genet 253:649–653

Boudreau E, Takahashi Y, Lemieux C, Turmel M, Rochaix JD (1997b) The chloroplast ycf3 and ycf4 open reading frames of *Chlamydomonas reinhardtii* are required for the accumulation of the photosystem I complex. EMBO J 16:6095–6104

Boynton JE, Gillham NW, Harris EH, Hosler JP, Johnson AM, Jones AR, Randolph-Anderson BL, Robertson D, Klein TM, Shark KB, Sanford JC (1988) Chloroplast transformation in *Chlamydomonas* with high velocity microprojectiles. Science 240:1534–1538

Breyton C, Tribet C, Olive J, Dubacq JP, Popot JL (1997) Dimer to monomer conversion of the cytochrome b6f complex. Causes and consequences. J Biol Chem 272:21892–21900

Burrows PA, Sazanov LA, Svab Z, Maliga P, Nixon PJ (1998) Identification of a functional respiratory complex in chloroplasts through analysis of tobacco mutants containing disrupted plastid ndh genes. EMBO J 17:868–876

Büttner M, Xie DL, Nelson H, Pinther W, Hauska G, Nelson N (1992) Photosynthetic reaction center genes in green sulfur bacteria and in photosystem 1 are related. Proc Natl Acad Sci USA 89:8135–8139

Cerutti H, Osman M, Grandoni P, Jagendorf AT (1992) A homolog of *Escherichia coli* RecA protein in plastids of higher plants. Proc Natl Acad Sci USA 89:8086–8072

Cramer WA, Soriano G M, Ponomarev M, Huang D, Zhang H, Martinez SE, Smith JL (1996) Some new structural aspects and old controversies concerning the cytochrome b_6f complex of oxygenic photosynthesis. Annu Rev Plant Physiol Plant Mol Biol 47:477–508

Crofts AR, Meinhardt SW, Jones KR, Snozzi M (1983) The role of the quinone pool in the cyclic electron transfer chain of *Rhodopseudomonas sphaeroides*: a modified Q-cycle. Biochim Biophys Acta 723:202–218

Cui L, Bingham SE, Kuhn M, Kass H, Lubitz W, Webber AN (1995) Site-directed mutagenesis of conserved histidines in the helix VIII domain of PsaB impairs assembly of the photosystem I reaction center without altering spectroscopic characteristics of P700. Biochemistry 34:1549–1558

Degli-Esposti M, De Vries S, Crimi MAG, Patarnello T, Meyer A (1993) Mitochondrial cytochrome-b-evolution and structure of the protein. Biochim Biophys Acta 1143:243–271

Delosme R, Béal D, Joliot P (1994) Photoacoustic detection of flash-induced charge separation in photosynthetic systems. Spectral dependence of the quantum yield. Biochim Biophys Acta 1185:56–64

Delosme R, Olive J, Wollman F-A (1996) Changes in light energy distribution upon state transitions: an in vivo photoacoustic study of the wild-type and photosynthesis mutants from *Chlamydomonas reinhardtii*. Biochim Biophys Acta 1273:150–158

Diaz-Quintana A, Leibl W, Bottin H, Sétif P (1998) Electron transfer in photosystem I reaction centers follows a linear pathway in which iron-sulfur cluster FB is the immediate electron donor to soluble ferredoxin. Biochemistry 37:3429–3439

Drepper F, Hippler M, Nitschke W, Haehnel W (1996) Binding dynamics and electron transfer between plastocyanin and photosystem I. Biochemistry 35:1282–1295

Drescher A, Ruf S, Calsa T Jr, Carrer H, Bock R (2000) The two largest chloroplast genome-encoded open reading frames of higher plants are essential genes. Plant J 22:97–104

Dunn P, Gray J (1988) Localisation and sequence of the gene for the 8 kDa subunit of photosystem I in pea and wheat chloroplast DNA. Plant Mol Biol 11:311–319

Endo T, Shikanai T, Takabayashi A, Asada K, Sato F (1999) The role of chloroplastic NAD(P)H dehydrogenase in photoprotection. FEBS Lett 457:5–8

Evans MC, Reeves SG, Cammack R (1974) Determination of the oxidation-reduction potential of the bound iron-sulphur proteins of the primary electron acceptor complex of photosystem I in spinach chloroplasts. FEBS Lett 49:111–114

Finazzi G, Buschlen S, de Vitry C, Rappaport F, Joliot P, Wollman FA (1997) Function-directed mutagenesis of the cytochrome b6f complex in *Chlamydomonas reinhardtii*: involvement of the cd loop of cytochrome b6 in quinol binding to the Q(o) site. Biochemistry 36:2867–2874

Fischer N, Stampacchia O, Redding K, Rochaix JD (1996) Selectable marker recycling in the chloroplast. Mol Gen Genet 251:373–380

Fischer N, Setif P, Rochaix JD (1997) Targeted mutations in the psaC gene of *Chlamydomonas reinhardtii*: preferential reduction of FB at low temperature is not accompanied by altered electron flow from photosystem I to ferredoxin. Biochemistry 36:93–102

Fischer N, Hippler M, Setif P, Jacquot JP, Rochaix JD (1998) The PsaC subunit of photosystem I provides an essential lysine residue for fast electron transfer to ferredoxin. EMBO J 17:849–858

Fischer N, Setif P, Rochaix JD (1999) Site-directed mutagenesis of the PsaC subunit of photosystem I. F(b) is the cluster interacting with soluble ferredoxin. J Biol Chem 274:23333–23340

Golbeck JH, Bryant DA (1991) Photosystem I. In: Lee CP (ed.) Current topics in bioenergetics: light driven reactions in bioenergetics, vol 16. Academic Press, New York, pp 83–177

Goldschmidt-Clermont M (1991) Transgenic expression of aminoglycoside adenyl transferase in the chloroplast: a selectable marker for site-directed transformation of Chlamydomonas. Nucleic Acids Res 19:4083–4089

Hager M, Biehler K, Illerhaus J, Ruf S, Bock R (1999) Targeted inactivation of the smallest plastid genome-encoded open reading frame reveals a novel and essential subunit of the cytochrome b6f complex. EMBO J 18:5834–5842

Hallahan BJ, Purton S, Ivison A, Wright D, Evans MCW (1995) Analysis of the proposed Fe–S_X binding region of Photosystem 1 by site directed mutation of PsaA *in Chlamydomonas reinhardtii.* Photosynth Res 46:257–264

Hamel P, Olive J, Pierre Y, Wollman FA, de Vitry C (2000) A new subunit of cytochrome b6f complex undergoes reversible phosphorylation upon state transition. J Biol Chem 275:17072–17079

Hayashida N, Matsubayashi T, Shinozaki K, Sugiura M, Inoue K, Hiyama T (1987) The gene for the 9 kd polypeptide, a possible apoprotein for the iron- sulfur centers A and B of the photosystem I complex, in tobacco chloroplast DNA. Curr Genet 12:247–250

Heathcote P, Williams-Smith DL, Sihra CK, Evans MC (1978) The role of the membrane-bound iron-sulphur centres A and B in the photosystem I reaction centre of spinach chloroplasts. Biochim Biophys Acta 503:333–342

Hippler M, Redding K and Rochaix JD (1998) *Chlamydomonas* genetics, a tool for the study of bioenergetic pathways. Biochim Biophys Acta 1367:1–62

Hoj PB, Svendsen I, Scheller HV, Moller BL (1987) Identification of a chloroplast-encoded 9-kDa polypeptide as a 2[4Fe–4S] protein carrying centers A and B of photosystem I. J Biol Chem 262:12676–12684

Horváth EM, Peter SO, Joet T, Rumeau D, Cournac L, Horváth G, Kavanagh TA, Schäfer C, Medgyesy P (2000) Targeted inactivation of the plastid ndhB gene in tobacco results in an enhanced sensitivity of photosynthesis to moderate stomatal closure. Plant Physiol 123:1337–1349

Howe G, Merchant S (1992) The biosynthesis of membrane and soluble plastidic c-type cytochromes of *Chlamydomonas reinhardtii* is dependent on multiple common gene products. EMBO J 11:2789–2801

Katoh A, Lee K-S, Fukuzawa H, Ohyama K, Ogawa T (1996) cemA homologue essential to CO_2 transport in the cyanobacterium Synechocystis PCC6803. Proc Natl Acad Sci USA 93:4006–4010

Kofer W, Koop H-U, Wanner G, Steinmüller K (1998) Mutagenesis of the genes encoding subunits A, C, H, I, J and K of the plastid NAD(P)H-plastoquinone-oxidoreductase in tobacco by polyethylene glycol-mediated plastome transformation. Mol Gen Genet 258:166–173

Kowallik KV, Stoebe B, Schaffran I, Kroth-Pancic P, Freier U (1995) The chloroplast genome of a chlorophyll a+c-containing alga, *Odontella sinensis.* Plant Mol Biol Rep 13:336–342

Mannan RM, He WZ, Metzger SU, Whitmarsh J, Malkin R, Pakrasi HB (1996) Active photosynthesis in cyanobacterial mutants with directed modifications in the ligands for two iron-sulfur clusters on the PsaC protein of photosystem I. EMBO J 15:1826–1833

Martinez SE, Huang D, Szczepaniak A, Cramer WA, Smith JL (1994) Crystal structure of chloroplast cytochrome f reveals a novel cytochrome fold and unexpected heme ligation. Structure 2:95–105

Martinez SE, Huang D, Ponomarev M, Cramer WA, Smith JL (1996) The heme redox center of chloroplast cytochrome f is linked to a buried five-water chain. Protein Sci 5:1081–1092

Mitchell P (1976) Possible molecular mechanism of the protonmotive function of cytochrome systems. J Theor Biol 62:327–367

Monod C, Takahashi Y, Goldschmidt-Clermont M, Rochaix JD (1994) The chloroplast ycf8 open reading frame encodes a photosystem II polypeptide which maintains photosynthetic activity under adverse growth conditions. EMBO J 13:2747–2754

Murata N (1969) Control of excitation transfer in photosynthesis. Light-induced decrease of chlorophyll a fluorescence related to photophosphorylation system in spinach chloroplasts. Biochim Biophys Acta 172:242–251

Naver H, Scott MP, Golbeck JH, Moller BL, Scheller HV (1996) Reconstitution of barley photosystem I with modified PSI-C allows identification of domains interacting with PSI-D and PSI-A/B. J Biol Chem 271:8996–9001

Naver H, Scott MP, Golbeck JH, Olsen CE, Scheller HV (1998) The eight-amino acid internal loop of PSI-C mediates association of low molecular mass iron-sulfur proteins with the P700-FX core in photosystem I. J Biol Chem 273:18778–18783

Newman SM, Gillham NW, Harris EH, Johnson AM, Boynton JE (1991) Targeted disruption of chloroplast genes in *Chlamydomonas reinhardtii*. Mol Gen Genet 230:65–74

Nitschke W, Feiler U, Rutherford AW (1990) Photosynthetic reaction center of green sulfur bacteria studied by EPR. Biochemistry 29:3834–3842

Ohyama K, Fukuzawa H, Kohchi T, Shirai H, Sano T, Sano S, Umesono K, Shiki Y, Takeuchi M, Chang Z, Aota S-I, Inokuchi H, Ozeki H (1986) Chloroplast gene organization deduced from complete sequence of liverwort *Marchantia polymorpha* chloroplast DNA. Nature 322:572–574

Pierre Y, Breyton C, Kramer D, Popot JL (1995) Purification and characterization of the cytochrome b_6/f complex from *Chlamydomonas reinhardtii*. J Biol Chem 270:29342–29349

Ponamarev MV, Cramer WA (1998) Perturbation of the internal water chain in cytochrome f of oxygenic photosynthesis: loss of the concerted reduction of cytochrome f and b_6. Biochemistry 37:17199–17208

Redding K, MacMillan F, Leibl W, Brettel K, Hanley J, Rutherford AW, Breton J, Rochaix JD (1998) A systematic survey of conserved histidines in the core subunits of Photosystem I by site-directed mutagenesis reveals the likely axial ligands of P700. EMBO J 17:50–60

Reith M, Munholland J (1995) Complete nucleotide sequence of the Porphyra purpurea chloroplast genome. Plant Mol Biol Rep 13:333–335

Rochaix JD (1997) Chloroplast reverse genetics: new insights into the function of plastid genes. Trends Plant Sci 2:419–425

Rochaix JD, Fischer N, Hippler M (2000) Chloroplast site-directed mutagenesis of photosystem I in *Chlamydomonas*: electron transfer reactions and light sensitivity. Biochimie 82: 635–646

Rodday SM, Webber AN, Bingham SE, Biggins J (1995) Evidence that the FX domain in photosystem I interacts with the subunit PsaC: site-directed changes in PsaB destabilize the subunit interaction in *Chlamydomonas reinhardtii*. Biochemistry 34:6328–6334

Rolland N, Dorne A-J, Amoroso G, Sültemeyer DF, Joyard J, Rochaix JD (1997) Disruption of the plastid ycf10 open reading frame affects uptake of inorganic carbon in the chloroplast of *Chlamydomonas*. EMBO J 16:6713–6726

Ruf S, Kössel H, Bock R (1997) Targeted inactivation of a tobacco intron-containing open reading frame reveals a novel chloroplast-encoded photosystem I-related gene. J Cell Biol 139:95–102

Ruf S, Biehler K, Bock R (2000) A small chloroplast-encoded protein as a novel architectural component of the light-harvesting antenna. J Cell Biol 149:369–377

Ruffle SV, Sayre RT (1999) Functional analysis of photosystem II. In: Rochaix JD, Goldschmidt-Clermont M, Merchant S (eds) The molecular biology of chloroplast and mitochondria in Chlamydomonas. Kluwer, Dordrecht, pp 287–322

Sainz G, Carrell CJ, Ponamarev MV, Soriano GM, Cramer WA, Smith JL (2000) Interruption of the internal water chain of the cytochrome f impairs photosynthetic function. Biochemistry 39:9165–9173

Sasaki Y, Sekiguchi K, Nagano Y, Matsuno R (1993a) Chloroplast envelope protein encoded by chloroplast genome. FEBS Lett 316:93–98

Sasaki Y, Hakamada K, Suama Y, Nagano Y, Furusawa I, Matsuno R (1993b) Chloroplast-encoded protein as a subunit of acetyl-CoA carboxylase in pea plant. J Biol Chem 268:25118–25123

Sazanov LA, Burrows PA, Nixon PJ (1998) The chloroplast Ndh complex mediates the dark reduction of the plastoquinone pool in response to heat stress in tobacco leaves. FEBS Lett 429:115–118

Schubert W, Klukas O, Krauss N, Saenger W, Fromme P, Witt HT (1997) Photosystem I of *Synechococcus elongatus* at 4 Å resolution: comprehensive structure analysis. J Mol Biol 272:741–769

Schwarz Z, Kössel H (1979) Sequencing of the 3'-terminal region of a 16 S rRNA gene from Zea mays chloroplast reveals homology with *E. coli* 16 S rRNA. Nature 279:520–522

Schwarz Z, Kössel H (1980) The primary structure of 16 S rDNA from Zea mays chloroplast is homologous to *E. coli* 16 S rRNA. Nature 283:739–742

Serino G, Maliga P (1998) RNA polymerase subunits encoded by the plastid rpo genes are not shared with the nucleus-encoded plastid enzymes. Plant Physiol 117:1165–1170

Shikanai T, Endo T, Hashimoto T, Yamada Y, Asada K, Yokota A (1998) Directed disruption of the tobacco ndhB gene impairs cyclic electron flow around photosystem I. Proc Natl Acad Sci USA 95:9705–9709

Shinozaki K, Ohme M, Tanaka M, Wakasugi T, Hayashida N, Matsubayashi T, Zaita N, Chunwongse J, Obokata J, Yamaguchi-Shinozaki K, Ohto C, Torazawa K, Meng BY, Sugita M, Deno H, Kamogashira T, Yamada K, Kusuda J, Takaiwa F, Kato A, Tohdoh N, Shimada H, Sugiura M (1986) The complete nucleotide sequence of the tobacco chloroplast genome: its gene organization and expression. EMBO J 5:2043–2049

Smart LB, Warren PV, Golbeck JH, McIntosh L (1993) Mutational analysis of the structure and biogenesis of the Photosystem I reaction center in the cyanobacterium *Synechocystis* sp. PCC 6803. Proc Natl Acad Sci USA 90:1132–1136

Soriano GM, Ponamarev G-S, Tae G-S, Cramer WA (1996) Effect of the interdomain basic region of cytochrome f on its redox reactions in vivo. Biochemistry 35:14590–14598

Soriano GM, Ponamarev MV, Piskorowski R, Cramer WA (1998) Identification of the basic residues of cytochrome f responsible for electrostatic docking interactions with plastocyanin in vitro: relevance to the electron transfer reaction in vivo. Biochemistry 37:15120–15128

Spreitzer RJ (1999) Genetic engineering of rubisco. In: Rochaix JD, Goldschmidt-Clermont M, Merchant S (eds) The molecular biology of chloroplast and mitochondria in *Chlamydomonas*. Kluwer, Dordrecht, pp 515–527

Stirewalt VL, Michalowski CB, Löffelhardt W, Bohnert HJ, Bryant DA (1995) Nucleotide sequence of the cyanelle genome from *Cyanophora paradoxa*. Plant Mol Biol Rep 13:327–332

Strotmann H, Shavit N, Leu S (1999) Assembly and Function of the Chloroplast ATP Synthase. In: Rochaix JD, Goldschmidt-Clermont M, Merchant S (eds) The molecular biology of chloroplast and mitochondria in *Chlamydomonas*. Kluwer, Dordrecht, pp 455–500

Svab Z, Maliga P (1993) High-frequency plastid transformation in tobacco by selection for a chimeric aadA gene. Proc Natl Acad Sci USA 90:913–917

Svab Z, Hajdukiewicz P, Maliga P (1990) Stable transformation of plastids in higher plants. Proc Natl Acad Sci USA 87:8526–8530

Takahashi Y, Goldschmidt-Clermont M, Soen SY, Franzen LG, Rochaix JD (1991) Directed chloroplast transformation in *Chlamydomonas reinhardtii*: insertional inactivation of the psaC gene encoding the iron sulfur protein destabilizes photosystem I. EMBO J 10:2033–2040

Takahashi Y, Goldschmidt-Clermont M, Rochaix JD (1992) Directed mutagenesis of chloroplast gene psaC in *Chlamydomonas reinhardtii*. In: Murata N (ed) Research in photosynthesis, vol III. Kluwer, Dordrecht, pp 359–366

Takahashi Y, Rahire M, Breyton C, Popot J-L, Joliot P, Rochaix JD (1996) The chloroplast ycf7 (petL) open reading frame of *Chlamydomonas reinhardtii* encodes a small functionally important subunit of the cytochrome b6f complex. EMBO J 15:3498–3506

Vassiliev IR, Jung YS, Smart LB, Schulz R, McIntosh L, Golbeck JH (1995) A mixed-ligand iron-sulfur cluster (C556SPaB or C565SPsaB) in the Fx-binding site leads to a decreased quantum efficiency of electron transfer in photosystem I. Biophys J 69:1544–1553

Vassiliev IR, Jung YS, Yang F, Golbeck JH (1998) PsaC subunit of photosystem I is oriented with iron-sulfur cluster F(B) as the immediate electron donor in ferredoxin and flavodoxin. Biophys J 74:2029–2035

Vener AV, Van Kann PJ, Gal A, Andersson B, Ohad I (1995) Activation/deactivation cycle of redox-controlled thylakoid protein phosphorylation. Role of plastoquinol bound to the reduced cytochrome bf complex. J Biol Chem 270:25225–25232

Vener AV, Van Kann PJ, Rich PR, Ohad I, Andersson B (1997) Plastoquinol at the quinol oxidation site of reduced cytochrome bf mediates signal transduction between light and protein phosphorylation: thylakoid protein kinase deactivation by a single-turnover flash. Proc Natl Acad Sci USA 94:1585–1590

Webber AN, Bingham SE (1999) Structure and function of photosystem I. In: Rochaix JD, Goldschmidt-Clermont M, Merchant S (eds) The molecular biology of chloroplast and mitochondria in *Chlamydomonas*. Kluwer, Dordrecht, pp 323–348

Webber AN, Gibbs PB, Ward JB, Bingham SE (1993) Site-directed mutagenesis of the photosystem I reaction center in chloroplasts. The proline-cysteine motif. J Biol Chem 268:12990–12995

Webber AN, Su H, Bingham SE, Kass H, Krabben L, Kuhn M, Jordan R, Schlodder E, Lubitz W (1996) Site-directed mutations affecting the spectroscopic characteristics and midpoint potential of the primary donor in photosystem I. Biochemistry 35:12857–12863

Wollman FA (1999) The structure, function and biogenesis of cytochrome b_6/f complexes. In: Rochaix JD, Goldschmidt-Clermont M, Merchant S (eds) The molecular biology of chloroplast and mitochondria in *Chlamydomonas*. Kluwer, Dordrecht, pp 459–476

Xia D, Yu CA, Kim H, Xia JZ, Kachurin AM, Zhang L, Yu L, Deisenhofer J (1997) Crystal structure of the cytochrome bc1 complex from bovine heart mitochondria. Science 277:60–66

Xie Z, Merchant S (1996) The plastid-encoded ccsA gene is required for heme attachment to chloroplast c-type cytochromes. J Biol Chem 271:4632–4639

Xie Z, Culler D, Dreyfuss BW, Kuras R, Wollman F-A, Girard-Bascou J, Merchant S (1998) Genetic analysis of chloroplast c-type cytochrome assembly in *Chlamydomonas reinhardtii*: one chloroplast locus and at least four nuclear loci are required for heme attachment. Genetics 148:681–692

Yokoi F, Vassileva A, Hayashida N, Torazawa K, Wakasugi T, Sugiura M (1990) Chloroplast ribosomal protein L32 is encoded in the chloroplast genome. FEBS Lett 276:88–90

Yu L, Vassiliev IR, Jung YS, Bryant DA, Golbeck JH (1995) Modified ligands to FA and FB in photosystem I. II. Characterization of a mixed ligand [4Fe-4S] cluster in the C51D mutant of PsaC upon rebinding to P700-Fx cores. J Biol Chem 270:28118–28125

Zhang Z, Huang L, Shulmeister VM, Chi Y-I, Kim KK, Hung L-W, Crofts A, Berry AE, Kim SH (1998) Electron transfer by domain movement in cytochrome bc1. Nature 392:677–684

Zhao J, Li N, Warren PV, Golbeck JH, Bryant DA (1992) Site-directed conversion of a cysteine to aspartate leads to the assembly of a [3Fe–4S] cluster in PsaC of photosystem I. The photoreduction of FA is independent of FB. Biochemistry 31:5093–5099

Zito F, Finazzi G, Joliot P, Wollman FA (1998) Glu78, from the conserved PEWY sequence of subunit IV, has a key function in cytochrome b6f turnover. Biochemistry 37:10395–10403

Zito F, Finazzi G, Delosme R, Nitschke W, Picot D, Wollman FA (1999) The Q_O site of the cytochrome b_6/f complexes controls the activation of the LHC kinase. EMBO J 18:2961–2969

Prof. Dr. Ralph Bock
Albert-Ludwigs-Universität Freiburg
Institut für Biologie III
Schänzlestraße 1
79104 Freiburg i. Br., Germany
Tel.: +49–761–2032721
Fax: +49–761–2032745
e-mail: bock@biologie.uni-freiburg.de
and
Westfälische Wilhelms-Universität Münster
Institut für Biochemie und Biotechnologie der Pflanzen
Hindenburgplatz 55
48143 Münster, Germany
Phone: +49–251–83-24790/91
Fax: +49–251–83-28371
e-mail: rbock@uni-muenster.de

Dr. Michael Hippler
Albert-Ludwigs-Universität Freiburg
Institut für Biologie II
Schänzlestraße 1
79104 Freiburg i. Br., Germany
Fax: +49–761–2032601

Present address:
Friedrich-Schiller-Universität Jena
Institut für Allgemeine Botanik
Dornburger Straße 159
07743 Jena, Germany
Tel.: +49-3641-949237
Fax: +49–3641–949232
e-mail: m.hippler@uni-jena.de

Communicated by K. Esser

Molecular Cell Biology: Mechanisms and Regulation of Protein Import into the Plant Cell Nucleus

By Vera Hemleben, Katrin Hinderhofer, and Ulrike Zentgraf

1 Introduction

Plant cells contain three genetically active compartments embedded in the cytoplasm which are acting in strictly regulated cooperation (Sitte 1998). Translation of proteins from organelle-encoded genes occurs in the respective semiautonomous plastid or mitochondrial compartments (stroma and matrix, respectively). Nuclear-encoded genes are transcribed in the cell nucleus, and posttranscriptionally modified mRNA, tRNA and preformed ribosomes are transported to the cytoplasm. The nuclear mRNA is translated into polypeptides either directly at free cytoplasmic ribosomes or at ribosomes associated with the rough endoplasmic reticulum (ER). Proteins necessary for the correct functioning of the cell have to be transported to the respective locations. Characteristic signal- or transit-peptides at the N-terminus of specific proteins mediate the correct transport to the plasma membrane or into the chloroplasts and mitochondria; other characteristic short motifs were found in proteins targeted to the vacuole, peroxisomes or into the cell nucleus.

The nucleus is a definitive structure of the eukaryotic cell, comprising two bilamellar membranes, the inner and outer nuclear membranes, which separate the cytoplasmic and the nucleoplasmic compartments; communication between these two is essential for proper functioning of the cell. Nuclear-targeted, karyophilic molecules, such as proteins translated in the cytoplasm and needed inside the nucleus (e.g. histones, ribosomal proteins, DNA and RNA polymerizing enzymes, DNA and RNA modifying enzymes, transcription factors etc.) are transported through the nuclear pore complexes (NPC). Therefore, a very frequent highly regulated molecular in and out traffic occurs through the nuclear membranes which are fused at specific sites where NPCs are formed (Forbes 1992; Davis 1995; Heese-Peck and Raikhel 1998a; Goldberg et al. 1999). For this translocation process, different specific mechanisms were developed.

Although much is known about NPC structure, only some of the approx. 100 different proteins forming this complex structure are characterized for animals and yeast (for reviews see Forbes 1992; Rout and

Progress in Botany, Vol. 63

Wente 1994; Davis 1995; Doye and Hurt 1997; Heese-Peck and Raikhel 1998a), and for plants, much less is known. However, for the elucidation of the different transport mechanisms, characterization and functional analysis of the NPC proteins is necessary (Heese-Peck and Raikhel 1998a).

Here, we intend to focus mainly on the nuclear import of proteins through the NPC. Again, many more details are known about this process in animal or yeast cells (for reviews see Nigg 1997; Görlich 1998; Mattaj and Engelmeier 1998; Görlich and Kutay 1999), and the data will be shortly summarized. Nuclear localization signals (NLS) present in karyophilic proteins were first recognized and characterized for the animal protein nucleoplasmin or the viral SV 40 T antigen proteins (for reviews see Dingwall and Laskey 1991; Garcia-Bustos et al. 1991; Hicks and Raikhel 1995b; Zentgraf and Hemleben 1996). In plants, initially the transfer of T-DNA from *Agrobacterium* into the nucleus of plant cells mediated by bacterial proteins encoded by the Ti-plasmid (Vir E and D) was investigated in more detail, and it was detected that the Vir proteins involved in this process carry a NLS-like motif (Citovsky et al. 1992; Zupan et al. 2000).

Since in vitro and in vivo systems were developed for plants, especially using GUS (β-glucuronidase) or GFP (green fluorescent protein) fusion proteins to visualize the transport of proteins into the nucleus (Harter et al. 1994; Hicks et al. 1996; Merkle et al. 1996; Smith et al. 1997; Chytilova et al. 2000), more information has been gained on plant nuclear protein import which will be summarized here. Previously, we reported on plant signal transduction processes through the cell (Zentgraf and Hemleben 1996) and on nuclear gene regulation by transcription factors (Zentgraf et al. 1998). Here, we concentrate mostly on components of signal transduction cascades, induced by exogenous (light, stress, pathogens, phytohormones, other signal molecules) or endogenous factors interacting indirectly or directly with those of the replication or transcription machinery which have to be transported to the nucleus and, therefore, have to be highly regulated and fine-tuned by specific mechanisms.

2 Plant Nuclear Pore Structure

The structure of the nuclear pore complex (NPC), its architecture and composition, has been intensively investigated in amphibian cells using different electron microscopic techniques and seems to be conserved among eukaryotes (Davis 1995). The NPC is a proteinaceous supramolecular structure of about 125 MDa exhibiting an overall cylindrical shape with a diameter of about 120 nm and a height of about 70 nm. It is

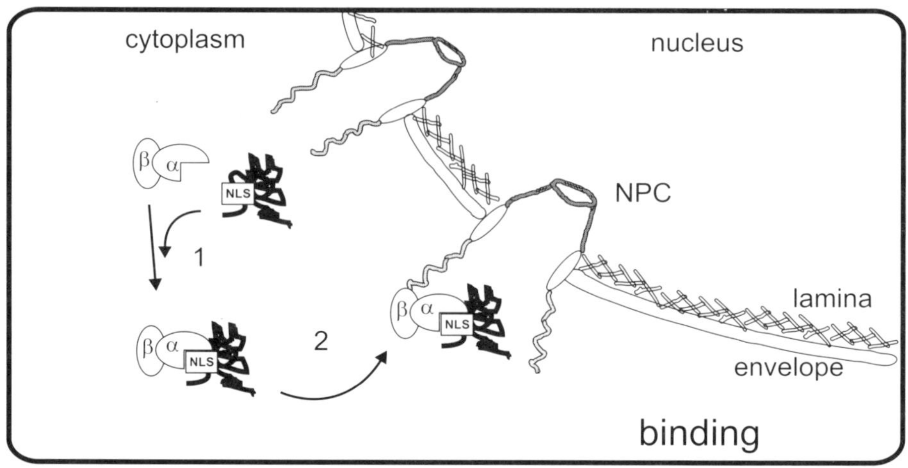

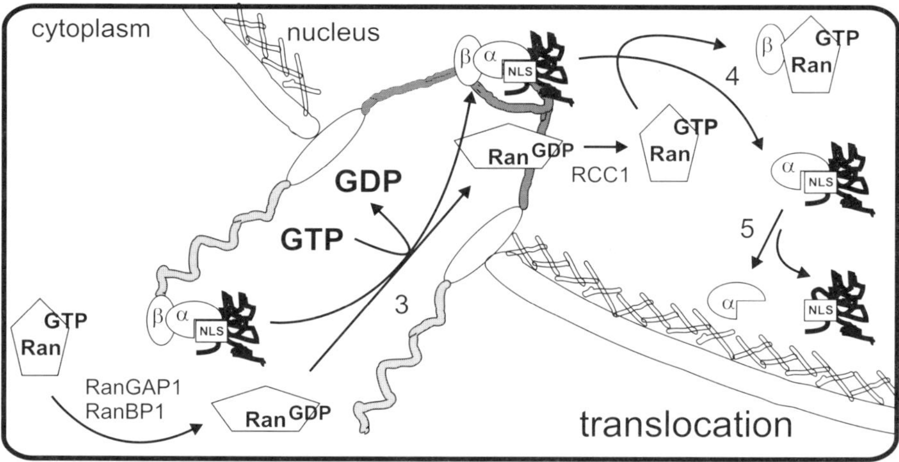

Fig. 1. Schematic drawing of the NLS protein import as a two-step process.The NLS docking step is mediated by the IMPα/β heterodimer. The initial cytoplasmic step is the binding of the import substrate to the IMPα/β heterodimer, where the α subunit provides the NLS binding site (*1*). Subsequently, the β subunit mediates the docking of the trimeric complex to the cytoplasmic periphery of the nuclear pore complex (*2*). How the trimeric complex is exactly translocated through the nuclear pore complex (*NPC*) and which NPC proteins are involved is still unknown, but translocation requires RanGDP and GTP (*3*). After translocation the trimeric complex sticks to the nucleoplasmic side of the NPC where the binding of nuclear RanGTP to the IMPβ subunit terminates import and releases the IMPα/NLS protein complex into the nucleoplasm (*4*). The NLS protein is released from IMPα (*5*). To date, homologues to IMPα and β, to Ran and RanBP have been characterized in plants. (Modified after Görlich 1998; Heese-Peck and Raikhel 1998a; Smith and Raikhel 1999)

a modular structure consisting of a series of stacked rings with associated fibers. It shows an eight-fold rotational symmetry of eight identical units. The spoke ring complex forms a central channel of up to 26 nm and eight smaller channels each with a diameter of 9 nm. Within this central channel, the hour-glass-shaped transporter is located which has been proposed to be involved in the active transport of macromolecules into and out of the nucleus. On the nucleoplasmic face, the spoke ring complex is connected with the nucleoplasmic ring. From the outer periphery of the nucleoplasmic ring, eight filaments extend into the nucleoplasm and terminate in a distal ring to form the nuclear basket. On the cytoplasmic face, the spoke ring complex is connected with the cytoplasmic ring consisting of the star ring, the thin ring and the cytoplasmic subunits. Eight cytoplasmic fibrils are anchored to the cytoplasmic subunits and appear to serve as docking sites for import substrates (for reviews see Davis 1995; Goldberg et al. 1999). The yeast NPCs appear to have a similar morphology to the amphibian NPCs but seem to be smaller (Rout and Blobel 1993). Morphological studies revealed that the NPCs from mono- and dicotyledonous plants also show the typical eight-fold symmetry and the modular structure of stacked rings connected to cytoplasmic fibbers and the nucleoplasmic basket (Fig. 1).

Approximately 30–40% of the genes encoding the NPC proteins of animal and yeast have been identified to date. Quantitative STEM (scanning transmission electron microscopy revealed that the amphibian NPC appears to consist of at least 100 different proteins while the yeast NPC may contain about 80 different proteins (Reichelt et al. 1990; Rout and Blobel 1993). The known NPC components have been identified using a variety of molecular and genetic approaches (for reviews see Forbes 1992; Davis 1995; Rout and Wente 1994; Doye and Hurt 1997). An excellent overview of the NPC proteins isolated for vertebrates and yeast is given by Heese-Peck and Raihkel (1998a). Although the plant NPCs have been described morphologically for many years, little is known about plant NPC components. A 100 kDa nucleoporin-like protein has been identified in carrot suspension cells using an antibody against the yeast NPC protein Nsp1p (Scofield et al. 1992). In plants, there are some indications that the components recognizing functional NLSs may be located at the plant NPC. In tobacco, proteins at the NPC are detected by wheat germ agglutinin (WGA) which is a lectin that specifically binds to GlcNAc residues and to glycosylated proteins. Several vertebrate NPC proteins are modified with a single O-GlcNAc. Sugar analyses of the plant glycans with terminal GlcNAc demonstrated that they consist of oligosaccharides of more than five GlcNAc (Fig. 2). This novel oligosaccharide modification may confer different properties to the NPC proteins of the plant (Heese-Peck et al. 1995). Although the significance of the GlcNAc modification of many NPCs is not understood, this modification has been very useful to identify and isolate NPC proteins.

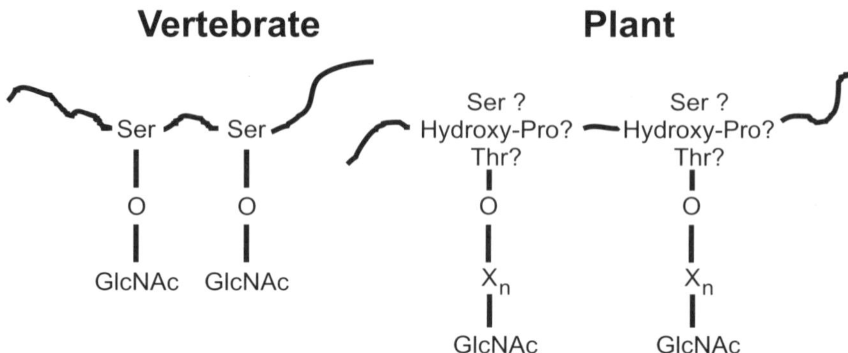

Fig. 2. Glycosylation of vertebrate and plant NPCs. O, O-linked oligosaccharides; X_n unknown saccharides equal to or larger than 5 GlcNAc. (Modified after Heese-Peck and Raikhel 1998a)

In tobacco, several NPC proteins modified by an O-linked oligosaccharide with a terminal GlcNAc have been identified by affinity chromatography using the lectin from *Erythrina crystagalli.* This purification scheme led to the characterization of three terminal GlcNAc proteins, gp33, gp40 and gp65. Subsequently, the gene encoding gp40 was isolated after peptide sequencing. It has 28–34% amino acid identity to aldose-1-epimerase from bacteria which has not been isolated from higher eukaryotes (Heese-Peck and Raihkel 1998b).

All O-GlcNAc NPC proteins characterized so far contain the short amino acid sequence motif FxFG. This motif is repeated and localized within the domain of the O-GlcNAc modification (Rout and Wente 1994; Davis 1995). Specific antibodies against the FxFG domain inhibit nuclear import in Xenopus oocytes indicating that this motif might play an important role in nuclear import (Dabauvalle et al. 1988). Several vertebrate NPC proteins containing the FxFG motif bind the importin β (IMPβ) domain of the NLS-dependent import receptor complex indicating that these NPC proteins might function as docking sites for the NLS-dependent import machinery (see Fig. 1 and below; Radu et al. 1995). In yeast, a second repeated motif (GLFG) has been identified which binds the yeast homologue of IMPβ, Kap95p (Rout and Wente 1994; Davis 1995; Iovine et al. 1995; Doye and Hurt 1997). In addition, several yeast and vertebrate NPC proteins contain Cys_2–Cys_2 type zinc-finger motifs that are proposed to bind DNA or coiled-coiled domains that may allow the formation of filamentous structures. NPC proteins with putative leucine zipper motifs might be involved in a protein-protein interaction during NPC assembly or substrate transport. Transmembrane domains of different NPC proteins have been suggested to facilitate anchoring of the NPC to the nuclear envelope. Although impressive progress has been made in identification, studies on functional domains, and localization in the mammalian and yeast systems, much more information is needed to resolve the exact function of the NPC. However, we are still at the beginning of identifying and understanding the function of the NPC of plants.

3 Mechanisms of protein import

a) Import in Yeast and Vertebrates

The best-characterized nucleoplasmic transport process at present is the nuclear import of proteins in vertebrate and yeast cells. The development of an in vitro system based on permeabilized mammalian cells (Adam et al. 1990) and further investigations have resulted in the purification, cloning and functional characterization of the soluble transport factors playing a key role in the classical nuclear import (for reviews see Nigg 1997; Görlich 1998; Mattaj and Engelmeier 1998; Görlich and Kutay 1999). Nuclear import of proteins can be considered as a two step process. First, the karyophilic protein binds to an import receptor via an NLS to form an import complex which then binds to the nuclear pore complex. Subsequently, the import-complex is translocated through the nuclear pore. This translocation appears to be the energy-consuming step.

GTPase Ran is the first factor which has been identified (Melchior et al. 1993; Moore and Blobel 1993). On the basis of its unusual property of binding tightly to nickel-agarose, IMPα was purified (Görlich et al. 1994). Subsequently, the β subunit of importin was stoichiometrically co-purified with IMPα by an anti-IMPα immunoaffinity column (Görlich et al. 1995).

Proteins larger than the exclusion limit (about 40 kDa) of the NPC must possess an NLS to be imported into the nucleus. First, the cytosolic NLS receptor IMPα binds the NLS of its cargo-protein, then the N-terminal IMPβ-binding domain of IMPα associates with IMPβ to form the trimeric import complex which than docks onto the cytoplasmic surface of the nuclear pore complex via IMPβ. The formation of this complex requires the presence of the small GTPase Ran in its GDP-bound form in the cytoplasm (Görlich et al. 1996). However, the actual mechanism of the passage through the NPC is still a black box. The transfer requires GTP. The role of Ran and GTP hydrolysis by Ran during translocation is still unclear. Translocation into the nucleus is terminated at the nuclear side of the pore complex by the disassembly of the trimeric NLS/IMPα/IMPβ complex. Binding of RanGTP to IMPβ dissociates the complex and releases IMPα together with its ligand into the nucleoplasm whereas IMPβ remains bound to the nucleoplasmic face of the pore complex. After IMPα and IMPβ have delivered their cargo into the nucleus, they need to return to the cytoplasm to mediate the next round of import.

The RanGTPase system is at the heart of the nuclear machinery and consists of Ran itself, the chromatin-bound guanine exchange factor RCC1, the Ran GTPase-activating protein Ran GAP1, the Ran-binding protein RanBP1, NTF2 (nuclear transport factor) and IMPβ transport receptors which are the regulatory targets of Ran. Ran itself is predominantly localized in the nucleus. RCC1, the major nucleotide exchange factor of Ran, is chromatin-bound and generates RanGTP inside the nucleus. In contrast, the RanGTPase-activating protein RanGAP1, which stimulates the conversion of RanGTP to the GDP-bound form, is excluded from the nucleus and depletes RanGTP from the cytoplasm. This asymmetric distribution of Ran, RCC1, RanGAP1 and Ran BP results in a steep gradient of RanGTP across the nuclear envelope with a high concentration in the nucleus and a very low level in the cytoplasm (Görlich 1998; Görlich and Kutay 1999).

b) Nuclear Localization Sequences

In yeast and vertebrate cells, the NLS is well defined. The majority of imported proteins contain either a monopartite NLS composed of a short stretch of basic amino acids or a bipartite NLS consisting of two basic domains separated by a spacer region (Dingwall and Laskey 1991; Garcia-Bustos et al. 1991). In addition, the IMPβ binding domain of IMPα and the M9 domain confer nuclear import, however, the M9 domain also confers export and is recognized by transportin, an IMPβ-like protein of a different nuclear import pathway (Merkle and Nagy 1997).

Three classes of NLS have been characterized in plants (Table 1; for reviews see Hicks and Raikhel 1995a; Zentgraf and Hemleben 1996). Beside the monopartite and the bipartite NLS found in vertebrates and yeast, the MAT α2-like NLS originally identified in the yeast mating-type factor MAT α2 is also functional in plants (Hicks et al. 1995; Merkle and Nagy 1997). Radiolabeled NLS-peptides or analyses of the localization of reporter protein fusions characterized the function of NLS motifs in plants (Hicks and Raikhel 1995a; Hicks et al. 1995). As the NLSs from animal and yeast are also functional in plants and vice versa, it can be expected that the import apparatus is highly conserved among eukaryotic organisms.

Table 1. Examples for nuclear localization signals (NLS) in plants

Class	NLS	Protein	Organism
Bipartite	**KKRAR**LVRNRESAQLS**RQRKK**	TGA-1B[a]	*Nicotiana tabacum*
SV40-like	MSE**RKRREK**L	R (NLS-M)[b]	*Zea mays*
Mat α2-like	MISEAL**RKAIGKR**	R (NLS-C)[b]	*Zea mays*

The one-letter amino acid code is used, the amino acids important for nuclear import are printed in bold
[a]van der Krol and Chua (1991)
[b]Shieh et al. (1993).

c) In Vitro Import Systems in Plants

The first reported in vitro import system for plants was an antibody co-translocation assay developed for evacuolated permeabilized protoplasts of parsley. After permeabilization of the cells, cytosolic proteins are allowed to leak out whereas large molecules such as antibodies can enter the cells. In the in vitro import assays, antibodies against proteins that are expected to be transported into the nucleus can be added to the permeabilized cells. After complex formation between antigen and antibody, import can be measured by protease protection assay of the imported (now intranuclear) antibody by immunoblot analysis (Harter et al. 1994). Using this assay, the light-dependent phosphorylation and ATP-dependent translocation of a G-box binding transcription factor was demonstrated. However, this method was indirect, so recently, in vitro import systems directly measuring NLS mediated protein import were developed for evacuolated permeabilized tobacco protoplasts (Hicks et al. 1996; Merkle et al. 1996). In this system, fluorescently-labeled NLS substrates like GFP-fusion proteins were added to the cells which had been permeabilized either by adding low amounts of Triton X-100 (Merkle et al. 1996) or by mild osmotic shock treatment (Hicks et al. 1996), and import was measured by fluorescence microscopy. The in vitro nuclear import in plants was found to be rapid and NLS-dependent. In consistence with animals cells, competition assays revealed that the nuclear import in plant cells is also receptor-mediated. In addition, nuclear uptake is inhibited by non-hydrolyzable guanosine nucleotide analogues also indicating the requirement for GTP and GTP hydrolysis. In contrast to yeast and animal cells, no addition of cytosolic fractions is required for protein import in plant cells, all import factors appear to be retained in the permeabilized protoplasts (Hicks et al. 1996; Merkle et al. 1996; Smith et al. 1997). The addition of crude cytosolic fractions isolated from plant cells even inhibits NLS protein import and binding to purified nuclei. This inhibition is caused by the low-*Mr* fraction of the cytosol. Obviously, two low-*Mr* proteins, p10 and the viral protein R (Vpr), are regulators of nuclear import in vertebrates and yeast (Smith and Raihkel 1999).

As plants are sessile and cannot escape from their environment, they have evolved mechanisms to adapt to environmental stress e.g. low temperature, light or heat. In contrast to the animal in vitro import system, plant in vitro import can also occur at 4 °C which might, in vivo, be important for low temperature growth. Interestingly, import into chloroplasts and mitochondria also occurs in the cold (Knorpp et al. 1994; Leheny and Theg 1994). Another difference between plant and animal import systems is that animal nuclear import can be completely blocked by wheat germ agglutinin whereas the plant in vitro system is not inhibited. However, vertebrate NPC proteins are modified with a single O-linked GlcNAc residue, whereas plant NCP proteins show complex sugar modifications (see Fig. 2; Heese-Peck et al. 1995). Therefore there seem to be plant-specific import mechanisms, besides the common features of NLS-dependent, receptor-mediated GTP and GTP-hydrolysis-dependent import mechanisms. (for review see Smith and Raikhel 1999).

d) Characterization of Plant Import Factors

Isolation and characterization of Ran, importins and other proteins essential for import in animal and yeast was achieved by the re-addition of cytosolic fractions to the in vitro import systems. As the plant in vitro systems developed to date are not dependent on the addition of cytoplasmic fractions, they cannot be used for the isolation of factors required for protein import. Nevertheless, several plant cDNAs have been isolated that encode proteins with significant homology to some of the well characterized import factors from vertebrates and yeast (see Fig. 1). In *Arabidopsis thaliana,* At-IMPα is encoded by a small gene family (Hicks et al. 1996; Smith et al. 1997; Schledz et al. 1998); it recognizes three classes of NLS and localizes to the cytoplasm and nucleus, but also to the nuclear envelope in a fashion more similar to IMPβ in mammalian cells (Hicks et al. 1996; Smith et al. 1997). At-IMPα binds to the NLS with a high affinity even in the absence of IMPβ and was able to accumulate nuclear proteins in the in vitro system to levels comparable with those accumulated by mouse IMPα/β heterodimers (Hübner et al. 1999). Thus, At-IMPα shows unique properties in being able to fulfil both NLS recognition and nuclear import and might, therefore, represent a new type of import receptor. In contrast, IMPα of rice recognizes only the mono- and bipartite NLSs and is able to cooperate with mouse IMPβ to mediate protein import in the vertebrate system (Jiang et al. 1998a). In addition, two rice IMPβ homologues have been isolated and characterized to be functional in vitro (Jiang et al. 1998b) indicating that the conserved α/β heterodimer pathway also exists in plants. In rice leaves, importin α transcription is down-regulated by light (Shoji et al. 1998). This might be due to the need for an accelerated import of nuclear proteins during dark periods.

Ran homologues have been characterized for several plants including *Arabidopsis* (Haizel et al. 1997), tomato (Ach and Gruissem 1994), tobacco (Merkle et al. 1994), *Vicia faba* (Saalbach and Christov 1994), and *Lotus japonicus* (Borg et al. 1997). The *Schizosaccharomyces pombe* Ran mutant *pim1* was functionally complemented by overexpression of tomato or tobacco Ran cDNA indicating that the plant homologues have similar functions as the yeast Ran proteins (Ach and Gruissem 1994; Merkle et al. 1994). This is also supported by the inhibitory effect of the non-hydrolyzable guanine nucleotide analogues in the in vitro import systems. Using the yeast two-hybrid system, two interacting partners for the Arabidopsis Ran homologue AtRan1 have been characterized, AtRanBP1a and AtRanBP1b. The two plant RanBP1s are very similar and contain the conserved "Ran binding domain"; both show the similar ubiquitous expression pattern as the Arabidopsis Ran homologues (Haizel et al. 1997). However, their actual role and their localization still have to be elucidated.

Double-labeling immunofluorescence studies with the confocal laser-scanning microscope revealed that most cytoplasmic IMPα coaligned with microtubules and microfilaments in tobacco protoplasts (Smith and Raihkel 1998). Depolymerization of the cytoskeleton disrupted the cytoskeleton-like pattern of IMPα. In vitro analyses showed that IMPα interacts with the microtubules and the microfilaments in an NLS-dependent manner indicating that this interaction could be essential for protein transport from the cytoplasm to the nucleus (Smith and Raihkel 1998). However, depolymerization of the microfilaments causes IMPα accumulation in the nucleus suggesting that the microfilaments are involved in the retaining of IMPα in the cytoplasm (Smith and Raihkel 1998). Obviously, IMPα contains highly hydrophobic *armadillo* repeats which were shown to be involved in protein-protein interactions of other proteins with the cytoskeleton (Barth et al. 1997). From these data and observations in other systems, a working model was developed (Smith and Raihkel 1999) suggesting that IMPα forms transport-competent complexes on microfilaments. In the next step, the assembled complexes are loaded onto microtubules for transport to the NPC which could be mediated by a microtubule motor protein. Thus, this transport step would precede the transport through the NPC.

4 Regulated Nuclear Import of Proteins

The import of regulatory proteins from the cytoplasm to the nucleus must be tightly controlled because regulatory proteins are required in the nucleus only at very specific developmental steps, in response to environmental signals, and in specific tissues. Therefore, the translocation process can be seen as an additional important step in the signal transduction pathway connecting intra- and extracellular stimuli with gene expression. Here, we will address plant regulatory proteins and their regulation upon nuclear import.

a) Transcription Factors

Opaque 2

The first regulatory protein found was Opaque 2 (O2) from maize. It belongs to the basic-region leucine-zipper (bZIP) protein family.

This transcription factor family consists of the bZIP domain, a DNA-binding and dimerization domain. The motif comprises an approximately 25-amino acid stretch, rich in basic residues (basic region) adjacent to a sequence with a leucine at every seventh residue in over three to six repeat units (leucine zipper; Landschulz et al. 1988). The amino acid sequence of the basic region specifies the target nucleotide sequence and the leucine zipper, folded in the α-helical structure, is responsible for the dimer formation (for reviews see Meshi and Iwabuchi 1995; L. Liu et al. 1999). The target of these factors are defined ACGT-containing elements (ACEs). The specificity of DNA recognition by bZIP proteins is determined by the sequences flanking the ACGT core (Foster et al. 1994).

O2 is localized in the nucleus in maize endosperm tissue and in transformed tobacco plants (Varagona et al. 1991), and it has two NLSs (Varagona et al. 1992). The first NLS (NLS A), inefficient in directing a reporter protein to the nucleus, is located at the N-terminal portion of the O2 protein between two acidic domains and has the structure of an SV40-type NLS. However, this region including another 13 amino acids may have a bipartite structure which was found to import better than the previously identified NLS. The more efficient NLS (NLS B) is located in the basic DNA-binding domain and has a bipartite structure (Varagona et al. 1992; Varagona and Raikhel 1994). Further analysis in transiently transformed onion epidermal cells revealed that both NLSs are necessary for efficient nuclear targeting of O2 fused to GUS as a reporter (Varagona and Raikhel 1994). Protein cross-linking experiments showed that NLS binding proteins which are located to the NPC are specifically associated with the NLS B (Hicks and Raikhel 1995a). Although the NLS B is located in the DNA-binding domain, it was possible to show, in the maize mutant o2-676, that the nuclear targeting function of this domain is independent of DNA binding (Varagona and Raikhel 1994).

Homeotic gene products

The homeotic APETALA3 (AP3) and PISTILLATA (PI) proteins are required to specify petal and stamen identities in the flower of Arabidopsis. Although both proteins contain multiple NLS consensus sequences, transient transformation assays of either AP3-GUS or PI-GUS fusion proteins demonstrated that neither AP3 nor PI fusion protein can be localized in the nucleus independently. However, co-expression resulted in nuclear localization of GUS activity (McGonigle et al. 1996). The N-terminal region of both proteins is responsible for the nuclear localization. It has been suggested that AP3 and PI form hetero-dimeric complexes which generate or expose a colocalization signal required for nuclear cotranslocation.

Pleiotropic regulatory locus 1 protein

Pleiotropic regulatory locus 1 (PRL1) encodes a regulatory protein containing seven tryptophan-aspartic acid WD-40 repeats that share homology with many WD proteins which perform different functions in eukaryotes (for reviews see Neer et al. 1994; Smith et al. 1999). The WD-40 repeats of PRL1 are distinct from those of other WD proteins indicating that PRL1 represents a novel class of regulatory factors. It functions as a pleiotropic regulator of glucose and hormone response in Arabidopsis (for review see Smeekens 2000). PRL1 is imported into the nucleus, but is also detected in association with membrane fractions consisting of fragments of nuclear envelope and endoplasmic reticulum. In a yeast two-hybrid screen, it could be demonstrated that PRL1 interacts with an Arabidopsis IMPα nuclear import receptor (Nemeth et al. 1998). A putative SV40-type NLS was mapped to the C-terminal part of PRL1.

Abscisic acid-responsive protein

The maize abscisic acid-responsive protein (RAB17) is located in the nucleus and in the cytoplasm of embryo cells. It is phosphorylated by casein kinase 2 (CK2) in vitro. Qualitative differences in the phosphorylation states of the protein can be found between the two subcellular compartments. The highest phosphorylated forms of RAB17 were more abundant in the cytoplasm than in the nucleus. In the phosphorylated form, RAB17 is able to interact with synthetic NLS peptides (Goday et al. 1994). Since RAB17 has similar domains of a mammalian NLS binding phosphoprotein, it is suggested that it plays a role in nuclear protein transport during stress conditions. Translational fusions of RAB17 to the GUS reporter protein demonstrated that a region containing the putative consensus site for CK2 phosphorylation is necessary for nuclear import of the fusion protein (Jensen et al. 1998). These results suggest that phosphorylation of RAB17 by CK2 is an important step for its nuclear localization, either by facilitating binding to specific proteins or as a direct part of the nuclear targeting apparatus.

Heat shock factor

The expression of heat shock (HS) genes is induced by heat stress and also by various conditions causing the accumulation of abnormal denatured proteins. The regulation of this heat shock response involves interaction between specific receptor elements in the promoter of HS genes, heat shock elements (HSEs), and the corresponding transcription factor (HSF); (for reviews see Becker and Craig 1994; Wu 1995; Schöffl et al. 1999). HSFs are characterized by an N-terminal DNA-binding domain with a central helix-turn-helix motif, an adjacent domain with heptad hydrophobic repeats involved in oligomerization, and a C-terminal activator domain (for reviews see Scharf et al. 1994; Nover and Scharf 1997). In addition to the oligomerization of the HSFs, nuclear import is an essential part of the stress-induced activation of HSFs. Almost all HSFs contain two basic amino acid residues (K/R motifs) which are potential NLS. Formally, both (K/R1 and K/R2) might be considered as bipartite NLS. In tomato three HSFs (HSFA1, HSFA2, and HSFB1) are known. Mutations of the NLS of HSFA1 and HSFA2 demonstrated that only one of the two motifs (K/R2) is functional as NLS (Lyck et al. 1997), whereas mutations in the K/R1 region affected DNA binding but do not interfere with nuclear import. Although K/R2 of HSFB1 differs in sequence, it is also important for nuclear import.

In transient expression assays using tobacco protoplasts, HSFA1 appeared to be distributed between cytoplasm and nucleus under control conditions, whereas during HS, an exclusive nuclear staining with antibodies generated against HSFA1 was found. In the recovery phase, a considerable part of HSFA1 returned to the cytoplasm. Interestingly, HSFA2 stayed in the cytoplasm under all conditions, whereas a C-terminal truncated version was constitutively in the nucleus (Lyck et al. 1997). Co-expression of HSFA2 with HSFA1 in tobacco protoplasts demonstrates that HSFA1 facilitates the nuclear import of

HSFA2 (Scharf et al. 1998). Obviously, the intramolecular shielding of the NLS in the wild-type HSFA2 can be released by interaction with the constitutively expressed HSFA1 which reflects the normal situation in tomato cells. The presence of HSFB1 did not alter the intracellular distribution of HSFA2, although HSFB1 is constitutively found in the nucleus. Co-immunoprecipitation and yeast two-hybrid tests demonstrated direct physical interaction of HSFA1 with HSFA2 forming relatively stabile hetero-oligomers. It is speculated that in the cytoplasm, HSFA2 is in a form with a shielded NLS motif, whereas the transport-competent form may represent hetero-oligomers with one or two subunits of HSFA1.

b) Protein Import During Light Regulation

Environmental variations, such as temperature, nutrition, and light, are monitored by plants with several sensory systems. Light is one of the most import factors affecting plant development. To monitor changes in the temporal and spatial pattern of the light environment, plants have evolved different classes of photoreceptors: red/far-red light photoreversible phytochromes, UV-A photoreceptors (cryptocromes, CRYs, and phototropin), and UV-B photoreceptors (for reviews see Kendrick and Kronenberg 1994; Ballare 1999; Briggs and Huala 1999; Cashmore et al. 1999). Phytochromes (PHY) are the best characterized photoreceptors. They are soluble chromoproteins which exist as dimers covalently linked to their chromophore (Furuya and Song 1994). Biologically-active phytochrome exists in two different photoconvertible forms, the red-light-absorbing (inactive) form (Pr) and the far-red-light-absorbing (active) form (Pfr). Five genes encoding phytochome proteins (PHYA to PHYE) have been described in *Arabidopsis thaliana* (Mathews and Sharrock 1997). PHYA is the sensor for very low fluence responses (VLFR) and for absorption of continuous far-red light, whereas PHYB is responsible for the photoperception of red light and functions as a classical red/far-red light reversible molecular switch (Furuya and Schäfer 1996). Purified oat PHYA shows autophosphorylation, and also phosphorylates other proteins in vivo. The autophosphorylation is slightly higher in the Pfr than in the Pr form (Fankhauser et al. 1999). Photosignal perception by the receptor is followed by conformational changes and activates an, as yet, almost unknown pathway leading to changes in the expression of light-regulated genes (Quail et al. 1995). Important known components involved in the regulation of photomorphogenesis in higher plants are illustrated in Fig. 3 (for reviews see Nagy and Schäfer 1999; Nagy and Schäfer 2000).

Interestingly, light-driven entrainment of the plant circadian clock is mediated by phytochromes and CRYs (Somers et al. 1998; for review see Cashmore et al. 1999). Moreover, expression of the PHYB gene is regulated by the circadian clock (Kozma-Bognar et al. 1999). Recently, GIGANTEA (GI), a nuclear protein involved in PHYB signaling, has been identified in Arabidopsis (Huq et al. 2000). This is consistent with the proposition that GI is involved in controlling light signaling to the circadian clock (Park et al. 1999). This protein has an atypical bipartite NLS, comprising four typical basic clusters where the distance between pairs of these clusters is much longer in comparison to typical bipartite NLS (Huq et al. 2000).

Regulation of photomorphogenesis and physiological responses are achieved by light quality-dependent nuclear transport of photoreceptors PHYA and PHYB. They both contain multiple NLS-like motifs in the C-terminal region. Fusion proteins containing the C-terminal part of PHYB and GUS as a reporter are localized in the nucleus of transgenic Arabidopsis (Sakamoto and Nagatani 1996). Moreover, nuclei prepared from light-grown Arabidopsis seedlings contained higher amounts of immunodetectable PHYB than nuclei from dark-adapted seedlings. Full-length PHYB-GFP fusion protein was translocated to the nucleus upon red light. It could be demonstrated that PHYB-GFP is functionally active by using PHYB-GFP to complement an Arabidopsis PHYB-deficient mutant in vivo (Yamaguchi et al. 1999). Kircher and coworkers (1999a) confirmed this latter observation and showed that PHYA-GFP and PHYB-GFP fusion proteins are photobiologically active and imported into the nuclei of transgenic tobacco in a light quality- and quantity-dependent manner. Nuclear import of Arabidopsis PHYB and tobacco PHYB-GFP displayed features of R/FR-reversible low fluence response (LFR). In contrast, nuclear transport of rice PHYA-GFP is mediated by VLFR in tobacco, whereas nuclear translocation of Arabidopsis PHYA-GFP is regulated by high irradiation response (HIR) and VLFR in transgenic Arabidopsis (Kim et al. 2000). This reflects the fluence requirements previously found for PHYA- and PHYB-mediated physiological responses. Gil and coworkers (2000) could demonstrate that the inhibition of hypocotyl elongation of tobacco seedlings induced by short red light-correlates with enhanced nuclear import of PHYB-GFP. The functionality of this fusion protein was confirmed by complementing the phenotype of the PHYB-deficient *Nicotiana plumbaginifolia hlg2* mutant. It has even been demonstrated that PHYB binds reversibly to G-box-bound transcription factor PIF3 (phytochrome-interacting factor 3) specifically in the Pfr form (Martinez-Garcia et al. 2000). Since a chromophore-deficient form of PHYB-GFP remains in the cytosol, only functional photoreceptors seem to obtain nuclear import competence (Kircher et al. 1999a). It is being discussed that in the Pr form, the NLS of the phytochromes is masked intramolecularly and/or bound to a hypothetical retention factor thereby abolishing nuclear uptake. Noteworthy, light-driven accumulation of PHYB-GFP and PHYA fusion proteins was accompanied by spot/speckle formation (Kircher et al. 1999a; Yamaguchi et al. 1999; Gil et al. 2000). These are rounded structures evenly distributed throughout the nucleus, but their biological relevance remains unclear.

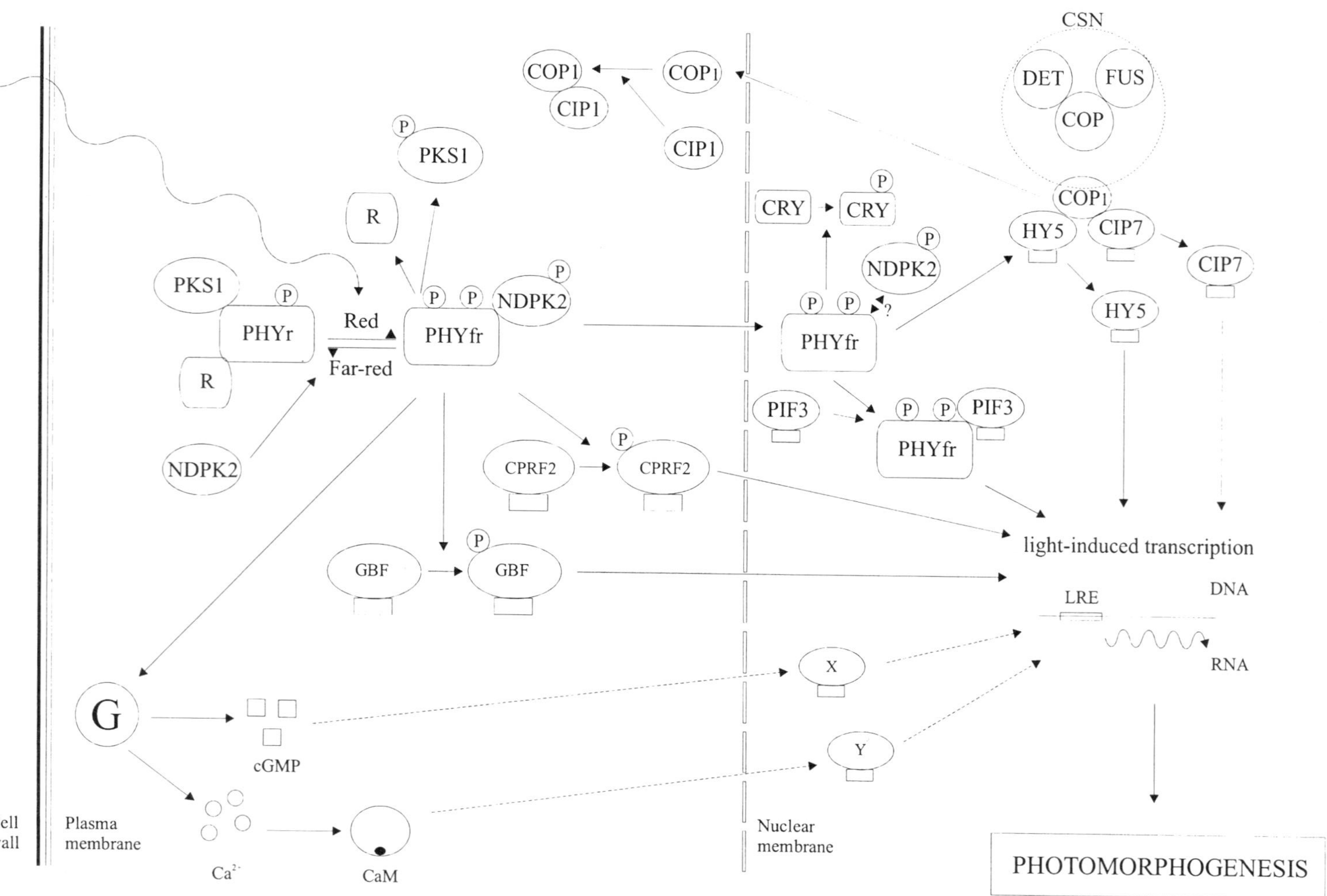
Light
CSN
DET
FUS
COP
COP1
CIP1
PKS1
P
R
CRY
HY5
CIP7
NDPK2
PHYr
Red
Far-red
PHYfr
?
PIF3
CPRF2
GBF
light-induced transcription
LRE
DNA
RNA
X
Y
G
cGMP
Cell
wall
Plasma
membrane
Ca2+
CaM
Nuclear
membrane
PHOTOMORPHOGENESIS

There are new insights into the nuclear import of the other photoreceptors. CRY1-GFP fusion protein localizes to the nucleus if transiently expressed in onion epidermal cells (see Fig. 3; Cashmore et al. 1999). Two independent groups found that CRY2 bears a basic bipartite NLS in the C-terminal domain (Guo et al. 1999; Kleiner et al. 1999). Immunoblot analysis and expression of fusion proteins with reporter genes demonstrated that CRY2 is a nuclear located protein. There is no obvious light regulation for the nuclear compartmentation (Guo et al. 1999). Interestingly, *Arabidopsis* CRY1 and CRY2 interact in vitro with PHYA and are phosphorylated by PHYA-associated kinase activity (Ahmad et al. 1998).

In addition to the photoreceptors there are other components involved in photomorphogenesis. The genes encoding these factors belong to the constitutive photomorphogenetic/de-etiolated/fusca (COP/DET/FUS) group and are thought to be members of large multiprotein complexes that repress directly, or indirectly, expression of light-inducible genes (Misera et al. 1994; Kwok et al. 1995). These are therefore negative regulators of photomorphogenesis or activators of skotomorphogenesis in darkness. COP1 plays an essential role in this repression (for reviews see Terzaghi and Cashmore 1995b; Holm and Deng 1999; Deshaies and Meyerowitz 2000). The nucleo-cytoplasmic partitioning of COP1 and that of PHYA and PHYB appears to be drastically different. COP1, when fused to GUS, is localized in the nucleus in darkness and disappears after light treatment, from hypocotyl cells of *Arabidopsis* seedlings and epidermal cells of onion bulbs (von Arnim and Deng 1994; von Arnim et al. 1997). Changes in the nuclear GUS-COP1 level in response to dark-light transition, quantitatively correlate with differential effects of light expo-

---

Fig. 3. Protein import during light regulation in higher plants. Light absorption (red light) changes the conformation of phytochromes (PHYr to PHYfr). Light-induced conformation of PHY is accompanied by autophosphorylation of the photoreceptors and phosphorylation of phytochrome kinase substrate 1 (PKS1), nucleoside diphosphate kinase 2 (NDPK2; Choi et al. 1999; Ogura et al. 1999), and other proteins such as the transcription factors CPRF2 and CBF (*P* phosphorylated residues). Phosphorylation abolishes cytosolic retention (*R* putative retention factor) of the photoreceptors and transcription factors and triggers their translocation to the nucleus. The nuclear localized CRYs are phosphorylated. GBFs interact with promoters of light-regulated genes. PHYfr leads, by an as yet unknown mechanism, to the activation of a trimeric GTP-binding protein (*G*) that is followed by altered levels of cyclic GMP (cGMP) and calmodulin/Ca^{2+} (CaM). Modulation of these second messengers then results in activation of putative transcription factors (*X* and *Y*) which are possible components of transcription complexes responsible for the expression of light-induced genes. Phytochrome imported into the nucleus in the Pfr form leads to the disassembly of the CSN: COP1 relocates to the cytoplasm where it binds to CIP1 associated to the cytoskeleton. Subsequently, transcription factors (such as long hypocotyl 5, HY5, and CIP7) are recruited and together with PIF3, associated with PHYfr, induce transcription of light-regulated genes in the form of active transcription complexes interacting with light-regulating *cis*-acting elements (LRE) of the target genes (for reviews see Whitelam and Halliday 1999; Nagatani 2000). *Dashed lines* and *question marks* indicate hypothetical and unknown steps, respectively. (Modified after Nagy and Schäfer 2000)

sure on Arabidopsis hypocotyl development. In contrast, the GUS-COP1 is constitutively nuclear in root cells, which is consistent with the role of COP1 in suppressing chloroplast development in Arabidopsis roots under light conditions (Deng and Quail 1992). These results suggest that its nucleoplasmic partitioning is regulated by light in a cell type-specific manner (von Arnim and Deng 1994). In addition, the activity of COP1 depends on the integrity of the NLS (Stacey et al. 2000). Interestingly, the subcellular localization of a mammalian homologue that contains all structural features present in Arabidopsis COP1 (AtCOP1) can be regulated in plant cells by light in a fashion similar to that of AtCOP1, and the expression of the N-terminal part of mammalian COP1 in Arabidopsis resulted in a hyperphotomorphogenic phenotype (Wang et al. 1999). Nuclear accumulation of COP1 in the dark depends on several genes, including DET1, COP8, 9, and 10, and FUS4, 5, 6, 11, and 12 (von Arnim et al. 1997). Most of these genes encode subunits of a nuclear protein complex called the COP9 complex or signalosome, which has now been renamed COP9 signalosome (CSN; Deng et al. 2000).

Redistribution of GUS-COP1 protein to the cytosol by light is mediated by multiple photoreceptors of the phytochrome and cryptochrome families in a complementary and cooperative way depending on the wavelength of light (Osterlund and Deng 1998). In vitro, COP1 interacts with COP1-interactive protein 1 (CIP1) which is associated with the cytoskeleton in hypocotyl and cotyledon cells but not in root cells. Therefore, CIP1 may act as a cell type-specific cytoplasmic sequester for COP1 and may be involved in light control of COP1 activity by affecting its nucleocytoplasmic partitioning (Matsui et al. 1995). GUS and GFP fusion protein experiments showed that the COP1 protein contains a cytoplasmic localization signal (CLS) motif that mediates the cytoplasmic localization and a classical bipartite NLS. Both domains act constitutively if tested independently. The light regulation of COP1 subcellular localization is achieved by combining them (Stacey et al. 1999). The NLS activity of COP1 is not regulated by phosphorylation of the phosphorylation sites adjacent to the NLS, but it is still possible that phosphorylation at other residues modulate NLS activity (Stacey et al. 2000). It has also been demonstrated that a C-terminal fragment of COP1, including a WD-40 repeat, is constitutively localized in the nucleus but not required for the light-regulated partitioning of COP1. This domain is involved in the repression of photomorphogenesis in Arabidopsis (McNellis et al. 1994; Stacey et al. 1999). More detailed deletion analysis of COP1 revealed that 58 amino acids overlapping the CLS and a putative α-helical coiled-coil domain that has been implicated in COP1 dimerization are responsible for the targeting of COP1 to subnuclear foci (speckles). This domain is referred to as a subnuclear localization signal (SNLS; Stacey and von Arnim 1999). In addition, the integrity of the WD-40 domain is important for localization to the foci, although alone, it is insufficient for targeting to nuclear foci. The observed correlation between loss of nuclear foci and loss of function after mutation of the WD-40 repeat is certainly consistent with a putative functional role of the foci (Stacey and von Arnim 1999; Stacey et al. 1999). Interestingly, the localization of COP1 to nuclear foci is reminiscent of those seen for phytochromes, an intriguing result given that COP1 nuclear localization is negatively regulated by phytochromes.

The last group of proteins involved in this process are G-box-binding transcription factors (GBFs). They belong to the plant bZIP proteins.

Various members of the G-box binding transcription factor family do interact specifically with promoters of light-regulated genes (for reviews see Terzaghi and Cashmore 1995a; Fankhauser and Chory 1997). G-box (CCACGTGG) binding activity has been localized not only in the nucleus but also in the cytosol of parsley protoplasts, although the specific DNA binding activity was much higher in the nuclear compartment. The DNA binding activity of the cytosolic G-box binding factors is modulated by light and phosphorylation/dephosphorylation activities. This correlates with the investigated cotranslocation of an antibody against GBF1 from Arabidopsis to the nucleus, upon white light, in an in vitro system indicating that the light-modulated and phosphorylation-dependent changes might lead to an enhanced nuclear translocation competence of the cytosolic GBFs (Harter et al. 1994). Similarly, immunoblot analysis of nuclear and cytoplasmic fractions from Arabidopsis and soybean cell cultures showed that over 90% of the detected proteins are in the cytoplasm. The same proportion of the G-box binding affinity was cytoplasmic.

Histochemical localization and cellular fractions of GUS fusion proteins expressed in soybean protoplasts have identified differences in the extent to which individual members of GBF show cytoplasmic localization. GBF2 is predominantly cytoplasmic only in dark-grown cells, in cells cultured under blue light it is mostly nuclear-localized. Red light has no effect which suggests that this import is specifically mediated by a blue light receptor. GBF4 always localizes to the nucleus. In contrast, GBF1 remained in the cytosol even under blue light conditions. Deletion of amino acids 112–164 of GBF1 resulted in enhanced nuclear localization, which suggests that this region is required for cytoplasmic retention of GBF1. Interestingly, a putative CK2 site lies in this region (Terzaghi et al. 1997). Since GBF1 has been shown to be a substrate for CK2 (Klimczak et al. 1992, 1995) phosphorylation of this site might also be involved in cytoplasmic retention. UV light also leads to an import of GBF2 but not of GBF1 from Arabidopsis into the nucleus of evacuolated parsley protoplasts (Kircher et al. 1998).

Another group of ACE-binding proteins isolated by South-western screening are the Common Plant Regulatory Proteins (CPRFs) in parsley (Weisshaar et al. 1991). In evacuolated protoplasts, recombinant full-length CPRF1 and CPRF2 were imported into the nucleus after UV irradiation. In contrast, full-length CPRF4a remained in the cytosol, while a CPRF4a fragment was detectable in the nuclear fraction. CPRF1 and CPRF2 may be retained in protein complexes in the cytosol until a light stimulus selectively releases specific factors that are then imported into the nucleus (Kircher et al. 1998). Experiments with CPRF-specific antisera in electrophoretic mobility supershift assays indicate that the three members of the CPRF family in parsley exhibit a differential intracellular distribution in dark-kept cells with CPRF1 exclusively localized in the nucleus, CPRF4 found in both compartments, and CPRF2 is retained in the cytosol, but translocated into the nucleus in response to a 30-min UV-containing white light irradiation. In addition, immunolocalization assays combined with confocal microscopy showed that CPRF2 was actively moving from the cytosol into the nucleus. The efficiency of the nuclear translocation of CPRF2 seems to be dependent on the light quality, which indicates that PHYA and PHYB are the main receptors involved in the nuclear translocation process. Two N-terminal domains were shown to be responsible for the cytosolic retention of CPRF2 in the darkness (Kircher et al. 1999b). The retention domain number 1 has strong homology to the α-helical cytoplasmic retention domain of the mammalian heat shock factor 2 (Sheldon and Kingston 1993). The second domain is acidic and contains a CK2 phosphorylation site. Interestingly, CPRF2 is rapidly phosphorylated in response to light, and since it correlates well with the phytochrome-mediated import of CPRF2 from the cytoplasm to the nucleus, it might be involved in the regulation of nuclear import (Wellmer et al. 1999).

Interestingly, both transcription factors involved in photomorphogenesis have putative CK2 phosphorylation domains, which seem to be necessary for their phytochrome-mediated import from the cytoplasm to the nucleus. CK2 is a serine/threonine protein kinase that is present in all eukaryotic cells examined to date. Subcellular localization studies of CK2 were carried out in maize embryo sections using anti-CK2 antibodies (Peracchia et al. 1999). CK2 protein could be detected in the cytoplasm and in the nucleus of several cell types. Further analysis, using CK2 deletions fused to GUS transiently expressed in onion epidermal cells, identified one SV40-type NLS. This region is highly conserved in other CK2 proteins.

Light-mediated changes in the different regulatory proteins suggest the involvement of IMPα. It has been found that the transcription of IMPα from rice is down-regulated by light in leaves. Thus, it is suggested that more IMP α genes might be necessary to perform both light-regulated accumulation and decrease of the nuclear proteins (Shoji et al. 1998).

c) Nuclear Protein Import During Pathogen/Plant Interactions

During their life span, plants are continuously affected by microorganisms, like bacteria or fungi – either pathogenic or symbiotic –, by virus infection or by attack from higher organisms feeding on them. Various sophisticated morphological and/or molecular mechanisms were developed against pathogenic or parasitic organisms which can induce a defense reaction of the plant. Signal transduction cascades through the cell, which subsequently lead to a quick answer (cell wall strengthening, oxidative burst etc.) and end in local and/or systemic synthesis of defense-associated proteins were characterized. These processes are accompanied by the import of components into the nucleus. In some cases either viral or bacterial DNA are transported mediated by proteins carrying NLS motifs, whereas viroid RNA molecules are suggested to be imported into the plant cell nucleus without any protein function (Woo et al. 1999).

During infection of plant DNA viruses which are replicated within the cell nucleus, an uptake step into the host nucleus is essential. Maize streak virus (MSV), a member of the family Geminiviridae (genus *Mastrevirus*), has a circular, single-stranded (ss) monopartite DNA genome (Lasarowitz 1988) and replicates in the cell nucleus via a double-stranded (ds) intermediate which also acts as a template for bi-directional transcription (Wright et al. 1997). Two of the four encoded viral proteins, the movement protein (MP) and the coat protein (CP), are required for systemic infection of the virus and disease development (Lasarowitz et al. 1989). Since CP is necessary for virus assembly and virus particles are present in the nucleus, it could be assumed that CP must enter the plant nucleus, and actually a bipartite NLS-like sequence is present at the N-terminal region of the MSV CP (H. Liu et al. 1999). Recombinant CP co-microinjected with single- or double-stranded virus DNA into maize or tobacco epidermal cells were accumulated in the nucleus indicating that MP facilitates the rapid transport of viral DNA into the nucleus (Liu et al. 1999). This is in agreement with the earlier finding for the BV1 protein, of some geminiviruses with a bipartite genome (species of the genus *Begomovirus*), which is able to bind single-stranded DNA and is accumulated in the nucleus. Here, the nuclear accumulation is supported by two NLS motifs located within the 113 N terminal amino acids of the BV1 protein (Pascal et al. 1994; Sanderfoot et al. 1996). CPs of Begomoviruses have been shown to contain several NLS sequences (Kunik et al. 1998; Qin et al. 1998; Unseld et al. 2001). In the case of African cassava mosaic virus (ACMV) GFP fusion experiments identified a tripartite NLS at the N-terminus and another NLS at the C-terminus of the CP. The central domain facilitates both, nuclear import and export of the fusion protein (Unseld et al. 2001).

Viroids are formed by circular rod-shaped RNA molecules which specifically infect plant cells and are replicated within the cell nucleus (recent reviews: Diener 1995, 1999). Therefore, after infection the RNA molecules have to be translocated into the nucleus. The process was investigated by following the import of infectious, fluorescein-labeled potato spindle tuber viroid RNA in permeabilized tobacco protoplasts (Woo et al. 1999). Import was not inhibited by GTP-gamma-S or GTP-ß-S, known inhibitors of the Ran-GTPase-coupled cycle of the import machinery (see Fig. 1). The results obtained indicate a sequence or structural transport motif of the viroid RNA and exclude a cytoskeleton-dependent uptake by a specific receptor on the nuclear membrane.

The plant parasitic bacterium known for its use in gene transfer, *Agrobacterium tumefaciens*, naturally induces the formation of callus cells or tumors (crown gall) by transferring a specific part of the Ti-plasmid, the T-DNA, through the plant cell into the nucleus where it is integrated into the genome, mostly of dicotyledonous plants. Transfer is initiated and mediated by bacterial virulence proteins encoded in the Ti-plasmid at the *Vir* region and produced within the bacteria and regulated by small phenolic compounds excreted by the plant (for review see Zupan et al.

2000). For the nicking and cutting processes of the T-DNA, VirD1 is necessary. The transfer intermediate assumed to be formed by the single stranded T-DNA/VirD2 protein complex is transported into the nucleus with the help of the VirE2 protein (Zupan et al. 1996). Interestingly, the protein components of this complex supporting or facilitating the import actually contain a bipartite NLS. The behavior of the two karyophilic proteins VirD2 and VirE2 were also tested in animal cells: VirD2 was targeted into the nucleus of Xenopus oocytes and *Drosophila* (Guralnick et al. 1996) or human cells (Relic et al. 1998), whereas VirE2 was only directed to the animal nucleus if one amino acid residue within the NLS was modified (Guralnick et al. 1996). The latter observation suggests a slightly different import machinery of animal and plant cells.

Other plant/bacteria pathogenic interactions are accompanied by specific nuclear import during the hypersensitive response (HR; Wengelnik et al. 1999, and references therein). Most effective defense reactions in plants are often mediated by specific resistance (R) gene products that are able to detect specific pathogen races through recognition of avirulence (Avr) proteins, in a gene-to-gene manner (reviewed in Feys and Parker 2000). Avirulence genes were named for the elicitation of plant disease resistance by the pathogen harboring the genes. For plant/bacteria interactions, however, it was found that the main function of the Avr proteins from the perspective of the bacterium is their role in virulence. Therefore, many of the avirulence factors are dual acting proteins (reviewed in White et al. 2000); they elicit resistance in one context and are involved in the virulence of the bacterium on susceptible host plants.

As a model system for plant/bacteria interactions, the Gram-negative bacterium *Xanthomonas campestris* pv. *vesicatoria* with its host plants pepper (*Capsicum annuum*) and tomato (*Lycopersicon esculentum*), has been extensively studied (Bonas et al. 2000). Stimulated by exogenous factors this interaction is controlled in *X. campestris* pv. *vesicatoria* by proteins encoded in the *hrp* (hypersensitive response and pathogenicity) gene region, some of which are predicted to encode components of a type III protein secretion (Sec) pathway. Similar to the Vir region in *A. tumefaciens*, the *hrp* operon is regulated by two gene products, HrpX and HrpG. HrpX is a protein of the AraC family and activates transcription of the operons *hrp*B to *hrp*F (Wengelnik and Bonas 1996). Expression of *hrp*X and *hrp*A depends on HrpG which is assumed to act as a response regulator protein of two-component signaling pathways (Wengelnik et al. 1996); mutations in *hrp*G result in constitutive expression of all *hrp* genes (Wengelnik et al. 1999). The putative regulator protein HrpG shows high similarity to Vir G in *A. tumefaciens*, which is the key molecule of a two-component signaling system activating the Vir region and therewith initiating and supporting T-DNA transport (reviewed in Zupan et al. 2000).

Since some of the Hrp proteins are components of the protein secretory type III pathway (Mudgett et al. 2000), it was suggested that defense reactions inducing Avr proteins encoded by *avirulence* (*avr*) genes are transported into the plant cell mediating the HR response and stimulating the expression of defense molecules. For the *X. campestris* pv. *vesicatoria*/plant interaction the avirulence factor AvrBs3 appears to be transported into the plant cell *via* the secretory type III pathway. AvrBs3 exhibits an interesting structure: At the N terminus three NLS are present, and an internal repeated region (17.5×34 aa) is characteristic for this protein. It is assumed that after transport into the cytoplasm, interactions with the resistance gene product Bs3, found only in resistant plants (Bonas 2000), occur, and after import into the plant nucleus or mediated by a signal cascade induced by this interaction, plant HR and production of defense proteins is stimulated (Van den Ackervelden et al 1996). Eukaryotic protein signatures were found in AvrBs3 that are required for its function in the host cell: The AvrBs3 C-terminal region contains an acidic transcription activation domain (AD) that is essential for HR induction and that can be functionally replaced by a heterologous sequence from *Herpes simplex* virus. As an indication for nuclear import, a yeast two-hybrid screening allowed the isolation of two pepper IMPα proteins which interact with an NLS in the AvrBs3 C-terminus suggesting that AvrBs3 is translocated into the nucleus and operates directly at the level of activation of host-gene expression (Bonas and coworkers, in prep.).

In the interaction of the rice pathogen *Xanthomonas oryzae* pv. *oryzae* and rice (*Oryza sativa*) a similar process has been observed. The avirulence factor AvrXa7 interacting with the plant resistance gene product Xa7 is encoded by a member of the avrBs3 avirulence gene family which are found in a variety of *Xanthomonas* strains (Yang et al. 2000); similar to AvrBs3 it also contains a central repeated structure of 25.5 direct repeat units, three C-terminal NLS motifs, and at the C-terminus, an activation domain. Avirulence and virulence specificities are associated with the central repeat region supporting the dual function of these Avr proteins. Both activities were abolished by mutations in the three NLS motifs, again similar to AvrBs3, and could be restored by addition of the NLS sequence of SV40 T-antigen. Avirulence activity was also associated with the presence of an intact acidic AD at the C-terminus. Interestingly, in gel shift assays a specific binding of AvrXa7 to double-stranded DNA could be shown indicating a putative role in transcriptional activation during the plant defense response.

For plant-fungal pathogen interactions (Feys and Parker 2000) the importance of regulated import of activating proteins is less clear. In the model system for non-host resistance, a signal transduction pathway mediated by MAP kinases is involved. Here, the last step of transcriptional activation of defense response genes can be assumed to occur *via* phosphorylation of factors by a specific MAP kinase which is targeted to the nucleus (Ligterink et al. 1997).

5 Conclusions and Perspectives

The general structure and mechanisms of plant NPC and transport of proteins into the cell nucleus appear to be relatively similar to those in animal or yeast cells. Components of the plant NPC have to be further characterized for a complete understanding of similarities among the different eukaryotic kingdoms. However, there are also indications that plants show the unique feature of an IMPβ-independent translocation process. In addition, plant IMPα interacts with the cytoskeleton so the import complexes are probably transported along the microfilaments. In some cases it is known that phosphorylation is necessary for nuclear import. However, the fact that MAP kinases are found in the nucleus suggests that phosphorylation may also take place inside the nucleus. Nevertheless, the exact mechanism and the factors involved in NLS-dependent import still have to be elucidated. Interaction between the common translocation components (importins, RanGTPase etc.) and the meanwhile characterized specifically transported proteins carrying respective NLSs has often not been proved yet.

Acknowledgements. The authors thank Prof. U. Bonas for providing unpublished results and Prof. E. Schäfer for correction of Fig. 3.

References

Ach RA, Gruissem W (1994) A small GTP-binding protein from tomato suppresses a *Schizosaccharomyces pombe* cell-cycle mutant. Proc Natl Acad Sci USA 91:5863–5867

Adam SA, Marr RS, Gerace L (1990) Nuclear protein import in permeabilized mammalian cells requires soluble cytoplasmic factors. J Cell Biol 11:807–816

Ahmad M, Jarillo JA, Smirnova O, Cashmore AR (1998) The CRY1 blue light photoreceptor of *Arabidopsis* interacts with phytochrome A in vitro. Mol Cell 1:939–948

Ballare CL (1999) Keeping up with the neighbours: phytochrome sensing and other signalling mechanisms. Trends Plant Sci 4:97–102

Barth AL, Nathke IS, Nelson WJ (1997) Cadherins, catenins and APC protein: interplay between cytoskeletal complexes and signaling pathways. Curr Opin Cell Biol 9:683–690

Becker SJC, Craig EA (1994) Heat-shock proteins as molecular chaperones. Eur J Biochem 219:11–23

Bonas U (2000) High-resolution genetic mapping of the pepper resistance locus *Bs3* governing recognition of the *Xanthomonas campestris* pv. *vesicatora* AvrBs3 protein. Theor Appl Genet 101:255–263

Bonas U, Van den Ackerveken G, Büttner D, Hahn K, Marois E, Nennstiel D, Noël L, Rossier O, Szurek B (2000) How the bacterial plant pathogen *Xanthomonas campestris* pv. *vesicatora* conquers the host. Mol Plant Pathol 1:73–76

Borg S, Brandstrup B, Jenson TJ, Poulsen C (1997) Identification of new protein species amount 33 different small GTP-binding proteins encoded by cDNAs from *Lotus japonicus*, and expression of corresponding mRNAs in developing root nodules. Plant J 11:237–250

Briggs WR, Huala E (1999) Blue-light photoreceptors in higher plants. Annu Rev Cell Dev Biol 15:33–62

Cashmore AR, Jarillo JA, Wu YJ, Liu D (1999) Cryptochromes: blue light receptors for plants and animals. Science 284:760–765

Choi G, Yi H, Lee J, Kwon YK, Soh MS, Shin B, Luka Z, Hahn TR, Song PS (1999) Phytochrome signalling is mediated through nucleoside diphosphate kinase2. Nature 401:610–613

Chytilova E, Macas J, Siwilinska E, Rafelski SM, Lambert GM, Galbraith DW (2000) Nuclear dynamics in *Arabidopsis thaliana.* Mol Biol Cell 11:2733–2741

Citovsky V, Zupan J, Warnick D, Zambryski P (1992) Nuclear localization of Agrobacterium VirE2 protein in plant cells. Science 256:1802–1805

Dabauvalle MC, Benevente R, Chaly N (1988) Monoclonal antibodies to a *Mr* 68,000 pore complex glycoprotein interfere with nuclear protein uptake in *Xenopus* oocytes. Chromsoma 97:193–197

Davis LI (1995) The nuclear pore complex. Annu Rev Biochem 64:865–896

Deng XW, Quail PH (1992) Genetic and phenotypic characterization of *cop1* mutants of *Arabidopsis thaliana.* Plant J 2:83–95

Deng XW, Dubiel W, Wei N, Hofmann K, Mundt K, Colicelli J, Kato J, Naumann M, Segal D, Seeger M, Glickman M Chamovitz DA, Carr A (2000) Unified nomenclature for COP9 signalosome an its subunits: an essential regulator of development. Trends Genet 16:202–203

Deshaies RJ, Meyerowitz E (2000) COP1 patrols the night beat. Nat Cell Biol 2:E102-E104

Diener TO (1995) Viroids and the nature of viroid diseases. Arch Virol Suppl 15:203–220

Diener TO (1999) Origin and evolution of viroids and viroid- like satellite RNAs. Virus Genes 11:119–131

Dingwall C, Laskey RA (1991) Nuclear targeting sequences-a consensus? Trends Biochem Sci 16:478–481

Doye V, Hurt EC (1997) From nucleoporins to nuclear pore complexes. Curr Opin Cell Biol 9:401 411

Fankhauser C, Chory J (1997) Light control of plant development. Annu Rev Cell Dev Biol 13:203–229

Fankhauser C, Yeh KC, Lagarias JC, Zhang H, Elich TD, Chory J (1999) PKS1, a substrate phosphorylated by phytochrome that modulates light signaling in *Arabidopsis.* Science 284:1539–1541

Feys BJ, Parker JE (2000) Interplay of signaling pathways in plant disease resistance. Trends Genet 16:448–445

Forbes DJ (1992) Structure and function of the nuclear pore complex. Annu Rev Cell Biol 8:495–527

Foster R, Izawa T, Chua NH (1994) Plant bZIP proteins gather at ACGT elements. FASEB J 8:192–200

Furuya M, Schäfer E (1996) Photoperception and signalling of induction reactions by different phytochromes. Trends Plant Sci 1:301–307

Furuya M, Song PS (1994) Assembly and properties of holophytochrome. In: Kendrick RE, Kronenberg GHM (eds) Photomorphogenesis in plants. Kluwer, Dordrecht, pp 105–134

Garcia-Bustos J, Heitman J, Hall MN (1991) Nuclear protein localization. Biochem Biophys Acta 1071:83–101

Gil P, Kircher S, Adam E, Bury E, Kozma-Bognar L, Schäfer E, Nagy F (2000) Photocontrol of subcellular partitioning of phytochrome-B: GFP fusion protein in tobacco seedlings. Plant J 22:135–145

Goday A, Jensen AB, Culianez-Marcia FA, Alba MM, Figueras M, Serratosa J, Torrent M, Pages M (1994) The maize abscisic acid-responsive protein Rab17 is located in the nucleus and interacts with nuclear localization signals. Plant Cell 6:351–360

Görlich D (1998) Transport into and out of the cell nucleus. EMBO J 17:2721–2727
Görlich D, Kutay U (1999) Transport between the cell nucleus and the cytoplasm. Annu Rev Cell Dev Biol 15:607–660
Görlich D, Prehn S, Laskey RA, Hartmann E (1994) Isolation of a protein that is essential for the first step of nuclear protein import. Cell 79:767–778
Görlich D, Kostka S, Kraft R, Dingwall C, Laskey RA, Hartmann E, Prehne S (1995) Two different subunits of importin cooperate to recognize nuclear localization signals and bind them to the nuclear envelope. Curr Biol 5:383–392
Görlich D, Pante N, Kutay U, Aebi U, Bischoff FR (1996) Identification of different roles for RanGDP and RanGTP in nuclear protein import. EMBO J 15:5584–5594
Goldberg MW, Cronshaw JM, Kiseleva E, Allen TD (1999) Nuclear-pore-complex dynamics and transport in higher eukaryotes. Protoplasma 209:144–156
Guo H, Duong H, Ma N, Lin C (1999) The *Arabidopsis* blue light receptor cryptochrome 2 is a nuclear protein regulated by a blue light-dependent post-transcriptional mechanism. Plant J 19:279–287
Guralnick B, Thomsen G, Citovsky V (1996) Transport of DNA into the nuclei of Xenopus oocytes by a modified VirE2 protein of Agrobacterium. Plant Cell 8:363–373
Haizel T, Merkle T, Pay A, Fejes E, Nagy F (1997) Characterization of proteins that interact with GTP-bound form of the regulatory GTPase Ran in *Arabidopsis*. Plant J 11:93–103
Harter K, Kircher S, Frohnmeyer H, Krenz M, Nagy F, Schäfer E (1994) Light-regulated modification and nuclear translocation of cytosolic G-box binding factors in parsley. Plant Cell 6:545–559
Heese-Peck A, Raikhel NV (1998a) The nuclear pore complex. Plant Mol Biol 38:145–162
Heese-Peck A, Raikhel NV (1998b) A glycoprotein modified with terminal N-acetylglucosamine and localized at the nuclear rim shows sequence similarity to aldose-1-epimerases. Plant Cell 10:599–612.
Heese-Peck A, Cole RN, Borhsenious ON, Hart GW, Raikhel NV (1995) Nuclear pore complex proteins from higher plants are modified by novel *O*-linked oligosaccharides. Plant Cell 7:1459–1471
Hicks GR, Raikhel NV (1995a) Nuclear localization signal binding proteins in higher plant nuclei. Proc Natl Acad Sci USA 92:734–738
Hicks GR, Raikhel NV (1995b) Protein import into the nucleus: an integrated view. Annu Rev Cell Biol 11:155–188
Hicks, GR, Smith HMS, Shieh M, Raikhel NV (1995) Three classes of nuclear import signals bind to plant nuclei. Plant Physiol 107:1055–1058
Hicks GR, Smith HMS, Lobreaux S, Raikhel NV (1996) Nuclear import in permeabilised protoplasts from higher plants has unique features. Plant Cell 8:1337–1352
Holm M, Deng XW (1999) Structural organization and interactions of COP1, a light-regulated developmental switch. Plant Mol Biol 41:151–158
Hübner S, Smith HMS, Hu W, Chan CK, Rihs HP, Paschal BM, Raihkel NV, Jans DA (1999) Plant importin α binds nuclear localization sequences with high affinity and can mediate nuclear import independent of importin β. J Biol Chem 274:22610–22617
Huq E, Tepperman JM, Quail PH (2000) GIGANTEA is a nuclear protein involved in phytochrome signaling in *Arabidopsis*. Proc Natl Acad Sci USA 97:9789–9794
Iovine MK, Watkins JL, Wente SR (1995) The GLFG repetitive region of the nucleoporin Nup116p interacts with Kap95p, an essential yeast nuclear import factor. J Cell Biol 131:1699–1713
Jensen AB, Goday A, Figueras M, Jessop AC, Pages M (1998) Phosphorylation mediates the nuclear targeting of the maize Rab17 protein. Plant J 13:691–697
Jiang CJ, Imamoto N, Matsuki R, Yoneda Y, Yamamoto N (1998a) Functional characterization of a plant importin α homologue – nuclear localization signal (NLS)-selective

binding and mediation of nuclear import of NLS proteins in vitro. J Biol Chem 272:24083–24087

Jiang CJ, Imamoto N, Matsuki R, Yoneda Y, Yamamoto N (1998b) In vitro characterization of rice importin beta 1: molecular interaction with nuclear transport factors and mediation of nuclear protein import. FEBS Lett 437:127–130

Kendrick RE, Kronenberg GHM (1994) Photomorphogenesis in plants. Kluwer, Dordrecht

Kim L, Kircher S, Toth R, Adam E, Schäfer E, Nagy F (2000) Light-induced nuclear import of phytochrome-A:GFP fusion proteins is differentially regulated in transgenic tobacco and *Arabidopsis*. Plant J 22:125–133

Kircher S, Ledger S, Hayashi H, Weisshaar B, Schäfer E, Frohnmeyer H (1998) CPRF4a, a novel plant bZIP protein of the CPRF family: comparative analyses of light-dependent expression, post-transcriptional regulation, nuclear import and heterodimerisation. Mol Gen Genet 257:595–605

Kircher S, Kozma-Bognar L, Kim L, Adam E, Harter K, Schäfer E, Nagy F (1999a) Light quality-dependent nuclear import of the plant photoreceptors phytochrome A and B. Plant Cell 11:1445–1456

Kircher S, Wellmer F, Nick P, Rügner A, Schäfer E, Harter K (1999b) Nuclear import of the parsley bZIP transcription factor CPRF2 is regulated by phytochrome photoreceptors. J Cell Biol 144:201–211

Kleiner O, Kircher S, Harter K, Batschauer A (1999) Nuclear localization of the Arabidopsis blue light receptor cryptochrome 2. Plant J 19:289–296

Klimczak LJ, Schindler U, Cashmore AR (1992) DNA binding activity of the *Arabidopsis* G-box binding factor GBF1 is stimulated by phosphorylation by casein kinase II from broccoli. Plant Cell 4:87–98

Klimczak LJ, Collinge MA, Farini D, Giuliano G, Walker JC, Cashmore AR (1995) Reconstitution of *Arabidopsis* casein kinase II from recombinant subunits and phosphorylation of transcription factor GBF1. Plant Cell 7:105–115

Knorpp C, Hugosson M, Sijoling S, Eriksson AC, Glaser E (1994) Tissue-specific differences of the mitochondrial protein import machinery: in vitro import, processing and degradation of the pre-fb subunit of the ATPase in spinach leaves and root mitochondria. Plant Mol Biol 26:571–579

Kozma-Bognar L, Hall A, Adam E, Thain SC, Nagy F, Millar AJ (1999) The circadian clock controls the expression pattern of the circadian input photoreceptor, phytochrome B. Proc Natl Acad Sci USA 96:14652–14657

Kunik T, Palanichelvam K, Czosnek H, Citovsky V, Gafni Y (1998) Nuclear import of the capsid protein of tomato yellow leaf curly virus (TYLCV) in plant and insect cells. Plant J 13:393–399

Kwok SF, Piekos B, Misera S, Deng XW (1995) A complement of ten essential and pleiotropic *Arabidopsis COP/DET/FUS* genes is necessary for repression of photomorphogenesis in darkness. Plant Physiol 110:731–742

Landschulz WH, Johnson PF, McKnight SL (1988) The leucine zipper: a hypothetical structure common to a new class of DNA-binding proteins. Science 240:1759–1764

Lasarowitz SG (1988) Infectivity and complete nucleotide sequence of the genome of a South African isolate of maize streak virus. Nucleic Acids Res 16:229–249

Lasarowitz SG, Pinder AJ, Damsteegt VD, Rogers SG (1989) Maize streak virus genes essential for systemic spread and symptom development. EMBO J 8:1023–1032

Leheny EA, Theg SM (1994) Apparent inhibition of chloroplast protein import by cold temperature is due to energetic considerations, not to membrane fluidity. Plant Cell 6:427–437

Ligterink W, Kroj T, zur Nieden U, Hirt H (1997) Receptor-mediated activation of a MAP kinase in pathogen defense of plants. Science 276:2054–2057

Liu H, Boulton MI, Thomas CL, Prior DAM, Oparka KJ, Davies JW (1999) Maize streak virus coat protein is karyophilic and facilitates nuclear transport of viral DNA. Mol Plant-Microbe Interact 12:894–900

Liu L, White MJ, MacRae TH (1999) Transcription factors and their genes in higher plants. Functional domains, evolution and regulation. Eur J Biochem 262:247–257

Lyck R, Harmening U, Höhfeld I, Treuter E, Scharf KD, Nover L (1997) Intracellular distribution and identification of the nuclear localization signals of two plant heat-stress transcription factors. Planta 202:117–125

Martinez-Garcia JF, Huq E, Quail PH (2000) Direct targeting of light signals to a promoter element-bound transcription factor. Science 288:859–863

Mathews S, Sharrock RA (1997) Phytochrome gene diversity. Plant Cell Environ 20:666–671

Matsui M, Stoop CD, von Arnim AG, Wei N, Deng XW (1995) *Arabidopsis* COP1 protein specifically interacts in vitro with a cytoskeleton-associated protein, CIP1. Proc Natl Acad Sci USA 92:4239–4243

Mattaj IW, Engelmeier L (1998) Nucleoplasmic transport: the soluble phase. Annu Rev Biochem 67:265–306

McGonigle B, Bouhidel K, Irish VF (1996) Nuclear localization of the *Arabidopsis* APATALA3 and PISTILLATA homeotic gene products depends on their simultaneous expression. Genes Dev 10:1812–1821

McNellis TW, von Arnim AG, Araki T, Komeda Y, Misera S, Deng XW (1994) Genetic and molecular analysis of an allelic series of *cop1* mutants suggests functional roles for the multiple protein domains. Plant Cell 6:487–500

Melchior F, Paschal B, Evans E, Gerace L (1993) Inhibition of nuclear protein import by nonhydrolyzable analogs of GTP and identification of the small GTPase Ran/TC4 as an essential transport factor. J Cell Biol 135:1457–1470

Merkle T, Nagy F (1997) Nuclear import of proteins: putative import factors and development of in vitro import systems in higher plants. Trends Plant Sci 2:458–464

Merkle T, Haizel T, Matsumoto T, Harter K, Dallmann G, Nagy F (1994) Phenotype of the fission yeast cell cycle regulatory mutant pim1-46 is suppressed by a tobacco cDNA encoding a small, Ran-like GTP-binding protein. Plant J 6:555–565

Merkle T, Leclerc D, Marshallsay C, Nagy F (1996) A plant in vitro system for nuclear import of proteins. Plant J 10:1177–1186

Meshi T, Iwabuchi M (1995) Plant transcription factors. Plant Cell Physiol 36:1405–1420

Misera S, Müller AJ, Weiland-Heidecker U, Jürgens G (1994) The *FUSCA* genes of *Arabidopsis*: negative regulators of light responses. Mol Gen Genet 244:242–252

Moore MS, Blobel G (1993) The GTP-binding protein Ran/TC4 is required for protein import into the nucleus. Nature 365:143–148

Mudgett MB, Chesnokova O, Dahlbeck D, Clark E, Bonas U, Staskawicz BJ (2000) Molecular signals required for type III secretion and translocation of the *Xanthomonas campestris* AvrBs2 protein to pepper plants. Proc Natl Acad Sci USA 97:13324–13329

Nagatani A (2000) Lighting up the nucleus. Science 288:821–822

Nagy F, Schäfer E (1999) Phytochromes, pif3 and light signaling go nuclear. Trends Plant Sci 4:125–126

Nagy F, Schäfer E (2000) Nuclear and cytosolic events of light-induced, phytochrome-regulated signalling in higher plants. EMBO J 19:157–163

Neer EJ, Schmidt CJ, Nambudripad R, Smith TF (1994) The ancient regulatory-protein family of WD-repeat proteins. Nature 371:297–300

Nemeth K, Salcher K, Putnoky P, Bhalerao R, Koncz-Kalman Z, Stankovic-Stangeland B, Bako L, Mathur J, Ökresz L, Stabel S, Geigenberger P, Stitt M, Redei GP, Schell J, Koncz C (1998) Pleiotropic control of glucose and hormone responses by PRL1, a nuclear WD protein, in *Arabidopsis*. Genes Dev 12:3059–3073

Nigg EA (1997) Nucleoplasmic transport: signals, mechanisms and regulation. Nature 386:779–787

Nover L, Scharf KD (1997) Heat stress proteins and transcription factors. Cell Mol Life Sci 53:80–103

Ogura T, Tanaka N, Yabe N, Komatsu S, Hasunuma K (1999) Characterization of protein complexes containing nucleoside diphosphate kinase with characteristics of light signal transduction through phytochrome in etiolated pea seedlings. Photochem Photobiol 69:397–403

Osterlund MT, Deng XW (1998) Multiple photoreceptors mediate the light-induced reduction of GUS-COP1 from Arabidopsis hypocotyl nuclei. Plant J 16:201–208

Park DH, Somers DE, Kim YS, Choy YH, Lim HK, Soh MS, Kim HJ, Kay SA, Nam HG (1999) Control of circadian rhythms and photoperiodic flowering by the *Arabidopsis* GIGANTEA gene. Science 285:1579–1582

Pascal E, Sanderfoot AA, Ward BM, Medville R, Turgeon R, Lasarowitz SG (1994) The geminivirus BR1 movement protein binds single- and double-stranded DNA and localizes to the nucleus. Plant Cell 6:995–1006

Peracchia G, Jensen AB, Culianez-Macia FA, Grosset J, Goday A, Issinger OG, Pages M (1999) Characterization, subcellular localization and nuclear targeting of casein kinase 2 from *Zea mays*. Plant Mol Biol 40:199–211

Qin S, Ward BM, Lazarowitz, SG (1998) The bipartite geminivirus coat protein aids BR1 function in viral movement by affecting the accumulation of viral single-stranded DNA. J Virol 72:9247–9256

Quail PH, Boylan MT, Parks BM, Short TW, Xu Y, Wagner D (1995) Phytochromes: photosensory perception and signal transduction. Science 268:675–680

Radu A, Blobel G, Moore MS (1995) Identification of a protein complex that is required for nuclear protein import and mediates docking of the import substrate to distinct nucleoporins. Proc Natl Acad Sci USA 92:1769–1773

Reichelt R, Holzenburg A, Buhle EL, Jarnik M, Engel A, Aebi U (1990) Correlation between structure and mass distribution of the nuclear pore complex, and distinct pore complex components. J Cell Biol 110:883–894

Relic B, Andjelkovic M, Rossi L, Nagamine Y, Hohn B (1998) Interaction of the DNA modifying proteins VirD1 and VirD2 of *Agrobacterium tumefaciens*: analysis by subcellular localization in mammalian cells. Proc Natl Acad Sci USA 95:9105–9110

Rout MP, Blobel G (1993) Isolation of the yeast nuclear pore complex. J Cell Biol 123:771–783

Rout MP, Wente SR (1994) Pores for thought: nuclear pore complex proteins. Trends Cell Biol 4:357–365

Saalbach G, Christov V (1994) Sequence of a plant cDNA from *Vicia faba* encoding a novel Ran-related GTP-binding protein. Plant Mol Biol 24:969–972

Sakamoto K, Nagatani A (1996) Nuclear localization activity of phytochrome B. Plant J 10:859–868

Sanderfoot AA, Ingham DJ, Lazarowitz SG (1996) A viral movement protein as a nuclear shuttle. Plant Physiol 110:23–33

Schäfer E, Marchal B, Marme D (1972) In vivo measurements of phytochrome photostationary state in far-red light. Photochem Photobiol 15:457–464

Scharf KD, Materna T, Treuter E, Nover L (1994) Heat stress promoters and transcription factors. In: Nover L (ed) Plant promoters and transcription factors. Springer, Berlin Heidelberg New York, pp 125–162

Scharf KD, Heider H, Höhfeld I, Lyck R, Schmidt E, Nover L (1998) The tomato Hsf system: HsfA2 needs interaction with HsfA1 for efficient nuclear import and may be localized in cytoplasmic heat stress granules. Mol Cell Biol 18:2240–2251

Schledz M, Leclerc D, Neuhaus G, Merkle T (1998) Characterization of four cDNAs encoding different importin alpha homologs from *Arabidopsis*. Plant Physiol 116:868

Schöffl F, Prändl R, Reindl A (1999) Molecular responses to heat stress. In: Shinozaki K, Yamaguschi-Shinozaki K (eds) Molecular responses to cold, drought, heat and salt stress in higher plants. RG Landes Company, Austin, TX, pp 81–98

Scofield GN, Beven AF, Shaw PJ, Doonan JH (1992) Identification and localization of a nucleoporin-like protein component of the plant nuclear matrix. Planta 187:414–420

Sheldon LA, Kingston RE (1993) Hydrophobic coiled-coil domains regulate the subcellular localization of the human heat shock factor 2. Genes Dev 7:1549–1558

Shieh MW, Wessler SR, Raikhel NV (1993) Nuclear targeting of the maize R protein requires two nuclear localization sequences. Plant Physiol 101:353–361

Shoji K, Iwasaki T, Matsuki R, Miyao M, Yamamoto N (1998) Cloning of a cDNA encoding an importin α and down-regulation of the gene by light in rice leaves. Gene 212:279–286

Sitte P (1998) Facts and concepts in cell compartmentation. Prog Bot 59:3–45

Smeekens S (2000) Sugar-induced signal transduction in plants. Annu Rev Plant Physiol Plant Mol Biol 51:49–81

Smith HMS, Raikhel NV (1998) Nuclear localization signal receptor importin α associates with the cytoskeleton. Plant Cell 10:1791–1799

Smith HMS, Raikhel NV (1999) Protein targeting to the nuclear pore. What can we learn from plants? Plant Physiol 119:1157–1163.

Smith HMS, Hicks GR, Raikhel NV (1997) Importin α *from Arabidopsis thaliana* is a nuclear import receptor that recognises three classes of import signals. Plant Physiol 114:411–417

Smith TF, Gaitatzes C, Saxena K, Neer EJ (1999) The WD repeat: a common architecture for diverse functions. Trends Biochem Sci 24:181–185

Somers DE, Devil PF, Kay SA (1998) Phytochromes and cryptochromes in the entrainment of the Arabidopsis circadian clock. Science 282:1488–1490

Stacey MG, von Arnim AG (1999) A novel motive mediates the targeting of the *Arabidopsis* COP1 protein to subnuclear foci. J Biol Chem 274:27231–27236

Stacey MG, Hicks SN, von Arnim AG (1999) Discrete domains mediate the light-responsive nuclear and cytoplasmic localization of *Arabidopsis* COP1. Plant Cell 11:349–363

Stacey MG, Kopp OR, Kim TH, von Arnim AG (2000) Modular domain structure of Arabidopsis COP1. Reconstitution of activity by fragment complementation and mutational analysis of a nuclear localization signal in plants. Plant Physiol 124:979–989

Terzaghi WB, Cashmore AR (1995a) Light-regulated transcription. Annu Rev Plant Physiol Plant Mol Biol 46:445–474

Terzaghi WB, Cashmore AR (1995b) Seeing the light in plant development. Curr Biol 5:466–468

Terzaghi WB, Bertekap RL, Cashmore AR (1997) Intracellular localization of GBF proteins and blue light-induced import of GBF2 fusion proteins into the nucleus of cultured *Arabidopsis* and soybean cells. Plant J 11:967–982

Unseld S, Höhnle M, Ringel M, Frischmuth T (2001) Subcellular targeting of the coat protein of African cassava mosaic geminivirus. Virology (in press)

Van den Ackerveken G, Marois E, Bonas U (1996) Recognition of the bacterial avirulence protein AvrBs3 occurs inside the host plant cell. Cell 87:1307–1316

Van der Krol AR, Chua NH (1991) The basic domain of plant B-ZIP proteins facilitates import of a reporter protein into plant nuclei. Plant Cell 3:667–675

Varagona MJ, Raikhel NV (1994) The basic domain in the bZIP regulatory protein Opaque2 serves two independent functions: DNA binding and nuclear localization. Plant J 5:207–214

Varagona MJ, Schmidt RJ, Raikhel NV (1991) Monocot regulatory protein Opaque-2 is localized in the nucleus of maize endosperm and transformed tobacco plants. Plant Cell 3:105–113

Varagona MJ, Schmidt RJ, Raikhel NV (1992) Nuclear localization signal(s) required for nuclear targeting of the maize regulatory protein Opaque-2. Plant Cell 4:1213–1227

Von Arnim AG, Deng XW (1994) Light inactivation of *Arabidopsis* photomorphogenic repressor COP1 involves a cell-specific regulation of its nucleocytoplasmic partitioning. Cell 79:1035–1045

Von Arnim AG, Osterlund MT, Kwok SF, Deng XW (1997) Genetic and developmental control of nuclear accumulation of COP1, a repressor of photomorphogenesis in *Arabidopsis*. Plant Physiol 114:779–788

Wang HY, Kang DM, Deng XW, Wei N (1999) Evidence for functional conservation of a mammalian homologue at the light-responsive plant protein COP1. Curr Biol 9:711–714

Weisshaar B, Armstrong GA, Block A, da Costa e Silva O, Hahlbrock K (1991) Light-inducible and constitutively expressed DNA-binding proteins recognizing a plant promoter element with functional relevance in light responsiveness. EMBO J 10:1777–1786

Wellmer F, Kircher S, Rügner A, Frohnmeyer H, Schäfer E, Harter K (1999) Phosphorylation of the parsley bZIP transcription factor CPRF2 is regulated by light. J Biol Chem 274:29476–29482

Wengelnik K, Bonas U (1996) HrpXv, an AraC-type regulator, activates expression of five out of six loci in the hrp cluster of *Xanthomonas campestris* pv. *vesicatoria*. J Bacteriol 178:3462–3469

Wengelnik K, Van den Ackerveken, Bonas U (1996) HrpG, a key hrp regulatory protein of *Xanthomonas campestris* pv. *vesicatoria* is homologous to two-component response regulators. Mol Plant-Microbe Interact 9:704–712

Wengelnik K, Rossier O, Bonas U (1999) Mutations in the regulatory gene *hrp*G of *Xanthomonas campestris* pv. *vesicatoria* result in constitutive expression of all *hrp* genes. J Bacteriol 181:6828–6831

White FF, Yang B, Johnson LB (2000) Prospects for understanding avirulence gene function. Curr Opin Plant Biol 3:291–298

Whitelam GC, Halliday KJ (1999) Photomorphogenesis: phytochrome takes a partner! Curr Biol 9:R225-R227

Woo Y-M, Itaya A, Owens RA, Tang L, Hammond R, Chou H-C, Lai MMC, Ding B (1999) Characterization of nuclear import of potato spindle tuber viroid RNA in permeabilized protoplasts. Plant J 17:627–635

Wright EA, Heckel T, Groenendijk J, Davies JW, Boulton MI (1997) Splicing features in maize streak virus virion- and complementary-sense gene expression. Plant J 12:1285–1297

Wu C (1995) Heat shock transcription factors: structure and regulation. Annu Rev Cell Dev Biol 11:441–469

Yamaguchi R, Nakamura M, Mochizuki N, Kay SA, Nagatani A (1999) Light-dependent translocation of a phytochrome B-GFP fusion protein to the nucleus in transgenic Arabidopsis. J Cell Biol 145:437–445

Yang B, Zhu W, Johnson LB, White FF (2000) The virulence factor AvrXa7 of *Xanthomonas oryzae* pv. *oryzae* is a type III secretion pathway-dependent nuclear-localized double-stranded DNA-binding protein. Proc Natl Acad Sci USA 97:9807–9812

Zentgraf U, Hemleben V (1996): Molecular cell biology: signal transduction in plants. Prog Bot 57:218–234

Zentgraf U, Velasco R, Hemleben V (1998) Molecular cell biology: different transcriptional activities in the nucleus. Prog Bot 59:131–168

Zupan J, Muth TR, Draper O, Zambryski P (2000) The transfer of DNA from *Agrobacterium tumefaciens* into plants: a feast of fundamental insight. Plant J 23:11–28

Zupan JR, Citovsky V, Zambryski P (1996) *Agrobacterium* VirE2 protein mediates nuclear uptake of single-stranded DNA in plant cells. Proc Natl Acad Sci USA 93:2392–2397

Prof. Dr. Vera Hemleben
Dr. Katrin Hinderhofer
Dr. Ulrike Zentgraf
ZMBP
(Zentrum für Molekularbiolobgie der Pflanzen)
Allgemeine Genetik
Auf der Morgenstelle 28
72076 Tübingen, Germany
Tel.: +49-07071-2976146
Fax: +49-07071-295042
e-mail: vera.hemleben@uni-tuebingen.de

Genetics of Phytopathology: Pathogenicity Factors and Signal Transduction in Plant-pathogenic Fungi

By Bettina Tudzynski and Paul Tudzynski

1 Introduction

The studies on molecular aspects of host-pathogen interaction in plant-pathogenic fungi are still booming. However, the new trends and focuses in this research area outlined in the last review of this series (Tudzynski and Tudzynski 1999) are still valid and – in spite of a vast number of publications which appeared in this area in the last two years – there are only a few really new aspects. Therefore, we will only present a short update of the last review here and refer to the detailed description of methods and trends and the tables of putative pathogenicity factors presented there. We will discuss in more detail the growing areas of signaling and transmembrane transport.

There are some recent books and conference proceedings which cover the field of molecular phytopathology, e.g., *Biology of Plant-Microbe Interactions* (de Wit et al. 1999) and *Fungal Pathology* (Kronstad 2000); several excellent recent reviews deal with various aspects of host pathogen interaction, e.g., rust fungi (Staples 2000); molecular diagnostics (Martin et al. 2000); horizontal gene transfer and evolution (Rosewich and Kistler 2000); mitochondrial DNA and hypovirulence (Bertrand 2000); cell biology and plant pathogen interactions (Heath 2000a); detoxification of plant antimicrobial agents (Osbourn et al. 1998); gene expression in haustoria of rust fungi (Hahn 2000); avirulence genes (Gabriel 1999); the hypersensitive response (Richael and Gilchrist 1999); and a new, promising European journal covering this area has been introduced: *Molecular Plant Pathology*.

2 Functional Analysis of Potential Pathogenicity Determinants

In the last two reviews of this series (Tudzynski and Tudzynski 1997, 1999) lists of functionally analyzed potential pathogenicity determinants have been presented. Here, we decided to refrain from presenting a further, updated list but would like to refer the interested reader to the published material and discuss only a few examples of interesting new aspects in the various stages of interactions/types of factors. Functional analyses of specific genes which are suspected to play a role in these various processes are now possible in a broad range of fungi, since transformation systems have been developed and optimized for most

Progress in Botany, Vol. 63

relevant fungi. These analyses are also possible for the first time in the biotrophic Barley powdery mildew *Erysiphe* (*Blumeria*) *graminis* (Chaure et al. 2000), a real breakthrough. In addition, several interesting genes have been directly identified via loss of function in insertional mutagenesis programs (see below).

a) Early Stages of Infection

"The first touch", i.e., early recognition events in plant-microbe interaction has been studied in detail in several interaction systems (summarized, e.g., by Heath 2000b): Bagga and Straney (2000) report on the induction of germination of conidia in *Nectria haematococca* by pea flavonoids (the same that induce *nod* gene expression in pea-specific rhizobia) and they show that obviously (possibly via inhibition of cAMP phosphodiesterase) cAMP is involved in this process. The latter has also been shown by Hall and Gurr (2000) for the differentiation of appressorial germ tubes, and the involvement of cAMP signaling pathways in various interactions is now evident (see below). In *Magnaporthe grisea* the calmodulin gene (*cam*) was shown to be expressed in early stages of conidia differentiation; self-inhibitor studies and concanavalin A-mediated inhibition of surface attachment indicate that for both, appressorium formation and *cam* expression, attachment to the surface of the plant is essential (Liu and Kolattukudy 1999). The finding of Nielsen et al. (2000) that conidia of *Blumeria graminis* can take up anionic low-molecular-weight compounds very rapidly (within 30 min) from plant surfaces sheds new light on the possible mechanisms of host recognition and induction of extracellular matrix release in this fungus. Obviously, "the first touch" initiating the fungus attack program is a much shorter moment than expected, and essential decisions (on both sides) are already made in the first few minutes of contact. There are now several reports on early expressed fungal genes that might be essential for manipulating the host's early defense; e.g., in *Colletotrichum gloeosporioides*, a gene (*CgDN3*) was identified, the product of which is obviously essential to prevent a hypersensitive response (HR)-like response in its host *Stylosenthes guianensis.* A cgDN3-deletion mutant is not able to penetrate, though colonization of tissue after wounding is normal, indicating an early action of this protein (Stephenson et al. 2000).

b) Cell Wall Degrading Enzymes

As documented in the previous review, up to now, only few functional analyses of cell wall-degrading enzymes (CWDE) have indicated an important role of single enzymes in pathogenicity, probably due to the complexity and redundancy of these enzyme systems. So far only pectin-degrading enzymes have been shown to be important, one example is an endo-polygalacturonase (PG) gene from *Botrytis cinerea*, the deletion of which reduced virulence on tomato (tenHave et al. 1998). In an elegant control experiment this result could be substantiated by modifying the plant partner: expression of a polygalacturonase inhibitory protein (from pear) resulted in a comparable reduction of virulence of *Botrytis cinerea* on these transgenic plants as had been observed with PG mutants on wild-type tomato (Powell et al. 2000).

The important role of pectin degradation has been further confirmed in other systems: Yakobey et al. (2000) could show that heterologous expression of a pectate lyase from *Colletotrichum gloeosprioides* in *Colletotrichum agna* led to increased virulence of transformants on watermelon. In *Claviceps purpurea*, the ergot fungus, replacement of two closely linked polygalacturonase genes resulted in drastic reduction of pathogenicity on rye (Oeser et al. 2001). On the other hand, just as an example for several papers, disruption of an *in planta* expressed exo-PG gene in *Fusarium oxysporum* did not result in any phenotype (Garcia-Maceira et al. 2000).

In the most thoroughly investigated system regarding CWDE, *Cochliobolus carbonum*, John Walton's group took an alternative approach to determine the role of CWDEs: they cloned an ortholog of the yeast SNF1 gene (encoding a protein kinase involved in carbon-catabolite repression), *ccsnf1*. Disruption of this gene resulted in a significant reduction of expression of several CWDE genes (coding e.g., for β-1,3-glucanases, xylanases, pectinases, and an α-arabinosidase) and in a significantly reduced number of spreading lesions. This interesting result now allows the inverse functional approach: increasing the expression of single genes in these mutants can define the role of specific enzymes/enzyme groups (Tonukari et al. 2000).

Two recent reports support the view that - apart from enzymatic effects - mechanic force could be relevant for penetration: Pryce-Jones et al. (1999) estimated (by cytorrhysis and plasmolysis experiments) the maximum turgor pressure in ripe appressoria of *E. graminis* to reach values of 2–4 Mpa. Bechinger et al. (1999) used an optical method to determine the invasive force exerted by appressoria of *Colletotrichum graminicola*: it was found to be as high as 17 μ-newtons, requiring an appressorium pressure of about 5 Mpa - extreme values, which should be sufficient to break the plant's cuticle.

c) Overcoming the Host's Chemical Defense

The role of enzymes detoxifying the host's chemical defense compounds, phytoalexins or phytoanticipines, has been studied in detail in several systems; however, in contrast to the early report of Bowyer et al. (1995) about the essential role of avenacinase in the colonization of *Avena* roots by *Gaeumannomyces graminis*, so far no other example with such a clear-cut role of the detoxifying enzyme could be detected. Especially the role of enzymes degrading α-tomatin was studied in several systems: disruption of the α-tomatinase gene in *Septoria lycopersici* did not influence virulence of the fungus on tomato (Martin-Hernandez et al. 2000); the same gene has been expressed in a field isolate of *Botrytis cinerea* (M3) lacking tomatinase activity and being non-pathogenic on tomato: the transformants showed α-tomatinase activity (with the *Septoria lycopersici* degradation mode, which is different from that of *Botrytis cinerea*: Quidde et al. 1998), but they showed no increased virulence on tomato, indicating that non-pathogenicity of M3 on tomato is not caused by the lack of α-tomatinase activity only (M. Ribbert, K.M. Weltring, A.E. Osbourn, P. Tudzynski, unpubl.). In *Fusarium oxysporum*, the gene *FoTom1* coding for a tomatinase was cloned and expressed in *Escherichia coli; the enzyme shows no sequence homology to other published saponinases. The gene is expressed in planta* in roots and stems throughout the complete disease cycle of the fungus; a functional analysis by gene inactivation is still lacking (Roldán-Arjona et al. 1999). Several other reports have been published on the role of detoxifying enzymes, e.g., on pisatin demethylases (Delserone et al. 1999) and cyanide hydratases (Wang et al. 1999; Sexton and Howlett 2000), however, no additional arguments for an essential role of these enzymes in pathogenesis have been raised. As outlined in more detail in Section 5 of this chapter, recent analyses indicated that degradation of chemical defense compounds is only an accompanying process to the far more important export though the membrane, mediated by special transporter systems.

3 "Genomics" and "Black Box" Approaches

The number of fungal genomes for which a sequencing program was initiated is rapidly increasing (Prade 1998). Among them are several phytopathogenic species, e.g., *Ustilago maydis*, *Magnaporthe grisea*, *Botrytis cinerea*. However, only very limited data are available so far in the public domain, most of the research is done (or sponsored) by companies aiming at the identification of potential new targets for fungicides. A new quality has been introduced in these "genomics" approaches by automated functional analyses, i.e., concomitant knock out of ORFs detected by genomic sequences (e.g., by serial transposon tagging of cosmids in E*scherichia coli,* Hamer et al. 2000) and systematic testing of mutants for biochemical, morphological and pathogenic properties.

These analyses could be of high value for a detailed understanding of the complex interactions of pathogenic fungi and their hosts, especially, a comparison of various systems could help to identify common pathogenicity strategies; this would require, however, that these data will be made fully available to the scientific community!

The amount of EST (expressed sequence tags, short random cDNA sequences) data of pathogenic fungi available to the public is increasing; some EST analyses have been evaluated in publications, e.g., in *Botryits cinerea* (Levis et al. 2000), in *Septoria tritici* (Keon et al 2000), *Phytophthora infestans* (Kamoun et al. 1999), *Erysiphe graminis* (Thomas et al. 2000), *Fusarium venenatum* (Rey et al. 2000), and *Claviceps purpurea* (Oeser et al. 2001). In all these cases only 50–60% of the sequences had significant homologies to genes with a known function, indicating a high "sleeping" potential of genes which might be important for pathogenicity.

The availability of whole sets of *in planta* expressed genes also allows the establishment of expression patterns, time- and tissue-specific, (see e.g., Oeser et al. 2001).

In addition to the EST analysis, differential cDNA screenings are still en vogue in plant pathogens, though the more classical differential screening techniques are being replaced more and more by PCR-based procedures; e.g., von den Biezen et al. (2000) used cDNA-AFLP (amplified fragment length polymorphism) to isolate genes of *Peronospora parasitica* expressed on *Arabidopsis thaliana*.

Also, random insertional mutagenesis techniques have been broadly applied for the analysis of host-pathogen interaction. Restriction enzyme-mediated-integration (REMI) has been successfully used in several systems, although this method has severe pitfalls like gross rearrangements, deletions, a high percentage of non-tagged mutants and multiple integration events (see recent reviews by Kahmann and Basse 1999; Maier and Schaefer 1999). Thon et al. (2000) screened a REMI library of *Colletotrichum graminicola* and could identify two pathogenicity mutants among 660 transformants. In *Magnaporthe grisea* the first membrane receptor involved in pathogenicity (Pth11p) was identified in an REMI approach, which yielded altogether 18 pathogenicity mutants of 5538 transformants (DeZwaan et al. 1999; see below). Redmann et al. (1999) used the REMI technique to isolate endophyte-like mutants (able to colonize host tissue without inducing disease symptoms) from *Colletotrichum magna*. Since in several other systems, REMI obviously is not the method of choice, (see above), alternative tagging systems have been developed. In a non-REMI insertional mutagenesis, Dufresne et al. (2000) identified a Gal4-like regulatory gene in *Colletotrichum lindemuthianum* involved in the switch between biotrophic and necrotrophic phases in this fungus. This approach had yielded altogether nine less-pathogenic mutants from 1200 transformants, among them a strain with a tag in the *clk1* gene coding for a putative serine/threonine kinase (Dufresne et al. 1998; see Table 1).

Recently, the first transposon-based mutation systems have also been described; Migheli et al. (2000) used the *Impala* transposon in *Fusarium oxysporum* to generate tagged pathogenicity mutants: 2 out of altogether 746 strains in which transposition of *Impala* was detected showed complete loss of pathogenicity on melon. In *Magnaporthe grisea* a comparable tagging system has been developed based on the same transposon (Villalba et al. 2000). Also, the *Agrobacterium*-mediated transformation, which only recently has been introduced into the repertoire of fungal molecular genetics (Dunn-Coleman and Wang 1998), will probably be highly useful for the generation of insertional mutant libraries: in *Botrytis cinerea* the system was successfully used to generate mutants with single copy random integrations (Linnemannstöns and Tudzynski, unpubl.).

4 Signal Transduction

Considerable progress has been made in elucidating signaling components that are involved in processes of pathogen differentiation in response to distinct host cues. In phytopathogenic fungi, the prototypical G protein/adenylate cyclase/cAMP-dependent protein Kinase A-activating circuit functions simultaneously with multiple mitogen-activated protein kinase cascades to ensure a rapid cellular response to environmental signals.

Despite the rapidly increasing number of cloned genes and their characterization, the initial events of substrate sensing and transduction of extracellular signals into an intracellular signal are still poorly understood. The binding of signal ligands to cell-surface receptors triggers a conformational change of heterotrimeric G proteins by dissociation of the Gα subunit from the ßγ subunits both activating or inhibiting appropriate target effectors such as protein kinases, adenylate cyclases, phospholipases, and ion channels (Kronstadt 1997). Recently, the first pathogenicity-related receptor gene has been cloned from *Magnaporthe grisea*. In addition, new insights into the frequent interactions between cAMP-signaling and mitogen-activated protein kinase (MAPK) pathways have provided better understanding of the complexity of regulatory processes including control of differentiation, sexual development, and virulence. In this chapter, we will briefly highlight recent data on signal transduction components in selected fungi and their role in pathogenicity.

Table. 1. Genes involved in signaling pathways in phytopathogenic fungi

Fungal species	Gene	Gene product	Virulence of mutants	References
G protein-coupled receptors				
Magnaporthe grisea	*pth11*	Transmembrane receptor?	Impaired appressorium maturation, disturbed host surface recognition	De Zwaan et al. (1999)
G protein subunits				
Cryphonectria parasitica	*cpg-1*	Gα subunit	Loss of virulence	Choi et al. (1995); Gao and Nuss (1998)
	cpg-2	Gα subunit	No effect	
	cpgb-1	Gβ subunit	Reduction in virulence	Kasahara and Nuss (1997)
Magnaporthe grisea	*magA*	Gα subunit	No effect	Liu and Dean (1997)
	magB	Gα subunit	Reduction of appressorium formation	Liu and Dean (1997)
	magC	Gα subunit	No effect	Liu and Dean (1997)
Cochliobolus heterostrophus	*cga1*	Gα subunit	Reduction of appressorium formation, but no effect on virulence	Horwitz et al. (1999)
Ustilago maydis	*gpa1*	Gα subunit	No effect	Regenfelder et al. (1997)
	gpa2	Gα subunit	No effect	Regenfelder et al. (1997)
	gpa3	Gα subunit	No tumor formation	Regenfelder et al. (1997)
	gpa4	Gα subunit	No effect	Regenfelder et al. (1997)

Table. 1. (continued)

Fungal species	Gene	Gene product	Virulence of mutants	References
Ustilago hordei	*fil1*	Gα subunit	Not yet clear	Lichter and Mills (1997)
Botrytis cinerea	*bcg1*	Gα subunit	Penetration but no secondary lesions on bean and tomato leaves	Schulze Gronover et al. (2001)
	bcg2	Gα subunit	Attenuation of infection process	Schulze Gronover et al. (2001)
Colletotrichum trifolii	*ctg-1*	Gα subunit	Decrease in spore germination and appressorium formation reduced pathogenicity	Truesdell et al. (2000)
cAMP signaling pathway				
Magnaporthe grisea	*mac1*	Adenylate cyclase	Loss of pathogenicity, defects in growth and mating	Choi and Dean (1997)
	sum1	Regulatory subunit of PKA	Restoration of growth and appresorium – defective phenotype in *mac1* mutants	Adachi and Hamer (1998)
	cpka	Catalytic subunit of PKA	No effect on appressorium formation, but no penetration	Mitchell and Dean (1995); Xu et al. (1997)
Ustilago maydis	*uac1*	Adenylate cyclase	Unable to colonize host plant, filamentous growth	Barrett et al. (1993)
	ubc1	Regulatory subunit of PKA	No tumor formation by homozygous *ubc1*$^-$ mutants	Gold et al. (1994); Gold et al. (1997)
	adr1	Catalytic subunit of PKA	Loss of pathogenicity, filamentous phenotype	Dürrenberger et al. (1998)
	uka1	Catalytic subunit of PKA	Little effect on morphogenesis, mating and virulence	Dürrenberger et al. (1998)

Colletotrichum trifolii	*ct-PKAR*	Regulatory subunit of PKA	?	Yang and Dickman (1999a)
Colletotrichum trifolii	*ct-PKAC*	Catalytic subunit of PKA	No penetration, infection only after artificial wounding	Yang and Dickman (1999b)
Erysiphe graminis	*Eg-cPKA*	Catalytic subunit of PKA	Role of PKA in conidial differentiation	Hall et al. (1999)
MAPK cascade				
Colletotrichum gloeosporioides	*Cg MEK1*	MAPKK	No appressoria, loss virulence	Kim et al. (2000)
Magnaporthe grisea	*pmk1*	MAPK	No appressoria	Xu and Hamer (1996)
Magnaporthe grisea	*mps1*	MAPK	Essential for appressorium penetration, loss of pathogenicity in mutants	Xu et al. (1998)
Magnaporthe grisea	*osm1*	MAPK	No effect	Dixon et al. (1999)
Ustilago maydis	*fuz7*	MAPKK	No tumor formation	Banuett and Herskowitz (1994)
Ustilago maydis	*kpp2*	MAPK	Attenuation in cell fusion, induction of pheromone-responsive genes and pathogenicity	Müller et al. (1999)
Ustilago maydis	*ubc3 (=kpp2)*	MAPK	Attenuation in cell fusion, induction of pheromone-responsive genes and pathogenicity	Mayorga and Gold (1998)
Ustilago maydis	*ubc5(=fuz7)*	MAPKK	Reduced pathogenicity	Andrews et al. (2000)
Ustilago maydis	*ubc4*	MAPKKK	Reduced pathogenicity	Andrews et al. (2000)
Cochliobolus heterostrophus	*chk1*	MAPK	No appressoria, reduction in pathogenicity	Lev et al. (1999)

Table. 1. (continued)

Fungal species	Gene	Gene product	Virulence of mutants	References
Botrytis cinerea	*bmp1*	MAPK	No penetration into plant tissue, loss of pathogenicity	Zheng et al. (2000)
Pyrenophora teres	*ptk1*	MAPK	Reduced conidiation no infection, no colonization after wounding	Ruiz-Roldán and Schäfer (2000)
Claviceps purpurea	*cpmk1*	MAPK	Loss of pathogenicity	G. Mey, B. Oeser, P. Tudzynski (unpubl.)
	cpmk2	MAPK	Loss of pathogenicity	G. Mey, B. Oeser, P. Tudzynski (unpubl.)
Other protein kinases				
Colletotrichum lindemuthianum	*clk1*	Serine/threonine protein kinase	Loss of pathogenicity	Dufresne et al. (1998)
Ustilago maydis	*ukc1*	Serine/threonine protein kinase	Reduced of pathogenicity	Dürrenberger and Kronstad (1999)
Cochliobolus carbonum	*ccSNF1*	SNF-like protein kinase	Down-regulation of catabolite-repressed wall-degrading enzymes; reduction in pathogenicity	Tonukari et al. (2000)

a) G Protein-Coupled Receptors

Fungi undergo specific differentiation and developmental processes in response to distinct physical and chemical environmental signals. All these events start with an initial "recognition phase" in which specific receptors play an important role by detecting surface structures or ligands and transducing this information to one or more downstream signaling pathways.

In the past 10 years, the pheromone signaling pathway has been well characterized in yeasts (Kujan 1993; Leberer et al. 1997) and some basidiomycetes, and several genes encoding pheromones receptors have been cloned (Wendland et al. 1995; O'Shea et al. 1998; Olesnicky et al. 1999). Pheromone receptors couple to a heterotrimeric G protein to effect intracellular signaling through a MAPK cascade, leading to the induction of genes required for mating (Herskowitz 1995; Leberer et al. 1997).

Only recently, the first pathogenicity-related transmembrane receptor protein-encoding gene, *pth11*, has been cloned by an REMI-approach from the rice blast fungus *Magnaporthe grisea* (De Zwaan et al. 1999). The mutation of the *pth11* gene was responsible for a 99% loss of pathogenicity of the fungus.

The predicted secondary structure of Pth11p suggested that it is an integral membrane protein which was demonstrated by construction of a Pth11-GFP gene fusion vector. Eukaryotic serpentine receptors have typically seven transmembrane domains (Bockaert and Pin 1999), whereas Pth11p appears to have nine, suggesting an atypical structure of this receptor protein.

Exogenous cellular second messengers, such as cAMP, suppressed defects associated with *pth11* mutants, suggesting that Pth11p mediates appressorium formation by activating intracellular cAMP production. Cutin monomers and other inductive substrate cues as well as hydrophobic surfaces were shown to be initial signals for appressorium formation and differentiation (De Zwaan et al. 1999).

b) Heterotrimeric GTP-Binding Proteins (G Proteins)

The importance of heterotrimeric G proteins in regulating diverse processes such as differentiation, mating, and pathogenicity in plant-pathogenic fungi has been demonstrated, following the cloning and disruption of a number of $G\alpha$ subunit-encoding genes (for review see Tudzynski and Tudzynski 1999). Recently, some new $G\alpha$-encoding genes (Table 1) have been cloned and studied in detail, e.g., *ctg-1* from *Colletotrichum trifolii*, the causal agent of alfalfa anthracnose (Truesdell et al. 2000), and *cga1* from the corn pathogen *Cochliobolus heterostrophus* (Horwitz et al. 1999; Table 1). Both genes were grouped into the class of $G\alpha_i$ subunits together with *cpg-1* from *Cryphonectria parasitica* (Choi et al. 1995) and *magB* from *Magnaporthe grisea* (Liu and Dean 1997) on the basis of characteristic sequence motifs, e.g., for potential N-myristoylation sites.

Replacement of *ctg-1* with a null allele resulted in transformants whose conidia fail to germinate, demonstrating the requirement of *ctg-1* for a very early stage in the pathogenic life cycle of *Colletotrichum trifolii* (Truesdell et al. 2000). *Cga1* mutants of *Cochliobolus heterostrophus* had a reduced ability to form appressoria on glass surfaces and corn leaves, but nevertheless caused lesions on corn plants (Horwitz et al. 1999). The signal transduction pathway, represented by the corresponding $G\alpha$ subunit, CGA1, appears to be involved in mating (mutants were female sterile) and/or appressorium formation.

In the gray mould *Botrytis cinerea*, two $G\alpha$ protein encoding genes, *bcg1* and *bcg2*, were cloned and functionally characterized (Schulze Gronover et al. 2001). Both genes, *bcg1*, belonging to the $G\alpha_i$ class, and *bcg2*, which was grouped together with *magC* from *Magnaporthe grisea* (Liu and Dean 1997), are expressed *in planta* at very early stages of infection. Knock-out-mutants for both genes caused similar primary necrosis lesions as the wild type in the first hours of infection on bean and tomato leaves. However, after 2 days, no further development was observed for the lesions caused by the *bcg1* mutants. *Bcg2*-mutants are able to produce spreading secondary lesions although with significantly reduced speed in comparison to the wild type.

Interestingly, knock out mutants for *cpg1*, *ctg-1*, *cga1*, and *bcg1*, all belonging to the class of $G\alpha_i$ proteins, showed a number of phenotypic changes, including a reduced growth rate and altered colony morphology. Recently, it has been shown for the *Aspergillus nidulans FadA* that fungal $G\alpha_i$ proteins may regulate the chitin content of the cell wall, the cell wall porosity and susceptibility to osmotin (Coca et al. 2000). This could be the reason for altered colony morphology in $G\alpha_i$ null mutants of several fungi.

For a better understanding of the effect of G proteins on processes of cellular development and pathogenesis, modifications of functional domains were introduced into several $G\alpha_i$-encoding genes, including *cpg-1* (Gao and Nuss 1998), *magB* (Fang and Dean 2000), and *gpa3* from *Ustilago maydis* (Krüger et al. 2000).

On *magB* from *Magnaporthe grisea* (Fang and Dean 2000), site-directed mutagenesis has been recently performed by introducing the point mutations $magB^{G42R}$ and $magB^{G203R}$. The conversion of glycine 42 to arginine disrupted the endogenous GTPase activity and led to a constitutively active G protein signaling. This mutation resulted in a 95% reduction in conidiation, repression of sexual reproduction, appressorium formation on both hydrophilic and hydrophobic surfaces, and smaller necrotic lesions on susceptible rice plants. Interestingly, the *magB* null mutants exhibited similar phenotypes, including reduced conidiation, sexual reproduction, and virulence (Fang and Dean 2000). $G\alpha$ heterocomplexes have been shown to be functionally active in null $G\alpha$ mutants (Yang and Borkovich 1999). Thus, it is possible that $G\alpha$ in *Magnaporthe grisea* remains active in both null and constitutively active magB^{G42R} mutants.

In magB^{G42R} mutants, the putative negative regulator, the $G\alpha$ subunit, is presumably unable to release itself from the $G\alpha$ subunit. Therefore, the phenotype of these mutants is very similar to that of the wild type.

In *Ustilago maydis* moderate activation of the cAMP signaling pathway by introducing a *gpa3*^{Q206L} mutation resulted in a drastically reduced amount of fungal material in the plant tumors and an arrest of fungal development within the plant (Krüger et al. 2000).

In *Cryphonectria parasitica*, mutation of the putative myristoylation site resulted in a significant increase of the CPG-1 accumulation, suggesting a possible role of this motif for the post-transcriptional regulation, whereas mutation of the putative palmitoylation site was found to alter the cellular localization of the protein (Gao and Nuss 1998).

Thus, results from site-directed mutagenesis experiments provide deeper insight into the complexity of the regulation of developmental processes in fungi.

Another interesting and new aspect of the recent research is the obvious stimulatory regulation of adenylate cyclase by several fungal $G\alpha_i$ subunits in contrast to mammalian systems, where $G\alpha_i$ proteins inhibit the activity of adenylate cyclase. Beside GNA-1 from *Neurospora crassa* (Ivey et al. 1999) also MAGB and BCG1 appear to be distinct from mammalian $G\alpha_i$ family members, since feeding of cAMP led to the reversion of appressorium development in *Magnaporthe grisea* (Liu and Dean 1997) and fully recovered the wild-type colony morphology in *Botrytis cinerea bcg1* mutants (Schulze Gronover et al. 2001).

c) cAMP Signaling Pathways

The cAMP signaling pathway in phytopathogenic fungi has been analyzed in some detail in the past years. It has been shown that it plays a crucial role during pathogenic development. Fungal strains in which cAMP signaling is blocked at different levels are disturbed at distinct stages of the infection process *in planta* (see Tudzynski and Tudzynski 1999).

Several new components of cAMP signaling pathway have been cloned recently. Their characterization supports the suggestion that especially the early infection stages such as conidial germination, appressorium formation and penetration, require an intact cAMP signaling pathway.

In *Magnaporthe grisea*, pathogenic wild-type strains were shown to have much higher cAMP-dependent protein kinase (PKA) activity during germination of condia and appressorium formation on hydrophobic surfaces. Transformants lacking the *cpkA* gene, encoding the catalytic subunit of a PKA, did not show protein kinase activity under the same conditions and produced only small non-functional appressoria (Kang et al. 1999).

One of the best-studied plant pathogens is *Ustilago maydis*, the agent causing corn smut disease. In this fungus, the cAMP pathway is needed not only for the early stages of infection, but also for subsequent fungal development *in planta* (Krüger et al. 2000). The components of the cAMP pathway involved in pathogenicity are the activating Gα subunit, *Gpa3*, the adenylate cyclase, *uac1*, and the regulatory and catalytic subunits of the protein kinase A (PKA), *ubc1*, and *adr1*, respectively (Table 1). Besides the activation of the cAMP pathway by a gpa3^{Q206L} point mutation (constitutively active *Gpa3*), a mutation in the regulatory subunit of the PKA (permanently active PKA) also influences tumor morphology and fungal development *in planta* (Krüger et al. 2000). However, the analysis of mutants with rather subtle changes in the activity of the cAMP pathway resulted in the suggestion that the distinct stages of development *in planta* are regulated by light changes in cAMP level.

Much progress has been achieved in molecular cloning and characterization of cAMP signaling components in *Colletotrichum trifolii* causing alfalfa anthracnose. Recently, the genes coding for the regulatory and catalytic subunits of the PKA, *ct-PKAR and ct-PKAC*, respectively, have been cloned (Yang and Dickman 1999a,b). However, although the gene *ct-PKAR* fully restored the wild-type characteristics of a *Neurospora crassa mcb* mutant, defective in the regulatory subunit of PKA, an overexpression of the gene by the *Aspergillus nidulans gpd* promoter did not affect growth and pathogenicity (Yang and Dickman 1999a). Disruption mutants were not described.

On the other hand, *ct-PKAC* disruption mutants were unable to infect intact alfalfa plants, but were able to colonize host tissue after artificial wounding. These data suggest that PKA has an important role in regulating the penetration into the plant surface (Yang and Dickman 1999b).

In *Magnaporthe grisea*, the turgor generation in appressoria by accumulating molar concentrations of glycerol is a prerequisite for the infection of rice leaves. Recently, it could be shown that the compartmentalization and rapid degradation of storage carbohydrate (glycogen) and lipid reserves is under genetic control of the *cpkA/sum1*-encoded PKA holoenzyme (Thines et al. 2000).

A catalytic subunit of PKA was recently cloned as an expressed sequence tag from the causal agent of barley powdery mildew, the obligate biotroph, *Erysiphe* (*Blumeria*) *graminis* f. sp. *hordei* (Hall et al. 1999). In contrast to *Magnaporthe grisea* and *Colletotrichum trifolii*, appressorial differentiation in the conidia of this obligate biotrophic pathogen is not induced by a single cAMP-mediated signal, such as contact with a hydrophobic surface (Lee and Dean 1993) or host cutin-derived compounds (Gilbert et al. 1996), but requires a complex series of external signals (Hall and Gurr 2000). Both cAMP and 8-Br-cAMP are able to activate and inactivate PKA activity during the appressorial differentiation demonstrating different requirements for cAMP signaling during the differentiation process.

d) MAP Kinases

In *Ustilago maydis*, the transcription factor, *Prf1*, plays a central role in pathogenicity and mating by connecting the pheromone-signaling pathway with the cAMP signaling pathway. *Prf1* gene expression is regulated by internal cAMP levels. However, the activation of *Prf1* via the cAMP pathway does not allow cell fusion in the absence of pheromone stimulation postulating the existence of a second pathway leading to the activation of *Pfr1* (Krüger et al. 1998; Hartmann et al. 1999). The elimination of putative MAP kinase sites in the *Pfr1* protein affected its function during mating supporting the suggestion of participation of a MAP kinase in activation of *Pfr1* (Müller et al. 1999). A gene, *Kpp2*, with significant similarity to fungal MAP kinases such as *pmk1* from *Magnaporthe grisea* (Xu and Hamer 1996) was cloned by PCR and characterized. Disruption mutants produce less pheromone, cannot react to pheromone stimulation and therefore, cannot produce conjugation tubes. On the other hand, ΔKpp2 mutants show the same response to external cAMP as wild-type strains demonstrating that Kpp2 is not an integral component of the known cAMP cascade (Müller et al. 1999; Fig. 1). Interestingly, ΔKpp2 mutants showed a significant reduction in their ability to induce plant tumors, and constitutive expression of *prf1* in haploid solo-pathogenic ΔKpp2 mutants did not positively affect pathogenic development reflecting the role of *Kpp2* in transmitting signals resulting in pathogenic development.

The same MAPK gene was cloned by complementation of one class of *uac1* suppressor mutants, and named *ubc3* (Mayorga and Gold 1998, 1999). Recently, the genes *ubc4* and *ubc5* have been cloned by complementation of other groups of *ubc* mutants. They encode a MAP kinase kinase kinase (MAPKKK) and a MAP kinase kinase (MAPKK), respectively (Andrews et al. 2000). Interestingly, *ubc5* was shown to be identical with *fuz7* (Banuett and Herskowitz 1994), whereas the *ubc4* gene is the most upstream member of the pheromone-responsive MAPK cascade in *Ustilago maydis*. This functional cascade is required for the filamentous phenotype of the *uac1* mutant, and the members of this cascade show important but variable roles in virulence and pathogenicity (Andrews et al. 2000). A general scheme of proposed signaling processes

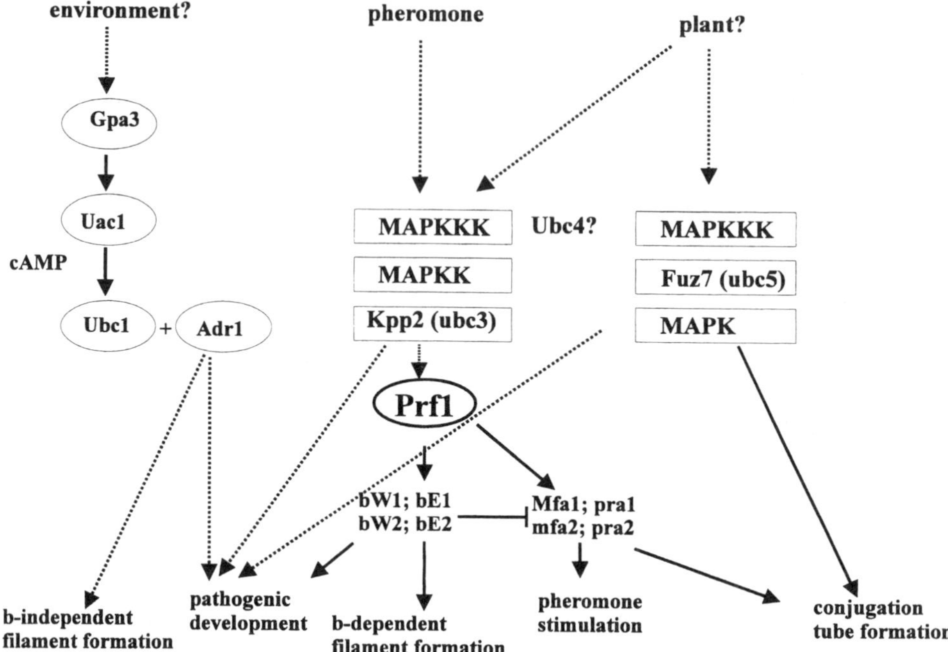

Fig. 1. Proposed signaling processes during mating and pathogenic development in *Ustilago maydis*. Different signaling inputs (*top*) are transmitted by three different cascades, eliciting various cellular processes (*bottom*). *Broken arrows* indicate missing components. (After Müller et al. 1999). The recently cloned genes are included

during mating and pathogenic development in *Ustilago maydis* is shown in Fig. 1.

In *Magnaporthe grisea*, besides *pmk1*, which was shown to be involved in appressorium formation and pathogenicity (Xu and Hamer 1996), two other MAPK genes were cloned: *mps1* (Xu et al. 1998) and *osm1* (Dixon et al. 1999). Disruption mutants of *osm1* were not affected in their pathogenicity whereas *mps1* plays an essential role in appressorium penetration; Δ*mps1* mutants totally lost their pathogenicity. Recently, *pmk1*-analogues were also cloned from the corn pathogen *Cochliobolus heterostrophus* (*chk1*, Lev et al. 1999), the gray mould *Botrytis cinerea* (*bmp1*, Zheng et al. 2000), *Colletotrichum gloeosporioides* (*cgMEK1*, Kim et al. 2000) and the barley pathogen *Pyrenophora teres* (*ptk1*, Ruiz-Roldán and Schäfer 2000). In *Claviceps purpurea*, two MAPK-genes were identified, showing significant homology to *pmk1* and *mps1* from *Magnaporthe grisea*, respectively (*cpmk1/cpmk2*; G. Mey, B. Oeser, P. Tudzynski, unpubl.; Table 1). Δ*chk1* mutants do not conidiate, are not able to produce appressoria and have a significantly reduced ability to infect corn leaves. In addition, when both mating partners lack a functional copy of *chk1*, they are not able to cross (Lev et al. 1999). The authors suppose that appressorium formation might depend on a cAMP signaling pathway through CGA1, together with the MAPK pathway through *Chk1*, as proposed for *Magnaporthe grisea* (Choi and Dean 1997). In *Botrytis cinerea*, *Δbmp1* mutants produced normal conidia and mycelia but were non-pathogenic. Germinating conida failed to penetrate and macerate plant tissues (Zheng et al. 2000). Disruption of *CgMEK1*, resulted in the loss of its ability to form appressoria in response to host signals and loss of virulence. In *Pyrenophora teres* disruption of *ptk1* led to reduced sporulation, no penetration and no colonization of host tissue after wounding, a rather drastic phenotype (Ruiz-Roldán and Schäfer 2000). In *Claviceps purpurea* mutants in both MAPK genes show significant reduction in pathogenicity; the *cpmk2*-mutant is also severely impaired in vegetative growth and sporulation (G. Mey, B. Oeser, P. Tudzynski, unpubl.).

All these observations show that MAPKs have diverse functions in fungal pathogenesis. Interestingly, most of the mutants in *pmk1*-homologous genes show defects in very early stages of infection, also those which form no (or not always) appressoria like *Botrytis cinerea* and *Claviceps purpurea*. So far, only in the *Claviceps purpurea* system an *mps1*-homologous MAPK gene was identified, indicating a similar organization of pathogenicity-related signal chains as in *Magnaporthe grisea*, whereas in *Cochliobolus heterostrophus chk1* mutants show an intermediate phenotype, indicating a different organization of MAPK cascades. Heterologous complementation experiments will be helpful to analyze the different functions of these kinases; preliminary data indicate that the *Claviceps purpurea* genes complement (only) the corresponding *Magnaporthe grisea* mutants, indicating a high degree of conservation between these two non-closely related pathogens (G. Mey, pers. comm.).

e) Interconnections Between cAMP and MAPK Signaling Pathways

From the recent detailed analyses of signaling processes in fungi it is becoming clear now that interactions frequently exist between cAMP signaling and MAPK signaling pathways. In *Ustilago maydis*, the fusion of two compatible signaling haploid cells that differ at the a and b mating type loci, is a prerequisite for mating and pathogenicity. In order to investigate post-fusion events, solo-pathogenic haploid and diploid strains that do not need to fuse with a mating partner to cause disease were constructed. In such strains it was shown that compounds from both the cAMP and MAPK signaling pathways are necessary for pathogenic development. It is well known that mutations in *gpa3*, *uac1*, *adr1* and *ubc1* affect the pathogenicity in *Ustilago maydis*. However, recent data demonstrated that strains in which signal-transmitting components of the pheromone MAPK cascade, such as *prf1* and *kpp2* (*ubc3*), are deleted became non-pathogenic or attenuated in pathogenicity. This is consistent with recent studies that have shown that *fuz7* (*ubc5*) also affect pathogenicity in solo-pathogenic strains (for review see Kronstad et al. 1998; Kahmann et al. 1999).

f) Other Protein Kinases Involved in Pathogenicity

Searching for additional catalytic subunits of PKAs in *Ustilago maydis*, Dürrenberger and Kronstad (1999) cloned a new protein kinase-encoding gene, *ukc1*, by PCR. This gene belongs to the family of Ser/Thre kinase genes and is related to, but distinct from the PKAs. *Ukc1* mutants became highly pigmented and are defective in their ability to cause disease on corn seedlings (Table 1).

As mentioned above (Sect. 1.b), an alternative approach for the functional analysis of the role of cell wall degrading enzymes led to the identification of a protein kinase involved in pathogenicity in *Cochliobolus carbonum*: since most of the wall-degrading enzymes are under negative control of carbon catabolite repression, a homologue to a yeast protein kinase gene, SNF1, required for depression of catabolite repressed genes, was cloned from *Cochliobolus carbonum* (Tonukari et al. 2000). *Cc SNF1*-mutants have reduced virulence on maize resulting from downregulation of several wall-degrading enzymes (see above).

5 Transporter Genes

Plant pathogenic fungi are constantly challenged by the presence of antifungal compounds produced by the host plants: preformed saponins and pathogen-induced phytoalexins. During evolution, fungi have developed several mechanisms to cope with this chemical threat. One mechanism to protect themselves is the detoxification of those plant defense compounds by specific saponinases (Osbourn et al. 1998) or phytoalexin-detoxifying enzymes, such as pisatin demethylase, *pda*, or maackiain-detoxifying enzyme, *mak1* (Enkerli et al. 1998; VanEtten et al. 1998). Another protection mechanism is the reduction of the concentration of toxic compounds in fungal cells by various families of integral membrane transporters. The most important families are the ATP-binding cassette (ABC) and the major facilitator superfamily (MFS) of transporters. In past years, studies on the function of transporters in plant pathogens became hot topics because they can function in the secretion of phytotoxic pathogenicity factors and in the protection against plant defense compounds as well (Del Sorbo et al. 2000).

Recently, a gene coding for an ABC transporter has been cloned from *Magnaporthe grisea* by insertional mutagenesis and shown to be essential during rice infection (Urban et al. 1999). Abc1 mutants arrest growth and die shortly after penetrating either rice or barley epidermal cells.

The ABC transporter genes *Bc atrA* and *Bc atrB* have been cloned from *Botrytis cinerea*. Targeted gene disruption of *Bc atrA* did not show a clear phenotype with regard to fungicide sensitivity and virulence on several host plants. However, disruption of *Bc atrB* causes increased sensitivity to the grape phytoalexin resveratrol and phenylpyrrole fungicides and lower virulence on grapevine leaves (Schoonbeek et al. 2001).

In the wheat pathogen *Mycosphaerella graminicola*, five ABC transporter-encoding genes, *Mg atr1-Mg atr5*, were cloned. Two of them, *Mg atr1* and *Mg atr 2*, are upregulated by plant secondary metabolites, such as eugenol, psoralen and reserpine (Zwiers and De Waard 2000). The authors suggest a possible role in pathogenicity although targeted gene disruption experiments failed until now.

Of the two ABC transporter genes from the crucifer pathogen *Leptosphaeria maculans*, *LMABC1* and *LMABC2*, the first is thought to be involved in defense of *L. maculans* against phytoalexins. So far, strains disrupted in either of the two genes have not been obtained. Therefore, the role in secretion of the *Leptosphaeria* phytotoxin sirodesmin and/or the defense against plant antifungal compounds is not yet clear (Taylor and Condie 1999).

Several MFS transporter-encoding genes are possibly involved in secretion of phytotoxins. The MFS transporter CFP from *Cercospora kikuchii* secretes the polyketide cercosporin. Disruption mutants do not produce the phytotoxin, are sensitive to this compound and display a reduced virulence (Callahan et al. 1999). *Tri12*, an MFS transporter of *Fusarium sporotrichoides*, secretes the trichothecene mycotoxins. *Tri12* mutants produce less toxins and are more sensitive to their own product (Alexander et al. 1999).

Transcription of *mfs-mdr1* from the potato pathogen *Gibberella pulicaris* is strongly induced by the potato sesquiterpenoid phytoalexin rishitin. Its role in protection of the fungus against rishitin is under investigation (Fleißner and Weltring, pers. comm.).

All these recent data have unequivocally demonstrated that MFS and ABC transporters may act as virulence factors of plant pathogens protecting the organisms against plant defense compounds and their own phytotoxic compounds as well.

6 Perspectives

The molecular analyses of the interaction of plant pathogenic fungi and their host plants in the past years have deepened significantly our understanding of the highly complex mechanisms underlying these pathogenic processes, especially with respect to the signaling events involved. Still, there remain more open than solved problems, and especially in the very early events, the exact nature of the signals and the receptors involved are unknown. However, the rapid progress especially in the "genomics" and random mutagenesis approaches will fill the gaps rather soon. Apart from forming the basis for efficient plant protection strategies in the near future, these investigations have yielded (and will yield) invaluable insights into the mechanisms of cell-cell interaction in eukaryotes which have an impact far beyond the field of molecular phytopathology.

Acknowledgements. We thank Ms. A. Kammerahl for typing this manuscript, and Dr. G. Mey, C. Schulze-Gronover and Dr. P. Linnemannstöns for discussion. The unpublished experimental work mentioned in the text, which was performed in our lab, was supported by the Deutsche Forschungsgemeinschaft (Bonn) and the European Community (TMR-network "CEREPAT").

References

Adachi K, Hamer JE (1998) Divergent cAMP signalling pathways regulate growth and pathogenesis in the rice blast fungus *Magnaporthe grisea*. Plant Cell 10:1361–1373

Alexander NJ, McCormick SP, Hohn TM (1999) TRI12, a trichothecene efflux pump from *Fusarium sporotrichioides*: gene isolation and expression in yeast. Mol Curr Genet 261:977–984

Andrews DL, Egan JD, Mayorga ME, Gold SE (2000) The *Ustilago maydis* ubc4 and ubc5 genes encode members of a MAP kinase cascade required for filamentous growth. Mol Plant-Microbe Interact 13:781–786

Bagga S, Straney D (2000) Modulation of cAMP and phosphodiesterase activity by flavonoids which induce spore germination of *Nectria haematococca MPVI* (*Fusarium solani*). Physiol Mol Plant Pathol 56:51–62

Banuett F, Herskowitz I (1994) Identification of fuz7, a *Ustilago maydis* MREK/MAPKK homolog required for *a*-locus-dependent and -independent steps in the fungal life cycle. Genes Dev 8:1367–1378

Bechinger C, Giebel KF, Schnell M, Leiderer P, Deising HB, Bastmeyer M (1999) Optical measurements of invasive forces exerted by appressoria of a plant pathogenic fungus. Science 285:1896–1899

Bertrand H (2000) Role of mitochondrial DNA in the senescence and hypovirulence of fungi and the potential for plant disease control. Annu Rev Phytopathol 38:397–422

Bockaert J, Pin JP (1999) Molecular tinkering of G protein-coupled receptors: an evolutionary success. EMBO J 18:1723–1729

Bowyer P, Clarke BR, Lunness P, Daniels MJ, Osbourn AE (1995) Host range of a plant pathogenic fungus is determined by a saponin detoxifying enzyme. Science 267:371–374

Callahan TM, Rose MS, Meade MJ, Ehrenshaft M, Upchruch RG (1999) GFP, the putative cercosporin transporter of *Cercospora kikuchii*, is required for wild type cercosporin production, resistance, and virulence on soybean. Mol Plant-Microbe Interact 12:901910

Chaure P, Gurr SJ, Spanu P (2000) Stable transformation of *Erysiphe graminis*, an obligate pathogen of barley. Nat Biotechnol 18:205–207

Choi GH, Chen BS, Nuss DL (1995) Virus mediated or transgenic suppression of a G protein alpha subunit and attenuation of fungal virulence. Proc Natl Acad Sci USA 92:305–309

Choi W, Dean RA (1997) The adenylate cyclase gene MAC1 of *Magnaporthe grisea* controls appressorium formation and other aspects of growth and development. Plant Cell 9:1973–1983

Coca MA, Damsz B, Yun, D-J, Hasegawa PM, Bressan RA, Narasimhan ML (2000) Heterotrimeric G-proteins of a filamentous fungus regulate cell wall composition and susceptibility to a plant PR-5 protein. Plant J 22:61–69

Delserone LM, McCluskey K, Mathews DE, VanEtten HD (1999) Pisatin demethylation by fungal pathogens and nonpathogens of pea: association with pisatin tolerance and virulence. Physiol Mol Plant Pathol 55:317–326

Del Sorbo G, Schoonbeek H-J, De Waard MA (2000) Fungal transporters involved in efflux of natural toxic compounds and fungicides. Fungal Gen Biol 30:1–15

De Wit PJGM, Bisseling T, Stiekema W (1999) Biology of plant-microbe interactions, vol 2. Proc 9th Int Congr on Molecular Plant-Microbe Interactions, Amsterdam, 25–30 July

DeZwaan TM, Carroll AM, Valent B, Sweigard JA (1999) *Magnaporthe grisea* Pth11p is a novel plasma membrane protein that mediates appressorium differentiation in response to inductive substrate cues. Plant Cell 11:2013–2030

Dixon KP, Xu J-R, Smirnoff N, Talbot NJ (1999) Independent signaling pathways regulate cellular turgor during hyperosmotic stress and appressorium-mediated plant infection by *Magnaporthe grisea*. Plant Cell 11:2045–2058

Dufresne M, Bailey JA, Dron M, Langin T (1998) CLK1, a serine/threonine protein kinase-encoding gene, is involved in pathogenicity of *Colletotrichum lindemuthianum* on common bean. Mol Plant Microbe Interact 11:99–108

Dufresne M, Perfect S, Pellier A-L, Bailey JA, Langin T (2000) A gal4-like Protein is involved in the switch between biotrophic and necrotrophic phases of the infection process of *Colletotrichum lindemuthianum* on common bean. Plant Cell 12:1579–1589

Dunn-Coleman N, Wang H (1998) Agrobacterium T-DNA: a silver bullet for filamentous fungi? Nat Biotechnol 16:817–842

Dürrenberger F, Kronstad J (1999) The *ukc1* gene encodes a protein kinase involved in morphogenesis, pathogenicity and pigment formation in *Ustilago maydis*. Mol Gen Genet 261:281–289

Enkerli J, Bhatt G, Covert SF (1998) Maackiain detoxification contributes to the virulence of *Nectria haematococca* MPVI on chickpea. Mol Plant-Microbe Interact 11:317–326

Fang EGC, Dean RA (2000) Site-directed mutagenesis of the magB gene affects growth and development in *Magnaporthe grisea*. Mol Plant-Microbe Interact 13:1214–1227

Gabriel DW (1999) Why do pathogens carry avirulence genes? Physiol Mol Plant Pathol 55:505–214

Gao S, Nuss DL (1998) Mutagenesis of putative acylation sites alters function, localization, and accumulation of a Giα subunit of the chestnut blight fungus *Cryphonectria parasitica*. Mol Plant-Microbe Interact 11:1130–1135

Garcia-Maceira FI, Di Pietro A, Roncero MIG (2000) Cloning and disruption of pgx4 encoding an in planta expressed exopolygalacturonase from *Fusarium oxysporum*. Mol Plant Microbe Interact 13:359–365

Gilbert RD, Johnson AM, Dean RA (1996) Chemical signals responsible for appressorium formation in the rice blast fungus *Magnaporthe grisea*. Physiol Mol Plant Pathol 48:335–346
Hahn M (2000) The rust fungi: cytology, physiology and molecular biology of infection. In: Kronstad JW (ed) Fungal pathology. Kluwer, Dordrecht, pp 267–306
Hall AA, Gurr SJ (2000) Initiation of appressorial germ tube differentiation and appressorial hooking: distinct morphological events regulated by cAMP signalling in *Blumeria graminis* f. sp. *hordei*. Physiol Mol Plant Pathol 56:39–46
Hall AA, Bindslev L, Rouster J, Rasmussen SW, Oliver RP, Gurr SJ (1999) Involvement of cAMP and protein kinase A in conidial differentiation by *Erysiphe graminis* f. sp. *hordei*. Mol Plant-Microbe Interact 12:960–968
Hamer JE, Hamer L, Woessner J, Adachi K, Tanzer M, Montenegro V, Ramamurthy L, Page A (2000) Gene function analysis in fungi. 5th. Eur Conf on Fungal Genetics, Arcachon, 26–29 March
Hartmann HA, Krüger J, Lottspeich F, Kahmann R (1999) Environmental signals controlling sexual development of the corn smut fungus *Ustilago maydis* through the transcriptional regulator Prf1. Plant Cell 11:1293–1306
Heath MC (2000a) Advances in imaging the cell biology of plant-microbe interactions. Annu Rev Phytopathol 38:443–460
Heath MC (2000b) The first touch. Physiol Mol Plant Pathol 56:49–50
Herskowitz I (1995) MAP kinase pathways in yeast: for mating and more. Cell 80:187–197
Horwitz BA, Sharon A, Shun-Wen L, Ritter V, Sandrock TM, Yoder OC, Turgeon BG (1999) A G protein Alpha subunit from *Cochliobolus heterostrophus* involved in mating and appressorium formation. Fungal Gen Biol 26:19–32
Ivey FD, Yang Q, Borkovich KA (1999) Positive regulation of adenylyl cyclase activity by a $G\alpha_i$ homolog in *Neurospora crassa*. Fungal Gen Biol 26:48–61
Kahmann R, Basse C (1999) REMI (restriction enzyme mediated integration) and its impact on the isolation of pathogenicity genes in fungi attacking plants. Eur J Plant Pathol 105:221–229
Kahmann R, Basse C, Feldbrügge (1999) Fungal-plant signalling in the *Ustilago maydis*-maize pathosystem. Curr Opin Microbiol 2:647–650
Kamoun S, Hraber P, Sobral B, Nuss D, Govers F (1999) Initial assessment of gene diversity for the oomycete pathogen *Phytophthora infestans* based on expressed sequences. Fungal Genet Biol 28:94–106
Kang SH, Khang CH, Lee YH (1999) Regulation of cAMP-dependent protein kinase during appressorium formation in *Magnaporthe grisea*. FEMS Microb Lett 170:419–423
Kasahara S, Nuss DL (1997) Targeted disruption of a fungal G-protein beta subunit results in increased vegetative growth but reduced virulence. Mol Plant Microbe Interact 10:984–993
Keon J, Bailey A, Hargreaves J (2000) A group of expressed cDNA sequences from the wheat fungal leaf blotch pathogen, *Mycosphaerella graminicola* (*Septoria tritici*). Fungal Genet Biol 29:118–133
Kim YK, Kawano T, Li D, Kolattukudy PE (2000) A mitogen-activated protein kinase kinase required for induction of cytokinesis and appressorium formation by host signals in the conidia of *Colletotrichum gloeosporioides*. Plant Cell 12:1331–1343
Kronstad J, De Maria A, Funnell D, Laidlaw RD, Lee N, Moniz de Sá M, Ramesh M (1998) Signaling via cAMP in fungi: interconnections with mitogen-activated protein kinase pathways. Arch Microbiol 170:395–404
Kronstad JW (1997) Virulence and cAMP in smuts, blasts and blights. Trends Plant Sci 2:193–199
Kronstad JW (2000) Fungal pathology. Kluwer, Dordrecht
Krüger J, Loubradou G, Regenfelder E, Hartmann A, Kahmann R (1998) cAMP complements the mating defect of a G protein mutant in *U. maydis* via regulation of pheromone gene expression. Mol Gen Genet 260:193–198

Krüger J, Loubradou G, Wanner G, Regenfelder E, Feldbrügge M, Kahmann R (2000) Activation of the cAMP pathway in *Ustilago maydis* reduces fungal proliferation and teliospore formation in plant tumors. Mol Plant-Microbe Interact 13:1034–1040

Kujan J (1993) The pheromone response pathway in *Saccharomyces cerevisiae*. Annu Rev Genet 27:147–179

Leberer E, Thomas DY, Whiteway M (1997) Pheromone signalling and polarized morphogenesis in yeast. Curr Opin Genet Dev 7:59–66

Lee HY, Dean RA (1993) cAMP regulates infection structure formation in the plant pathogenic fungus *Magnaporthe grisea*. Plant Cell 5:693–700

Lev S, Sharon A, Hadar R, Ma H, Horwitz BA (1999) A mitogen-activated protein kinase of the corn leaf pathogen *Cochliobolus heterostrophus* is involved in conidiation, appressorium formation, and pathogenicity: diverse roles for mitogen-activated protein kinase homologs in foliar pathogens. Proc Natl Acad Sci USA 96:13542–13547

Levis C, Bitton F, Fortini D, Pradier JM, Brygoo G, Brygoo Y (2000) Analyse of *Botrytis cinerea* expressed sequence tags: gene expression level and codon usage. XIIth Int *Botrytis* Symp Reims, 3–7 July

Lichter A, Mills D (1997) Fil1, a G-protein alpha subunit that acts upstream of CAMP and is essential for dimorphic switching in haploid cells of *Ustilago hordei*. Mol Gen Genet 256:426–435

Liu S, Dean RA (1997) G protein (subunit genes control growth, development and pathogenicity of *Magnaporthe grisea*. Mol Plant-Microbe Interact 10:1075–1086

Liu Z-M, Kolattukudy PE (1999) Early expression of the calmodium gene, which precedes appressorium formation in *Magnaporthe grisea*, is inhibited by self inhibitors and requires surface attachment. J Bacteriol 181:3571–3577

Maier FJ, Schäfer W (1999) Mutagenesis via insertional or restriction enzyme-mediated-integration (REMI) as a tool to tag pathogenicity related genes *in planta* pathogenic fungi. Biol Chem 380:855–864

Martin RP, James D, Lévesque CA (2000) Impacts of molecular diagnostic technologies on plant disease management. Annu Rev Phytopathol 38:207–240

Martin-Hernandez AM, Dufresne V, Hugouvieux MR, Osbourn AE (2000) Effects of targeted replacement of the tomatinase gene on the interaction of *Septoria lycopersici* with tomatoes. Mol Plant Microbe Interact 13:1301–1311

Mayorga ME, Gold SE (1998) Characterization and molecular genetic complementation of mutants affecting dimorphism in the fungus *Ustilago maydis*. Fungal Gen Biol 24:364–376

Mayorga ME, Gold SE (1999) A MAP kinase encoded by the *ubc3* gene of *Ustilago maydis* is required for filamentous growth and full virulence. Mol Microbiol 34:485–497

Migheli Q, Steinberg C, Davière J-M, Olivan C, Gerlinger C, Gautheron N, Alabouvette C, Daboussi M-J (2000) Recovery of mutants impaired in pathogenicity after transposition of *Impala* in *Fusarium oxysporum* f. sp. *melonis*. Phytopathology 90:1279–1284

Mitchell TK, Dean RA (1995) The cAMP-dependent protein kinase catalytic subunit is required for appressorium formation and pathogenesis by the rice blast pathogen *Magnaporthe grisea*. Plant Cell 7:1869–1878

Müller P, Aichinger C, Feldbrügge M, Kahmann R (1999) The MAP kinase Kpp2 regulates mating and pathogenic development in *Ustilago maydis*. Mol Microbiol 34:1007–1017

Nielsen KA, Nicholson RL, Carver TLW, Kunoh H, Oliver RP (2000) First touch: an immediate response to surface recognition in conidia of *Blumeria graminis*. Physiol Mol Plant Pathol 56: 63–70

Oeser B, Tenberge KB, Moore S, Mihlan M, Heidrich PM, Tudzynski P (2001) Pathogenic development of *Claviceps purpurea*. In: Osiewacz H (ed) Molecular biology of fungal development. Marcel Dekker, New York (in press)

Olesnicky NS, Brown AJ, Dowell SJ Casselton LA (1999) A constitutively active G-protein-coupled receptor causes mating self-compatibility in the mushroom *Coprinus*. EMBO J 18:2756–2763

Osbourn AE, Melton RE, Wubben JP, Flegg LM, Oliver RP, Daniels MJ (1998) Saponin detoxification and fungal pathogenesis. In: Kohmoto K, Yoder OC (eds) Molecular genetics of host-specific toxins in plant diseases, vol 13. Kluwer, Dordrecht, pp 309–315

O'Shea SF, Chaure PT, Halsall JH, Olesnicky NS, Leibbrandt A, Connerton IF, Casselton LA (1998) A large pheromone and receptor gene complex determines multiple *B* mating type specificities in *Coprius cinereus*. Genetics 148:1081–1090

Powell ALT, vanKan JAL, tenHave A, Visser J, Greve LC, Bennett AB, Labavitch JM (2000) Transgenic expression of pear PGIP in tomato limits fungal colonization. Mol Plant Microbe Interact 13:942–950

Prade RA (1998) Fungal genomics – one per week. Fungal Genet Biol 25:76–78

Redman RS, Ranson JC, Rodriguez RJ (1999) Conversion of the pathogenic fungus *Colletotrichum magna* to a nonpathogenic, endophytic mutualist by gene disruption. Mol Plant-Microbe Interact 12:969–975

Pryce-Jones E, Carver T, Gurr SJ (1999) The roles of cellulase enzymes and mechanical force in host penetration by *Erysiphe graminis* f. sp. *hordei*. Physiol Mol Plant Pathol 55:175–182

Quidde T, Osbourn AE, Tudzynski P (1998) Detoxification of α-tomatine by *Botrytis cinerea*. Physiol Mol Plant Pathol 52:151–165

Regenfelder E, Spellig T, Hartmann A, Lauenstein S, Bölker M, Kahmann R (1997) G proteins in *Ustilago maydis*: transmission of multiple signals? EMBO J 16:1934–1942

Rey MW, Nelson BA Bernauer S, Berka RM (2000) The *Fusarium venenatum* EST project: analysis of 8,000 ESTs. 5. Eur Conf on Fungal Genetics, Arcachon

Richael C, Gilchrist D (1999) The hypersensitive response: a case of hold or fold? Physiol Mol Plant Pathol 55:5–12

Roldán-Arjona T, Pérez-Espinosa A, Ruiz-Rubio M (1999) Tomatinase from *Fusarium oxysporum* f. sp. *lycopersici* defines a new class of saponinases. Mol Plant-Microbe Interact 12:852–861

Rosewich UL, Kistler HC (2000) Role of horizontal gene transfer in the evolution of fungi. Annu Rev Phytopathol 38:325–364

Ruiz-Roldán MC, Schäffer W (2000) *ptk1*, amitogen-activated protein kinase gene, is involved in conidia building and pathogenicity of *Pyrenophora teres* on barley. 5th Eur Conf Fungal Genetics Arcachon, 26-29 March

Schoonbeek H, Del Sorbo G, De Waard MA (2001) The ABC transporter BcatrB affects the sensitivity of *Botrytis cinerea* to the phytoalexin resveratrol and the fungicide fenpiclonil. Mol Plant Microbe Interact 14:562–571

Schulze Gronover C, Kasulke D, Tudzynski P, Tudzynski B (2001) The role of G protein α subunits in the infection process of the gray mould fungus *Botrytis cinerea*. Molec Plant-Microbe Interact (in press)

Sexton AC, Howlett BJ (2000) Characterization of a cyanide hydratase gene in the phytopathogenic fungus *Leptosphaeria maculans*. Mol Gen Genet 263:463–470

Staples RC (2000) Research on the rust fungi during the twentieth century. Annu Rev Phytopathol 38:49–70

Stephenson SA, Hatfield JH, Rusu AG, Maclean DJ,. Manners JM (2000) CgDN3: an essential pathogenicity gene of *Colletotrichum gloeosporioides* necessary to avert a hypersensitive-like response in the host *Stylosanthes gulanensis*. Mol Plant Microbe Interact 13:929–941

Taylor JL, Condie J (1999) Characterization of ABC transporters from the fungal phytopathogen *Leptosphaeria maculans*. Proc 9th Int Congr on Molecular Plant Microbe Interactions, p 73

tenHave A, Mulder W, Visser J, vanKan JAL (1998) The edopolygalacturonase gene *Bcpg1* is required for full virulence of *Botrytis cinerea*. Mol Plant Microbe Interact 11:1009–1016

Thines E, Weber RWS, Talbot NJ (2000) MAP kinase and protein kinase A-dependent mobilization of triacylglycerol and glycogen during appressorium turgor generation by *Magnaporthe grisea.* Plant Cell 12:1703–1718

Thomas SW, Rasmussen SW, Glaring MA, Rouster JA, Christiansen SK, Oliver RP (2001) Gene identification in the fungal pathogen *Blumeria graminis* by expressed sequence tag analysis. Proc Natl Acad Sci USA (in press)

Thon MR, Nuckles EM, Vaillancourt LJ (2000) Restriction enzyme-mediated integration used to produce pathogenicity mutants of *Colletotrichum graminicola.* Mol Plant-Microbe Interact 13:1356–1365

Tonukari NJ, Scott-Craig JS, Walton JD (2000) The *Cochliobolus carbonum SNF1* gene is required for cell wall-degrading enzyme expression and virulence on maize. Plant Cell 12:237–247

Truesdell GM, Zhonghui Y, Dickman MB (2000) A Gα subunit gene from the phytopathogenic fungus *Colletotrichum trifolii* is required for conidial germination. Physiol Mol Plant Pathol 56:131–140

Tudzynski P, Tudzynski B (1997) Genetics of plant pathogenic fungi. Prog Bot 59:169–193

Tudzynski P, Tudzynski B (1999) Phytopathogenic fungi: genetic aspects of host-pathogen interaction. Prog Bot 61:119–147

Urban M, Bhargava T, Hamer JE (1999) An ATP-driven efflux pump is a novel pathogenicity factor in rice blast disease. EMBO J 18:512–521

Van den Biezen EA, Juwana H, Parker JE, Jones JDG (2000) cDNA-AFLP display for the isolation of *Peronospora parasitica* genes expressed during infection in *Arabidopsis thaliana.* Mol Plant-Microbe Interact 13:895–898

VanEtten H, Jorgensen S, Enkerli J, Covert SF (1998) Inducing the loss of conditionally dispensable chromosome in *Nectria haematococca* during vegetative growth. Curr Genet 33:299–303

Villalba F, Lebrun M-H, Hua-Van A, Daboussi M-J, Grosjean-Cournoyer M-C (2000) Transposon gene tagging in the rice blast fungus *Magnaporthe grisea* using impala, a Tc1-mariner element from *Fusarium oxysporum.* 5th Eur Conf on Fungal Genetics, Arcachon, 26-29 March

Wang P, Sandrock RW, VanEtten HD (1999) Disruption of the cyanide hydratase gene in *Gloeocercospora sorghi* increases its sensitivity to the phytoanticipin cyanide but does not affect its pathogenicity on the cyanogenic plant sorghum. Fungal Gen Biol 00:126–134

Wendland J, Vaillancourt LJ, Hegner J, Lengeler KB, Laddison KJ, Specht CA, Raper Ca, Kothe E (1995) The mating-type locus B-Alpha-1 of *Schizophyllum-commune* contains a pheromone receptor gene and putative pheromone genes. EMBO J 14:5271–5278

Xu JR, Hamer JE (1996) MAP-kinase and cAMP signaling regulate infection structure formation and pathogenic growth in the rice blast fungus *Magnaporthe grisea.* Genes Dev 10:2696–2706

Xu JR, Urban M, Sweigard J, Hamer J (1997) The CPKA gene of *Magnaporthe grisea* is essential for appressorial penetration. Mol Plant Microbe Interact 10:187–194

Xu JR, Staiger CJ, Hamer JE (1998) Inactivation of the mitogen-activated protein kinase Mps1 from the rice blast fungus prevents penetration of host cells but allows activation of plant defense responses. Proc Natl Acad Sci USA 95:12713–12718

Yang Q, Borkovich KA (1999) Mutational activation of a G alpha(i) causes uncontrolled proliferation of aerial hyphae and increased sensitivity to heat and oxidative stress in *Neurospora crassa.* Genetics 151:107–117

Yang Q, Bieszke JA, Borkovich KA (2000) Differential complementation of a *Neurospora crassa* $G\alpha_i$ mutation using mammalian Gα protein genes. Mol Gen Genet 263:712–721

Yang Z, Dickman MB (1999a) Molecular cloning and characterization of *Ct-PKAR*, a gene encoding the regulatory subunit of cAMP-dependent protein kinase in *Colletotrichum trifolii*. Arch Microbiol. 171:249–256

Yang Z, Dickman MB (1999b) *Colletrotrichum trifolii* mutants disrupted in the catalytic subunit of cAMP-dependent protein kinase are nonpathogenic. Mol Plant-Microbe Interact 12:430–439

Zheng L, Campbell M, Murphy J, Lam S, Xu J-R (2000) The BMP1 gene is essential for pathogenicity in the gray mold fungus *Botrytis cinerea*. Mol Plant-Microbe Interact 13:724–732

Zwiers L-H, De Waard MA (2000) Characterization of the ABC transporter genes *MgAtr1* and *MgAtr2* from the wheat pathogen *Mycosphaerella graminicola*. Fungal Gen Biol 30:115–125

Priv. Doz. Dr. Bettina Tudzynski
Prof. Dr. Paul Tudzynski
Institut für Botanik
Westf. Wilhelms-Universität Münster
Schlossgarten 3
48149 Münster, Germany
Tel.: +49-0251-83 2 48 01/83 2 49 98
Fax: +49-0251-83 2 16 01
e-mail: Bettina.Tudzynski@uni-muenster.de
e-mail: tudzyns@uni-muenster.de
http://www.uni-muenster.de/Biologie/botanik/Tudzynsk.htm

Key Genes of Crop Domestication and Breeding: Molecular Analyses

By Günter Theißen

1 Introduction

The life of human beings depends on a sufficient supply with fruits, grains and vegetables, which are consumed either directly, or fed to livestock. Without crop plants such as rice, maize, wheat, tomato, potato, beans, apples and so on, human civilization as we know it would not exist. The importance of crop plants for human culture thus can hardly be overestimated. Most calories consumed by humans and livestock derive from cereals, the three globally most important of which are wheat (*Triticum aestivum*), rice (*Oryza sativa*) and maize (*Zea mays* ssp. *mays*).

The domestication of all major crop plants occurred during a relatively brief and recent period in human history, roughly about 10.000 years ago. During this time, ancient agriculturists brought about dramatic changes in plant form and physiology, the result of which is sometimes collectively called the 'domestication syndrome' of cultivated plants (Koornneef and Stam 2001). Selection for larger seeds, reduced articulation of the mature inflorescence, day length-insensitive flowering, and a reduction of thorns or prickles eventually led to the derivation of extremely valuable crops out of low-yielding or even agronomically almost useless wild ancestors (Paterson et al. 1995; Koornneef and Stam 2001). Later phases of plant breeding by more systematic procedures led to major refinements in crop plant structure and function ; an endeavour which is still going on.

Since plant form and function are largely under genetic control, the question arises as to which genetic changes were involved in the domestication and breeding processes of crops. Agriculturists selected the seed of preferred forms and culled out seed of undesired phenotypes to produce each subsequent generation of crop plants. Favoured alleles at loci-controlling traits of interest thus increased in frequency, eventually reaching fixation, while alleles that bring about undesired plant properties were removed from the crop plant gene pool. But how exactly are genotype and phenotype related in traits of agronomic importance?

Progress in Botany, Vol. 63

First of all, how many loci are involved in the control of agronomic traits? This question could already be answered in a number of cases by examination of segregating progenies produced from hybrids within and between plant species (Gottlieb 1984). Evidence was provided that the 'classical' components of agricultural yield, especially those of dimensions, weight and number, are usually governed by multiple gene systems, or Quantitative Trait Loci (QTLs), while many other morphological characters, particularly those of changed structure, shape or architectural orientation, are controlled by just one or two gene loci (Gottlieb 1984). The exciting finding that the domestication of sorghum (*Sorghum bicolor*), rice and maize resulted from independent mutations at orthologous genetic loci suggested that a few genes with large effects determined some of the agronomic key traits (Paterson et al. 1995), but did not reveal the molecular nature of these loci.

So what then is the structure and function of these QTLs or genes, and which changes did they undergo during domestication and breeding? During the last few years, the first 'domestication genes' and 'breeding genes' could be cloned and characterised at the molecular level, so that these questions can be answered now in a few precedent cases. In the following some of the recent breakthroughs in understanding are outlined. A summary of the genes of interest is provided in Table 1.

2 *TB1*, a Key Gene of Maize Architecture

During the domestication of maize (*Zea mays* ssp. *mays*), one of the most rapid and drastic morphological changes known for plants occurred. Especially the female inflorescence of maize, the ear, can be regarded as an unparalleled morphological novelty. Nowhere else in the plant kingdom has a similar structure been found, not even in the direct ancestor of maize, which is a wild Mexican grass known as teosinte (*Zea mays* ssp. *parviglumis*) (Doebley 1990). However, the plant habit also changed dramatically during maize domestication.

Beadle (1980), Doebley (1992), and Szabo and Burr (1996) noticed that the impressive morphological differences between teosinte and maize can be boiled down to just five key traits. Four of these, concerning the number and arrangement of spikelets and kernels, and the structure of the glumes, are restricted to the female inflorescence and are thus of direct agronomic importance (Doebley et al. 1990; Doebley 1992). The fifth trait refers to overall plant architecture: teosinte plants have elongated primary lateral branches that are tipped by male inflorescences, so that teosinte plants have a bushy appearance; in maize, the lateral branches are short and are terminated by female inflorescences.

Table 1. Key genes of plant domestication or breeding

Gene[a]	Gene family[b]	Fct.[c]	Mutation type[d]	Mutation site[e]	Crop[f]	Trait of interest[g]	Reference
BoCAL	MADS	TF	Loss-of-fct.	Coding region	Cauliflower	Curd development	Kempin et al. (1995)
FUL	MADS	TF	Ectopic expr.[h]	Promoter region[h]	Canola	Indehiscent fruits	Ferrándiz et al. (2000)
fw2.2	*RAS*-like	G	Weaker expr.	Promoter region	Tomato	Increased fruit weight	Frary et al. (2000)
J	MADS	TF	Loss-of-fct.	Prom. reg., 5′-UTR	Tomato	Stemless fruits	Mao et al. (2000)
RHT	*GAI*-like	TF	Altered fct.	Coding region	Wheat	Reduced plant height	Peng et al. (1999)
SHP	MADS	TF	Loss-of-fct.[h]	Arbitrary[h]	Canola	Indehiscent fruits	Liljegren et al. (2000)
TB1	TCP	TF	Stronger expr.	Promoter region	Maize	Apical dominance	Doebley et al. (1997)

[a]Abbreviations for gene names used: *BoCAL*, *Brassica oleracea CAULIFLOWER*; *FUL*, *FRUITFULL*; *J*, *JOINTLESS*; *RHT*, *Reduced height*; *SHP*, *SHATTERPROOF*; *TB1*, *Teosinte branched1*.
[b]Abbreviations used: *GAI*-like, *Arabidopsis Gibberellin Insensitive*-like; MADS, MADS-box gene (MCM1-AGAMOUS-DEFICIENS-SRF-like); TCP, TB1-CYC-PCF-like.
[c]Fct., function: means putative general basic function: G, GTP-binding protein; TF, transcription factor.
[d]Abbreviations used: expr., expression; fct., function.
[e]Abbreviations used: Prom. reg., promoter region; 5′-UTR, 5′-untranslated region.
[f]Botanical names: canola, *Brassica napus*, *Brassica rapa*; cauliflower, *Brassica oleracea* ssp. *botrytis*; maize, *Zea mays* ssp. *mays*; tomato, *Lycopersicon esculentum*; wheat, *Triticum aestivum*.
[g]Only primary traits are listed here. Indirectly, they may lead to other traits, e.g., increased yield.
[h]May be realised in the future by transgenic technology or marker assisted breeding.

Despite the dramatic morphological difference between teosinte and maize, it is clear now that maize was derived from teosinte by human selection in middle America just about 7500 years ago. It is thus not surprising that the overall genetic difference between both plants is very small. By analysing an F_2 population, which had been derived from a teosinte-maize cross, Beadle (1980) could already show that the number of genes controlling the key differences between teosinte and maize is approximately five. Using Quantitative Trait Loci (QTL) mapping, Doebley and his coworkers were able to localise these loci on the first five chromosomes of the maize genome (Doebley et al. 1990; Doebley and Stec 1991, 1993; Doebley 1992).

With five genomic loci and five key traits it was tempting to speculate that each of these loci controls one trait. Unfortunately, things turned out to be much more complicated, since the relationship between *Zea* genotypes and phenotypes is heavily influenced by polygenic and pleiotropic effects (Doebley 1992). For example, the average length of internodes in the primary lateral branches is heavily influenced not only by a QTL on the long arm of maize chromosome 1, but also by other loci on chromosomes 3 and 5; and the QTL on chromosome 1 influences not only primary branch length, but also, e.g., inflorescence sex (Doebley and Stec 1993; Doebley et al. 1995). A couple of years ago, genetic complementation tests indicated that there is a major locus representing the QTL on 1L, namely the gene *Teosinte branched1* (*TB1*) (Doebley et al. 1995). This identified the *TB1* gene as a major contributor to changes of maize plant habit during domestication. The maize plant homozygous mutant for *tb1* has long lateral branches tipped by tassels at upper nodes of the main culm, thus resembling teosinte in plant architecture.

Just a few years ago Doebley and coworkers managed to clone the *TB1* gene by transposon tagging and to determine the expression pattern of the gene (Doebley et al. 1997). This revealed that the maize and teosinte alleles have similar spatial expression patterns in axillary primordia, but the maize allele is expressed at about twice the level of the teosinte allele. The pattern of *TB1* expression and the structure of *tb1* mutant plants suggests that *TB1* acts by repressing the growth of axillary structures and supports the formation of female inflorescences. In maize, the relatively strong *TB1* gene expression obviously prevents the outgrowth of buds at lower nodes. Therefore, maize does not form elongated primary branches tipped by male inflorescences. This is in contrast to the situation in teosinte, in which *TB1* is expressed at a lower level and long primary branches are formed.

The wild-type *TB1* gene may be involved in the toesinte plant's response to varying environmental conditions, by producing either long (good conditions) or short branches (poor conditions, such as strong competition from surrounding plants, shade, restricted moisture). Thus maize domestication may have involved a change at the *TB1* locus to produce short branches under all environmental conditions (Doebley et al. 1995).

Despite the significant influence of the *TB1* gene on plant architecture, only subtle differences between the maize and teosinte alleles were found. Sequencing of the *TB1* gene from quite a number of different maize and teosinte varieties revealed that the effects of selection were limited to the gene's regulatory region and cannot be found in the protein-coding region (Wang et al. 1999). This further corroborates the notion that changes in gene expression, not changes in the properties of the encoded protein, were responsible for a key step of maize domestication.

But what are the basic functions of this protein in biochemical or biophysical terms? Surprisingly, sequence comparisons revealed that *TB1* shares significant similarity with *CYCLOIDEA* (*CYC*) and *DICHOTOMA* (*DICH*), two genes controlling zygomorphy (bilateral symmetry) in the flowers of snapdragon (*Antirrhinum majus*) (Luo et al. 1996, 1999; for a review, see Theißen 2000a). *TB1*, *CYC* and two DNA-binding proteins from rice, *PCF1* and *PCF2*, are the founding members of the TCP family of putative transcription factors (Cubas et al. 1999). Although at first glance the developmental processes controlled by *TB1* on the one hand, and *CYC* and *DICH*, on the other, appear to be very different, there may be common themes. All three genes are involved in the control of the growth of meristems forming axillary structures (flowers or side branches), probably by affecting cell division (Cubas et al. 1999). Thus *TB1* may encode a transcription factor which activates or represses target genes such that cell division is affected, thus controlling side branch development.

3 Cloning of the 'Green Revolution' Genes

During the so called Green Revolution in the 1960s and 1970s, the bread wheat (*Triticum aestivum*) grain yields of the world increased significantly because farmers adopted new varieties and cultivation methods. These new varieties are shorter, which means that they increase grain yield at the expense of straw biomass. In addition, they are more resistant to damage by wind and rain. The new wheat varieties are short because they respond abnormally to the plant hormone gibberellin, an essential regulator of plant growth which, e.g., positively influences stem and leaf elongation. The reduced response to gibberellin by the new wheat varieties is conferred by mutant dwarfing alleles at one of two *Reduced height-1* (*RHT-B1* or *RHT-D1*) loci (Peng et al. 1999).

Peng et al. (1999) used a candidate gene approach for molecular cloning of the 'Green Revolution' genes. They employed a part of the sequence of the – *nomen est omen*! – *Arabidopsis Gibberellin Insensitive* (*GAI*) gene to identify an expressed sequence tag (EST) from rice (*Oryza sativa*) by database searches. This EST was then used to isolate homologous cDNA and genomic clones from wheat and maize. Sequence analysis of several mutant alleles confirmed that the wheat *RHT-B1* and *RHT-D1* and the maize *DWARF-8* (*D8*) genes had been cloned (Peng et al. 1999). Comparative sequence analysis and mapping in colinear regions of the maize, wheat and rice genomes strongly suggests that *RHT-B1*, *RHT-D1* and *D8* are orthologs of the *GAI* gene, meaning that these genes originated from a common ancestor gene by speciation events which separated the lineages that led to *Arabidopsis*, maize and rice.

Orthologs are often interpreted as "the same genes in different species", although, by definition, orthology refers to gene genealogy and implies neither sequence identity nor functional similarity. In fact, however, due to functional conservation, many orthologs have very similar functions, as is certainly true for the dwarfing genes considered here.

Analysis of the *GAI*, *RHT-1*, and *D8* sequences revealed that these genes encode proteins that encode nuclear transcription factors and contain an SH2-like domain in the C-terminal part of the proteins (Peng et al. 1999). Such domains bind tyrosine-phosphorylated polypeptides and are associated with phosphotyrosine signalling in animals. Thus phosphotyrosine signalling may be involved in gibberellin-mediated plant growth regulation. The *RHT* genes and *D8* are defined by an allelic series of semi-dominant altered- or gain-of-function (rather than recessive loss-of-function) mutations that confer differing severities of dwarfism. In line with this there is evidence that the mutated genes still encode active gene products. Peng et al. (1999) showed for different mutant alleles of these genes, that they encode proteins that are altered in a conserved domain in the N-terminal regions of the proteins which is very likely involved in gibberellin signalling. In contrast, putative *Arabidopsis gai* and wheat *rht-B1* loss-of-function derivatives, which are probably unable to produce any proteins, confer a tall, gibberellin-responsive, rather than dwarf, gibberellin-resistant phenotype (Peng et al. 1999).

What do these findings tell us about the molecular mechanism by which the 'Green Revolution' genes confer their effects? The loss-of-function phenotypes indicate that the dwarfing genes considered here function as growth repressors. Their action is obviously opposed by gibberellin, an effect which is conferred by the N-terminal region of the dwarfing gene products. If this region is mutated, the dwarfing proteins are relatively insensitive to the effects of gibberellin. This means that they may have changed into constitutive growth repressors and explains why the respective mutant alleles are dominant.

So what is the wild-type function of the 'Green Revolution' genes? It is well known that gibberellin elicits plant responses in a dose-dependent fashion, which led Peng et al. (1999) to suggest that one of the functions of *RHT-1*, *D8* and *GAI* may be to modulate the gibberellin dose-response.

Does this knowledge provide us with novel tools of agronomic importance? The answer is probably yes! Peng et al. (1999) generated transgenic Basmati 370 rice plants containing a suitable mutant *GAI* allele. These plants showed reduced response to gibberellin and a dwarfed phenotype, suggesting that mutant *GAI* orthologs could indeed be employed to change plant height and increase yield by transgenic technology. Will gene technology thus enable a second 'Green Revolution'? What could be achieved?

Basmati rice, for example, looks good, cooks well and has a wonderful aroma. The plants are tall, however, and have weak culms. Wind and rain, therefore, cause considerable yield losses and a reduction in grain quality. For this and other reasons, Basmati rice is so expensive that many people cannot afford it. Conventional breeding methods to reduce the height of Basmati rice failed, because of a loss of the unique characters for which it is valued. However, introduction of a single, dominant dwarfing gene may suffice to generate a 'Green Revolution' version of Basmati rice, as suggested by the data of Peng et al. (1999), which may enable considerable increases in Basmati rice production.

Like the wheat *Rht-1* and maize *d8* mutants, *dwarf1* (*d1*) mutants from rice (*Oryza sativa*) have been classified as gibberellin-insensitive dwarf mutants. Interestingly, however, isolation by a map-based cloning approach revealed that *D1* does not encode a putative transcription factor orthologous to *RHT*-1 from wheat, *D8* from maize or *GAI* from *Arabidopsis*, but the α-subunit of a GTP-binding protein (G protein) (Ashikari et al. 1999). It could well be, therefore, that a GTP-binding protein is involved in gibberellin signal transduction, although other functions cannot be excluded so far. Will G proteins become a second class of 'Green Revolution' genes?

4 *fw2.2*, a Quantitative Trait Locus Important for the Evolution of Tomato Fruit Size

The fruit weight of tomato (*Lycopersicon esculentum*) is a very complex character that is influenced by many developmental and environmental processes (Gottlieb 1984). Accordingly, the genetic basis of fruit weight is everything but simple. A genome-wide scan, carried out by Tanksley and coworkers during the 1990s identified at least 28 QTLs controlling the different fruit weight between wild (*Lycopersicon pennellii*) and cultivated (*Lycopersicon esculentum*) tomato (Doebley 2000; Frary et al. 2000). One of these QTLs, termed *fw2.2*, changes fruit weight by up to 30% and appears to have been crucial for a key step during tomato domestication, because all wild *Lycopersicon* species that were tested contain small-fruit alleles at *fw2.2*, whereas modern tomato cultivars have large-fruit alleles. By using a map-based approach, Tanksley and coworkers were able to clone and characterise the gene responsible for the QTL effect (Frary et al. 2000). Genetic complementation analysis in transgenic plants was used to corroborate that the right gene was indeed isolated: when transformed into large-fruited cultivars, a cosmid derived from the *fw2.2* region of a small-fruited wild tomato species reduced fruit size by the expected percentage (Frary et al. 2000).

The gene responsible for the QTL effect was given the imaginative name *ORFX*. Expression analyses revealed that *ORFX* is transcribed only very weakly in all floral organs (sepals, petals, stamens and carpels), the highest level being in carpels, which give rise to fruits. Small-fruited, nearly isogenic lines showed higher levels of expression than large-fruited ones, and a lower number of cells in the carpel, while cell size was found unchanged (Frary et al. 2000).

Evaluation of the conceptual ORFX amino acid sequence suggested that it adopts a structure similar to the human oncogene RAS protein (c-H-*ras* p21), with an overall shape of heterotrimeric GTP-binding proteins. Sequence comparisons between different large- and small-fruit *ORFX* alleles, together with the expression studies, suggest that changes upstream of the coding sequence, in the putative promoter region, were responsible for the changes in fruit weight, although the importance of some potential amino acid substitutions could not be completely excluded (Frary et al. 2000). Taken together, the data currently available suggest that ORFX is a negative regulator of cell division in the carpel of tomato. Changes in the gene's promoter region may have led to weaker expression, and thus more cell divisions, resulting in more carpel cells and thus higher fruit weight.

5 *JOINTLESS*, a MADS-Box Gene Which Controls Tomato Flower Abscission Zone Development

Not only fruit weight, but also fruit abscission is an important trait in tomato. Generally, abscission is a process in plants whereby organs such as flowers, fruits or leaves are shed. This may happen during normal developmental processes, or in response to damage or stress. Shedding occurs by separation of cells in anatomically distinct regions of the plant, the so called abscission zones, due to enzymatic degradation of the middle lamella between cells in the abscission zone.

Normal tomato plants have abscission zones at the midpoint of the pedicels that carry flowers and fruits. In *jointless*, however, a tomato mutant known for decades, the formation of pedicel abscission zones is completely suppressed (Mao et al. 2000). The lack of abscission zones on pedicels of *jointless* mutants yields 'stemless' tomato fruits, which aids mechanical harvesting and prevents physical wounding during transportation. Therefore, the *jointless* mutation is widely used in the processing tomato industry and is thus of considerable agronomic value.

Mao et al. (2000) used a map-based cloning approach to isolate a candidate for the *JOINTLESS* gene. Complementation of the *jointless* mutant with the candidate gene, and antisense suppression experiments yielding the *jointless* mutant phenotype were used to confirm that *JOINTLESS* really had been cloned (Mao et al. 2000). Sequence comparisons identified *JOINTLESS* as a MADS-box gene (Mao et al. 2000).

MADS-box genes encode putative transcription factors. They are defined by the presence of a highly conserved DNA sequence, termed the MADS-box, which encodes the DNA-binding domain of the respective MADS-domain transcription factors (for a recent review about MADS-box genes, see Theißen et al. 2000). 'MADS' is an acronym for the four founder proteins MCM1 (from brewer's yeast, *Saccharomyces cerevisiae*), AGAMOUS (from *Arabidopsis*), DEFICIENS (from *Antirrhinum*), and SRF (from human), on which the definition of this gene family is based. In flowering plants, MADS-box genes control many developmental processes during both vegetative and reproductive growth (Theißen et al. 2000). MADS-box genes are especially well known for their role in the specification of floral organ identity (for recent reviews, see Theißen 2001a; Theißen and Saedler 2001).

Phylogeny reconstructions revealed that *JOINTLESS* is a member of a novel subfamily of MADS-box genes (Mao et al. 2000), termed *STMADS11*-like genes (Becker et al. 2000). Members of that gene subfamily are typically expressed in vegetative rather than floral organs. *STMADS11*-like genes have already been found in a number of flowering plant species, but also in a gymnosperm, indicating that the first gene subfamily member existed more than 300 million years ago (Becker et al. 2000). Thus *JOINTLESS* orthologs should be present in most, if not all flowering plant species, and it is tempting to suggest that these may become valuable tools to manipulate fruit abscission. But things may be more complicated. The closest relative of *JOINTLESS* known so far is *SHORT VEGETATIVE PHASE* (*SVP*) from *Arabidopsis*, a gene which encodes a repressor of the floral transition rather than a product involved in flower or fruit abscission (Hartmann et al. 2000). Thus *STMADS11*-like genes may be involved in quite a diversity of developmental processes, including the floral transition, some of which may be of agronomic interest (Theißen 2001b).

6 *BoCAL*, a MADS-Box Gene Key to the Evolution of the Cauliflower Curd

Among the most divergent structures that characterise varieties within a single plant species are the vegetables that have been developed from

Brassica oleracea. Cabbage has a high number of leaves overlapping its terminal meristem, kohlrabi has a swollen corm-like stem, kale has a fleshy marrow stem, Brussels sprouts have enlarged axillary buds, and cauliflower, 'romanesco' and broccoli have thickened inflorescences and fleshy flower buds (Gottlieb 1984). The classical white semi-spherical curd of cauliflower (*Brassica oleracea* ssp. *botrytis*) consists of a dense mass of developmentally arrested inflorescence meristems, only a small fraction of which eventually develop into floral primordia and flowers. This suggests that the transition from inflorescence to floral meristem identity is disturbed in cauliflower.

In *Arabidopsis*, expression of the MADS-box genes *APETALA1* (*AP1*) and *CAULIFLOWER* (*CAL*) is required for the transition from inflorescence to floral meristems and the specification of floral meristem identity. In plants which are mutant for both genes, a dense mass of inflorescence meristems develops similarly to the cauliflower curd (Kempin et al. 1995). This raised the intriguing possibility that the cauliflower curd is caused by the mutation of two MADS-box genes.

Molecular genetic analyses indeed provided evidence that the *CAL* ortholog of cauliflower, termed *BoCAL*, has a premature termination codon and thus encodes a prematurely terminated protein which may be impaired in its function (Kempin et al. 1995). With the ortholog of *AP1*, termed *BoAP1*, the situation appears to be more complicated. As a matter of fact, there are two *BoAP1* copies in the *Brassica* genome, only one of which encodes a prematurely terminated protein in the case of cauliflower (Lowman and Purgganan 1999). Whether the *BoAP1* function is impaired in cauliflower is unclear, however. In any case, the available evidence strongly suggests that changes at least one (*BoCAL*) or even two (*BoAP1*) loci encoding MADS-box genes were of great importance during the origin of cultivated cauliflower.

7 *SHP* and *FUL* in Canola: Shatterproof Fruits by Design?

Thus far we have considered genes which were of importance during plant domestication and breeding processes in the past. In the last example we will have a look at a class of genes which may play a role during future plant breeding. The trait of interest here is fruit dehiscence, or pod shatter.

Pod shatter is an important agronomic trait in a number of crop plants. Depending on the crop, yields might be increased by making seedpods weaker or stronger. In the case of cotton (*Gossypium*), for example, harvesting might be easier if the seedpods are more fragile. In contrast, strong, shatterproof seeds are desirable in oilseed crops such as canola (*Brassica napus*, *B. rapa*), where pod shatter before harvest can cause considerable yield losses (up to 50% under adverse weather conditions) (Moffat 2000). Indehiscent pods permit harvest to continue over longer time periods and helps to increase the amount of seed actually obtained.

If there is sufficient genetic diversity in a crop variety, shatter-resistant pods may be obtained by simple selection schemes and classical breeding, as already demonstrated many decades ago for lupins (*Lupinus luteus*) by von Sengbusch (1934). A better understanding of the molecular genetic mechanisms underlying pod shatter, however, could enable more direct approaches to generate shatter-resistant crop plants.

Fortunately, the currently most intensively studied plant model system, the tiny weed *Arabidopsis thaliana*, is relatively closely related to the oilseed crop canola. *Arabidopsis* is typical of many of the approx. 3000 species of the mustard family (Brassicaceae) in that it produces dry, dehiscent fruits, i.e. fruits that burst open at maturity to release its seeds. Since *Arabidopsis* is amenable to efficient molecular genetic techniques, including reverse genetics and transformation, it could be used as a starting point for unravelling the molecular genetic basis of pod shatter.

Arabidopsis fruits are composed of two valves (carpel walls) separated by a thin structure called the replum (Ferrándiz et al. 1999). At the boundary between the valve and the replum, a narrow band of cells develops into the dehiscence zone. Separation of the cells from one another in the dehiscence zone late in fruit development leads to valve detachment from the replum and thus allows seed dispersal (Ferrándiz et al. 1999). Lignification of valve margin cells adjacent to the dehiscence zone and of an internal valve cell layer, very likely contributes to fruit dehiscence.

By systematically employing the methods of reverse genetics, Yanofsky and his coworkers have generated *Arabidopsis* plants in which two very similar and functionally redundant MADS-box genes, termed *SHATTERPROOF1* (*SHP1*) and *SHATTERPROOF2* (*SHP2*) have both lost their function (Liljegren et al. 2000). In mature fruits of the double mutants the dehiscence zones are absent and, therefore, the fruits fail to dehisce. The lignification of valve margin cells is reduced in these mutants. The mutant phenotype is in line with the expression pattern of the *SHP* genes, which are transcribed in narrow stripes just before the valve margin is distinct. This expression pattern at the valve margins is maintained after fertilisation of the flower, suggesting that the *SHP* genes function both to specify the identity of the margin and to control dehiscence zone development in the mature fruit (Liljegren et al. 2000).

Yanofsky and colleagues have complemented their loss-of-function analyses by examining transgenic plants that constitutively express the *SHP* genes under the control of the cauliflower mosaic virus (CaMV) 35S promoter, and by employing putative downstream targets of *SHP1* and *SHP2* transcription factors as molecular markers to monitor the cellular differentiation of valve margins. The data obtained are all compatible with the view that in wild type plants, *SHP1* and *SHP2* control proper development of the dehiscence zone cells and of adjacent cells at the valve margin of *Arabidopsis* fruits (Liljegren et al. 2000).

Yanofsky and coworkers also reported evidence that another MADS-box gene, termed *FRUITFULL* (*FUL*), interacts antagonistically with the *SHP* genes during valve margin development (Ferrándiz et al. 2000b). Plants ectopically expressing *FUL* under the control of the CaMV 35S promoter resemble *shp* loss-of-function plants in that their fruits become indehiscent and fail to disperse the seeds normally. In wild-type plants *FUL* expression in the valves is required for valve cell differentiation and expansion after fertilisation (Gu et al. 1998).

From an evolutionary point of view, it is interesting to note that *FUL* as well as the *SHP* genes are very close relatives of genes involved in the specification of floral meristem or organ identity, respectively (Theißen 2000b). *FUL* may still function as a floral meristem identity gene during early stages of flower development (Ferrándiz et al. 2000a). The available evidence documents the involvement of new genes, such as the *SHP* s and *FUL*, generated by gene duplication, sequence divergence and fixation, in the evolution of novel reproductive devices, i.e., fruits, during the course of flowering plant evolution (Theißen 2000b).

How could the novel insights into the molecular genetics of fruit dehiscence in *Arabidopsis* (Ferrándiz et al. 2000b; Liljegren et al. 2000) be used to generate shatter-resistant canola? *Arabidopsis* and canola are so closely related that knocking out the function of the canola *SHP* ortholog, or overexpressing the *FUL* gene, may result in shatter-resistant canola fruits. The respective plants might be generated by genetic engineering. Especially the ectopic expression of *FUL* could be a fast way to generate shatter-resistant oilseed plants (Ferrándiz et al. 2000b). However, given the current reluctance to accept the introduction of transgenically modified crops, one may prefer to apply alternatives which do not involve transgenic technology. Plants could be randomly mutagenised, and high yielding, *shp* loss-of-function plants may be selected via marker-assisted breeding, during the course of which, *shp* loss-of-function alleles are brought into the genetic background of canola elite lines. Alternatively, high-yielding lines ectopically expressing *FUL* may be developed in analogous ways, although such gain-of-function mutants may be more difficult to find, since the occurrence of gain-of-function mutations is less likely than that of loss-of-function mutations due to the molecular mechanisms involved.

8 Concluding Remarks

We have seen that the first plant 'domestication genes' or 'breeding genes' could be analysed at the molecular level (summarised in Table 1).

Although the number of case examples is certainly still too small to allow statistically sound general conclusions, a critical discussion in the light of contemporary theories may be allowed.

Not long ago, Doebley and Lukens (1998) published the interesting hypothesis that changes in the *cis*-regulatory elements of transcriptional regulators provides a predominant mechanism for the generation of novel phenotypes in plants. Does this hold for the often dramatic changes that occurred during plant domestication?

We have seen that single loci can have major effects on traits of agronomic interest, and that the vast majority of the genes cloned so far indeed encode transcription factors (Table 1), although different kinds of biases introduced by the methods of gene identification and mapping (Barton and Turelli 1989) and during molecular cloning may have influenced the picture. But cases like the *BoCAL* and the *RHT* genes exemplify that both, loss- as well as gain-of-function (or altered function) effects can be obtained by mutations in the coding region. Thus the relative importance of changes at *cis*-regulatory elements, which were obviously essential in some exciting cases (*TB1* from maize and *fw2.2* from tomato, to mention the two most spectacular and instructive examples considered here), remains unclear, and its critical assessment simply awaits a significantly higher number of case studies.

However, whatever may turn out to be the most important molecular genetic mechanisms behind key phenotypic changes during crop plant domestication, it is yet another question whether the same principles also apply to the origin of natural plant variability. Are polygenes, rather than single Mendelian loci, of greater importance here than in the case of artificially selected crop plants? Or are single genes that control plant development, most of which encode transcription factors, also important for the 'natural' evolution of plant structure (Theißen and Saedler 1995; Theißen et al. 2000)? Answering questions like these will be the key to our understanding of the origin of biodiversity on this planet.

Acknowledgements. I would like to thank Heinz Saedler and the members of my group for continuous support during recent years.

References

Ashikari M, Wu J, Yano M, Sasaki T, Yoshimura A (1999) Rice gibberellin-insensitive dwarf mutant gene *Dwarf1* encodes the α-subunit of GTP-binding protein. Proc Natl Acad Sci USA 96:10284–10289

Barton NH, Turelli M (1989) Evolutionary quantitative genetics: how little do we know? Annu Rev Genet 23:337–370

Beadle GW (1980) The ancestry of corn. Sci Am 242:96–103

Becker A, Winter K-U, Meyer B, Saedler H, Theißen G (2000) MADS-box gene diversity in seed plants 300 million years ago. Mol Biol Evol 17:1425–1434

Cubas P, Lauter N, Doebley J, Coen E (1999) The TCP domain: a motif found in proteins regulating plant growth and development. Plant J 18:215–222

Doebley J (1990) Molecular evidence and the evolution of maize. Econ Bot 44 (Suppl 3):6–27

Doebley J (1992) Mapping the genes that made maize. Trends Genet 8:302–307

Doebley J (2000) A tomato gene weighs in. Science 289:71–72

Doebley J, Lukens L (1998) Transcriptional regulators and the evolution of plant form. Plant Cell 10:1075–1082

Doebley J, Stec A (1991) Genetic analysis of the morphological differences between maize and teosinte. Genetics 129:285–295

Doebley J, Stec A (1993) Inheritance of the morphological differences between maize and teosinte: comparison of results for two F_2 populations. Genetics 134:559–570

Doebley J, Stec A, Wendel J, Edwards M (1990) Genetic and morphological analysis of a maize-teosinte F_2 population: implications for the origin of maize. Proc Natl Acad Sci USA 87:9888–9892

Doebley J, Stec A, Gustus C (1995) *teosinte branched 1* and the origin of maize: evidence for epistasis and the evolution of dominance. Genetics 141:333–346

Doebley J, Stec A, Hubbard L (1997) The evolution of apical dominance in maize. Nature 386:485–488

Ferrándiz C, Pelaz S, Yanofsky MF (1999) Control of carpel and fruit development in *Arabidopsis*. Annu Rev Biochem 68:321–354

Ferrándiz C, Gu Q, Martienssen R, Yanofsky MF (2000a) Redundant regulation of meristem identity and plant architecture by *FRUITFULL*, *APETALA1* and *CAULIFLOWER*. Development 127:725–734

Ferrándiz C, Liljegren SJ, Yanofsky MF (2000b) Negative regulation of the *SHATTERPROOF* genes by FRUITFULL during *Arabidopsis* fruit development. Science 289:436–438

Frary A, Nesbitt TC, Frary A, Grandillo S, van der Knaap E, Cong B, Liu J, Meller J, Elber R, Alpert KB, Tanksley SD (2000) *fw2.2:* a quantitative trait locus key to the evolution of tomato fruit size. Science 289:85–88

Gottlieb LD (1984) Genetics and morphological evolution in plants. Am Nat 123:681–709

Gu Q, Ferrándiz C, Yanofsky MF, Martienssen R (1998) The *FRUITFULL* MADS-box gene mediates cell differentiation during *Arabidopsis* fruit development. Development 125:1509–1517

Hartmann U, Höhmann S, Nettesheim K, Wisman E, Saedler H, Huijser P (2000) Molecular cloning of *SVP*: a negative regulator of the floral transition in *Arabidopsis*. Plant J 21:351–360

Kempin SA, Savidge B, Yanofsky MF (1995) Molecular basis of the *cauliflower* phenotype in *Arabidopsis*. Science 267:522–525

Koornneef M, Stam P (2001) Changing paradigms in plant breeding. Plant Physiol 125:156–159

Liljegren SJ, Ditta GS, Eshed Y, Savidge B, Bowman JL, Yanofsky MF (2000) *SHATTERPROOF* MADS-box genes control seed dispersal in *Arabidopsis*. Nature 404:766–770

Lowman AC, Purugganan MD (1999) Duplication of the *Brassica oleracea APETALA1* floral homeotic gene and the evolution of domesticated cauliflower. J Hered 90:514–520

Luo D, Carpenter R, Vincent C, Copsey L, Coen E (1996) Origin of floral asymmetry in *Antirrhinum*. Nature 383:794–799

Luo D, Carpenter R, Copsey L, Vincent C, Clark J, Coen E (1999) Control of organ asymmetry in flowers of *Antirrhinum*. Cell 99:367–376

Mao L, Begum D, Chuang H-W, Budiman MA, Szymkowiak EJ, Irish EE, Wing RA (2000) *JOINTLESS* is a MADS-box gene controlling tomato flower abscission zone development. Nature 406:910–913

Moffat AS (2000) Can genetically modified crops go 'greener'? Science 290:253–254

Paterson AH, Lin Y-R, Li Z, Schertz KF, Doebley JF, Pinson SRM, Liu S-C, Stansel JW, Irvine JE (1995) Convergent domestication of cereal crops by independent mutations at corresponding genetic loci. Science 269:1714–1718

Peng J, Richards DE, Hartley NM, Murphy GP, Devos KM, Flintham JE, Beales J, Fish LJ, Worland AJ, Pelica F, Sudhakar D, Christou P, Snape JW, Gale MD, Harberd NP (1999) 'Green revolution' genes encode mutant gibberellin response modulators. Nature 400:256–261

Szabó VM, Burr B (1996) Simple inheritance of key traits distinguishing maize and teosinte. Mol Gen Genet 252:33–41

Theißen G (2000a) Evolutionary developmental genetics of floral symmetry: the revealing power of Linnaeus' monstrous flower. Bioessays 22:209–213

Theißen G (2000b) Shattering developments. Nature 404:711–713

Theißen G (2001a) Development of floral organ identity: stories from the MADS house. Curr Opin Plant Biol 4:75–85

Theißen G (2001b) *SHATTERPROOF* oil seed rape: a *FRUITFULL* business? MADS-box genes as tools for crop plant design. Biotech News Int 6:13–15

Theißen G, Saedler H (1995) MADS-box genes in plant ontogeny and phylogeny: Haeckel's 'biogenetic law' revisited. Curr Opin Genet Dev 5:628–639

Theißen G, Saedler H (2001) Floral quartets. Nature 409:469–471

Theißen G, Becker A, Di Rosa A, Kanno A, Kim JT, Münster T, Winter K-U, Saedler H (2000) A short history of MADS-box genes in plants. Plant Mol Biol 42:115–149

Wang R-L, Stec A, Hey J, Lukens L, Doebley J (1999) The limits of selection during maize domestication. Nature 398:236–239

Von Sengbusch R (1934) Lupinen mit nichtplatzenden Hülsen. Züchter 6:1–5

Prof. Dr. Günter Theißen
Max-Planck-Institut für Züchtungsforschung
Abteilung Molekulare Pflanzengenetik
Carl-von-Linné-Weg 10
50829 Köln, Germany
Tel.: +49-221-5062-122
Fax: +49-221-5062-113
e-mail: theissen@mpiz-koeln.mpg.de

Physiology

Redox Regulation in Oxigenic Photosynthesis

By Karl-Josef Dietz, Gerhard Link, Elfriede K. Pistorius, Renate Scheibe

1 Introduction

Photosynthetic cells use the energy from absorbed light to synthesize partially or fully reduced organic compounds such as carbohydrates, fatty acids and amino acids from oxidized substrates. A series of elaborate redox reactions links the fundamental events of light absorption to the ultimate supply of reduced assimilates within the cell and also for export to sink tissues. When one looks from a modeling point of view, three sets of parameters appear to be of interest and – if possible – should be controlled in order to optimize the process of photosynthesis, to minimize the waste of energy and to prevent the development of damage (Fig. 1). These three sets of parameters are (1) the input parameters, mainly the incident photon flux density, other environmental parameters and gene activity, (2) the output parameters, particularly the accumulated amount of assimilate products and the energy status, and (3) process parameters, the redox state of intermediate reactions which are of key importance for the whole process.

1. Input parameters
   The absolute amount of light quanta absorbed in the antenna is not necessarily a useful parameter since the energy input has to be related to the energy consumption in the whole process of assimilate synthesis. Under conditions of sufficient availability of oxidized substrates and fast growth, energy conversion can proceed at a high rate and there is a need for efficient light absorption to sustain high rates of photosynthesis. Conversely, under stress conditions, the major portion of absorbed energy will have to be dissipated and light absorption should be minimized. Consequently, the balance between absorbed light on the one hand and energy use on the other hand is determined in the photosynthetic apparatus by measuring the state of reactions which, in the reaction sequence are 'downstream' of the absorption and charge separation events. Either the state of redox reactions as discussed below or the degree of thylakoid energization as reflected in the luminal proton concentration of the thylakoids are sensed and transformed to appropriate adaptive responses such as

Progress in Botany, Vol. 63

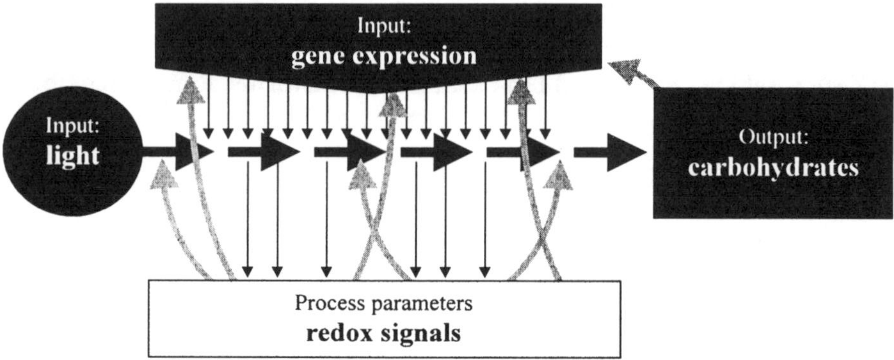

Fig. 1. A simplified model of photosynthesis. The input parameters considered here are light and the expression of photosynthetic genes needed for the buildup and maintenance of the photosynthetic apparatus. Other input parameters as for instance the CO_2-availability are considered constant. The output of the system are sugars and other assimilates. The complexity of the photosynthetic process requires feed-back and feed-forward control mechanisms (*gray arrows*). System parameters, particularly as defined by the redox states of key reactions, are used as signals reporting on the efficiency of the process and to tune signaling pathways, which in turn adjust the activity of individual reactions for optimization of the whole system performance

increased non-photochemical quenching. At low luminal pH the violaxanthin deepoxidase is activated and zeaxanthin and antheraxanthin are formed from violaxanthin. Incorporation of these 'quenching' carotenoids in the antenna increases the non-radiative dissipation of excitation energy (Demmig-Adams and Adams 1992). In addition, the pH-gradient also affects non-photochemical quenching independently of the zeaxanthin cycle (Havaux et al. 2000). Gene expression and the build-up and maintenance of the photosynthetic apparatus represent another input parameter which is not measured as a direct system parameter but indirectly on the basis of the process performance employing parameters such as the state of redox reactions (see below). It should be noted that the distinction between the redox state of the plastoquinone pool and the transthylakoid pH gradient as a primary signal for acclimatory responses is not possible yet on the basis of the available results (Mullineaux et al. 2000).

2. Output parameters
Plant cells measure their carbohydrate status and adjust their metabolism accordingly, mainly on the level of gene expression. This phenomenon is referred to as 'sugar sensing' and has received considerable attention during the last years (Farrar et al. 2000). In photosynthetic cells, high sugar contents induce global changes in gene expression which result in a decrease of the photosynthetic capacity (Pego et al. 2000). This regulation apparently involves various signaling mechanisms and can be considered as typical feedback inhibi-

tion, the response of which is manifested only slowly within several hours to days (Farrar et al. 2000).

3. Process parameters
   A time scale of hours to days is too slow and not sufficient to respond appropriately, when photosynthetic cells are subjected to fast changing environmental parameters which immediately affect the balance between energy input and assimilate output. The cells must counteract such sudden imbalances otherwise they may induce damage at sensitive targets of the photosynthetic machinery long before appreciable changes in sugar accumulation occur. The main mechanism of damage involves production of reactive oxygen species (ROS) by consecutive reduction of dioxygen and subsequent oxidative inactivation and destruction of proteins and lipids. ROS production occurs when more light energy is absorbed than can be converted into assimilates. For example such a situation occurs under drought stress, when the photon fluence rate is high and the CO_2 availability is low, and can easily be illustrated by measuring chlorophyll a fluorescence quenching parameters (Dietz et al. 1985). The reduction state of the photosynthetic electron transport chain is usually inversely related to the photosynthetic yield and directly related to the production of ROS, and thus provides important redox information on the system. The information on the state of redox reactions is then transduced to regulatory responses as will be outlined below.

During the last decade, a large body of information has accumulated on the central importance of redox reactions in the control of photosynthesis. A single review cannot cover the whole field of redox-control in photosynthesis exhaustively. Therefore, the aim of this review is to provide an introduction for the reader into the topic of redox-dependent regulation in the chloroplast and the cytoplasm as related to oxigenic photosynthesis using selected examples from higher plants, also highlighting the advantage of employing the prokaryotic system of cyanobacteria to deepen our understanding of redox control in oxigenic photosynthesis.

2 Redox Signals from the Photosynthetic Electron Transport Chain

Figure 2 summarizes our present knowledge on redox-related signals which may originate from the photosynthetic electron transport chain. The signal may be related to the redox state of an individual component of the electron transport chain (PQ, Cyt b559), to the concentration or the redox potential of a redox active metabolite (GSH, ascorbate, NADPH) or to the concentration of reactive oxygen species ($O_2^{\bullet-}$, H_2O_2).

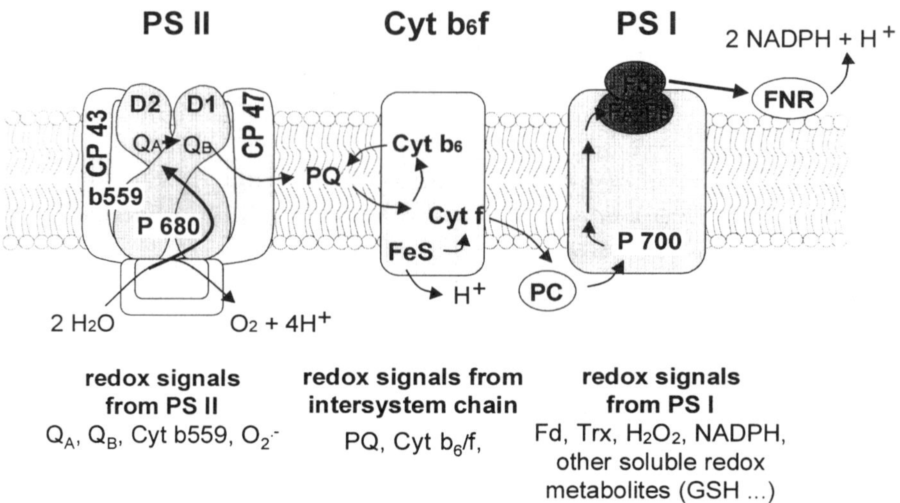

Fig. 2. Schematic depiction of redox-signals originating from the photosynthetic electron transport chain. Various types of redox signals may originate either from photosystem II, the intersystem electron transport chain or from PS I. Other redox-active metabolites can be synthesized or reduced from NADPH

The signals must be detected by sensor molecules and transduced to the appropriate response. Particularly, knowledge of the nature of the sensing and transduction mechanisms is only slowly emerging. Examples of genetic and bioinformatic approaches will be given in Sections 3 and 7. Much more is known about the redox-induced responses which cover a broad range of mechanisms from fast and direct biochemical regulation of enzymes to the expressional activation of gene networks. Four major fields of redox control will be addressed in the following. Due to a lack of space, reference will frequently be given to concise review articles and only occasionally to original articles. At least partly, the selection of the references in this rapidly growing field must reflect an arbitrary choice or a personal opinion.

3 Short-Term Redox Control of Primary Metabolism

Plants in their natural environment experience continuous changes in light intensity, CO_2 concentration, nutrient availability, and temperature. All these factors have great impact on the velocities of basic metabolic reactions of a photoautotrophic organism. In the long run, plants possess the ability to adapt their structure and their enzymatic equipment to the growth conditions. However, the time- and energy-requiring processes at the level of gene expression and translation are not suited to adjust the metabolism to short-term changes as often encountered by a leaf in a canopy during a sunny day with wind and clouds permanently affecting the rates of photosynthesis and transpiration. Using reversible redox-changes on target enzymes, plants possess a flexible mechanism to cope with changing electron flow and metabolism, and thus can maintain homeostasis over a wide range of conditions (Scheibe 1996; Kelly 1999, 2000).

a) Redox Potentials of Intracellular Proteins

As the intracellular milieu is generally kept in a reduced state, so are the protein thiols. This is the prerequisite for most intracellular proteins to be in their functional state. In extracellular compartments, as in the apoplast or in the endoplasmic reticulum (ER) when proteins are to be secreted, permanent disulfide bridges are formed for higher stability in an oxidized environment.

As an exception, the chloroplast compartment with its extreme redox conditions houses the light/dark-modulated enzymes that undergo a continuous redox change at special regulatory thiol groups. Since oxidized and reduced enzyme forms exhibit distinct properties, a change in the steady-state ratios of both forms is immediately reflected as changed enzyme activity causing a changed flux. The electron flow required for this system of continuous reduction and reoxidation is generated in the light through photosynthetic electron flow and is mediated via the ferredoxin/thioredoxin system.

The components of this system are well characterized. The 1-electron transfer from the highly electronegative iron-sulfur cluster in ferredoxin changes to 2-electron/2-proton steps in the form of protein thiols/disulfides in subunit B of the ferredoxin/thioredoxin reductase (FTR) and is further channeled to thioredoxins m and f that reduce the various target enzymes (Ruelland and Miginiac-Maslow 1999; Schürmann and Jacquot 2000). The more "classical" targets are the Calvin-cycle enzymes (Martin et al. 1999) NADP-glyceraldehyde 3-phosphate dehydrogenase (NADP-GAPDH), fructose 1,6-bisphosphatase (FBPase), sedoheptulose 1,7-bisphosphatase (SBPase), phosphoribulokinase (PRK), the key enzyme of the oxidative pentose phosphate pathway glucose 6-phosphate dehydrogenase (G6PDH) that is inactivated upon reduction, NADP-malate dehydrogenase (NADP-MDH) as part of the malate valve, and CF_1 that is reduced in its γ subunit. When electron flow stops, these enzymes are converted to their oxidized forms due to the fact that their redox potential is rather negative (Hirasawa et al. 1999, 2000) and the presence of O_2.

New developments in the elucidation of the structures responsible for such posttranslational mechanisms have lately been reviewed in various articles. Most exciting was the resolution of crystal structures of the oxidized FBPase (Chiadmi et al. 1999) and of NADP-MDH (Carr et al. 1999; Johansson et al. 1999). Although the earlier approaches to identify the regulatory cysteines and the mechanism of inactivation had led to rather good models, the structures gave interesting clues. For NADP-MDH, the oxidized C-terminal sequence was identified to have an autoinhibitory effect by binding of a glutamate residue to the active site residues, while the N-terminus stabilizes the inactive dimer (Ruelland et al. 1998). For FBPase, two of the three cysteines on the inserted sequence of the chloroplast enzyme are involved in the actual disulfide bridge that leads to a conformational change comparable to the allosteric change upon effector (AMP) binding to the non-plastid isoforms.

b) Interaction with Metabolism

As a reversible system of redox modification the thioredoxin system can function as an on/off switch. This is the case upon darkening when electron flow stops completely. In the light, however, fine-tuning at individual steps is required in order to adjust the fluxes not only to electron flow, but also to the metabolic situation which might vary even under constant light. This can be achieved by the fact that at steady state the redox potentials of the various enzymes are influenced by the presence of certain positive or negative effectors (Faske et al. 1995). This leads to a very flexible situation where metabolites act very efficiently to adjust the fluxes through the pathway under constant light (Holtgrefe et al. 1997).

Thus, a changed metabolite pool can act upon an enzyme activity by changing its activation state or its catalytic activity. This fact leads to a homeostatic compensation of the flux, as long as none of the enzymes become limiting under the given conditions (Fridlyand et al. 1999; Fridlyand and Scheibe 2000). It is feasible to believe that a prolonged period of imbalance due to a limitation will lead to the release of a signal affecting gene expression; sugar sensing and redox control being examples for such links between metabolism and the transcriptional level.

c) Are There More Redox-Controlled Enzymes?

For the enzymes mentioned above, there is considerable evidence that redox regulation is actually functioning although in some cases the redox change is not directly linked to the activity change. Even the oxidized forms of NADP-GAPDH and of FBPase can be activated at least <u>in vitro</u> (Reichert et al. 2000). Especially for NADP-GAPDH this can also happen inadvertently <u>in vitro</u> by dilution during activity determination, since activation is achieved upon dissociation of oligomeric forms. This makes it difficult to identify redox-regulated enzymes merely by determining their activation state. Good evidence for redox regulation requires the simultaneous redox changes at regulatory cysteines and of activity being dependent upon the redox situation both in vivo and in vitro. It is also not trivial to distinguish between regulation and oxidative damage that is repaired by reduction.

There are, however, various interesting new targets of redox regulation. Thus, the key enzyme of starch biosynthesis, **ADP-glucose pyrophosphorylase** turned out to be reductively activated in the presence of ADP-glucose (Fu et al. 1998; Preiss et al. 1999; Ballicora et al. 2000). Here, the regulatory disulfide bridge appears to be localized between two small subunits.

Rubisco activase in its larger isoform is also reductively activated at low ADP/ATP ratios (Zhang and Portis 1999). Thioredoxin f catalyzes this redox modification which is specific for the larger isoform, and the two participating Cys residues have been identified.

The plastid enzyme **acetyl-CoA carboxylase** has been identified to be light/dark-modulated only recently (Sasaki et al. 1997; Hunter and Ohlrogge 1998). In addition, the enzyme is activated by preincubation with acetyl-CoA. This key regulatory step in fatty-acid synthesis is thereby linked both to light and metabolism. Since the enzyme is composed of four different polypeptides, it is difficult to identify the regulatory Cys residues. The successful production of recombinant enzyme consisting of subunits α and β is therefore the prerequisite for further investigations (Kozaki et al. 2000).

There is evidence that a cytosolic enzyme, namely a **calmodulin-dependent NAD^+ kinase** responds to redox changes. Upon short-term salt stress, *Lycopersicon pimpinellifolium* cells are under oxidative stress leading to inactivation of this enzyme, which could be restored by dithiothreitol (DTT) treatment (Delumeau et al. 2000). Since the enzyme is localized in the cytosol, thioredoxin h and NADP-thioredoxin reductase should mediate electron transfer from NADPH. Changes in the cytosolic redox state, as triggered by NaCl treatment in this case, could be linked to other signal transduction systems.

The **hydrogenase** activity of the green alga *Scenedesmus obliquus* was shown to be inactivated by reduced thioredoxin f, but no sequence data are available yet at all (Wünschiers et al. 1999). This reductive inactivation might be of physiological relevance, since hydrogen production would compete for electrons during light-dependent CO_2 assimilation.

In mitochondria, the **2-oxoacid dehydrogenase** complex (Bunik et al. 1999) and the **alternative oxidase** (Vanlerberghe et al. 1998, 1999) are likely to be further targets of redox-modulation. As in the other cases, both the reduced state and metabolites (e.g. pyruvate) are required to obtain the active form.

These more recently discovered examples of non-plastid enzymes that respond to changes in the redox state open the possibility to find links between the chloroplast and other cell compartments and set the challenge to identify the signal transduction chains to these pathways as well as to gene expression in the nucleus.

4 Redox Regulation of Plastid Gene Expression

Chloroplasts also provide an ideal environment for redox regulatory mechanisms to adjust plastid gene expression both to the status of photosynthetic activity and the requirements of the cell. In fact the hypothesis was made that the presence of certain genes in the chloroplast or even the maintenance of the plastome as a subgenomic unit of the plant cells is required to allow for rapid redox response to photosynthetic needs (Allen 1993b). Redox regulation of plastid gene expression is not unique, as precedents for redox regulated gene expression exist both in bacteria (Demple 1998; Zheng et al. 1998) and eukaryotic nuclear systems (Sen and Packer 1996). Plastids contain their own genetic material and a complete set of proteins responsible for all steps in gene expression, ranging from transcription via RNA maturation to translation, and these processes take place in close physical proximity to photosynthesis (Bogorad and Vasil 1991; Sugita and Sugiura 1996). Together, this has stimulated ideas that the intraorganellar genetic system might be under redox control in a tightly interconnected fashion with the photosynthetic apparatus (Allen 1993a). What is the experimental evidence for this?

a) Multitude of Redox-Regulatory Mechanisms

Redox regulation of chloroplast gene expression has been detected in different species (algae, higher plants), stages and tissue types (seedlings, mature leaves), and under a variety of environmental conditions (light/dark, photostress, PS I/PS II photoacclimation). For more comprehensive reviews on the physiology of redox-dependent gene regulation in plants the reader is referred to Durnford and Falkowski (1997) and Link (2001).

As shown in Table 1, mechanisms that represent major steps in plastid gene expression, including transcription, posttranscriptional mechanisms at RNA level (5′ and 3′ processing, splicing), and translation (initiation and elongation), are all subject to redox regulation. In view of the fact that many of these findings are quite recent, it seems likely that this is only the 'tip of the iceberg' in a sense that only the more obvious effects have been analyzed thus far. More subtle (and more difficult to analyze) redox-controlled changes may still have been over-

Table 1. Evidence for redox regulation of chloroplast gene expression

Transcription	RNA Maturation	Translation	
Altered in vivo transcript levels of chloroplast genes in response to: redox-reactive reagents (diamide, H_2O_2, DTT, glutathione) chemicals that affect glutathione biosynthesis (NAC, BSO) photosynthetic electron transport inhibitors (DCMU, DBMIB) photostress conditions			Karpinski et al. (1997); Kettunen et al. (1997); Salvador and Klein (1999); S. Stratmann, T. Pfannschmidt and G. Link (unpubl.)
Altered transcription rates during in organello run-on experiments using chloroplasts from plants treated with PS I- or PS II-sensitizing light ± DCMU, DBMIB from plants under photostress conditions			Deng et al. (1989); Pfannschmidt et al. (1999a, b); Tullberg et al. (2000) Deng et al. (1989); Pfannschmidt et al. (1999a,b); Tullberg et al. (2000 E. Baena-Gonzalez, S. Baginsky, H. Summer, E.-M. Aro and G. Link (unpublished)
Chloroplast RNA polymerase (PEP-A) contains an associated protein kinase activity (ATP- and GTP-dependent) that responds to SH-group redox state In vitro transcription using chloroplast RNA polymerase is affected by kinase inhibitors and redox-reactive reagents			Baginsky et al. (1997); Baginsky et al. (1999); K. Ogrzewalla and G. Link (unpublished)

Table 1. (continued)

Transcription	RNA Maturation	Translation	
	Splicing of chloroplast precursor transcripts in vivo is affected by DCMU, DBMIB		Deshpande et al. (1997)
	Light-enhanced degradation of chloroplast transcripts is delayed by diamide or DCMU		Salvador and Klein (1999)
	RNA binding and processing activities of chloroplast 3′-UTR binding endonuclease p54 are modulated by (ATP-dependent) phosphorylation and SH-group redox state in vitro		Nickelsen and Link (1993); Liere and Link (1997)
	In vitro activity of a protein complex that binds to the 5′-UTR of plastid mRNA is affected by (ADP-dependent) phosphorylation and SH-group redox state:		Danon and Mayfield (1994a, b); Danon (1997); Kim and Mayfield (1997); Yohn et al. (1998a); Fong et al. (1999)
	Extracts from light-grown cells have higher binding activity than those from dark-grown cells		
	The DTNB-oxidized complex can be reactivated by DTT		
	RNA binding is decreased in extracts from a PSI mutant		

	The complex contains components related in sequence to protein disulfide isomerase (PDI) and poly(A)-binding proteins (PABPs), respectively		
		Light-dependent recruitment of plastid mRNA to polysomes and enhanced protein synthesis consistent with initiation control of translation. Amounts of newly synthesized proteins in isolated chloroplasts are affected by the redox state of vicinal SH-groups on the PDI component of the 5′UTR binding complex	Danon and Mayfield (1994b); Yohn et al. (1998b); Trebitsh et al. (2000)
		Protein synthesis in isolated chloroplast under pulse-chase conditions, in the presence of lincomycin (initiation inhibitor), nigericin, DCMU, and other inhibitors, photostress	Kuroda et al. (1996); Kettunen et al. (1997); Mühlbauer and Eichacker (1998); Zhang et al. (1999); Zhang et al. (2000)
		Evidence for elongation control of Chloroplast translation by redox poise and SH-group modification	
		Cross-linking of nascent chains points to redox control of posttranslational events, including precursor processing, and assembly of multi-protein complexes	Zhang et al. (2000)

DTT, dithiothreitol; NAC, N-acetyl-cysteine; BSO, buthionine sulfoximine; DCMU, dichlorophenyldimethylurea; DBMIB, dibromomethyl isopropylbenzoquinone; UTR, untranslated region; DTNB, dithiobis nitrobencoic acid.

looked. Furthermore, it has become increasingly clear that the noticeable impact of a certain mechanism (e.g., transcription *versus* translation) on the *steady-state* concentration of the final gene product seems to be highly variable, depending on the choice of organism, cell and tissue type, developmental and environmental situations. Redox control of translation (initiation) was established in work done with *Chlamydomonas reinhardtii* (Mayfield et al. 1995; Danon 1997), whereas redox regulation of transcription was studied in higher plants (Link 1996; Link et al. 1997). This makes it difficult to generalize the findings in each system and, despite technical challenges, it should be highly rewarding to raise comparative data both in lower and higher photosynthetic eukaryotes.

What are the redox signaling processes responsible for changes in plastid gene expression, and where do they originate? In part of the work listed in Table 1 redox control is operationally defined based on the criteria of interference by photosynthetic electron transport inhibitors (such as DCMU: dichlorophenyldimethylurea) with gene expression in vivo (e.g., Kuroda et al. 1996; Deshpande et al. 1997; Salvador and Klein 1999; Zhang et al. 2000) or in isolated organelles ('*in-organello*') (Pfannschmidt et al. 1999a,b; Zhang et al. 2000). The same or other workers listed in Table 1 (e.g. Danon and Mayfield 1994b; Baginsky et al. 1999; Fong et al. 2000; Trebitsh et al. 2000; Zhang et al. 2000) have also addressed the effectiveness of chemicals that affect the thiol group redox state of chloroplast proteins (e.g., DTT, glutathione, N-ethyl maleimide (NEM), diamide). The conclusions usually drawn by the DCMU-type experiments were that, if the inhibitor was effective, then photosynthetic electron transport would play a role in the regulation of plastid gene expression. In addition, considering the known mode of action of DCMU (Trebst 1980), it has been suggested that a direct outlet to gene expression might exist, which originates at or between PS II and cytochrome b_6f and involves the plastoquinone pool. On the other hand, experiments in which thiol regulation of gene expression was analyzed pointed to a signaling mechanism that acts downstream of PS I, possibly involving thioredoxins as the terminal components that connect to gene expression. Although the identity of the actual 'redox sensor(s)' for plastid gene regulation is still a matter of debate, the data obtained in the two different approaches are consistent with each other. Even if a sensor of photosynthetic electron transport is located close to PS I, rather than PS II, it would monitor the changes in electron flow resulting, e.g., from inhibition at PS II.

b) Transcription

Unlike earlier ideas adopted from bacterial two-component systems (Allen 1993b), the current picture is that the major chloroplast RNA polymerase contains an associated serine-specific protein kinase of the CK2 type (Maliga 1998; Baginsky et al. 1999; K. Ogrzewalla and G. Link, unpublished). This kinase, referred to as PTK (Plastid Transcription Kinase) is subject to SH-group regulation by glutathione, i.e., a major redox mediator in chloroplasts (Karpinski et al. 1997; Noctor and Foyer 1998). PTK controls chloroplast transcription via phosphorylation of sigma-like transcription factors and several other polypeptides that are associated with the plastid transcription apparatus (Baginsky et al. 1999). The polymerase-associated proteins include additional redox-relevant components such as Fe-SOD (Pfannschmidt et al. 2000). The identity of the full set of more than 15 polypeptides present in purified ('PEP-A'; Maliga 1998) chloroplast RNA polymerase preparations is currently under investigation using proteomics and reverse genetics techniques (K. Ogrzewalla, S. Jung, A. Sickmann, H.E. Meyer and G. Link, unpubl.). Interestingly, PTK is not only redox-responsive but also subject to phosphorylation by itself (Baginsky et al. 1997; Link et al. 1997). As protein phosphorylation is characteristic of the photosynthetic apparatus, and the kinases involved are beginning to become defined (Gal et al. 1997; Vener et al. 1998; Rintamäki et al. 2000), one current priority issue is to address the details of the signaling chain(s) between photosynthesis and gene expression. Likewise, as the PEP-A polymerase contains a component that is sequence-related to the 3'RNA-binding protein CSP41 (Yang et al. 1996), this may mean that the connections between plastid transcription and post-transcriptional processes may be much closer than anticipated, with a possibility for common redox-regulatory mechanisms.

c) Translation

Perhaps the most complete picture on redox-regulatory mechanisms in plastid gene expression for translation initiation has emerged in *Chlamydomonas*. The current models (Danon 1997; Bruick and Mayfield 1999; Trebitsh et al. 2000) are based on the properties of a 5'UTR mRNA-binding complex consisting of four proteins, two of which have been tentatively identified as a protein disulfide isomerase (cPDI; RB60) (Kim and Mayfield 1997) and a chloroplast member of the poly(A)-binding proteins (cPABP; RB47) (Yohn et al. 1998a), respectively. The in vitro properties of the purified complex and cloned recombinant proteins, combined with genetic evidence from photosynthesis-deficient mutants (Danon and Mayfield, 1994b; Yohn et al. 1998b), together with protein synthesis in isolated chloroplasts and in vivo experiments, have led to a detailed view that is briefly summarized (and simplified) as follows: A thiol redox signal reflecting the state of photosynthetic electron transport is first transmitted from thioredoxin to the cPDI component of the RNA-binding complex and from here to the cPABP protein. Only in its reduced form does the latter confer strong mRNA binding on the complex, which is a necessary prerequisite for efficient translation (initiation).

Despite the beauty and simplicity of this model, several aspects require additional comments. (1) Thiol redox regulation is not the only regulatory mechanism involved, but it acts in a concerted manner with protein phosphorylation. The kinase responsible is an ADP-dependent enzyme (Danon and Mayfield 1994a). This is in contrast to transcription, where the redox-regulated CK2-type kinase (PTK) uses ATP and GTP but not ADP as phosphodonor (Baginsky et al. 1999; K. Ogrzewalla and G. Link, unpubl.), indicating that the details of transcriptional and translational redox regulation differ considerably. (2) Although recombinant cPABP was clearly demonstrated to respond to reversible disulfide bond formation (Fong et al. 2000), evidence from work using the authentic purified protein complex suggests that redox-regulatory vicinal thiol groups are present exclusively on cPDI (Trebitsh et al. 2000). This apparent discrepancy is likely to be resolved once all components of the complex are cloned and thus available for reconstitution and mutational studies. (3) The translational initiation model does not exclude additional redox regulatory mechanisms during subsequent steps, including elongation, protein modification, breakdown, and assembly, and ample evidence for this indeed to be the case has recently accumulated (Kuroda et al. 1996; Kettunen et al. 1997; Mühlbauer and Eichacker 1998; Zhang et al. 1999, 2000). It can be anticipated that combined reverse genetics (Suzuki and Maliga 2000) and biochemical approaches will further help to clarify redox-regulatory mechanisms during translation as well as all other steps throughout chloroplast gene expression.

5 Redox Regulation in the Extrachloroplastic Compartment

Since chloroplasts are only partially autonomous, development and maintenance of the functional photosynthetic apparatus depend on gene expression in both the plastids and the nucleus. Accordingly, signals originating from the photosynthetic electron transport chain not only affect the gene expression of plastids as outlined in the previous section but also the nuclear gene expression. However, in addition to the importance of redox signals in the regulation of photosynthesis, redox-dependent changes in nuclear gene expression also relate to stress and

Table 2. Examples of genes whose expression changes in response to light and redox signals

Gene	Species	Function	Redox-related response of transcript amount	Reference
cab	*Dunaliella tertiolecta/Dunaliella salina*	Chlorophyll-a/b-binding protein (LHCP II)	Excitation pressure/redox state of plastoquinone	Escoubas et al. (1995); Maxwell et al. (1995)
sep	*Arabidopsis th.*	Stress enhanced proteins (pigment storage?)	High light/UV	Heddad and Adamska (2000)
elip	*Hordeum vulgare*	Early light-induced protein (pigment storage)	Light	Montane et al. (1997)
apx 1	*Arabidopsis th.*	Ascorbate peroxidase (cytosolic isoform)	Excess light, systemic signals	Karpinski et al. (1997); Karpinski et al. (1999)
apx2	*Arabidopsis th.*	Ascorbate peroxidase (cytosolic isoform)	Excess light, systemic signals	Karpinski et al. (1997); Karpinski et al. (1999)
gor 1	*Arabidopsis th.*	Glutathione reductase (cytosolic isoform)	Post excess light stress	Karpinski et al. (1997)
2-cp	*Arabidopsis th./Riccia fluitans*	Peroxiredoxin (chloroplast peroxide detoxification)	Downregulation at high ascorbate and -lesser extent- high glutathione	Baier and Dietz (1997); Horling et al. (2001)
sod	*Pinus sylvestris*	CuZn-superoxide dismutase	Decrease by GSH	Wingsle and Karpinski (1996)
chs	*Phaseolus vulgaris*	chalcone synthase (phenylpropanoid synthesis)	Stimulation by GSH	Wingate et al. (1988)
pal	*Phaseolus vulgaris*	Phenylalanine ammonium lyase (phenylpropanoid synthesis)	Stimulation by GSH	Wingate et al. (1988)

defense responses of plants. In fact, pathogenesis- and stress-related redox regulation has been known longer and in more detail than photosynthesis-related redox control (Foyer et al. 1997; Wingate et al. 1988). At least in part, both redox-dependent regulatory processes employ identical signal transduction elements such as reactive oxygen species, ascorbate and glutathione and are linked by cross-talk. In addition, signaling mechanisms exist which specifically and stress-independently link biochemical and genetic responses to the redox status of photosynthetic reactions. For the time being, the latter mechanisms may be considered to represent the photosynthesis-specific feedback mechanism. Conversely, the former signaling pathway is part of the general stress acclimation response. In the stress response, ROS are likely to serve as predominant redox signals and are synthesized in various reactions. For example ROS are liberated at the acceptor site of photosystem I, by reduced plastoquinone, by the plasma membrane NADH oxidase, by peroxisomal glycollate oxidase, lipoxigenases and other mechanisms (Foyer and Noctor 2000). Table 2 lists some nuclear genes whose expression is modified depending on redox and ROS signals. The gene products are either constituents of the photosynthetic machinery, enzymes of the antioxidant metabolism or part of the general plant defense. These genes respond depending on the redox state of the photosynthetic electron transport chain, to increased production of ROS or the pool size and redox poise of the low molecular antioxidants. The redox responses of the nuclear-encoded chlorophyll a/b binding protein of PS II and of two types of peroxide-processing antioxidant enzymes are described as examples in the following.

a) Changes in Nuclear Gene Expression Depending on the Redox State of the Photosynthetic Electron Transport Chain

As outlined before, redox signals derived from the photosynthetic electron transport chain allow to extract system information on the balance between excitation pressure and energy consumption prior to the establishment of oxidative stress and photoinhibition. Thus, in addition to their use in regulating biochemical reactions and chloroplast gene expression, changes in the redox state of the electron transport carriers also relate to alterations in nuclear gene expression (Escoubas et al. 1995; Maxwell et al. 1995; Karpinski et al. 1997). Using the green alga *Dunaliella tertiolecta*, Escoubas and coworkers (1995) investigated the expression of the nuclear encoded chlorophyll a/b-binding protein of PS II (cab) by run-on transcription of isolated nuclei. The transcriptional activity was high at low light and there was a concomitantly high oxidation state of the electron transport chain. Addition of two inhibitors that block the electron transport chain either upstream (DCMU) or downstream (DBMIB: dibromomethyl isopropylbenzoquinone) of the plastoquinone pool mimicked the low light-dependent induction of cab gene expression even in high light and the high light suppression in low light, respectively. The results implied that the redox state of the plastoquinone pool is implicated in the signaling pathway. Furthermore, inhibitors of protein phosphatases such as okadaic acid blocked the low light-induced upregulation of cab gene expression suggesting the involvement of reversible protein phosphorylation in redox signaling from the chloroplast. Using a slightly different approach, Maxwell et al. (1995) arrived at the same conclusions: Following the transfer of *Dunaliella salina* cells to growth conditions which established high or low excitation pressure, cab expression decreased or increased up to eightfold. The authors concluded that the redox poise of the intersystem electron transport chain regulated the expression of the cab genes and the chlorophyll contents of the cells. As stated above, the reduction state of the cytochrome b_6/f complex may also function as a redox signal (Pearson et al. 1993; Fujita et al. 1994). In these studies either cyanobacteria or isolated lettuce chloroplasts were used. Therefore, it is not possible to make a conclusion about the importance of signals from the cytochrome b_6/f complex for the regulation of nuclear gene expression. Although similar redox signals are likely to be used for the acclimatory regulation of photosynthesis in algae and higher plants, it is important to note that an imbalance between excitation energy and energy consumption does not induce identical responses in all plants and algae. For example evergreen plants and cereals activate distinct adaptive responses under conditions of excess light and low temperatures. Evergreen plants decrease their photosynthetic efficiency by reorganizing the light-harvesting complexes and increasing the dissipation of excess energy as heat (Huner et al. 1998). Conversely, cereals recover a high photosynthetic capacity at low growth temperatures by increasing the activity of enzymes of the Calvin cycle and sucrose synthesis (Strand et al. 1999). This mechanism allows the energy consumption to be balanced by increasing the activity of the strongly temperature-dependent 'dark' reactions with the energy supply by the rather temperature-independent photochemical processes. The signals involved in this adaptation are not known.

b) Redox Regulation of the Expression of Two Peroxide-Processing Enzymes, the Ascorbate Peroxidase and the Two-Cysteine Peroxiredoxin

When the imbalance between photon absorption and energy consumption in subsequent photosynthetic reactions is large, reactive oxygen species are synthesized by electron transfer to O_2 (Foyer and Noctor 2000). Reduction of dioxygen to superoxide at the acceptor site of PS I is called 'Mehler reaction' and is likely to be the most quantitatively important reaction in producing reactive oxygen species at the thylakoid membrane. Other components of the electron transport chain such as reduced plastoquinone, PS II reaction centers and iron-sulfur clusters of PS I can reduce O_2 also, but at much lower rates than the Mehler reaction. The relative contribution of the Mehler reaction to the photosynthetic electron transport rate appears to be below 10% in C3 and CAM plants, even under conditions of high light and water stress (Badger et al. 2000). A complex network of enzymes and low molecular weight antioxidants is present both in the chloroplasts and the cytosol to protect cell metabolism from oxidative damage (Noctor and Foyer 1998; Baier and Dietz 1998). However, in plants not adapted to high light previously, the antioxidant defense may be insufficient and overwhelmed under conditions of excess excitation. Thus, photodamage develops in non-acclimated leaves due to the toxicity of ROS on the one hand. On the other hand, however, ROS provide valuable information on the system performance. Their concentration is monitored and employed to trigger appropriate acclimation responses. Recently, a MAP kinase (mitogen activated kinase) was identified in plants, which was shown to be activated during oxidative stress (Kovtun et al. 2000) and could be part of a redox-dependent regulatory circuit. Evidence for the involvement of other typical signal transduction elements such as small GTP binding proteins has been provided, however the knowledge on redox- and ROS-signaling remains rudimentary (Mullineaux et al. 2000).

Ascorbate peroxidases (Apx) and 2-cysteine peroxiredoxins (2-CP) constitute important antioxidant enzymes. Apx detoxifies H_2O_2, 2-CP both H_2O_2 and alkyl hydroperoxide. These enzymes are good examples for distinct redox regulation of gene expression and will be briefly discussed in the following.

1. In *Arabidopsis*, Apx is present as a small gene family with two isoforms in the cytosol (Apx 1 and 2), two in the chloroplast (Apx 4 and 5) and one in the microbodies (Apx 3) (Mullineaux et al. 2000). Expression of Apx 1 and 2 is induced under excessive light, by treatment of the leaves with DBMIB and H_2O_2, and inhibited by DCMU. Thus the expression of the cytosolic Apx is linked both to the redox state of the photosynthetic intersystem electron transport chain and to the concentration of ROS. The inhibitory effect of DCMU cannot be overridden by treatment with H_2O_2 (Karpinski et al. 1999). This indicates the existence of at least two interfering signaling cascades which act on Apx expression. The expressional upregulation of Apx 1 and 2, and of catalase (cat 2) under excess light is part of an acclimation program which adapts the photosynthetic cells to high light conditions. Interestingly, when single leaves of a plant were treated with excessive light, other leaves also acclimated to the high light at positions distant to the site of treatment and not exposed to the ex-

cessive excitation regime. Thus, there exists an analogy to the systemic acquired resistance in the pathogen-host interaction where infection and the induction of the hypersensitive response at one site will induce resistance in the whole plant. Apparently, the acclimation to excess excitation energy and the antioxidant defense program are induced systemically (Karpinski et al. 1999). Although the nature of the systemic signal is not known yet, the authors observed considerable overlap in the signaling and genetic response between excess light stress and wounding and speculate on the involvement of H_2O_2 (Mullineaux et al. 2000).

2. Peroxiredoxins occur in all organisms and reduce a broad range of peroxides including H_2O_2, alkylhydroperoxides and, as shown recently for the bacterial peroxiredoxin, peroxinitrite (Chae et al. 1999; Bryk et al. 2000). The plant 2-cysteine peroxiredoxin (2-CP) occurs in the chloroplasts (Baier and Dietz 1997). Work with *Arabidopsis*-plants and the cyanobacterium *Synechocystis* with genetically modified contents of 2-CP has shown that the 2-CP protects the photosynthetic apparatus from oxidative damage (Klughammer et al. 1998; Baier and Dietz 1999; Baier et al. 2000). In young leaves and thalli of the liverwort *Riccia fluitans*, 2-CP was expressed at a high level. Treatment of the tissues with glutathione and, particularly strongly, with ascorbate decreased 2-CP expression (Baier and Dietz 1997; Horling et al. 2001). In *Riccia* thalli, externally added ascorbate at a concentration of 20 mM, inhibited the 2-CP expression fast and completely. The ascorbate effect was suppressed by addition of the protein kinase inhibitor staurosporine suggesting the involvement of a protein kinase in the redox signaling pathway to the nucleus.

Apx, cab and 2-CP represent three examples with opposite redox response. Cab expression is high under optimum growth conditions and decreases at a high reduction state of the plastoquinone pool. Cytosolic Apx expression is low and increases with electron pressure in the photosynthetic electron transport chain. H_2O_2 stimulates Apx expression. 2-CP expression is high and decreases with improved reduction state of the cells. Other examples of genetic changes induced by redox-active compounds could be given. For instance Wingate et al. (1988) observed massive and selective genetic changes upon addition of reduced glutathione (GSH) at low concentrations to a suspension of cultured cells of *Phaseolus*. Among the GSH-induced genes were the phenylalanine ammonia lyase and chalcone synthase which catalyze key steps in the phenylpropanoid biosynthetic pathway and are part of the general plant defense against pathogens and other causes of stress. Oxidized glutathione (GSSG), cysteine, ascorbate and dithiothreitol did not induce similar genetic changes. Transient changes in gene expression were seen at GSH concentrations as low as 10 μM. Apparently, relatively small changes in glutathione homeostasis, i.e., in concentration or redox state, are sufficient to induce major changes in nuclear gene expression. Such changes in the glutathione system may also occur under photoinhibitory conditions.

The selected examples show that a multiplicity of redox-dependent signaling pathways exists in plant cells which could be implicated in regulating photosynthesis-related changes in nuclear gene expression. Obviously, the question needs to be addressed which redox-related signals may be exchanged between the chloroplast and the cytosol.

c) Signals from the Chloroplast to the Cytoplasm

In order to alter extrachloroplastic processes such as nuclear gene expression, the redox status of components of the photosynthetic apparatus or of the chloroplast antioxidant system must be communicated to the cytosol either by signal elements which cross the inner and outer envelope or by signal transduction at the envelope membrane. A summary of possible but still partly theoretical mechanisms which could be involved in intercompartment signaling between the chloroplast and the cytosol was recently presented in this series (Baier and Dietz 1998). Little new information has been made available since then and the reader is referred to the appropriate section of that review. Since ascorbate was shown to be involved in regulating nuclear gene expression (Horling et al. 2001), transport of ascorbate may be part of an important redox signaling mechanism. The cytosolic and chloroplastic ascorbate pools are connected by an active but low affinity diffusion carrier which is likely to transmit changes in concentration and redox state from the chloroplast to the cytosol (Beck et al. 1983). Conversely, information on transport across the envelope of the other major low molecular weight antioxidant glutathione is not available, possibly due to a very low permeability and thus negative results in such experiments. Assuming the absence of a glutathione exchange mechanism between stroma and cytosol, the organellar and cytosolic glutathione pools would be uncoupled and only indirectly linked through ascorbate- and NAD(P)H-dependent reactions such as dehydroascorbate reductase and glutathione reductase (May et al. 1998). This implies (1) that the redox state of the subcellular glutathione pools allows independent redox sensing in the cytosol and stroma and (2) that glutathione does not serve as a mobile intercompartment redox signal. Diffusion of ROS, the activity of the malate valve which links the stromal $NADPH/NADP^+$-system to the cytosolic $NADH/NAD^+$-system (Fridlyand et al. 1998), transport of lipid degradation products or other as yet, unknown mechanisms may communicate redox signals from the stroma to the cytosol and affect nuclear gene expression.

d) The Inactivation of the Peroxisomal Catalase as a Case Study of Signal Transfer

Shang and Feierabend (1999) described an interesting observation related to signal transfer from the chloroplast to the cytosol and may be relevant for redox signaling. Photoinactivation of catalase by blue light absorbed by the heme group is a long-known phenomenon. With red light which is not active in heme excitation, inactivation of isolated catalase is not observed. However, high flux densities of red photons inactivate catalase in leaves. By reconstituting isolated intact or broken chloroplasts and purified catalase in vitro and additional experiments with electron acceptors and inhibitors of the photosynthetic electron transport chain, Shang and Feierabend (1999) showed that overreduction of PS II and the plastoquinone pool correlates with the red light-dependent inactivation of catalase. Furthermore, the radical scavenger Trolox fully suppressed the inactivation reaction. Addition of SOD or incubation at a low temperature decreased the red light-dependent inactivation of catalase. The authors suggest that the superoxide anion radical, possibly in its protonated form, escapes from the illuminated intact chloroplast and reacts with the catalase. Addition of DBMIB stimulated the inactivation, addition of electron acceptors such as methylviologen decreased the inactivation. Dissipation of the transthylakoid proton gradient did not prevent the chloroplast-mediated inactivation of the catalase. It was concluded that the inactivating compound originates from PS II or the plastoquinone pool. The Mehler reaction as an alternative source of catalase-inhibiting $O_2{\cdot}^-$ was excluded. This pattern of effects resembles the relationship between the reduction state of the photosynthetic electron transport chain and the induction of Apx expression (see above, Karpinski et al. 1997). For the time being, it could be hypothesized that the chloroplast-mediated inactivation of catalase is an indicator system for the transfer of a redox-active compound from the chloroplast to the cytosol which may also be relevant for redox signaling in nuclear gene expression. The increased rate of $HO_2{\cdot}$ release (protonated superoxide anion radical) from the chloroplast could cause oxidization of a transcription factor or of a cytosolic redox sensor and activate a signaling cascade which initiates downstream events.

6 Cyanobacteria as Oxigenic Photosynthetic Model Organisms to Study Redox Regulation

Cyanobacteria, being prokaryotic oxigenic photosynthetic organisms are remarkable for their ability to flourish in environments with widely fluctuating chemical and physical parameters, such as nutrient and water availability, light intensity and quality, temperature and osmotic conditions, contribute significantly to photosynthesis on our planet (Whitton and Potts 2000). Cyanobacteria are also of interest because of their considerable morphological and metabolic diversity, long evolutionary history (extending to at least 3500 Ma ago) (Schopf 2000), and their economic importance as a health food (Spirulina) (Vonshak 1997), as a source of polyaspartate, obtained from cyanophycin and representing a biodegradable plastic (Allen 1984; Simon 1987; Schamborn 1996), and as a source of a number of unusual primary and secondary metabolites with a wide range of biological and pharmacological activities (Falch 1996). Moreover, cyanobacteria have roles in oil biosynthesis and degradation (Radwan and Al-Hasan 2000).

Some recent reviews have covered the present knowledge on redox regulation in cyanobacteria (see e.g. Allen 1992, 1993b; Allen et al. 1995; Huner et al. 1996; Allen and Nilsson 1997; Link 2001). This section will focus on reviewing how bioinformatic methods, taking advantage of the availability of the entire nucleotide sequence of the unicellular cyanobacterium *Synechocystis* sp. strain PCC 6803 genome (subsequently called *Synechocystis* PCC 6803) (Kaneko et al. 1996; Kotani and Tabata 1998), were used in combination with genetic/physiological methods to contribute to our present knowledge of redox-mediated regulatory circuits in cyanobacteria. Cyanobacteria can be considered as good model organisms for at least two reasons. Firstly, findings about regulatory circuits from bacteria can be applied to cyanobacteria, being oxigenic photosynthetic prokaryotes, and thus provide the possibility of investigating how such regulatory circuits have been conserved or how they have been changed to fit the oxigenic photosynthetic life style. Secondly, comparisons can be made to see which cyanobacterial regulatory circuits are conserved or modified in eukaryotic photosynthetic organisms: algae and plants. With the entire nucleotide sequence of the *Arabidopsis thaliana* genome (nuclear, plastid and mitochondria genome) now being available (The *Arabidopsis* Genome Initiative 2000), an optimal basis for such comprehensive comparisons is given.

a) General Considerations About Adaptation Processes in Cyanobacteria

Cyanobacteria, as other oxigenic photosynthetic organisms, have developed a large range of effective mechanisms to recognize their environment, to use this information for initiating a cascade of morphological, physiological and molecular changes that lead to the adaptation and optimal use of the available resources (Anderson et al. 1995; Grossman et al. 1994; Bhaya et al. 2000; Mann 2000). The dominant nutritional mode of cyanobacteria is photoautotrophy. Growth occurs via the light-dependent fixation of CO_2 and the acquisition of simple inorganic nutrients. Variations in these factors constitute the primary environmental stimuli to which these organisms must adapt. Being very adaptable, cyanobacteria contain a large reservoir of genetic information encoding biochemical pathways to achieve optimal utilization of absorbed light energy, to avoid oxidative damage induced by excessive excitation (photoinhibition) and to effectively coordinate photosynthesis with the overall cellular metabolism. It is well documented that in cyanobacteria, as in other oxigenic photosynthetic organisms, many of these adaptation processes are under redox control (Allen 1992, 1993b; Allen et al. 1995; Huner et al. 1996; Allen and Nilsson 1997; Link 2001). As expected for a mainly photoautotrophic organism, the redox state of components of the thylakoid electron transport chain, especially the plastoquinone pool and/or the cytochrome b_6f-complex as well as the ferredoxin/thioredoxin system and possibly also PS II, has been implicated as a signal which regulates acclimation to light quantity/quality and nutrient status. In many adaptation processes, the primary target for modification is the photosynthetic/respiratory electron transport chain itself. It is well documented that in response to fluctuation in light intensity/quality the photosynthetic apparatus can be modified within minutes by a process called "state transitions" (Mullineaux and Allen 1990; van Thor et al. 1998). In addition, cyanobacteria are also capable of altering the phycobilisome structure (complementary chromatic adaptation) (Bhaya et al. 2000), the total amount of photosystems, and the ratio of PS I to PS II in response to changes in the light and/or nutrient environment (Fujita et al. 1994). Furthermore, cyanobacteria can modify the protein composition of PS II in high light to make it less susceptible to photoinhibition (Golden 1994). The altered redox state of components of the electron transport chain subsequently represents one major input signal for the regulatory circuits of cellular metabolism, especially C- and N-assimilation, thereby coupling cellular regulatory pathways controlling gene expression and enzyme activation to utilization of light energy (Gleason 1994; Bhaya et al. 2000; Mann 2000). Although there is no question about the importance of redox-dependent signaling pathways in adaptation processes of cyanobacteria, there is still little understanding of the signaling molecules that sense and link the redox state of the plastoquinone pool/cytochrome b_6/f-complex or of PS II with the modifying processes.

b) Two-Component Regulatory System: Histidine Kinase-Response Regulator

Protein phosphorylation is a common cellular response to external and internal signals. Protein kinases that catalyze the transfer of phosphate from ATP to serine, threonine, tyrosine or histidine residues are widely spread in all three kingdoms: eubacteria, archaebacteria and eukaryotes.

In bacteria, numerous sensory-response circuits operate by making use of a phosphorylation control mechanism referred to as the "two-component system". In *Escherichia coli*, for example, 40 such systems have been identified (Mizuno 1997). The two-component regulatory system generally consists of a sensor kinase and a response regulator. The sensor kinase has sensor and histidine phosphotransferase domains. The sensor domain recognizes the signal and autophosphorylates a histidine residue. The phosphoryl group is subsequently transferred to an aspartate residue of the cognate response regulator which is activated by the phosphorylation. The activated regulator will then either directly or indirectly regulate the expression of the target genes implying that the response regulator usually functions as a transcription factor. Some two-component systems utilize more than one histidine-kinase or response regulator (multi-step phosphorelay), and some single proteins contain both two-component elements (Parkinson and Kofoid 1992; Parkinson 1993; Chang and Stewart 1998; Fabret et al. 1999).

As in non-photosynthetic bacteria, also in cyanobacteria, numerous phosphorylation events as a response to external and/or internal signals have been described (Mann 1994; Allen and Nilssen 1997; Gal et al. 1997; Bhaya et al. 2000; Mann 2000). After the entire nucleotide sequence of the cyanobacterium *Synechocystis* PCC 6803 genome became available (Kaneko et al. 1996), information about two-component regulatory systems from non-photosynthetic bacteria has successfully been used to identify such putative systems in cyanobacteria. An extensive computer-aided similarity search was conducted for all open reading frames (ORFs) of the *Synechocystis* PCC 6803 genome showing that at least 80 ORFs (out of a total of 3168 ORFs representing 2.5%) exhibit a significant similarity to known two-component signal transducers from other bacterial species (Mizuno et al. 1996; Kotani and Tabata 1998). 26 ORFs were identified as a sensory kinase containing a transmitter, 38 ORFs as response regulators containing a receiver, and 16 ORFs as hybrid sensory kinases containing both the transmitter and receiver domain and in some cases an additional alternative transmitter. Further examination of the putative response regulators revealed that they can be classified into distinct subgroups: CheY-, OmpR-, NarL-, and PatA-subfamily. This evaluation also gave evidence, that the chromosomal positions of these 80 ORFs are scattered evenly over the entire genome of PCC 6803, while in *Escherichia coli* a cognate sensor-regulator pair in most cases is located in the same transcriptional unit. In *Synechocystis* PCC 6803, however, only 14 sets of signal transducers were considered to be located in close proximity (32 out of 80 ORFs). Recent nucleotide sequence evaluations revealed that cyanobacterial two-component systems have several unusual structures that point to more complex and sophisticated signaling circuits than those present in bacteria (Urao et al. 2000). Thus, cyanobacteria represent a good model of how bacteria-like two-component systems have been changed to fit the oxigenic photosynthetic life style.

Two recent papers show how this DNA sequence information was successfully used to identify two-component signal transducers and to elucidate their function. Suzuki et al. (2000) have insertionally inactivated all 43 genes of putative histidine kinases in *Synechocystis* PCC 6803 and so far identified 2 histidine kinases (Hik19: Sll0698 and Hik33: Sll1905) and an unusual response regulator (Rer1) as components of the pathway for perception and transduction of low-temperature in *Synechocystis* PCC 6803. Li and Sherman (2000) used the *Synechocystis* genome sequence information in combination with sequence comparisons to the photosynthesis response regulator and kinase genes *regA-regB* and *prrA-prrB* in *Rhodobacter capsulatus* and *Rhodobacter sphaeroides*, respectively (Zeilstra-Ryalls et al. 1998; Masuda et al. 1999). The products of these genes are a global signal transduction system involved in the anaerobic induction of many physiological processes including synthesis of light-harvesting, reaction center, and cytochrome components of the bacterial photosystem. With this approach, the genes *rppA*: *sll0797* and *rppB*: *sll0798* in *Synechocystis* PCC 6803 (both genes being present in the list compiled by Mizuno et al. 1996) were identified as a putative response regulator and histidine kinase. The Sherman group could show that in the RppA-free *Synechocystis* PCC6803 mutant, the PS II gene transcripts were highly upregulated relative to wild type under all redox conditions, whereas transcription of phycobilisome-related genes and PS I genes was decreased. The results were interpreted to suggest that RppA as a regulator of photosynthesis- and photopigment-related gene expression, is involved in the establishment of the appropriate stoichiometry between photosystems, and can sense changes in the plastoquinone redox poise.

In addition to the very recently, above-described members of a two-component sensory system in *Synechocystis* PCC 6803, a number of two-component modules have previously been identified by genetic/physiological methods and shown to be involved in a wide range of adaptive responses in *Synechocystis* PCC 6803 as well as in other cyanobacteria. The complete list of two-component modules identified so far, with the corresponding references is given in a recent review by Mann (2000; see Table 1 in this review). For example, such modules are involved in phycobilisome degradation and survival under nutrient-limited/high light conditions, complementary chromatic adaptation, heterocyst pattern formation, nitrogen fixation, CO_2 availability, phytochrome- and ethylene response-like signaling, herbicide resistance/chemical signaling, and phosphate availability.

Although the His-to-Asp phosphorelay system was initially considered to be a classical bacteria-like regulatory module, many two-component regulatory genes have been identified in higher plants (Alex and Simon 1994; Kennelly and Potts 1996). These code for histidine kinases, response regulators, and phosphorelay intermediates containing HPt domains. The first identified eukaryotic two-component system element was the *Arabidopsis thaliana* ETR1 (predicted hybrid kinase) involved in ethylene signaling. Subsequently, a number of such modules have been identified and shown to be involved in cytokinin signaling or to have a function as osmosensor or as a component of the circadian clock (Loomis et al. 1997; D'Agostino and Kieber 1999; The ***Arabidopsis*** Genome Initiative 2000; Urao et al. 2000). In this respect, it is also relevant to point out that portions of several putative gene products of sensor histidine kinase genes in *Synechocystis* PCC 6803 (e.g. Slr0473 and Sll1124) have been shown to possess a high degree of sequence-similarity to a chromophore-binding domain commonly found in phytochromes of green plants (Kehoe and Grossman 1996; Yeh et al. 1997; Hughes et al. 1997; Kotani and Tabata 1998). This suggests that, as expected, at least some two-component systems in plants have evolved from cyanobacterial systems.

c) Serine/Threonine and Tyrosine Kinases

Serine/threonine and tyrosine kinases as well as the corresponding phosphatases are abundant in plants (Hardii 1999). In the *Arabidopsis thaliana* nuclear genome 340 receptor-like kinase genes were identified. However, very little is known so far about their regulators and target proteins (The *Arabidopsis* Genome Initiative 2000). Originally, the serine/threonine kinases were considered to be classical "eukaryotic" protein kinases, but during recent years, serine/threonine kinases have also been identified in a wide range of prokaryotes (Kennelly and Potts 1996; Chang and Stewart 1998).

Evaluation of the nucleotide sequence of the *Synechocystis* PCC 6803 genome (Zhang et al. 1998) revealed that this cyanobacterium possesses at least seven serine/threonine kinases and seven serine/threonine and tyrosine phosphatases. Several genes encoding serine/threonine kinases or phosphatases in *Synechocystis* PCC 6803 are found in the same cluster as those encoding members of two-component modules. Since genes that are involved in the same cellular process in prokaryotes are frequently clustered or form an operon, it could be expected that at least some serine/threonine kinases or phosphatases may interact with two-component regulatory proteins encoded by the same gene cluster. The elucidation of such possible molecular interactions might provide new insights of how the non-photosynthetic bacteria-like signal transduction pathways have been altered in oxigenic photosynthetic organisms.

Adaptive responses in cyanobacteria in which monoester phosphorylation of proteins has so far been implicated are also summarized in Mann (2000) and include, for example, complementary chromatic adaptation, state transition, β-phycocyanin phosphorylation, heterocyst formation, and nitrate transport. The best characterized example of monoester phosphorylation in cyanobacteria is the phosphorylation of the P_{II} protein. The phosphorylation of the P_{II} protein is primarily determined by the N-status of the cell, but is also responsive to carbon availability and an imbalance in photosynthetic electron transport (Allen et al. 1985; Allen 1992; Forchhammer and Tandeau de Marsac 1995; H.-M. Lee et al. 1999).

d) Protein Domains with Putative Function in Redox-, Light-, Oxygen- and Energy-Sensing: PAS Domain-Containing Proteins

Although it is obvious that two-component regulatory systems play an important role in adaptation processes in cyanobacteria, the identity of the sensor detecting the redox state of the corresponding components in the thylakoid and/or cytoplasma membrane has remained unclear. In this respect recent results obtained about PAS containing proteins seem to be relevant (Zhulin et al. 1997; Zhulin and Taylor 1998; Taylor and Zhulin 1999). Again based on improved techniques for computer-assisted homology searches, a PAS domain superfamily of sensory transduction elements has been identified. The first proteins with PAS domains (PER, ARNT, SIM and phytochromes) were identified in eukaryotes and shown to be typically paired with a repeat domain and to be involved in protein-protein interactions. Where a function is known, PAS domains mostly sense redox potential, light, oxygen or cellular energy. To date, several different redox- and/or light-sensitive cofactors have been identified in PAS domains, including 4-hydroxy-cinnamoyl, heme, FAD/FMN, and 2Fe-2S centers. In prokaryotes, PAS domains are mostly input domains for sensor histidine kinases in two-component regulatory systems sensing redox changes in the electron transport system or overall cellular redox status. This suggests that the PAS domains might be the missing link. Well-characterized PAS domain containing proteins in microorganisms include the oxygen-sensing FixL (heme-containing) in *Sinorhizobium meliloti* and other rhizobial species, the redox sensing Aer (FAD containing) aerotaxis transducer in *E. coli*, and NifL (FAD-containing) regulating N_2 fixation in response to redox status in *Azotobacter vinelandii* (Zhulin and Taylor 1998). Although some of the proteins with PAS input domains have been clearly shown to be global regulators of metabolism, for the majority of PAS containing prokaryotic sequences, the function of the protein is so far only postulated or unknown.

Analysis of 11 completely sequenced microbial genomes (Zhulin and Taylor 1998; Taylor and Zhulin 1999) led to the identification of five species without a PAS domain. The other species had between 6–17 proteins with PAS domains, and the number of PAS domains per protein varied from one to six. This group showed that there was no correlation between the size of a bacterial genome and the total number of PAS domains present in the genome, while there was a positive correlation between the number of PAS domains and the number of electron transport-associated proteins in the species. As suggested by this group, this correlates quite well with the assumption that the primary role of PAS domains is in sensing redox potential, oxygen, or light. The species with the lowest incidence of electron transport proteins and the absence of PAS domains are animal parasites that live in an environment where they have little need for a complex electron transport system and redox sensing. In contrast, *Synechocystis* PCC 6803 whose survival is absolutely dependent on sensing light and the redox status of a complex photosynthetic/respiratory electron transport chain, is an organism with an abundance of PAS domains. In *Synechocystis* PCC 6803 17 genes that code for proteins with PAS domains (containing a total of 47 PAS domains) have been identified. A number of these proteins contain several PAS domains, either multiple copies of similar domains which might provide an advantage by amplifying the sensory signal (e.g. three N-terminal similar PAS domains in Sll0779) or domains that are unrelated and might provide the response to multiple input signals (e.g. six unrelated domains in Slr0222). Several of these proteins with PAS domains in *Synechocystis* PCC 6803 show similarity to histidine kinases, such as Slr1759, Sll1124, Slr2098, Slr0222, and Slr0311.

Results from the Grossman group (Schwarz and Grossman 1998; Dolganov and Grossman 1999; Grossman et al. 1999) suggest that such PAS domains could indeed be the redox sensing modules not only in non-photosynthetic bacteria but also in oxigenic photosynthetic cyanobacteria. This group recently presented results on a two-component system called NblS-NblR in *Synechococcus* PCC 7942. NblR was shown to be a response regulator that controls some of the general responses that occur during a number of different stress conditions. It is required for degradation of phycobilisomes and also appears to be necessary for controlling additional aspects of photosynthetic activity during both nutrient limitation and high light conditions. NblR appears to be controlled by NblS, a sensor histidine kinase that has a PAS domain. Preliminary results of this group suggest that this PAS domain binds a redox-sensing prosthetic group such as a flavin and thus is a good candidate for sensing the redox status of the cell when taking into consideration the above outlined criteria.

In this respect, the sensory histidine kinase genes *sasA* and *cikA* as circadian clock-related genes in *Synechococcus* PCC 7942 should also be mentioned. CikA is a novel member of the bacteriophytochrome family (Iwasaki and Kondo 2000).

In plants, well-characterized proteins containing PAS domains, are the phytochromes (absorbing red/far-red light) having an N-terminal chromophore-binding domain, two PAS domains, and a C-terminal serine/threonine kinase domain. Well characterized are also the blue light receptor phototropin (Nph1) involved in phototropism, which has two LOV domains (representing a subfamily of PAS domains – LOV referring to Light, Oxygen, Voltage) for FMN binding and a C-terminal serine/threonine kinase domain, and the cryptochrom (Cry1) containing pterin and FAD being involved in photomorphogenesis (Christie et al. 1999; Marwan 1999). Additional proteins containing the combination of PAS and kinase domains have been found in the nuclear genome of *Arabidopsis*, but the function so far is unclear (The ***Arabidopsis*** Genome Initiative 2000).

7 Redox Regulation as a Global Player in the Control of Plant Cell Development and Metabolism: An Outlook

During the last two decades, a large number of reports described aspects of the relationship between the redox state of the electron transport chain or subsequent reactions and the regulation of photosynthesis. During the last five years, redox control of plastid and nuclear gene expression has come into focus. Transgenic approaches were used to probe the significance of specific reactions in the acclimation response of photosynthetic cells to excess excitation energy. Sites of signal detection have been narrowed down, targets of redox-dependent regulation have been identified. Our review attempts to selectively describe this progress. Nevertheless, our understanding of redox signal perception and transduction is only slowly emerging. The availability of the entire nucleotide sequences of the *Synechocystis* PCC 6803 and *Arabidopsis thaliana* genomes provides an excellent opportunity to apply computer-aided homology searches for the identification of putative regulatory modules in organisms with oxigenic photosynthesis. Cyanobacteria can serve as model organisms. With the prediction that nine further genomes of cyanobacteria will be sequenced by the end of 2001, a multitude of information will be available to be used for elucidation of redox-mediated regulatory circuits by rapid experimental genetics. There are several advantages of using cyanobacteria for such investigations: Cyanobacteria are easy to handle and manipulate genetically. Moreover, in cyanobacteria, not having separate organelles for photosynthesis and respiration, the interrelationship between photosynthesis and respiration as well as the interrelationship between photosynthesis/respiration and the overall cellular metabolism is much more direct than in algae and plants. Therefore, major signal transduction pathways and the cross-talk between various regulatory circuits can be identified more easily. Finally, cyanobacteria represent a good model organism because of their great adaptability to a wide range of environmental conditions, their significant contribution to photosynthesis on earth, and their economic importance. With the entire nucleotide sequence of genomes being available for a fairly large number of heterotrophically growing bacteria, the cyanobacterium *Synechocystis* PCC 6803 (and soon from nine additional cyanobacteria) and now also for *Arabidopsis thaliana,* comprehensive comparisons are possible to see to what extent such redox-mediated regulatory circuits have been conserved or have been changed over evolution. However, the results obtained with cyanobacteria can only partly be extrapolated and have to be tested vigorously for applicability in higher plants. In addition, the complexity of eukaryotic cells with the chloroplast as a separate compartment requires communication between chloroplast and cytosol during development and adaptation to environmental conditions. The review shows that intercompartment signaling is still little understood. But major progress can be expected in the near future. Furthermore, the quantitative importance of redox control, i.e., the quantity and function of genes subjected to redox control in general and depending on photosynthetic reactions in particular, and the regulatory networks need to be investigated in plants, similarly to previous investigations in yeast and *Rhodobacter* for example (J. Lee et al. 1999; Oh and Kaplan 2000). Modern methods of transcriptome and proteome analysis will facilitate such investigations.

Acknowledgments. The authors gratefully acknowledge the support of their work on redox regulation in photosynthetic organisms by the Deutsche Forschungsgemeinschaft within the Forschergruppe FOR 387, projects 1, 2, 3 and 7. The authors are grateful to Prof. H. Rennenberg (University of Freiburg) for helpful discussion.

References

Alex LA, Simon MI (1994) Protein histidine kinases and signal transduction in prokaryotes and eukaryotes. Trends Genet 10:133–139

Allen JF (1992) Protein phosphorylation in regulation of photosynthesis. Biochim Biophys Acta 1098:275–335

Allen JF (1993a) Redox control of gene expression and the function of chloroplast genomes – a hypothesis. Photosynth Res 36:95–102

Allen JF (1993b) Redox control of transcription: sensors, response regulators, activators and repressors. FEBS Lett 332:203–207

Allen JF, Nilsson A (1997) Redox signaling and the structural basis of regulation of photosynthesis by protein phosphorylation. Physiol Plant 100:863–864

Allen JF, Sanders CE, Holmes NG (1985) Correlation of membrane protein phosphorylation with excitation energy distribution in the cyanobacterium *Synechococcus* 6301. FEBS Lett 193:271–275

Allen JF, Alexciev K, Hakansson G (1995) Regulation of redox signaling. Curr Biol 5:869–872

Allen MM (1984) Cyanobacterial cell inclusions. Annu Rev Microbiol 38:1–25

Anderson JM, Chow WS, Park Y-I (1995) The grand design of photosynthesis: acclimation of the photosynthetic apparatus to environmental cues. Photosynth Res 46:129–139

Badger MR, von Caemmerer S, Ruuska S, Nakano H (2000) Electron flow to oxygen in higher plants and algae: rates and control of direct photoreduction (Mehler reaction) and rubisco oxygenase. Philos Trans R Soc Lond B 355:1433–1446

Baginsky S, Tiller K, Link G (1997) Transcription factor phosphorylation by a protein kinase associated with chloroplast RNA polymerase from mustard (*Sinapis alba*). Plant Mol Biol 34:181–189

Baginsky S, Tiller K, Pfannschmidt T, Link G (1999) PTK, the chloroplast RNA polymerase-associated protein kinase from mustard (*Sinapis alba*), mediates redox control of plastid in vitro transcription. Plant Mol Biol 39:1013–1023

Baier M, Dietz K-J (1997) The plant 2-Cys peroxiredoxin BAS1 is a nuclear encoded chloroplast protein. Its expressional regulation, phylogenetic origin, and implications for its specific physiological function in plants. Plant J 12:179–190

Baier M, Dietz K-J (1998) The costs and benefits of oxygen in photosynthetic plant metabolism. Prog Bot 60 283–314.

Baier M, Dietz K-J (1999) Protective function of chloroplast 2-Cys peroxiredoxin in photosynthesis: Evidence from transgenic *Arabidopsis thaliana*. Plant Physiol 119:1407–1414.

Baier M, Noctor G, Foyer CH, Dietz KJ (2000) Antisense suppression of 2-Cys peroxiredoxin in *Arabidopsis thaliana* specifically enhances the activities and expression of enzymes associated with ascorbate metabolism, but not glutathione metabolism. Plant Physiol 124:823–832.

Ballicora MA, Frueauf JB, Fu Y, Schürmann P, Preiss J (2000) Activation of the potato ADP-glucose pyrophosphorylase by thioredoxin. J Biol Chem 275:1315–1320

Beck E, Burkert A, Hofmann M (1983) Uptake of L-ascorbate by intact spinach chloroplasts. Plant Physiol 73:41–45.

Bhaya D, Schwarz R, Grossman AR (2000) Molecular responses to environmental stress. In: Whitton BA, Potts M (eds) The ecology of cyanobacteria. Their diversity in time and space. Kluwer, Dordrecht, pp 397–442

Bogorad L, Vasil IK (1991) The molecular biology of plastids. Academic Press, San Diego

Bruick RK, Mayfield SP (1999) Light-activated translation of chloroplast mRNAs. Trends Plant Sci 4:190–195

Bryk R, Griffin P, Nathan C (2000) Peroxynitrite reductase activity of bacterial peroxiredoxins. Nature 407:211–215

Bunik V, Raddatz G, Lemaire S, Meyer Y, Jacquot J-P, Bisswanger H (1999) Interaction of thioredoxins with target proteins: role of particular structural elements and electrostatic properties of thioredoxins in their interplay with 2-oxoacid dehydrogenase complexes. Protein Sci 8:65–74

Carr PD, Verger D, Ashton AR, Ollis DL (1999) Chloroplast NADP-malate dehydrogenase: structural basis of light-dependent regulation of activity by thiol oxidation and reduction. Structure Folding Design 7:461–475

Chae HZ, Kang SW, Rhee SG (1999) Isoforms of mammalian peroxiredoxin that reduce peroxides in presence of thioredoxin. Methods Enzymol 300:19–226

Chang C, Stewart RC (1998) The two-component system. Regulation of diverse signaling pathways in procaryotes and eukaryotes. Plant Physiol 117:723–731

Chiadmi M, Navaza A, Miginiac-Maslow M, Jacquot J-P, Cherfils J (1999) Redox signalling in the chloroplast: structure of oxidized pea fructose-1,6-bisphosphate phosphatase. EMBO J 18:6809–6815

Christie JM, Salomon M, Nozue K, Wada M, Briggs WR (1999) LOV (light, oxygen, or voltage) domains of the blue-light photoreceptor phototropin (*nph1*): Binding sites for the chromophore flavin mononucleotide. Proc Natl Acad Sci USA 96:8779–8783

D'Agostino IB, Kieber JJ (1999) Phosphorelay signal transduction: the emerging family of plant response regulators. Trends Biochem Sci 24:452–456

Danon A (1997) Translational regulation in the chloroplast. Plant Physiol 115:1293–1298

Danon A, Mayfield SP (1994a) ADP-dependent phosphorylation regulates RNA-binding in vitro: implications in light-modulated translation. EMBO J 13:2227–2235

Danon A, Mayfield SP (1994b) Light-regulated translation of chloroplast messenger RNAs through redox potential. Science 266:1717–1719

Delumeau O, Renard M, Montrichard F (2000) Characterization and possible redox regulation of the purified calmodulin-dependent NAD^+ kinase from *Lycopersicon pimpinellifolium*. Plant Cell Environ 23:1267–1273

Demmig-Adams B, Adams III WW (1992) Photoprotection and other responses of plants to high light stress. Annu Rev Plant Physiol Plant Mol Biol 43:599–625

Demple B (1998) Signal transduction – a bridge to control. Science 279:1655–1656

Deng XW, Tonkyn JC, Peter GF, Thornber JP Gruissem W (1989) Post-transcriptional control of plastid mRNA accumulation during adaptation of chloroplasts to different light quality environments. Plant Cell 1:646–654

Deshpande NN, Bao Y, Herrin DL (1997) Evidence for light/redox-regulated splicing of psbA pre-RNAs in *Chlamydomonas* chloroplasts. RNA 3:37–48

Dietz K-J, Schreiber U, Heber U (1985) The relationship between the redox state of Q_A and photosynthesis in leaves at various carbon dioxide, oxygen and light regimes. Planta 166:219–226

Dolganov N, Grossman AR (1999) A polypeptide with similarity to phycocyanin α-subunit phycocyanin lyase involved in degradation of phycobilisomes. J Bacteriol 181:610–617

Durnford DG, Falkowski PG (1997) Chloroplast redox regulation of nuclear gene transcription during photoacclimation. Photosynth Res 53:229–241

Escoubas JM, Lomas M, LaRoche J, Falkowski PG (1995) Light intensity regulation of cab gene transcription is signaled by the redox state of the plastoquinone pool. Proc Natl Acad Sci USA 92:10237–10241

Fabret C, Feher VA, Hoch JA (1999) Two-component signal transduction in *Bacillus subtilis*: how one organism sees its world. J Bacteriol 181:1975–1983

Falch B (1996) Was steckt in Cyanobakterien? Pharm Unserer Zeit 25:311–321

Farrar J, Pollock C, Gallagher J (2000) Sucrose and the integration of metabolism in vascular plants. Plant Sci 154:1–11.

Faske M, Holtgrefe S, Ocheretina O, Meister M, Backhausen JE, Scheibe R (1995) Redox equilibria between the regulatory thiols of light/dark-modulated chloroplast enzymes and dithiothreitol: fine-tuning by metabolites. Biochim Biophys Acta 1247:135–142

Fong CL, Lentz A, Mayfield SP (2000) Disulfide bond formation between RNA binding domains is used to regulate mRNA binding activity of the chloroplast poly(A)-binding protein. J Biol Chem 275:8275–8278

Forchhammer K, Tandeau de Marsac N (1995) Functional analysis of the phosphoprotein P_{II} (*glnB* gene product) in the cyanobacterium *Synechococcus* sp. strain PCC 7942. J Bacteriol 177:2033–2040

Foyer CH, Noctor G (2000) Oxygen processing in photosynthesis: regulation and signaling. New Phytol 146:359–388

Foyer CH, Lopez-Delgado H, Dat JF, Scott IM (1997) Hydrogen peroxide- and glutathione-associated mechanisms of acclimatory stress tolerance and signaling. Physiol Plant 100:241–254

Fridlyand LE, Scheibe R (2000) Regulation in metabolic systems under homeostatic flux control. Arch Biochem Biophys 374:198–206

Fridlyand LE, Backhausen JE, Scheibe R (1998) Flux control of the malate valve in leaf cells. Arch Biochem Biophys 349:290–298

Fridlyand LE, Backhausen JE, Scheibe R (1999) Homeostatic regulation upon changes of enzyme activities in the Calvin cycle as an example for general mechanisms of flux control. What can we expect from transgenic plants? Photosynth Res 61:227–239

Fu Y, Ballicora MA, Leykam JF, Preiss J (1998) Mechanism of reductive activation of potato tuber ADP-glucose pyrophosphorylase. J Biol Chem 273:25045–25052

Fujita Y, Murakami A, Aizawa K, Ohki K (1994) Short-term and long-term adaptation of the photosynthetic apparatus: homeostatic properties of thylakoids. In: Bryant DA (ed) The molecular biology of cyanobacteria. Kluwer, Dordrecht, pp 677–692

Gal A, Zer H, Ohad I (1997) Redox-controlled thylakoid protein phosphorylation. News and views. Physiol Plant 100:869–885

Gleason FK (1994) Thioredoxins in cyanobacteria: Structure and redox regulation of enzyme activity. In: Bryant DA (ed) The molecular biology of cyanobacteria. Kluwer, Dordrecht, pp 715–729

Golden SS (1994) Light-responsive gene expression and the biochemistry of the photosystem II reaction center. In: Bryant DA (ed) The molecular biology of cyanobacteria. Kluwer, Dordrecht, pp 693–714

Grossman AR, Schaefer MR, Chiang GG, Collier JL (1994) The responses of cyanobacteria to environmental conditions: Light and nutrients. In: Bryant DA (ed) The molecular biology of cyanobacteria. Kluwer, Dordrecht, pp 641–675

Grossman AR, van Waasbergen L, Schwarz R, Dolganov N (1999) The acclimation of photosynthetic organisms to adverse environmental conditions. In: Börner T, Hess WR, Lockau W, Schuler G, Tittel C (eds) Fourth European Workshop on the Molecular Biology of Cyanobacteria, 15–17 Sept, Berlin, 1999, p 39

Hardii DG (1999) Plant protein serine/threonine kinases: classification and functions. Annu Rev Plant Physiol Plant Mol Biol 50:97–131

Havaux M, Bonfils J-P, Lütz C, Niyogi KK (2000) Photodamage of the photosynthetic apparatus and its dependence on the leaf developmental stage in the *npq*[1] *Arabidopsis* mutant deficient in the xanthophyll cycle enzyme violaxanthin de-epoxidase. Plant Physiol 124:273–284

Heddad M, Adamska I (2000) Light stress-regulated two-helix proteins in *Arabidopsis thaliana* related to the chlorophyll a/b-binding gene family. Proc Natl Acad Sci USA 97:3741–3746

Hirasawa M, Schürmann P, Jacquot J-P, Manieri W, Jacquot P, Keryer E, Hartman FC, Knaff DB (1999) Oxidation-reduction of chloroplast thioredoxins, ferredoxin:thioredoxin reductase, and thioredoxin *f*-regulated enzymes. Biochemistry 38:5200–5205

Hirasawa M, Ruelland E, Schepens I, Issakidis-Bourguet E, Miginiac-Maslow M, Knaff DB (2000) Oxidation-reduction properties of the regulatory disulfides of sorghum chloroplast nicotinamide adenine dinucleotide phosphate-malate dehydrogenase. Biochemistry 39:3344–3350

Holtgrefe S, Backhausen JE, Kitzmann C, Scheibe R (1997) Regulation of steady-state photosynthesis in isolated intact chloroplasts under constant light: responses of carbon fluxes, metabolic pools and enzyme-activation states to changes of electron pressure. Plant Cell Physiol 38:1207–1216

Horling F, Baier M, Dietz K-J (2001) The cellular redox poise regulates expression of the peroxide detoxifying chloroplast 2-Cys peroxiredoxin in the liverwort *Riccia fluitans*. Planta (in press)

Hughes J, Lamparter T, Mittmann E, Gartner W, Wilde A, Borner T (1997) A prokaryotic phytochrome. Nature 386:663.

Huner NPA, Maxwell DP, Gray GR, Savitch LV, Krol M, Ivanov AG, Falk S (1996) Sensing environmental temperature change through imbalances between energy supply and energy consumption: redox state of photosystem II. Physiol Plant 98:358–364

Huner NPA, Öquist G, Sarhan F (1998) Energy balance and acclimation to light and cold. Trends Plant Sci 3:224–330

Hunter SC, Ohlrogge JB (1998) Regulation of spinach chloroplast acetyl-CoA carboxylase. Arch Biochem Biophys 359:170–178

Iwasaki H, Kondo T (2000) The current state and problems of circadian clock studies in cyanobacteria. Plant Cell Physiol 41:1013–1020

Johansson K, Ramaswamy S, Saarinen M, Lemaire-Chamley M, Issakidis-Bourguet E, Miginiac-Maslow M, Eklund H (1999) Structural basis for light activation of a chloroplast enzyme: the structure of sorghum NADP-malate dehydrogenase in its oxidized form. Biochemistry 38:4319–4326

Kaneko T, Sato S, Kotoni H, Tanaka A, Asamizu E, Nakamura Y, Miyajima N, Hirosawa M, Sugiura M, Sasamoto S, Kimura T, Hosouchi T, Matsuno A., Muraki A, Nakazaki N, Naruo K, Okumura S, Shimpo S, Takeuchi C, Wada T, Watanabe A, Yamada M, Yasuda M, Tabata S (1996) Sequence analysis of the genome of the unicellular cyanobacterium *Synechocystis* sp. strain PCC 6803. II. Sequence determination of the entire genome and assignment of potential protein-coding regions. DNA Res 3:109–136

Karpinski S, Escobar C, Karpinska B, Creissen G, Mullineaux PM (1997) Photosynthetic electron transport regulates the expression of cytosolic ascorbate peroxidase genes in *Arabidopsis* during excess light stress. Plant Cell 9:627–640

Karpinski S, Reynolds H, Karpinski B, Wingsle G, Creissen G, Mullineaux P (1999) Systemic signaling and acclimation in response to excess excitation energy in *Arabidopsis*. Science 284:654–657

Kehoe DM, Grossman AR (1996) Similarity of a chromatic adaptation sensor to phytochrome and ethylene receptors. Science 273:1409–1412

Kelly GJ (1999) Photosynthesis. Carbon metabolism: in and beyond the chloroplast. In: Esser K, Kadereit JW, Lüttge U, Runge M (eds) Progress in Botany, vol 60. Springer, Berlin Heidelberg New York, pp 254–281

Kelly GJ (2000) Photosynthesis. Carbon metabolism from DNA to deoxyribose. In: Esser K, Lüttge U, Kadereit JW, Beyschlag W (eds) Progress in Botany, vol 62. Springer, Berlin Heidelberg New York, pp 238–265

Kennelly PJ, Potts M (1996) Fancy meeting you here! A fresh look at "prokaryotic" protein phosphorylation. J Bacteriol 178:4759–4764

Kettunen R, Pursiheimo S, Rintamäki E, Van Wijk KJ, Aro EM (1997) Transcriptional and translational adjustments of *psbA* gene expression in mature chloroplasts during photoinhibition and subsequent repair of photosystem II. Eur J Biochem 247:441–448

Kim JM, Mayfield SP (1997) Protein disulfide isomerase as a regulator of chloroplast translational activation. Science 278:1954–1957

Klughammer B, Baier M, Dietz K-J (1998) Inactivation by gene disruption of 2-cysteine-peroxiredoxin in *Synechocystis* sp PCC 6803 leads to increased stress sensitivity. Physiol Plant 104:699–706

Kotani H, Tabata S (1998) Lessons from sequencing of the genome of a unicellular cyanobacterium, *Synechocystis* sp. PCC 6803. Annu Rev Plant Physiol Plant Mol Biol 49:151–171

Kovtun Y, Chiu WL, Tena G, Sheen J (2000) Functional analysis of oxidative stress-activated mitogen activated protein kinase cascade. Proc Natl Acad Sci USA 97:2940–2945

Kozaki A, Kamada K, Nagano Y, Iguchi H, Sasaki Y (2000) Recombinant carboxyltransferase responsive to redox of pea plastidic acetyl-CoA carboxylase. J Biol Chem 275:10702–10708

Kuroda H, Kobashi K, Kaseyama H, Satoh K (1996) Possible involvement of a low redox potential component(s) downstream of photosystem I in the translational regulation of the D1 subunit of the photosystem II reaction center in isolated pea chloroplasts. Plant Cell Physiol 37:754–761

Lee H-M, Vázquez-Bermúdez MF, Tandeau de Marsac N (1999) The global nitrogen regulator NtcA regulates transcription of the signal transducer P_{II} (GlnB) and influences its phosphorylation level in response to nitrogen and carbon supplies in the cyanobacterium *Synechocococus* sp. strain PCC 7942. J Bacteriol 181:2697–2702

Lee J, Godon C, Lagniel G, Spector D, Garin J, Labarre J, Toledano MB (1999) Yap1 and Skn7 control two specialized oxidative stress response regulons on yeast. J Biol Chem 274:16040–16046

Li H, Sherman LA (2000) A redox-responsive regulator of photosynthesis gene expression in the cyanobacterium *Synechocystis* sp. strain PCC 6803. J Bacteriol 182:4268–4277

Liere K, Link G (1997) Chloroplast endoribonuclease p54 involved in RNA 3′-end processing is regulated by phosphorylation and redox state. Nucleic Acids Res 25:2403–2408

Link G (1996) Green life: control of chloroplast gene transcription. BioEssays 18:465–471

Link G (2001) Redox regulation of photosynthetic genes. In: Andersson B, Aro E-M (ed) Advances in photosynthesis, vol 7. Kluwer, Dordrecht,

Link G, Tiller K, Baginsky S (1997) Glutathione, a regulator of chloroplast transcription. In: Hatzios KK (ed) Regulation of enzymatic systems detoxifying xenobiotics in plants. Kluwer, Dordrecht, pp 125–137

Loomis WF, Shaulsky G, Wang N (1997) Histidine kinases in signal transduction pathways of eukaryotes. J Cell Sci 110:1141–1145

Maliga P (1998) Two plastid RNA polymerases of higher plants: an evolving story. Trends Plant Sci 3:4–6.

Mann NH (1994) Protein phosphorylation in cyanobacteria. Microbiology 140:3207–3215

Mann NH (2000) Detecting the environment. In: Whitton BA, Potts M (eds) The ecology of cyanobacteria. Their diversity in time and space. Kluwer, Dordrecht, pp 367–395

Martin W, Scheibe R, Schnarrenberger C (1999) The Calvin Cycle and its regulation. In: Leegood RC, Sharkey TD, von Caemmerer S (eds) Photosynthesis: physiology and metabolism. Kluwer, Dordrecht, pp 2–43

Marwan W (1999) Kryptochrome und LOV/PAS-Domänen-Proteine – vielseitige Regulatoren des Zellgeschehens. Biospektrum 6:443–448

Masuda S, Matsumoto Y, Nagashima KVP, Shimada K, Inoue K, Bauer CE, Matsuura K (1999) Structural and functional analyses of photosynthetic regulatory genes *regA* and *regB* from *Rhodovulum sulfidophilum*, *Roseobacter denitrificans*, and *Rhodobacter capsulatus*. J Bacteriol 181:4205–4215

Maxwell DP, Laudenbach DE, Huner NPA (1995) Redox regulation of light harvesting complex II and cab mRNA abundance in *Dunaliella salina*. Plant Physiol 109:787–795

May MJ, Vernoux T, Leaver C, Van Montagu M, Inzé D (1998) Glutathione homeostasis in plants: implications for environmental sensing and plant development. J Exp Bot 49:649–667

Mayfield SP, Yohn CB, Cohen A, Danon A (1995) Regulation of chloroplast gene expression. Annu Rev Plant Physiol Plant Mol Biol 46:147–166

Mizuno T (1997) Compilation of all genes encoding two-component phosphotransfer signal transducers in the genome of *Escherichia coli*. DNA Res 4:161–168

Mizuno T, Kaneko T, Tabata S (1996) Compilation of all genes encoding bacterial two-component signal transducers in the genome of the cyanobacterium, *Synechocystis* sp. strain PCC 6803. DNA Res 3:407–414

Montane MH, Dreyer S, Triantaphylides C, Kloppstech K (1997) Early light-inducible proteins during long-term acclimation of barley to photooxidative stress caused by light and cold: High level of accumulation by posttranscriptional regulation. Planta 202:293–302.

Mühlbauer SK, Eichacker LA (1998) Light-dependent formation of the photosynthetic proton gradient regulates translation elongation in chloroplasts. J Biol Chem 273:20935–20940

Mullineaux CW, Allen JF (1990) State 1-state 2 transitions in the cyanobacterium *Synechococcus* 6301 are controlled by the redox state of electron carriers between photosystems I and II. Photosynth Res 23:297–311

Mullineaux P, Ball L, Escobar C, Karpinski B, Creissen G, Karpinski S (2000) Are diverse signaling pathways integrated in the regulation of *Arabidopsis* antioxidant defence gene expression in response to excess excitation energy. Philos Trans R Soc Lond 355:1531–1540

Nickelsen J, Link G (1993) The 54 kDa RNA-binding protein from mustard chloroplasts mediates endonucleolytic transcript 3′ end formation in vitro. Plant J 3:537–544

Noctor G, Foyer CH (1998) Ascorbate and glutathione: keeping active oxygen under control. Annu Rev Plant Physiol Plant Mol Biol 49:249–279

Oh JI, Kaplan S (2000) Redox signaling: globalization of gene expression. EMBO J 19:4237–4247

Parkinson JS (1993) Signal transduction schemes of bacteria. Cell 73:857–871

Parkinson JS, Kofoid EC (1992) Communication modules in bacterial signaling proteins. Annu Rev Genet 26:71–112

Pearson CK, Wilson SB, Schaffer R, Ross AW (1993) NAD turnover and utilisation of metabolites for RNA synthesis in a reaction sensing the redox state of the cytochrome b_6/f complex in isolated chloroplasts. Eur J Biochem 218:397–404.

Pego J, Kortstee AJ, Huijser, Smeekens SCM (2000) Photosynthesis, sugars and the regulation of gene expression. J Exp Bot 51:407–416

Pfannschmidt T, Nilsson A, Allen JF (1999a) Photosynthetic control of chloroplast gene expression. Nature 397:625–628

Pfannschmidt T, Nilsson A, Tullberg A, Link G, Allen JF (1999b) Direct transcriptional control of the chloroplast genes *psbA* and *psaAB* adjusts photosynthesis to light energy distribution in plants. Biochem Mol Biol Int 48:271–276

Pfannschmidt T, Ogrzewalla K, Baginsky S, Sickmann A, Meyer HE, Link G (2000) The multisubunit chloroplast RNA polymerase A from mustard (*Sinapis alba* L.): integration of a prokaryotic core into a larger complex with organelle-specific functions. Eur J Biochem 267:253–261

Preiss J, Ballicora M, Fu Y (1999) Allosteric regulation and reductive activation of ADP glucose pyrophosphorylase. In: Bryant JA, Burrell MM, Kruger NJ (eds) Plant carbohydrate biochemistry, BIOS Scientific Publishers Ltd, Oxford, pp 103–125

Radwan SS, Al-Hasan RH (2000) Oil pollution and cyanobacteria. In: Whitton BA, Potts M (eds) The ecology of cyanobacteria. Their diversity in time and space. Kluwer, Dordrecht, pp 307–319

Reichert A, Baalmann E, Vetter S, Backhausen JE, Scheibe R (2000) Activation properties of the redox-modulated chloroplast enzymes glyceraldehyde 3-phosphate dehydrogenase and fructose-1,6-bisphosphatase. Physiol Plant 110:330–341

Rintamäki E, Martinsuo P, Pursiheimo S, Aro EM (2000) Cooperative regulation of light-harvesting complex II phosphorylation via the plastoquinol and ferredoxin-thioredoxin system in chloroplasts. Proc Natl Acad Sci USA 97:11644–11649

Ruelland E, Miginiac-Maslow M (1999) Regulation of chloroplast enzyme activities by thioredoxins: activation or relief from inhibition? Trends Plant Sci 4:136–141

Ruelland E, Johansson K, Decottignies P, Djukic N, Miginiac-Maslow M (1998) The autoinhibition of sorghum NADP malate dehydrogenase is mediated by a C-terminal negative charge. J Biol Chem 273:33482–33488

Salvador ML, Klein U (1999) The redox state regulates RNA degradation in the chloroplast of *Chlamydomonas reinhardtii*. Plant Physiol 121:1367–1374

Sasaki Y, Kozaki A, Hatano M (1997) Link between light and fatty acid synthesis: Thioredoxin-linked reductive activation of plastidic acetyl-CoA carboxylase. Proc Natl Acad Sci USA 94:11096–11101

Schamborn M (1996) Polyasparaginsäure. Nachr Chem Tech Lab 44:1167–1170

Scheibe R (1996) Die Regulation der Photosynthese durch das Licht. BiuZ 26:27–34

Schopf JW (2000) The fossil record: Tracing the roots of the cyanobacterial lineage. In: Whitton BA, Potts M (eds) The ecology of cyanobacteria. Their diversity in time and space. Kluwer, Dordrecht, pp 13–35

Schürmann P, Jacquot J-P (2000) Plant thioredoxin system revisited. Annu Rev Plant Physiol Plant Mol Biol 51:371–400

Schwarz R, Grossman AR (1998) A response regulator of cyanobacteria integrates diverse environmental signals and is critical for survival under extreme conditions. Proc Natl Acad Sci USA 95:11008–11013

Sen CK, Packer L (1996) Antioxidant and redox regulation of gene transcription. FASEB J 10:709–720

Shang W, Feierabend J (1999) Dependence of catalase photoinactivation in rye leaves on light intensity and quality and characterization of a chloroplast-mediated inactivation in red light. Photosynth Res 59:201–213.

Simon RD (1987) Inclusion bodies in the cyanobacteria: Cyanophycin, polyphosphate, polyhedral bodies. In : Fay P, van Baalen C (eds) The cyanobacteria. Elsevier, Amsterdam, pp 199–225

Strand A, Hurry V, Henkes S, Huner N, Gustafsson P, Gardestrom P, Stitt M (1999) Acclimation of *Arabidopsis* leaves developing at low temperatures. Increasing cytoplasmic volume accompanies increased activities of enzymes in the Calvin cycle and in the sucrose-biosynthesis pathway. Plant Physiol 119:1387–1397

Sugita M, Sugiura M (1996) Regulation of gene expression in chloroplasts of higher plants. Plant Mol Biol 32:315–326

Suzuki I, Los DA, Kanesaki Y, Mikami K, Murata N (2000) The pathway for perception and transduction of low-temperature signals in *Synechocystis.* EMBO J 19:1327–1334

Suzuki JY, Maliga P (2000) Engineering of the *rpl23* gene cluster to replace the plastid RNA polymerase alpha subunit with the *Escherichia coli* homologue. Curr Genet 38:218–225

Taylor BL, Zhulin IB (1999) PAS domains: internal sensors of oxygen, redox potential, and light. Microbiol Mol Biol Rev 63:479–506

The *Arabidopsis* Genome Initiative (2000) Analysis of the genome sequence of the flowering plant *Arabidopsis thaliana.* Nature 408:796–815

Trebitsh T, Levitan A, Sofer A, Danon A (2000) Translation of chloroplast *psbA* mRNA is modulated in the light by counteracting oxidizing and reducing activities. Mol Cell Biol 20:1116–1123

Trebst A (1980) Inhibitors in electron flow. Methods Enzymol 69:675–715

Tullberg A, Alexciev K, Pfannschmidt T, Allen JF (2000) Photosynthetic electron flow regulates transcription of the *psaB* gene in pea (*Pisum sativum* L.) chloroplasts through the redox state of the plastoquinone pool. Plant Cell Physiol 41:1045–1054

Urao T, Yamaguchi-Shinozaki K, Shinozaki K (2000) Two-component systems in plant signal transduction. Trends Plant Sci 5:67–74

Vanlerberghe GC, McIntosh L, Yip JYH (1998) Molecular localization of a redox-modulated process regulating plant mitochondrial electron transport. Plant Cell 10:1551–1560

Vanlerberghe GC, Yip JYH, Parsons HL (1999) In organello and in vivo evidence of the importance of the regulatory sulfhydryl/disulfide system and pyruvate for alternative oxidase activity in tobacco. Plant Physiol 121:793–803

Van Thor JJ, Mullineaux CW, Matthijs HCP, Hellingwerf KJ (1998) Light harvesting and state transitions in cyanobacteria. Bot Acta 111:430–443

Vener AV, Ohad I, Andersson B (1998) Protein phosphorylation and redox sensing in chloroplast thylakoids. Curr Opin Plant Biol 1:217–223

Vonshak A (ed) (1997) *Spirulina platensis* (*Arthrospira*): Physiology, cell-biology and biotechnology. Taylor & Francis Ltd, London

Whitton BA, Potts M (2000) Introduction to the cyanobacteria. In: Whitton BA, Potts M (eds) The ecology of cyanobacteria. Their diversity in time and space. Kluwer, Dordrecht, pp 1–11

Wingate VPM, Lawton MA, Lamb CJ (1988) Glutathione causes a massive and selective induction of plant defence genes. Plant Physiol 87:206 – 210

Wingsle G, Karpinski S (1996) Differential regulation by glutathione of glutathione reductase and CuZn-superoxide dismutase gene expression in *Pinus sylvestris* L. needles. Planta 198:151–157

Wünschiers R, Heide H, Follmann H, Senger H, Schulz R (1999) Redox control of hydrogenase activity in the green alga *Scenedesmus obliquus* by thioredoxin and other thiols. FEBS Lett 455:162–164

Yang JJ, Schuster G, Stern DB (1996) CSP41, a sequence-specific chloroplast mRNA binding protein, is an endoribonuclease. Plant Cell 8:1409–1420

Yeh K-C, Wu S-H, Murphy JT, Lagarias JC (1997) A cyanobacterial phytochrome two-component light sensory system. Science 277:1505–1507

Yohn CB, Cohen A, Danon A, Mayfield SP (1998a) A poly(A) binding protein functions in the chloroplast as a message-specific translation factor. Proc Natl Acad Sci USA 95:2238–2243

Yohn CB, Cohen A, Rosch C, Kuchka MR, Mayfield SP (1998b) Translation of the chloroplast psbA mRNA requires the nuclear-encoded poly(A)-binding protein, RB47. J Cell Biol 142:435–442

Zeilstra-Ryalls J, Gomelsky M, Eraso JM, Yeliseev A, O'Gara J, Kaplan S (1998) Control of photosystem formation in *Rhodobacter sphaeroides*. J Bacteriol 180:2801–2809

Zhang Ch-C, Gonzalez L, Phalip V (1998) Survey, analysis and genetic organization of genes encoding eukaryotic-like signaling proteins on a cyanobacterial genome. Nucleic Acids Res 26:3619–3625

Zhang LX, Paakkarinen V, Van Wijk KJ, Aro EM (1999) Co-translational assembly of the D1 protein into photosystem II. J Biol Chem 274:16062–16067

Zhang LX, Paakkarinen V, Van Wijk KJ, Aro EM (2000) Biogenesis of the chloroplast-encoded D1 protein: regulation of translation elongation, insertion, and assembly into photosystem II. Plant Cell 12:1769–1781

Zhang N, Portis AR Jr (1999) Mechanism of light regulation of Rubisco: a specific role for the larger Rubisco activase isoform involving reductive activation by thioredoxin-f. Proc Natl Acad Sci USA 96:9438–9443

Zheng M, Åslund F, Storz G (1998) Activation of the OxyR transcription factor by reversible disulfide bond formation. Science 279:1718–1721

Zhulin IB, Taylor BL (1998) Correlation of PAS domains with electron transport-associated proteins in completely sequenced microbial genomes. Mol Microbiol 29:1522–1523

Zhulin IB, Taylor BL, Dixon R (1997) PAS domain S-boxes in archaea, bacteria and sensors for oxygen and redox. Trends Biochem Sci 22:331–333

Prof. Dr. Karl-Josef Dietz
Stoffwechselphysiologie und Biochemie der Pflanzen
Fakultät für Biologie
Universität Bielefeld
33501 Bielefeld, Germany

Prof. Dr. Gerhard Link
Ruhr-Universität Bochum
Fakultät für Biologie, ND 2–72,
Arbeitsgruppe Pflanzliche Zellphysiologie
Universitätsstraße 150
44780 Bochum, Germany

Prof. Dr. Elfriede K. Pistorius
Zellphysiologie
Fakultät für Biologie
Universität Bielefeld
33501 Bielefeld, Germany

Prof. Dr. Renate Scheibe
Fachbereich Biologie/Chemie
FB 5/Pflanzenphysiologie
Universität Osnabrück
49069 Osnabrück, Germany

NO Production in Plants: Nitrate Reductase Versus Nitric Oxide Synthase

By Peter Rockel and Werner M. Kaiser

1 Introduction

Nitric oxide (NO) is an inorganic free radical that acts as a signalling molecule with multiple biological functions in vertebrates, including vasorelaxation, neurotransmission and modulation of the immune response. Due to its lipophilicity and being a small uncharged molecule it can easily diffuse through cell membranes but also through water phases. In spite of its reactivity, its lifetime in biological systems is in the range of 5–15 s (Lancaster 1997). Production and functions of NO have been intensively studied in animal physiology, often under clinical aspects. However, during recent years, NO has also gained increasing attention in plant research. Although information available at this stage is rather limited, it indicates already the potential of NO as a multifaceted signalling compound and as both a protective or toxic agent. Here, we will only briefly review some major functions of NO in plants, and will then summarize more recent insights into NO synthesis pathways in plants.

2 NO in Plants: Friend and Foe

Due to its reactivity, NO can rapidly interact with biological material. Generally, the preferred direct targets of NO in cells are proteins containing iron (II) porphyrins such as heme, and proteins with sulfhydryl (SH)-groups. It was first shown in animal cells that NO inhibits aconitase, which is converted into an mRNA-binding protein that regulates iron homeostasis. Recently, it has been demonstrated that NO donors also inhibit aconitase in tobacco (Navarre et al. 2000). NO may also block mitochondrial electron transport through cytochrome oxidase, thereby favouring electron flow through the alternative oxidase (AOX) pathway. In addition, induction of salicylic acid (SA) synthesis by NO (see below) also blocks mitochondrial electron transport and increases expression of AOX genes (Xie and Chen 1999, review by Murphy et al. 1999). The S-nitrosylation of SH-groups by NO can inhibit several SH-

Progress in Botany, Vol. 63

dependent enzymes like glutathione reductase and the reaction of NO with glutathione yields S-nitrosoglutathione which serves as a putative NO-storage (Kröncke et al. 1997).

NO can further influence plant growth and development. It stimulates seed germination and de-etiolation, inhibits hypocotyl elongation (Leshem and Haramaty 1996; Beligni and Lamattina 2000; Leshem et al. 2000) and causes accumulation of phytoalexins (Noritake et al. 1996).

Most importantly, NO appears as a second messenger in plant pathogen resistance (PR) (Pfeiffer et al. 1994; Delledonne et al 1998; Durner and Klessig 1999; Bolwell 1999, also cf. Fig. 1). When plants are exposed to pathogens or elicitors, an NO burst takes place, probably in parallel with (or even triggered by) the well-described oxidative burst (Dangl et

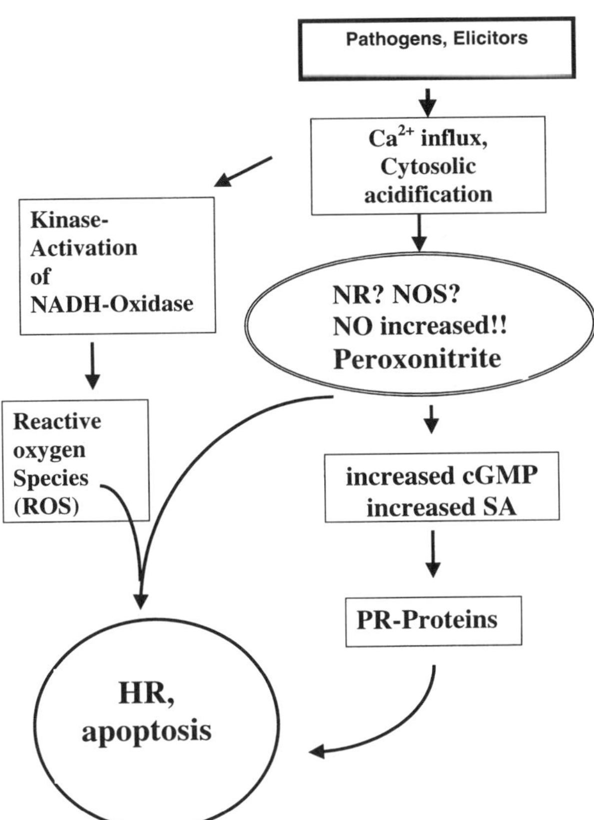

Fig. 1. Simplified scheme of the role of NO in plant pathogen defence (modified according to Dangl et al, 1996) *HR* Hypersensitive response; *NOS* nitric oxide synthase; *NR* nitrate reductase; *PR* pathogen resistance; *ROS* reactive oxygen species; *SA* salicylic acid

al. 1996; McDowell and Dangl 2000) which is mediated by a plasma membrane (PM)-anchored NADPH-oxidase. NO is accumulated during incompatible, but not during compatible plant pathogen interactions, and pharmacological inhibitors of nitric oxide synthase (NOS) ameliorate the establishment of plant disease resistance (Delledonne et al. 1998). NO treatment, transiently increasing cGMP through activation of guanylate cyclase (Durner et al. 1998), eventually also through inhibition of cyclic nucleotide phosphodiesterase, thereby activates phenylalaninelyase to increase salicylic acid levels. Subsequently, PR genes are induced, possibly via cyclic ADP-ribose, and thus activation of Ca^{2+} channels occurs (Klessig et al. 2000; cf. Fig. 1).

In context with the effects of NO on guanylate cyclase, it is also interesting that the well-known drug Sildenafil, which is the effective component of VIAGRA, has very impressive senescence-decelerating effects on plants and can keep cut flowers fresh for prolonged time periods, probably by affecting NO levels and subsequently cGMP-levels, which act on cGMP-gated ion channels (Leshem et al. 2000). This may suggest practical aspects for NO in horticulture.

It is puzzling that NO may also protect against cytotoxicity of reactive oxygen species (ROS) (Laxalt et al. 1997; Beligni and Lamattina 1999), with which NO may rapidly react forming peroxonitrite which reacts especially with sulfhydryl groups (for review see Grant and Loake 2000). Interestingly, in animal cells (neutrophils), the activity of superoxide (O_2^-) producing NADPH-oxidase is blunted by NO (Grant and Loake 2000). Further, NO may act as a potent inhibitor of lipid peroxidation, but can also inhibit many potential initiators of lipid peroxidation (Hogg and Kalyanaraman 1999). Lipid peroxidation, in elicited plants, produces jasmonic acid and induces salicylic acid production which then play a signalling role in the activation of plant defence responses (Mueller et al. 1993; Blee 1998; Klessig et al. 2000). Thus, in plants, NO itself may actually play a critical role in controlling the extent of cell death in the hypersensitive response (HR). Altogether, NO can undergo partly conflicting reactions, and it is not yet clear which role is dominant under which condition.

Plants can also be important sources for atmospheric NO in non-polluted areas. As mentioned above, plants can emit NO into purified air, and if no ozone is present NO is stable enough to be precisely measured by chemiluminescence in the ppt range. It is known that different nitrate-nourished plant species have a compensation point for NO uptake from the atmosphere, showing a net NO emission at low atmospheric NO concentrations (Rockel et al 1996; Wildt et al. 1997). The emission of nitric oxide (NO) from a variety of plant species was measured in a continuously stirred tank reactor. During daytime and at NO concentrations below 1 ppb in the chamber air, NO emissions were observed for all nitrate-nourished plant species studied (sunflower, spinach, rape, spruce, sugar cane, tobacco). A relation was found between the emission rates of NO during daytime and the uptake rates for CO_2. The ratios of the NO emission rates over those of CO_2 uptake were similar for all plants. Changes in the net rate of photosynthesis by variations of light intensity or changes of CO_2 concentrations correspondingly changed NO emission rates. From an assumed link between NO emissions and CO_2 uptake during daytime, the potential of vegetation to evolve NO was roughly estimated to be about 0.23 Tg (N) a^{-1} on a global scale. Strong NO emissions during the night were observed when the nitrate concentration in the nutrient solution was enhanced. This led to emissions of NO with flux densities comparable to the highest emission rates observed from soil.

3 How Is NO Produced in Plants?

a) By Nitric Oxide Synthase (NOS)

In vertebrates NO is produced mainly by the nitric oxide synthase (NOS, EC 1.14.13.39) family, which catalyses NO and L-citrulline formation from O_2, NADPH and L-arginine in the presence of the cofactor tetrahydrobiopterin.

NOS activity was also detected in higher plant extracts (Cueto et al. 1996; Ninnemann and Maier 1996; Ribeiro et al. 1999) and peroxisomes (Barroso et al. 1999). The measurement of NOS activity is usually based on a test using ^3H-labelled arginine. Using this test system, as well as immunoblot analysis, NOS was detected in isolated leaf peroxisomes of *Pisum sativum* L. and in chloroplasts, while no activity was found in isolated mitochondria (Barroso et al. 1999). The fluorescent NO-indicator DAF 2DA (diaminofluorescein diacetate) has been used to visualise NO production in animal cells (Kojima et al. 1998). It is specific, highly sensitive and does not react with nitrite or nitrate or reactive oxygen species. Meanwhile, it has also been applied to detect NO in plant tissues (Pedroso et al. 2000; Foissner et al. 2000). Pedroso et al., using leaves and callus of *Kalanchoe daigremontiana* and *Taxus brevifolia*, localised NO in the cytosol of epidermal cells, in chloroplasts of guard cells and leaf parenchyma cells. Vacuoles were not stained, therefore showing no evidence for vacuolar NO production. By using DAF 2DA in combination with real-time imaging by confocal laser scanning microscopy, it was shown that the fungal elicitor cryptogein (from *Phytophtora cryptogea*) caused a rapid (2-min) onset of NO production in epidermal cells from tobacco leaves. NO production was visualised in the cytosol, but also in chloroplasts and other organelles (Foissner et al. 2000). Further evidence for NOS-dependent NO production is based on the use of commercially available NOS inhibitors, which are usually structural substrate analogues. Whenever NO production can be inhibited by these compounds, it is concluded to be the result of NOS activity. However, we have to be aware that both techniques, DAF-fluorescence and the use of NOS inhibitors or NO donors, do not permit a quantification of NO production. DAF-fluorescence appears to be an excellent tool to localise sites of NO production at the tissue level and even at the subcellular level, but does not give clear quantitative data. NOS inhibitors and donors give only qualitative data anyway, and often it is unknown how much NO was actually present or was produced in the tissues. On the other hand, chemiluminescence detection of NO in the gas phase may be limited to situations where NO is not immediately trapped inside the tissue. This may be especially critical in the most interesting situation, when the hypersensitive response is evoked in incompatible plant pathogen interactions. Here, a time-parallel production of reactive oxygen species and of NO may not at all lead to net NO emission from the plants. Final proof for the presence of NOS genes in plants is still lacking, in spite of a meanwhile completed sequencing of the *Arabidopsis* genome.

b) By Nitrate Reductase (NR)

In contrast to vertebrates, plants (and microorganisms) can produce NO by pathways unrelated to NOS (Wojtaszek 2000). This was first observed by Klepper (1975) with soybean plants treated with photosynthetic inhibitor herbicides (Klepper 1978, 1979) or other chemicals (Klepper 1990, 1991) as well as under dark anaerobic conditions (Klepper 1987, 1990). It was suggested that this emission was due to chemical reactions of accumulated nitrite with plant metabolites, such as salicylate derivatives, or the chemical decomposition of HNO_2.

NO was observed to be the predominant compound evolved during a purged in vivo assay with soybean nitrate reductase derived from accumulated nitrite (Harper 1981). Results obtained with boiled leaflets (Harper 1981) and the mutant soybean line nr_1 (Nelson et al. 1983; Ryan et al. 1983; Dean and Harper 1986, 1988), further indicated that the enzymatic reaction of a constitutive NAD(P)H: nitrate-reductase is responsible for evolution of NO_x. Experiments with ^{15}N-labelled nitrate as substrate for nitrate reduction showed that NO_x is produced from ^{15}N-NO_3^- (Dean and Harper 1986). More recently, it has been shown that NR from maize could also produce NO from nitrite plus NADH, and that this reaction could be prevented by azide (Yamasaki et al. 1999; Yamasaki and Sakihama 2000). These authors also found an NO production with nitrate as substrate, but in that case with a time-lag. They concluded that the actual substrate for NR-dependent NO-production was nitrite, not nitrate.

Similar conclusions have been drawn recently in work done with unicellular green algae (Rai et al. 1999; Mallick et al. 2000a,b; Mallick and Mohn 2000) and also with prokaryotic cyanobacteria (Mallick et al. 1999). All these organisms produced and emitted significant quantities of NO, either directly from nitrite fed to the cells, or from nitrate. NO emission was always closely related to the nitrite concentration in the cells.

By high-sensitive chemiluminescence detection, we have recently investigated NO production and quantified its emission into the gas phase by sunflower plants (*Helianthus annuus* L.), by detached spinach leaves (*Spinacia oleracea* L.), by desalted spinach leaf extracts or by purified maize (*Zea mays* L.) nitrate reductase (NR, EC 1.6.6.1) (Rockel et al., submitted). As stated before, NO production by purified NR occurred immediately with nitrite (not nitrate) and NADH as substrates. Nothing is known as yet about the molecular details of that one-electron transfer reaction. However, maximum NO production rates at saturating NADH and nitrite concentrations were only about 1% of the NR capacity. This contrasts with previous results from Klepper (1990), who found an NO production rate (under "semi-in vitro" conditions with soybean leaves) almost as high as the rate of nitrate reduction. In our experiments, the Km for nitrite was relatively high (100 µM) compared to mean nitrite concentrations in illuminated leaves (10 µM). Importantly, NO production was competitively inhibited by physiological nitrate concentrations (Ki 50 µM). In crude extracts, NR can be inactivated by preincubation with MgATP in the presence of a protein phosphatase inhibitor (for review see Kaiser et al. 1999). The treatment phosphorylates NR on a serine residue in the hinge 1 region (serine 543 in spinach), and subsequently a 14-3-3 dimer binds to phospho-NR, which inactivates NR in the presence of divalent cations. Such NR inactivation in vitro completely inhibited NO production from nitrite plus NADH. On the other hand, preincubation with EDTA and/or 5′AMP, which activates NR, also gave high rates of NO production from nitrite plus NADH.

Nitrate fertilised plants or leaves can emit NO into purified air (Rockel et al. 1996). NO emission was lower in the dark than in the light (Fig. 2), but was generally only a small fraction of total NR activity in the tissue (about 0.01 to 0.1%)(Rockel et al., submitted). When NR was artificially activated by treatments like anoxia, by feeding uncouplers or AICAR [(5-amino-4-imidazolecarboxyamide ribonucleoside), a cell permeate 5′- AMP analogue], NO production by leaves was especially high in the dark. In these cases, leaves were accumulating nitrite to concentrations exceeding those in normal, illuminated leaves up to 200-fold (Table 1). This high nitrite accumulation reflected a strong imbalance between nitrite produced (by NR) and nitrite consumed (by NiR (nitrite reductase) in the plastids) under these special conditions.

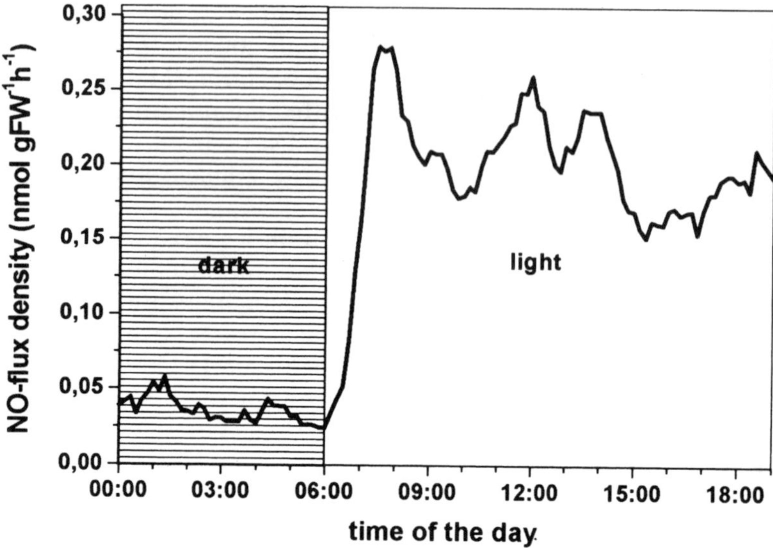

Fig. 2. NO-flux density (nmol NO g^{-1} FW h^{-1}) of sunflower plants in an exposure-chamber during a dark-light-transient (six plants in the chamber, light intensity 450 µmol m^{-2} s^{-1})

Further evidence for the high amounts of cytosolic nitrite required for high NO production rates was obtained from experiments with transgenic tobacco expressing an antisense NiR (Goshima et al. 1999; Rockel, unpubl. results). When these plants were kept under normal light/dark cycles, they accumulated high concentrations of nitrite in the light, and showed an NO emission (high in the light, low in the dark) which was up to three orders of magnitude higher than with the wild-type plants.

NO production by non-infected leaves was insensitive to nitric oxide synthase inhibitors. We concluded that in these cases, NO is produced in

Table 1. NO-emission of spinach leaves in an exposure chamber under different light and air conditions. NO-flux density (NO), actual nitrate reductase activity (NRA) and nitrite content of spinach leaves in light and dark (air) and dark/anaerobe (N_2). Samples for NRA and nitrite determination were taken after 60 min in the light, after a further 30 min in the dark and after 180 min in the dark + N_2 (mean of four samples for NRA and nitrite, for NO mean over 10 min around the moment of sample collection, errors are standard deviation)

	NO flux density (nmol g^{-1} FW h^{-1})	NRA (µmol g^{-1} FW h^{-1})	Nitrite (µmol g^{-1} FW h^{-1})
Light	0.52 ± 0.20	10.0 ± 2.0	0.026 ± 0.003
Dark	0.06 ± 0.04	5.1 ± 0.6	0.011 ± 0.001
Dark/anoxia	165.49 ± 7.27	19.1 ± 3.7	4.873 ± 0.420

variable quantities by NR, depending on at least four variables: (1) NR activity and activation state, (2) NiR activity, (3) cytosolic nitrite concentrations and (4) cytosolic nitrate concentrations. Indeed, under all conditions tested so far, increased NO emission was correlated with increased tissue nitrite concentrations (Rockel et al., submitted).

Thus, NR is a bifunctional enzyme:

it reduces nitrate to nitrite according to Eq. (1):

$$NO_3^- + NAD(P)H + H^+ \Rightarrow NO_2^- + NAD(P)^+ + H_2O \quad (1)$$

or it reduces (with a much lower capacity) nitrite to NO [(Eq. (2)]:

$$2\,NO_2^- + NAD(P)H + H^+ \rightarrow 2\,NO + NAD(P)^+\ 2\,OH^- \quad (2)$$

Higher plants contain not only a soluble NR localised in the cytosol of green and non-green tissues, but also a PM-bound enzyme of as yet unknown physiological relevance. This reduces nitrate preferentially with organic acids like succinate as the electron donor (Stöhr et al. 2000). In addition, Stöhr et al. (2001)) has recently found yet another PM-bound protein (NO-oxidoreductase, PI-NOR) that catalysed the reduction of nitrite to NO. The role of this PI-NOR is especially interesting in our context, since NADPH-oxidase, which is probably a major source of reactive oxygen species outside the chloroplast and which is triggered by elicitors and pathogens (cf. Fig. 1), is also located in the PM and therefore potentially in the neighbourhood of PM-NR. This may help to assure a concerted action of NO and reactive oxygen species (McDowell and Dangl 2000).

3 Conclusions and Future Aspects

The situation in plants, with respect to control of NO production in response to biotic or abiotic stress, is certainly complicated by the fact that they have, in contrast to animals, at least two different sources for NO, namely NOS and NR (including PM-bound NI-NOR).

If NO is really an obligatory intermediary signal in eliciting the hypersensitive response in plants attacked by pathogens, the requirement for NOS-derived NO is obvious: Plants growing in the absence of nitrate do not express NR, and indeed they do not emit NO normally (Rockel et al. 1996). This may be a frequent situation in horticulture, at least transiently, when the only fertiliser is ammonium nitrogen, or urea nitrogen. It may also be a natural situation in plants which grow on acid soils with low nitrification rates, or in plants which receive their nitrogen exclusively from symbiotic or associative N_2 fixation. Without the ability to induce NOS, all these plants would lack an important part of the signalling pathway leading to HR and to acquired resistance. They should be more accessible to biological attack, which has not been observed so far.

In that context it would be extremely helpful to have genetically transformed plants where NOS is repressed, e.g. by antisense approaches. However, this is impossible until the NOS gene(s) are identified. A possible indirect approach to the problem would be to degrade NO as rapidly as it is formed. This has already been achieved by treating plants with NO scavengers such as PTIO (carboxy-2-phenyl-4,4,5,5-tetramethylimidazoline-3-oxide-1-oxyl), which prevents the HR (Durner et al. 1998). Even more promising may be the use of transgenic plants expressing NO-degrading enzymes. Recently, *Arabidopsis* has been successfully transformed with a bacterial NO-oxidase (Jürgen Zeier, personal communication), but unfortunately no data on its HR have yet been obtained.

It will also be important to find out whether NR-derived NO may, at least partly, replace NO from NOS. If this is the case, one would expect NR to be upregulated or NiR to be downregulated during incompatible plant pathogen interactions. Here again, experimental evidence is yet completely lacking.

Acknowledgements. This work was supported by the DFG (Ka 456/12-1).

References

Barroso JB, Corpas FJ, Carreras A, Sandalio LM, Valderrama R, Palma JM, Lupianez JA, del Rio LA (1999) Localization of nitric-oxide synthase in plant peroxisomes. J Biol Chem 274:36729–36733

Beligni MV, Lamattina L (1999) Nitric oxide counteracts cytotoxic processes mediated by reactive oxygen species in plant tissues. Planta 208:337–344

Beligni MV, Lamattina L (2000) Nitric oxide stimulates seed germination and de-etiolation, and inhibits hypocotyl elongation, three light-inducible responses in plants. Planta 210:215–221

Blee E (1998) Phytooxylipins and plant defense reactions. Prog Lipid Res 37:33–72

Bolwell GP (1999) Role of active oxygen species and NO in plant defence responses. Curr Opin Plant Biol 2:287–294

Cueto M, Hernandez-Perera O, Martin R, Bentura ML, Rodrigo J, Lamas S, Golvano MP (1996) Presence of nitric oxide synthase activity in roots and nodules of *Lupinus albus*. FEBS Letters 398:159–164

Dangl JL, Dietrich RA, Richberg MH (1996) Death don't have no mercy: cell death programs in plant-microbe interactions. Plant Cell 8:1793–1807

Dean JV, Harper JE (1986) Nitric oxide and nitrous oxide production by soybean and winged bean during in vivo nitrate reductase assay. Plant Physiol 82:718–723

Dean JV, Harper JE (1988) The conversion of nitrite to nitrogen oxide(s) by the constitutive NAD(P)H-nitrate reductase enzyme from soybean. Plant Physiol 88:389–395

Delledonne M, Xia Y, Dixon RA, Lamb C (1998) Nitric oxide functions as a signal in plant disease resistance. Nature 394:585–588

Durner J, Klessig DF (1999) Nitric oxide as a signal in plants. Curr Opin Plant Biol 2:369–374

Durner, J, Wendehenne, D, Klessig DF (1998) Defence gene induction in tobacco by nitric oxide, cyclic GMP, and cyclic ADP-ribose. Proc Natl Acad Sci USA 95:10328–10333

Foissner I, Wendehenne D, Langebartels C, Durner J (2000) In vivo imaging of an elicitor-induced nitric oxide burst in tobacco. Plant J 23:817–824

Goshima N, Mukai T, Suemori M, Takahashi M, Caboche M, Morikawa H (1999) Emission of nitrous oxide (N_2O) from transgenic tobacco expressing antisense NiR mRNA. Plant J 19:75–80

Grant JJ, Loake GJ (2000) Role of reactive oxygen intermediates and cognate redox signalling in disease resistance. Plant Physiol 124:21–29

Harper JE (1981) Evolution of nitrogen oxide(s) during in vivo nitrate reductase assay of soybean leaves. Plant Physiol 68:1488–1493

Hogg N, Kalyanaraman B (1999) Nitric oxide and lipid peroxidation. Biochim Biophys Acta 1411:378–384

Kaiser WM, Weiner H, Huber SC (1999) Nitrate reductase in higher plants: A case study for transduction of environmental stimuli into control of catalytic activity. Physiol Plant 105:385–390

Klepper LA (1975) Evolution of nitrogen oxide gases from herbicide treated plant tissues. WSSA Abst 184:70

Klepper LA (1978) Nitric oxide (NO) evolution from herbicide-treated soybean plants. Plant Physiol Suppl 61:65

Klepper LA (1979) Nitric oxide (NO) and nitrogen dioxide emissions from herbicide-treated soybean plants. Atmos Environ 13:537–542

Klepper LA (1987) Nitric oxide emissions from soybean leaves during in vivo nitrate reductase assays. Plant Physiol 85:96–99

Klepper LA (1990) Comparison between NOx evolution mechanisms of wild-type and nr1 mutant soybean leaves. Plant Physiol 93:26–32

Klepper LA (1991) NOx evolution by soybean leaves treated with salicylic acid and selected derivatives. Pesticide Biochem Physiol 39:43–48

Klessig DF, Durner J, Noad R, Navarre DA, Wendehenne D, Kumar D, Zhou JM, Shah J, Zhang S, Kachroo P, Trifa Y, Pontier D, Lam E, Silva H (2000) Nitric oxide and salicylic acid signalling in plant defense. Proc Natl Acad Sci USA 97:8849–8855

Kojima H, Nakatsubo N, Kikuchi K, Kawahara S, Kirino Y, Nagoshi H, Hirata Y, Nagano T (1998) Detection and imaging of nitric oxide with novel fluorescent indicators: diaminofluoresceins. Anal Chem 70:2446–2453

Kröncke K-D, Fehsel K, Kolb-Bachofen V (1997) Nitric Oxide: Cytotoxicity versus cytoprotection - how, why, when and where. Nitric Oxide Biol Chem 1:107–120

Lancaster JR (1997) A tutorial on the diffusibility and reactivity of free nitric oxide. Nitric Oxide Biol Chem 1:18–30

Laxalt AM, Beligni MV, Lamattina L (1997) Nitric oxide preserves the level of chlorophyll in potato leaves infected by *Phytophtora infestans*. Eur J Plant Pathol 103:643–651

Leshem YY, Haramaty E (1996) The characterization and contrasting effects of the nitric oxide free radical in vegetative stress and senescence of *Pisum sativum* L. foliage. J Plant Physiol 148:258–263

Leshem YY, Huang JS, Tzeng DDS, Chou CC (2000) Nitric oxide in plants: occurrence, function and use. Kluwer, Dordrecht

McDowell JM, Dangl JL (2000) Signal transduction in the plant immune response. TIBS 25:79–82

Mallick N, Rai LC, Mohn FH, Soeder CJ (1999) Studies on nitric oxide (NO) formation by the green alga *Scenedesmus obliquus* and the diazotrophic cyanobacterium *Anabaena doliolum*. Chemosphere 39:1601–1610

Mallick N, Mohn FH, Rai LC, Soeder CJ (2000a) Evidence for the non-involvement of nitric oxide synthase in nitric oxide production by the green alga *Scenedesmus obliquus*. J Plant Physiol 156:423–426

Mallick N, Mohn FH, Rai LC, Soeder CJ (2000b) Impact of physiological stresses on nitric oxide formation by green alga, *Scenedesmus obliquus*. J Microbiol Biotechnol 10:300–306

Mallick N, Mohn FH (2000) Reactive oxygen species: response of algal cells. J Plant Physiol 157:183–193

Murphy AM, Chivasa S, Singh DP, Carr JP (1999) Salicylic acid-induced resistance to viruses and other pathogens: a parting of the ways? TIPS 4:155–160

Mueller MJ, Brodschelm W, Spannagle E, Zenk MH (1993) Signalling in the elicitation process is mediated through the octadecanoid pathway leading to jasmonic acid. Proc Natl Acad Sci USA 90:7490–7494

Navarre DA, Wendehenne D, Durner J, Noad R, Klessig DF (2000) Nitric oxide modulates the activity of tobacco aconitase. Plant Physiol 122:573–582

Nelson RS, Ryan SA, Harper JE (1983) Soybean-mutants lacking a constitutive nitrate reductase activity. I. Selection and initial plant characterization. Plant Physiol 72:503–509

Ninnemann H, Maier J (1996) Indications for the occurrence of nitric oxide synthases in fungi and plants and the involvement in photoconidiation of *Neurospora crassa*. Photochem Photobiol 64(2):393–398.

Noritake T, Kawakita K, Doke N (1996) Nitric oxide induces phytoalexin accumulation in potato tuber tissues. Plant Cell Physiol 37:113–116

Pedroso MC, Magalhaes JR, Durzan D (2000) A nitric oxide burst precedes apoptosis in angiosperm and gymnosperm callus cells and foliar tissues. J Exp Bot 51:1027–1036

Pfeiffer S, Janistyn B, Jessner G, Pichorner H, Ebermann R (1994) Gaseous nitric oxide stimulates guanosine-3′,5′-cyclic monophosphate (cGMP) formation in spruce needles. Phytochemistry 36(2):259–262

Rai LC, Mohn FH, Rockel P, Wildt J, Soeder CJ (1999) Formation of nitric oxide (NO) in nitrate-supplied suspensions of green algae (*Scenedesmus*). Algol Stud 93:119–130

Ribeiro EA Jr, Cunha FQ, Tamashiro WMSC, Martins IS (1999) Growth-phase dependent subcellular localization of nitric oxide synthesis in maize cells. FEBS Lett 445:283–286

Rockel P, Rockel A, Wildt J, Segschneider H-J (1996) Nitric oxide (NO) emission by higher plants. In: Van Cleemput O, Hofman G, Vermoesen A (eds) Progress in nitrogen cycling studies. Kluwer, Dordrecht, pp 603–606

Ryan SA, Nelson RS, Harper JE (1983) Soybean mutants lacking constitutive nitrate reductase activity. II. Nitrogen assimilation, chlorate resistance, and inheritance. Plant Physiol 72:510–514

Stöhr C, Wienkoop S, Ullrich WR (2000) Nitrate reductase in roots: Succinate - and NADH-dependent plasma membrane bound forms. Plant Soil (in press)

Stöhr C, Strube F, Marx G, Ullrich WR, Rockel P (2001) A plasma membrane-bound enzyme of tobacco roots catalyzes the formation of nitric oxide from nitrite. Planta 212:835–843

Wildt J, Kley D, Rockel A, Rockel P, Segschneider H-J (1997) Emission of NO from several higher plant species. J Geophys Res 102 (D5):5919–5927

Wojtaszek P (2000) Nitric oxide in plants To NO or not to NO. Phytochemistry 54:1–4

Xie Z, Chen Z, (1999) Salicylic acid induces rapid inhibition of mitochondrial electron transport and oxidative phosphorylation in tobacco cells. Plant Physiol. 120:217–225

Yamasaki H, Sakihama Y (2000) Simultaneous production of nitric oxide and peroxonitrite by plant nitrate reductase: in vitro evidence for the NR-dependent formation of active nitrogen species. FEBS Lett 4689:89–92

Yamasaki H, Sakihama Y, Takahashi S (1999) An alternative pathway for nitric oxide production in plants: new features of an old enzyme. Trends Plant Sci 4:128–129

Prof. Dr. Werner M. Kaiser
Julius-von-Sachs-Institut für Biowissenschaften
Lehrstuhl für Molekulare Pflanzenphysiologie und Biophysik
Julius-von-Sachs-Platz 2
97082 Würzburg, Germany
Tel.: +49-0931 888 6120
Fax: +49-0931 888 6148
e-mail: kaiser@botanik.uni-wuerzburg.de

Dr. Peter Rockel
Forschungszentrum Jülich GmbH
Abt.: ICG-6
52425 Jülich, Germany
Tel.: +49-02461 61 4830
Fax: +49-02461 612492
e-mail: P.Rockel@fz-juelich.de

Organismic Interactions and Plant Water Relations

By Rainer Lösch and Dirk Gansert

Plants normally function as physiological units that respond to ambient conditions regulating their water relations, mostly independent of other organisms that co-occur in the same habitat. However, water consumption from the soil and vapor transpiration into the atmosphere can influence local pedospheric and atmospheric conditions in a rather specific manner. Feedback can occur in this way, from a plant community as part of the biotic component of an ecosystem, to the abiotic parameters of the habitat. Competition for soil water reserves results from these interactions, and the canopy microclimate that influences transpiration rates can deviate drastically from the conditions that would control latent heat exchange of a solitary plant under the temperature and humidity conditions of the mixed layer of the atmosphere. Moreover, quite often plant water relations are influenced directly by other organisms that interact with a plant individual in mutualistic, parasitic or symbiotic ways. Other plant individuals, fungi and microorganisms as well as animals are such interacting partners. In the following overview of such relationships the focus will be on mutualistic effects on plant water supply by mycophyta and prokaryotic organisms, on phytopathogenic effects on uptake, transport and loss of water by the plant, and on host-parasite water relationships in xylem-sucking or tapping animal and plant parasites. The water relations of lichens as symbiotic organisms have been dealt with in previous reports on plant water relations.

1 Competitive and Mutualistic Effects on Water Relations and Metabolism of the Partners by Root-Associated Microorganisms

The great majority of plants possess mycorrhiza. The vesicular-arbuscular (VA) mycorrhiza, the ectomycorrhiza of conifers and various angiosperm tree families, and the ectendomycorrhiza of some Ericales families play an important role in the root water uptake of the higher plant (Safir 1987). Often roots with a rich ectomycorrhiza do not develop root hairs so exploitation of the soil water reserves is completely based

Progress in Botany, Vol. 63

on the mycorrhizal hyphae. These have diameters between 2 and 5 μm and are, therefore, much smaller than root hairs (10–20 μm diameter). Thus, they can penetrate very small soil pores (Reid 1979) and exploit the soil matrix quite efficiently. They can approach total hyphal lengths of up to 50 m cm^{-3} of soil (Allen 1991). If the hyphae are combined to hyphal strands, these "rhizomorphs" are particularly efficient in improving root water supply (Skinner and Bowen 1974; Duddridge et al. 1980; Foster 1981). Rhizomorphs are composed of three cell types, one type with thin walls containing cytoplasm, another thin-walled type is devoid of cell contents, and a ring of thick-walled small hyphae surrounds both cell types (Parke et al. 1983). It has been proven without any doubt, at least with conifer-associated ectomycorrhizae (Boyd et al. 1986), that a better water supply exists in mycorrhiza plants as compared with non-infected ones. Evidence for a direct improvement of the plant water supply by VA mycorrhiza comes from, e.g., decreased plant transpiration rates, if external hyphae are removed experimentally (Hardie 1985). Water uptake rates of mycorrhiza hyphae of up to 2.8×10^{-5} mg s^{-1}, estimated by Allen (1982, 1991) are sufficient to maintain normal plant water relations (Sánchez-Díaz and Honrubia 1994). However, doubts have been raised, for hydraulic reasons, about the high hyphal water flow rates needed in this case (Fitter 1985). In addition to an improved instantaneous water supply, the root mycorrhiza makes soil water reserves available that are inaccessible to a fungal-free fine root system (Bethlenfalvay et al. 1988). Moreover, many studies document an improved water flow through mycorrhizal plants as compared with non-infected ones (e.g. Allen et al. 1981; Cooper 1984; Augé et al. 1986a; Nelsen 1987; Augé 1989; Faber et al. 1991). One reason for better water uptake is due to an extended root system: In clover, greater lengths and diameters of mycorrhiza-infected roots result in a 26–86% increase in absorption surface compared with non-mycorrhizal root systems (Hardie and Leyton 1981). Similarly, root length of mycorrhizal *Leucaena* plants amounts to twice that of non-mycorrhizal plants (Huang et al. 1985). By contrast, Kothari et al. (1990) found reduced root lengths in mycorrhizal maize in comparison with controls. But this gramineous root system – structurally quite different from a dicot root system – supported doubled water uptake rates, possibly by reduced transport resistance of the roots.

Often an improved plant water status cannot be traced directly back to higher water uptake rates. Rather it results as a secondary effect of (1) altered root hydraulic conductivity (e.g. Graham and Syvertsen 1984; Andersen et al. 1988; Newman and Davies 1988). This could be based on a better water transfer across the root cortex as well as on a higher number of (meta-) xylem vessels (Daft and Okusanya 1973; Kothari et al. 1990). Very often (2) the supply of nutrients, in particular phosphate, is improved by a dense mycorrhiza (e.g. Safir et al. 1971, 1972; Nelsen and

Safir 1982; Buwalda et al. 1983; Koide 1985; Graham et al. 1987). As a result of this improved mineral nutrition, a better osmoregulation can occur (e.g. Hardie and Leyton 1981; Augé et al. 1986b; Augé and Stodola 1990). This osmoregulation is normally based (a) on higher cellular solute contents. However, studies of Augé and Stodola (1990) in mycorrhiza-infected and uninfected *Rosa hybrida* plants of similar size and adequate phosphorus nutrition point (b) to a higher symplastic to apoplastic water content relation in droughted plants under the fungal influence. Further, (c) in well-watered mycorrhiza plants, the bulk modulus of tissue elasticity can be lower than in mycorrhiza-free ones, so that the turgor loss with decreasing tissue water content proceeds less drastically. (3) Finally, mycorrhizal infection of roots can alter the phytohormonal balance of the host. A higher cytokinin activity in mycorrhiza plants as compared with uninoculated ones is most often reported in comparative studies on the phytohormonal physiology of the host (Allen et al. 1980; Edriss et al. 1984; Dixon et al. 1988; Baas and Kuiper 1989; Drüge and Schönbeck 1992). Also gibberellin activity is higher in mycorrhiza plants, while ABA activity is decreased (Allen et al. 1982). In all cases it is more a shift in balance of the phytohormonal activities that is decisive, than their changed absolute amounts. As a rule, in mycorrhizal plants the conducive hormones are favored at the cost of stress hormones (Levy and Krikum 1980).

As a consequence, vigor of mycorrhiza-infected plants is improved. The comparatively better performance of heavily mycorrhiza-infected plants is based on higher gas exchange rates resulting from higher leaf conductance (e.g. Allen 1982; Levy et al. 1983; Huang et al. 1985; Augé et al. 1986a; Brown and Bethlenfalvay 1987; Sánchez-Díaz et al. 1990; Drüge and Schönbeck 1992) and on better growth (Sieverding 1981; Abbott and Robson 1984). Under soil drought, mycorrhiza-infected roots cease growing later in time than uninfected ones mitigating the effects of water shortage. For all these reasons it is most often under drought conditions that mycorrhiza plants are superior to mycorrhiza-free ones (Bethlenfalvay et al. 1988; Sánchez-Díaz and Honrubia 1994); for well watered plants the differences are not as obvious. As a consequence, a good mycorrhiza colonization generally brings advantages to plants of drought-prone habitats, even if severe drought also reduces the fungal colonization, and different species responses must be taken into account as well (Boyle and Hellenbrand 1991: spruce responding better to mycorrhizal inoculation than pine; Lansac et al. 1995: different mycorrhiza-affected rooting of different Mediterranean shrubs). Under the combined influence of soil drought and competition from other species (Allen and Allen 1986: *Agropyron smithii*), an improved water and nutrient gain through mycorrhizae might become decisive for the success of a species in a particular habitat. Tripartite symbioses of legumes with root nodules and mycorrhiza show distinctly higher nitrogen and carbon gain and better survival under a drought than a two partner symbiosis (Peña et al. 1988; Sánchez-Díaz et al. 1990).

Perirhizal sheaths and the interstitial spaces of the soil matrix are crowded by free-living bacteria, blue-green algae, fungi (spores and their resting stages). This microbial community is the recruiting basis for mycorrhiza and other mutualistic partners of the higher plants. The free-living organisms interact with the roots exchanging ions and organic solutes with the reciprocal environment (Schönwitz and Ziegler 1989). Release and uptake of solutes take place in the moist environment of the soil solution and interacts by necessity with the local water potential gradients. This occurs as part of the normal soil-plant water exchange dynamics and must not be specified as a peculiar organism influence on plant water relations.

Besides mycorrhizal fungi, yet another group of true symbiotic microorganisms is recruited from the free-living microbial community and is specifically affected by the higher plant root water relations: the nodule-forming bacteria and actinomycetes. Hunt and Layzell (1993) reviewed findings on nodule gas exchange and nitrogenase activities, and only some additional citations shall be mentioned here to update the literature quoted there. Concerning the water stress effects upon these processes, Hunt and Layzell (1993) summarized an O_2 limitation to the energy requirements of the nitrogenase activity as a key process for a reduced nitrogen fixation under drought. While the enzyme itself is sensitive against O_2, the metabolism of functional nodules requires a high ATP production by oxidative respiration. Under a good water supply nodule respiration rates are five times higher than root respiration (Aguirreolea and Sánchez-Díaz 1989). Under water stress these rates are distinctly decreased (Sprent 1971). Also, the other extreme, waterlogging, is detrimental for high nitrogen fixation rates (usually measured as acetylene reduction) so that a nodule water potential equilibrium around −0.2 MPa is optimal (Huang et al. 1975a). Sheoran et al. (1988) assume that limited supplies of energy and carbon skeletons resulting from differently reduced enzyme activities of pidgeon pea nodules under water stress is the primary reason for reduced N_2 fixation rates. Findings by Huang et al. (1975b) on the inhibition of nodule acetylene reduction by heavily decreased photosynthesis under water stress point into the same direction. A direct breakdown of nitrogenase activity occurs under severe water stress. A transient increase of anaerobically induced enzyme activity in alfalfa nodules (Irigoyen et al. 1992b,c) could indicate decreased oxygen access to the nodule center at intermediate (75%) water content so that the effects of oxygen on nitrogenase activity do not occur yet. With nodule water contents below 60%, the transient hypoxic situation cannot be maintained further.

An altered oxygen availability within the nodule compartments delicately influences the balance of nodule carbon metabolism and N_2 fixation and, as a result, limits the whole process. Oxygen diffusion to the central zone of the bacterioids seems to be controlled by an aqueous barrier of cells and very small intercellular spaces soaked with water. Nodule oxygen permeability is changed if the length of the diffusion pathway is altered by water being lost or taken up (Hunt and Layzell 1993). An efficient control of the extent of this oxygen diffusion pathway could be based on osmoregulatory processes as assumed by various studies (e.g. Sheehy et al. 1983; Sheehy and Webb 1991; Hunt and Layzell 1991). Alternatively or additionally, the amount and activity of the O_2-affine leghemoglobin may be changed upon water stress (Khanna-Chopra et al.1984; Swaraj et al. 1986; Guerin et al. 1990, 1991; Irigoyen et al. 1992b). Osmoregulation upon nodule water shortage always occurs at higher water potentials than osmoregulation in leaves. This holds true for the accumulation of both soluble sugars and proline. The latter also functions as a stabilizer of pH and protein functionality (Irigoyen et al. 1992a). Nodules are supplied with water essentially from the higher plant roots (Sprent 1972b). Water diffusion through the nodule surface occurs mostly as water loss to the surroundings (Sprent 1972a). Determinate and indeterminate nodules differ somewhat in the pathway of diffusive water loss. In determinate nodules it occurs with a high priority across lenticels (Pankhurst and Sprent 1975; Sprent and Gallacher 1976; Aguirreolea et al. 1989).

Nitrogen-fixing nodules and mycorrhizal fungi are partners of symbiotic relationships of the higher plant, i.e., their function brings advantages for the carbon-autotrophic partner organism. Besides the *Rhizobium* and *Bradyrhizobium* bacteria and the *Frankia* actinomycetes of the nodules and the *Pisolithus*, *Rhizopogon*, *Endogone* and particularly *Glomus* fungi of the mycorrhiza, a great number of other mostly saprophytic fungi can grow in intimate contact with the plant roots. In some cases they function - as well as mycorrhiza - in improving the growth conditions of the higher plant. However, all intermediate effects are also possible with decidedly detrimental influences by the fungi upon the cormophytic plant. Then the fungus acts as a phytopathogen and even causes trees to die back by spreading into the host tissues. Some fungi are obligate parasites, but the majority are classified only as facultative parasites. In this case it depends on the higher plant's predisposition whether or not the mycobiont leads to symptoms of disease in the higher plant or even brings about the tree's decline and death. As a rule, tree colonization by facultative parasites develops into a disease only if the higher plant is additionally stressed by other abiotic or biotic stresses. Drought stress plays a very prominent role in such a predisposition to fungal attacks. Wargo (1996), when reviewing the predisposal of oaks to fungal-caused decline enumerates many papers about the effects of frost, drought, defoliation and other causes that predispose oak trees to fungal pathogens. Because the phenomenon of plant disease predisposition has already been reviewed several times during the last decades (Colhoun 1973; Schoeneweiss 1975, 1986; Boyer 1995) it shall not be discussed here with details from primary publications.

2 Effects on Host Plant Water Relations by Mutualistic and Pathogenic Fungi

a) Endophytes

Many mycophyta exist as neutral or even mutualistic parabionts on the surfaces and inside of most plants. Whether they influence the water relations of the host or, whether they themselves suffer from a bad water status in the surrounding tissue is mostly unknown in detail. Asymptomatous fungal endophytes of woody plants and grasses (*Neotyphodium*, *Epichloe*; asexual anamorphs: *Acremonium*) are intercellular inhabitants of the higher plant shoots. Their interactions with water relation parameters of the host were noted when endophyte-infected and -free populations of *Festuca arundinacea* (tall fescue), an important pasture grass of North America, were studied comparatively (e.g. Belesky et al. 1989; White et al. 1992; Bacon 1993; Richardson et al. 1993; Elmi and West 1995). From symposium reports dealing with endophyte-grass relationships (Quisenberry and Joost 1990; Bacon and White 1994) and a review update (Clay 1990), it can be generalized that (1) endophyte influences on host water relations were described as increased leaf resistance of the host (Arachevaleta et al. 1989). This, however, is not necessarily a consistent drought response (Elmi and West 1995). (2) Under desiccation, endophyte-colonized tall fescue leaf blades become rolled more rapidly than leaves of uninfected plants (Arachevaleta et al. 1989). Both features, stomatal response and leaf rolling, are typical drought-avoidance mechanisms. However, reports are more consistent about (3) a good osmoregulation of the leaf meristematic zones (Elmi et al. 1989; Richardson et al. 1992; Elmi and West 1995), a means of increasing drought tolerance. Endophytes are propagated vertically, i.e., via seeds of the host plant, or horizontally, from one individual host to the next. The first way is the normal one for grass endophytes; it seems to be linked with a mutualistic character of the host-endophyte relationships (Saikkonen et al. 1998). The horizontal infection pathway is the rule in woody plant endophytes and may occur sometimes with grass endophytes. In vertical propagule transmission, the fate of the endophytic hyphae completely depends on germination success of the host seeds. Vitality and infection intensity of horizontally spread fungal spores are heavily governed by the microclimatic conditions, humidity being the most prominent parameter of influence.

b) Phytopathogenic Fungi

Asymptomatic endophytes that rely on the horizontal transmission pathway share the dependency on air movement and humidity with externally visible fungal epiphytes and parasites of higher plants. Dix and Webster (1995) give an overview of the fungal colonization on higher plant surfaces and the seasonally and locally different environmental conditions that influence the spread of saprophytic or parasitic fungal communities on surfaces and inside tissues of living plants, of decomposing leaves, and of decaying wood. Some fungi can endure or even remain active growing under quite low water potentials, some are very sensitive against dry conditions, and all intermediate fungus-environment relationships can be found. Among economically important fungal parasites of crops *Phytophthora cinnamomi* and *Rhizoctonia solani* are most virulent under wet conditions, *Fusarium solani* and *F. roseum* are associated with dry soil. *Fusarium oxysporum* and *Verticillium albo-atrum* can live under very low water potentials, and only bring about host disease when the soil is wet (Cook 1973). Overviews about the potential effects of substrate water conditions on vitality and virulence of wood saprophytic and plant parasitic fungi are given, respectively, by Griffin (1977) and Cook and Papendick (1972). Many soil-borne pathogenic fungi are stimulated to grow by slightly lowered osmotic potentials, but not if the water potential is reduced due to matrical forces (Cook 1973).

Wood-rotting fungi, developed when the host is predisposed by drought (e.g. *Fomes annosus* on pines: Towers and Stambaugh 1968), increasingly deteriorate tree water supply by their influence on the root structures so that finally the saprophytic fungus inhabits a dead tree. Root decay also causes the cotton wilting disease brought about by *Phymatotrichum* (Olsen et al. 1983). Tree bark cancer develops best if bark moisture content is below 80% (Bier 1961 and related studies cited therein), but often stem wounding is required for the vigorous establishment of fungal stem parasites (e.g. Christ and Schoeneweiss 1975). Possibly saprophytic bark fungi that block the virulence of facultative parasites when living on lesion-free bark (Bier and Rowat 1962) rely on a high moisture content of the growing substrate (>80% rel. turgidity). Perhaps such interactions of protecting and damaging fungal parabionts on plant surfaces are also responsible for other aftereffects of drought on the intensity of plant diseases (e.g. *Cytospora* canker: Bertrand et al. 1976). On the other hand, the physiological status of the host can drastically promote the development of disease: stress-related changes of amino acid contents correlate with differences in *Hypoxylon* canker susceptibility of different *Populus* clones (Bélanger et al. 1990), and vessels embolized by drought can offer low resistance to hyphal growth and the pathological spread of this parasite that normally lives as an inconspicuous endophyte of many healthy trees (Vannini and Valentini 1994). As a rule, different plant species, even co-existing under the same habitat conditions, have different sensitivities to a phytopathogen attack (Dawson and Weste 1982). This may explain why most studies about these interactions remain more or less descriptive and are centered upon the phytopathogen response of the plants investigated under particular conditions.

Wilting diseases of higher plants (Dimond 1955; Mace et al. 1981) are caused by a great variety of phytopathogenic fungi (in particular *Fusarium* and *Verticillium* species) but also by bacteria. Wilting indicates a water imbalance of the plant that can result from different causes (Beckman et al. 1962): (1) Fungal or bacterial exotoxins (Scheffer 1976) act directly on leaf cells and disturb tissue water relations (Hancock 1981) – in this case the disturbed water relations are a secondary effect of the pathogen.

Duniway (1973) quotes net solute losses of *Pseudomonas syringae*-inoculated floating tobacco leaf disks as an example of a possible changed cell membrane permeability under the influence of the pathogen. Briggs et al. (1984) emphasize the occurrence of water potential imbalances between apoplast and symplastic compartments if *Helminthosporium* toxin causes membrane lesions. *Erysiphe graminis* infection of barley leaf epidermal cells brings about an altered plasmolysis form. Lee-Stadelmann et al. (1984) interpret this as an intensified adhesion between plasmalemma and cell wall, mediated by Ca^{2+} bridges between negative charges of plasmalemma and extrinsic proteins of the wall that are accumulated upon the fungal infection (Clarke et al. 1981).

Alternatively, wilting is caused directly by the pathogen-disturbed plant water relations. This could result from (2) an excessive water loss due to impairment of stomatal regulation and/or a changed cuticular vapor conductance, (3) an insufficient leaf and shoot water supply resulting from (a) submicroscopic plugging of intermicellar spaces of cell membranes by molecular products of the pathogen or host-pathogen interactions, (b) vascular plugging by the pathogens themselves or by pathogenic or pathogen-inducted products (e.g. slimes, gels, tyloses) or (c) disintegration of the anatomical structures necessary for the long-distance transport of water in roots and shoots. Situation 3c is the case in many root rot diseases.

Pathologically changed leaf conductance and transpiration rates are a frequently observed phenomenon in plants suffering from wilting diseases, mildew or rust infections. In plants with wilting diseases (Duniway 1973), transpiration rates often decline in comparison with the water loss of healthy plants as wilting symptoms develop (e.g. Beckman et al. 1962; Duniway 1971; Harrison 1971; Helms et al. 1971). This does not necessarily indicate a disturbed guard cell metabolism; rather it can result from extremely lowered plant water potentials (Duniway 1971). Rust infections most often increase transpirational water losses (e.g. Duniway and Durbin 1971a,b), while powdery mildew of leaves may or may not be accompanied by excessive water loss, depending on host-fungus combinations (Shtienberg 1992).

Peronospora-infected tobacco leaves have lower transpiration rates than non-infected leaves in the light, but have higher rates in the dark (Cruickshank and Rider 1961), *Phytophthora infestans* causes abnormal stomatal opening in invaded regions of potato leaves (Farrell et al. 1969), both characterizations quoted from Duniway (1973). According to our own measurements (Lösch, Heinrichs and Jungbluth, unpublished), the autumn mildew infection of oak leaves by *Microsphaera alphitoides* intensifies a senescence-dependent loss of stomatal regulation of leaf conductance, whereas the same seasonal phenomenon reduces the gas exchange of *Trifolium* leaves (mildew: *Erysiphe trifolii*) to half the rate measured with uninfected leaves.

High stomatal resistance could result from pathogen-impaired (inward directed) ion fluxes at the guard cell plasma membrane (e.g. Arntzen et al. 1973). This could occur via a changed phytohormonal balance: A reduced cytokinin concentration in the xylem sap due to *Phytophthora* infection could be coupled with an intensified delivery of abscisic acid to the leaves, and the stomata could close under the influence of this phytohormone (Cahill et al. 1986). Detailed studies about abscisic acid relations of such pathogen-infected plants are still missing. Increased transpiration rates of fungus-colonized leaves can result from reduced cuticular resistance, as was demonstrated with rust-infected bean plants (Sempio et al. 1966; Duniway and Durbin 1971a,b). In addition, toxins of the pathogens could also disturb the ion transfer balance at the guard cell plasma membrane in the opposite direction, viz. bringing about an extreme stomatal opening. In this respect fusicoccin, produced by the phytopathogenic fungus *Fusicoccum amygdali*, is well known (Turner and Granitti 1969; Chain et al. 1971). The toxic substance brings about hyperpolarization of membranes. At the guard cell plasmalemma the increased efflux of protons is coupled with an intensified K^+ influx so that the guard cells gain turgor in a very strong manner.

In wilting diseases it is often the transpiration stream through the xylem vessels that is blocked by phytopathogenic organisms. Primarily among the xylem plugging organisms, mycophyta must be enumerated (*Ceratocystis, Fusarium, Verticillium*), but bacteria (*Pseudomonas, Erwinia, Xylella*) and viruses can also attack the xylem. Several comprehensive reviews exist about xylem-limited bacteria (Davis et al. 1981; Hopkins 1989; Purcell and Hopkins 1996). These bacteria cause, among others, Pierce's disease of grapevines and Bermuda grass (Davis et al. 1978) and ratoon stunting disease of sugarcane (Teakle et al. 1973; Kao and Damann 1978). They are at the least involved in citrus blight and in leaf scorch of deciduous trees (Hearon et al. 1980), and they may be the cause of shoot bending below the flowers of cut roses (Van Doorn et al. 1989). Xylem-limited bacteria are propagated mostly by xylem-sucking insects; xylem infection by parasitic fungi occurs via shoot lesions, leaf scars and through rotting roots (*Phytophthora*: Duniway 1977). Due to the xylem-limited organisms the water transport capacity of infected roots and shoots can be reduced to a few percent of that encountered in healthy plants (Dimond 1970). Bacterial masses or fungal hyphae can simply clog the tracheal and tracheidal lumina. From the infection points, the pathogens are transported by the xylem stream over great distances. Vessel plugging increases the axial hydraulic resistance. Only in particular cases is it partially offset by a reduced root axial resistance if appropriate anatomical changes are elicited by the pathogen (Tissera and Ayres 1988).

Two days after inoculation with *Erwinia tracheiphila* Main and Walker (1971) found the petioles of cucumber leaves to be completely plugged. Duniway (1971b) reports about such wilting phenomena resulting from xylem blockade in *Fusarium*-infested tomatoes, El Mahjoub and Le Picard (1985) found similar vessel plugging in melon leaves.

Plants respond to an infection most often by increased production of tyloses (Robb et al. 1979). In addition to the clogging of vessels by the pathogens themselves and the tylose response by the plant, phytotoxic glycoproteins are produced by the infection. Moreover, cell walls are disintegrated by the bacterial enzymes so that the xylem strands become mucilaginous masses.

Glucans with molecular masses >30 kDa are produced, e.g., by endopectin lyase released from *Verticillium* pathogens in tomato petioles (Street and Cooper 1982). These colloidal polysaccharides accumulate at pit fields paralyzing any water movement through the shoots. This can also explain the breakdown of the hydraulic conductivity of the whole root system of sensitive *Eucalyptus* plants under *Phytophthora* infection, even if the pathogen itself remains restricted to limited root zone areas (Dawson and Weste 1984). The impairment of water flow through pit membranes by metabolites of pathogenic fungi without tamping the vessel lumina themselves seems to be sufficient to bring about branch drying and even whole tree mortality of *Cupressus sempervirens* (Madar et al. 1990). Citrus blight results from fibrous plugs in the vessels (Brlansky et al. 1985; Timmer et al. 1986; Beretta et al. 1988). Sánchez-Díaz and Aguirreolea (1993) point to the possibility of reduced surface tensions (due to metabolic products of the pathogens) that facilitate the development of vessel embolisms. Vessel walls of *Fusarium*-infected carnation stems are invaded by microhyphae and finally destroyed (Baayen and Elgersma 1985; Ouellette et al. 1999). The partly disintegrated walls become impregnated by electron-dense material that possibly is secreted from the fungal hyphae (Ouellette and Baayen 2000). But, vessel wall coatings due to *Fusarium* infection are also secreted from xylem parenchyma cells of the host, probably as a barrier against spread of the parasite (e.g. Jordan et al. 1988; Tessier et al. 1990; Shi et al. 1992). This function is, at least, ascribed to xylem parenchyma secretion products of *Verticillium*-infected plants (e.g. Robb et al. 1987; Newcombe and Robb 1988). These substances could act physically, blocking the fungal advance through pits and vessel end walls (Beckman et al. 1976), and they may also function as phytoalexins affecting the vitality of a pathogen chemically. Such a defense response of the host at the cellular level parallels the reaction zone formation of axial organs that is found in many trees where the heartwood is inhabited by wood-rotting fungi (e.g. Shigo 1984; Biggs 1987; Pearce 1990, 1991).

3 Water Relations of Parasitic Plants and Their Hosts

(Books and reviews: e.g. Kuijt 1969, 1977; Musselman 1980, 1987; Atsatt 1983; Calder and Bernhardt 1983; Ter Borg 1986; Weber and Forstreuter 1987; Press et al. 1990, 1999a; Stewart and Press 1990; Press and Whittaker 1993; Press and Graves 1995)

With more than 4000 species, parasitic plants make up almost 1% of all flowering plants (Press et al. 1999b). With respect to carbon acquisition one can distinguish holo- and hemiparasites. But all parasites take up water and mineral nutrients from their hosts. After successful contact of the seedling radicle of the parasite with the host tissue, either by the placement of seeds on host twigs by birds or after foraging growth of the parasite radicle up to host roots that is very often elicited and guided by specific chemicals (Press et al. 1990), haustorial initiation starts, triggered again by specific signal compounds exuded by host roots (Lynn and Chang 1990; Estabrook and Yoder 1998). The haustorium develops into the parasite organ for the uptake of water, ions and carbon compounds. Upon host contact the haustorial ontogeny of *Striga* involves the following steps (Press et al. 1990): cessation of radicle cells elongation – radial expansion of peripheral cells at the radicle tip – initiation of haustorial hairs – development of vesicle-rich, densely staining cells at the haustorial apex – penetration of host cortex and vascular system. Hyphal-like cells from the haustorium penetrate the host xylem aided by lytic enzymes and turgor pressure produced by the parasite. Once a vessel contact is achieved, the parasite haustorial cell differentiates into a transfer cell. Thereafter xylem-type wall thickenings develop, the wall labyrinth characteristic for transfer cells disintegrates, and a dead parasite vessel comes into contact with the dead host vessel (Dörr and Kollmann 1976; Press et al. 1990). There are exceptions to this generalized developmental pathway. In particular, not all mistletoes have direct contacts of host and parasite vessels. In this case, water and substance diffusion occurs across the apoplastic space of common cell walls, possibly facilitated by plasma tubules of the parenchyma cells (Lamont 1982; Alosi and Calvin 1985; Coetzee and Fineran 1987).

In order to divert water and solute flow from the host xylem to the parasite, an appropriate gradient of the water potential must exist. It results from high transpiration rates of the parasite and from high solute concentrations in its tissues. The parenchymatic tissues of parasites are often distinguished by remarkably high contents of inorganic solutes. In hemiparasites a phloem-based retranslocation of xylem-delivered potassium between parasite and host is not possible (Glatzel 1983); the parasite functions therefore as a potassium sink. In addition to such a passive accumulation nutrients may also be actively acquired, at least by various mistletoe parasites (Panvini and Eickmeier 1993). Schulze et al. (1984; see also Ehleringer et al. 1985) emphasize the need for nitrogen acquisition from the very diluted xylem sap by high transpirational water flow through mistletoes in order to attain vigorous growth. Mistletoe transpiration rates on a nitrogen-rich xylem sap from a leguminous (N_2-fixing) host proved indeed to be lower than mistletoe transpiration rates on a host with low nitrogen delivery (Schulze and Ehleringer 1984). Similarly, Gauslaa (1990) measured higher leaf conductance of the root hemiparasite *Melampyrum pratense* from oligotrophic growing sites if compared with plants from mesotrophic habitats. However, in some root hemiparasites, like *Rhinanthus minor*, transpiration may not be tightly coupled with nitrogen supply (Seel et al. 1983a), and a better nitrogen supply by N_2-fixing hosts may improve growth performance without significantly reducing transpiration rates (Radomiljac et al. 1999). Based on studies with host-free cultivated *Rhinanthus minor* plants Seel et al. (1993b) emphasize that phosphorus requirements rather than nitrogen demands may dominate xylem sap nutrient delivery to this parasite. While N (in the majority being supplied as amino acids and amides: Stewart et al. 1984) and P are metabolized in the parasite tissues, inorganic solutes not required for metabolic processes in the cytoplasm are accumulated in the cell vacuoles thereby depressing the osmotic potential. For the cell internal balance, symplastically compatible solutes become accumulated, like mannitol (e.g. Nour et al. 1984) and – seasonally – proline (Lösch, Bienert and Kiefer, unpubl.). The resulting water potential of the haustorial tissue is lower than that of the adjacent host tissue, often only as a result of synergistic effects of solute concentration, tissue elasticity and actual cell water content: pressure volume (pV)-analyses of tissue water relations in the *Cuscuta* haustorial region indicated that a water potential gradient from the host to the parasite exists neither at full turgor nor at the turgor loss point, but is present at intermediate water contents within the range of turgescence (Lösch et al. 1995). Similarly, tissue water potentials of the root hemiparasite *Santalum album* and its hosts diverge more and more during the course of a day, starting from similar predawn values (Radomiljac et al. 1999). Thus, at least during daytime hours a certain water deficit of the host-parasite system always guarantees that the transpiration stream is diverted, at least partly, from the host to the parasite xylem (e.g. Scholander et al. 1965; Fisher 1983).

The diurnal dynamics of host and parasite water potentials and the concomitant sapflow patterns are illustrated by Fig. 1: During a fair-weather period in May, *Viscum album* growing on *Malus transitoria* approaches minimal twig water potentials of –2 MPa while host twig water potentials, proximal to the haustorium insertion, never fall much below –1 MPa. Nocturnal recovery of water potentials approaches –0.5 MPa in both species. Both the lowest daytime values and highest nocturnal water potentials occur earlier in time in the host than in the parasite. Mistletoe sap-flow pattern more or less correlated with the prevailing radiation; the maximal sap flow rates correspond to transpiration rates of 12 mmol H_2O m^{-2} s^{-1} (Lösch and coworkers, unpubl.).

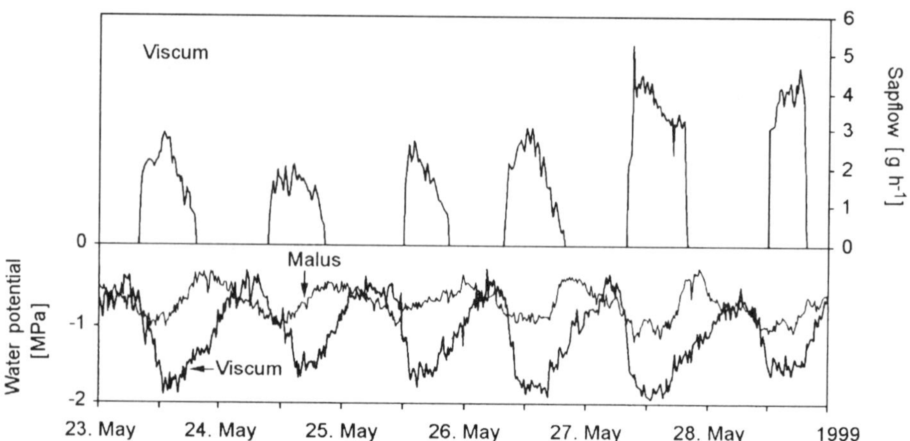

Fig. 1. Time courses of twig water potentials of *Viscum album* and its host, *Malus transitoria*, and of xylem sap flow of the mistletoe (Lösch, Hillebrand, Bienert and Stöhr, unpubl.). [In-situ water potential psychrometers (PWS Inc., Canada), Dynagage sap flow collars SGA 5 (Dynamax, USA), data loggers Squirrel 1250 (Grant, UK) and CR 10 (Campbell, UK); Botanical Garden Düsseldorf, 1999)]

Parasite transpiration rates are twice to nine-fold higher than those of the hosts (Ullmann et al. 1985; further examples, e.g. Davidson et al. 1989; Goldstein et al. 1989; Johnson and Choinski 1993). There are also, however, some studies emphasizing a distinctly higher transpiration of the host leaves than those of the parasite (e.g. Beserra de Oliveira et al. 1962; Hellmuth 1971; Küppers et al. 1992). Küppers (1992) interprets the deviation from the usual patterns of transpiration characteristics in the case of the host-parasite pair *Eucalyptus behriana/Amyema miquelii* by a low water, nitrogen and phosphorus availability in the particular habitat of this *Eucalyptus* species so that the normal mistletoe strategy of maximizing transpiration and by this way maximizing nitrogen gain would not be favourable. While, as a rule, water-use efficiency of the parasite is lower (mostly determined as very negative $\delta^{13}C$ values or calculation of the $\Delta\delta d^{13}C$ for the host-parasite pair: e.g. Ehleringer et al. 1985; Ehleringer et al. 1986), in this particular case the values of host and parasite do not differ much. The more often observed very high transpiration rates result from nearly functionless stomata in *Striga* (Press et al. 1987; Shah et al. 1987; Smith and Stewart 1990) and some holoparasites (e.g. *Cynomorium*: Fahmy 1993), and from generally wider stomatal apertures, equivalent to higher leaf conductance, in most of the other parasitic cormophytes investigated so far (e.g. Ullmann et al. 1985: 19 host-mistletoe pairs of Australia, porometric measurements; Press et al. 1987: 8 root hemiparasites from England, Scotland and Lappland, porometric measurements; Lüttge et al. 1998: 17 host-parasite pairs of the Brazilian cerrado, stomatal conductance inferred from carbon isotope analyses). Short-lived root parasites keep the stomata more or less open, also at nighttime (Press et al. 1987, 1993), whereas perennial root and stem hemiparasites close their stomata during the night. This occurs even in mistletoes parasitic on hosts performing Crassulacean acid metabolism (Schulze et al. 1991).

Stomata of hemiparasitic plants show the normal responses to environmental parameters and do not differ qualitatively in their metabolism from those of normal C3 plants. But the closing responses are always comparatively weak. El-Sharkawy et al. (1986) found the stomata of *Phthirusa pyrifolia* unable to respond to humidity changes, while the host, *Citrus reticulata*, responded sensitively. By contrast, Hollinger (1983) showed direct stomatal response of *Phoradendron villosum* to air humidity, and Ullmann et al. (1985) generally found decreased parasite leaf conductance with increasing leaf-to-air vapor pressure difference. Moreover, a certain coordination between host and parasite stomatal regulation could be deduced from these data. Such a concordance in stomatal behavior could be based on comparable independent responses of host and parasite to the actual environmental situation, or it could have been induced by an internal signal common to both plants. According to Fisher (1983) *Viscum album* is unable to synthesize abscisic acid (ABA), the phytohormone first of all associated with stomatal control, but Ihl et al. (1987) found much higher ABA concentrations in parasites as compared with their hosts, and De Bock and Fer (1992) documented an intensive ABA transfer from host tissue to the holoparasite *Cuscuta reflexa*, where the hormone plays an important role enhancing the phloem sugar transfer between host and parasite. The closing influence on stomata of ABA works via control of the guard cell potassium contents. Smith and Stewart (1990) demonstrated that *Striga* stomata, when isolated by low pH treatment of floating epidermal strips, respond rapidly, by closing, to low CO_2 concentrations, darkness and ABA, but are unable to do so – as is also the case in the intact *Striga* leaf – if incubated in the presence of high potassium concentrations. Obviously, the potassium accumulation of the xylem-tapping hemiparasites reduces osmoregulatory decrease of guard cell potassium concentrations, so that their turgor loss bringing up stomatal closure is much less pronounced than in the hosts.

4 Animal Xylem Sap Suckers and Herbivore Influences upon Plant Water Relations

Not only plant parasites, but also several animals use the xylem sap of host plants for their nutrition. In particular, in the insect order of homoptera (Cercopidae and Cicadidae, spittlebugs) many taxa are sucking on the plant xylem. They are equipped with a powerful cibarial pump and excrete fluids in much higher quantities than phloem-sucking insects (Press and Whitaker 1993). They puncture the xylem elements with a robust piercing proboscis. Plant structural defenses against the insect attacks are trichomes, a greater depth of xylem elements inside the shoot and a high tissue hardness (Hoffman and McEvoy 1986). If xylem strands are reached by the insects, the strength of the proboscis and the power of the cibarial pump are just sufficient to overcome the suction of the transpiration stream (Raven 1983). Stem water potentials below –1.5 to –2 MPa cannot be overcome (Andersen et al. 1992). Spittlebugs are polyphagous and move from plant to plant. Seasonally-changing host preferences can be paralleled with the amounts and composition of amino acids in the xylem sap (Brodbeck et al. 1990). More often, insect spittle is found on the shoots of herbs, but root sucking seems to be obligatory for the extremely long-lived *Magicicada* species (White and Strehl 1978). The animal parasites resorb the diluted nutrients. In particular, it is the amount of amides that determines the amount of xylem sap required by the insects (Horsfield 1977; Andersen et al. 1992). The xylem suckers have adapted their metabolism to compensate for the imbalance in amino acid composition of the sap relative to their physiological requirements and to specifically detoxify the few secondary plant substances ingested with the sap (Andersen et al. 1989). Xylem sucking can be classified therefore as a specialized type of herbivory. Additionally, the water demands of the insects are covered by the xylem sap (Press and Whitaker 1993). These demands are high, because spittlebugs are very sensitive to desiccation (Wiegert 1964; Whittacker 1970), and the viscous spittle on Cercopidae-infected plants essentially serves as protection against desiccation. Hourly feeding rates can exceed the insect body weight by an order of magnitude (Andersen et al. 1989), and the feeding rates can be adapted to diurnal fluctuations in xylem fluid chemistry (Brodbeck et al. 1993). This contrasts distinctly with the sap extraction of phloem feeders. However, at times, phloem-sucking aphids can also switch to xylem sap (Spiller et al. 1990). By this way they maintain their water balance in the face of the solute-concentrated phloem sap (xylem sap osmotic potential $\approx$ –0.2 MPa, phloem sap osmotic potential $\approx$ –1.5 MPa: Cull and Van Emden 1977), and several facultative xylem feeders exist among less specialized Auchenorrhyncha (Press and Whittacker 1993). The high amounts of water taken up by the obligate xylem suckers are passing a filtration process inside the insects, and the excess water is transferred directly to the ileum (Cheung and Marshall 1973). The detrimental effect to the plants of the xylem-sucking homoptera is obvious only in heavily infected plants if wilting occurs above the punctured xylem area. However, Wiegert (1964) emphasizes that the withdrawal of nitrogen compounds from the plant metabolism by the sucking animals is more aggravating than herbivore feeding on carbohydrate biomass. The ingestion of one calorie of xylem sap organic matter can thus equal a 5-cal loss due to a reduced biomass production and can make up, by this way, a significant part of the energy flow in ecosystems where xylem-feeders are abundant.

There is yet another phylum of animals, the nematodes, some of whose taxa rely on plant xylem fluid (Webster 1975; Melakeberhan and Webster 1993). By destroying plant root systems, some of them impede a sufficient water uptake and cause plant dieback (e.g. *Globodera rostochiensis*: Evans et al. 1977; Fatemy and Evans 1986). *Bursaphelenchus xylophilus* is a xylem destroying nematode that interrupts xylem water flow (Ikeda and Susaki 1984; Ikeda et al. 1990) and triggers xylem cavitation of pine trees (Kuroda et al. 1988; Kuroda 1989). Parasite infection leading to disturbed tree water relations is followed by damage to the photosynthetic apparatus (Melakeberhan et al. 1991), needle chlorosis and the death of weakened trees.

Herbivory can influence plant water relations by general weakening effects. Systemic defense responses exist in plants signaling wounding over long distances via the vessel system (Boari and Malone 1993; Malone 1993). Caterpillars biting the leaves brings thickness changes of distant, uninfected leaves that probably result from a hydraulic transmission of the disturbance (Alarcon and Malone 1993). Chemical elicitors may be carried by the mass flow that accompanies such hydraulic pulses (Malone et al. 1994). Activation of allelochemicals can be the consequence of the hydraulically propagated signal. On the other hand, the peculiar plant water status can also influence conditions and behavior of herbivores. Under drought conditions the amino acid composition of plants is changed, due to increased proline levels in the course of osmoregulatory processes. In an experiment, Haglund (1980) stimulated grasshopper herbivory in a Montana grassland by raising the proline and valine contents of the diet. He concluded that higher amino acid contents of the grasses under drought could favor orthoptera population outbreaks. This is in line with interpretations of insect outbreaks (after drought periods) being due to more available nutritious food (White 1976, 1984) that at the same time has lower levels of defensive compounds (Rhoades 1983). A comprehensive overview of this topic can be found in Barbosa and Schultz (1987). English-Loeb (1990) contrasts the enumeration of studies in favor of such a connection between drought, biomass nutrient content, and the population increase of herbivores by quoting studies negating such a connection. Taking into account reduction of biomass by drought stress, the concomitant increase of herbivore predators and food competitors, non-linear effects of the changed ecosystem water availability upon all the depending changes etc., it becomes evident, that monocausal interpretations will lead to unrealistic oversimplifications. Nevertheless, complex conceptual models for all these interrelations can be constructed, and for limited scenarios cause-and-effect relationships may be worked out. As qualitative statements Mattson and Haack (1987) summarize the following points that characterize the role of drought for outbreaks of plant-eating insects: (1) drought-stressed plants are behaviorally more attractive or acceptable for insects, (2) they are physiologically more suitable for them, (3) drought enhances insect detoxification systems and immunocompetence, (4) drought favors mutualistic microorganisms but not natural enemies of phytophagous insects and (5) drought might even induce genetic changes in the herbivorous insects.

Acknowledgements. We thank Mrs. S. Miljanovic and K. Kiefer for technical help.

References

Abbott LK, Robson AD (1984) The effect of VA mycorrhizae on plant growth. In: Powell C, Bagyaraj J (eds) VA mycorrhiza. CRC Press, Boca Raton, pp 113–130

Aguirreolea J, Sánchez-Díaz M (1989) CO_2 evolution by nodulated roots in *Medicago sativa* L. under water stress. J Plant Physiol 134:598–602

Aguirreolea J, Sánchez-Díaz M, Irigoyen J (1989) Measurement of leaf, nodule and root water potentials in alfalfa plants under water stress. J Plant Physiol 134:352–356

Alarcon J-J, Malone M (1993) Substantial hydraulic signals are triggered by leaf-biting insects in tomato. J Exp Bot 45:953–957

Allen EB, Allen MF (1986) Water relations of xeric grasses in the field: interaction of mycorrhizas and competition. New Phytol 104:559–571

Allen MF (1982) Influence of vesicular-arbuscular mycorrhizae on water movement through *Bouteloua gracilis* (H.B.K.) Lag ex Steud. New Phytol 91:191–196

Allen MF (1991) The ecology of mycorrhizae. Cambr Univ Pr, Cambridge

Allen MF, Moore TS, Christensen M (1980) Phytohormone changes in *Bouteloua gracilis* by vesicular-arbuscular mycorrhizae. I. Cytokinin increases in the host plant. Can J Bot 58:371–374

Allen MF, Smith WK, Moore TS, Christensen M (1981) Comparative water relations and photosynthesis of mycorrhizal and nonmycorrhizal *Bouteloua gracilis* H.B.K. Lag ex Steud. New Phytol 88:683–693

Allen MF, Moore TS, Christensen M (1982) Phytohormone changes in *Bouteloua gracilis* by vesicular-arbuscular mycorrhizae. II. Altered levels of gibberellin-like substances and abscisic acid in the host plant. Can J Bot 60:468–471

Alosi MC, Calvin CL (1985) The ultrastructure of dwarf mistletoe (*Arceuthobium* spp.) sinker cells in the region of the host secondary vasculature. Can J Bot 63:889–898

Andersen CP, Markhart AH, Dixon RK, Sucoff EI (1988) Root hydraulic conductivity of vesicular-arbuscular mycorrhizal green ash seedlings. New Phytol 109:465–471

Andersen PC, Brodbeck BV, Mizell RF (1989) Metabolism of amino acids, organic acids and sugars extracted from the xylem fluid of four host plants by adult *Homalodisca coagulata*. Entomol Exp Appl 50:149–159

Andersen PC, Brodbeck BV, Mizell RF (1992) Feeding by the leafhopper, *Homalodisca coagulata*, in relation to xylem fluid chemistry and tension. J Insect Physiol 38:611–622

Arachevaleta M, Bacon CW, Hoveland CS, Radcliff DE (1989) Effect of tall fescue endophyte on plant response to environmental stress. Agron J 81:83–90

Arntzen CJ, Haugh MF, Bobick S (1973) Induction of stomatal closure by *Helminthosporium maydis* pathotoxin. Plant Physiol 52:569–574

Atsatt PR (1983) Host-parasite interactions in higher plants. In: Lange OL, Nobel PS, Osmond CB, Ziegler H (eds) Physiol Plant Ecol III, Responses to the chemical and biological environment. Encycl Plant Phys New Ser 12C. Springer, Berlin Heidelberg New York, pp 519–535

Augé RM (1989) Do VA mycorrhizae enhance transpiration by affecting host phosphorus content? J Plant Nutr 12:743–753

Augé RM, Stodola AJW (1990) An apparent increase in symplastic water contributes to greater turgor in mycorrhizal roots of droughted *Rosa* plants. New Phytol 115:285–295

Augé RM, Schekel KA, Wample RL (1986a) Greater leaf conductance of well-watered VA mycorrhizal rose plants is not related to phosphorus nutrition. New Phytol 103:107–116

Augé RM, Schekel KA, Wample RL (1986b) Osmotic adjustment in leaves of VA mycorrhizal and nonmycorrhizal rose plants in response to drought stress. Plant Physiol 82:765–770

Baas R, Kuiper D (1989) Effects of vesicular-arbuscular mycorrhizal infection and phosphate on *Plantago major* ssp. *pleiosperma* in relation to internal cytokinin concentrations. Physiol Plant 76:211–215
Baayen RP, Elgersma DM (1985) Colonization and histopathology of susceptible and resistant carnation cultivars infected with *Fusarium oxysporum* f. sp. *dianthi*. Neth J Plant Pathol 91:119–135
Bacon CW (1993) Abiotic stress tolerances (moisture, nutrients) and photosynthesis in endophyte-infected tall fescue. Agric Ecosyst Environ 44:123–141
Bacon CW, White JF (eds) (1994) Biotechnology of endophytic fungi of grasses. CRC Press, Boca Raton
Barbosa P, Schultz JC (eds) (1987) Insect outbreaks. Acad Pr, San Diego
Beckman CH, Brun WA, Buddenhagen IW (1962) Water relations in banana plants infected with *Pseudomonas solanacearum*. Phytopathology 52:1144–1148
Beckman CH, Vandermolen GE, Mueller WC, Mace ME (1976) Vascular structure and distribution of vascular pathogens in cotton. Physiol Plant Pathol 9:87–94
Bélanger RR, Manion PD, Griffin DH (1990) Amino acid content of water-stressed plantlets of *Populus tremuloides* clones in relation to clonal susceptibility to *Hypoxylon mammatum* in vitro. Can J Bot 68:26–29
Belesky DP, Stringer WC, Hill NS (1989) Influence of endophyte and water regime upon tall fescue accessions. I. Growth characteristics. Ann Bot 63:495–503
Beretta MJG, Brlansky RH, Lee RF (1988) A comparison of histochemical staining reactions of the xylem occlusions in trees affected by citrus blight and decline. Plant Dis 72:1058–1060
Bertrand PF, English H, Uriu K, Schick FJ (1976) Late season water deficits and development of *Cytospora* canker in French Prune. Phytopathology 66:1318–1320
Beserra de Oliveira JG, Marques Válio IF, Felippe GM, De Campos SM (1962) Balanço d' água do hemiparasito *Struthanthus vulgaris* Mart.: I. Estudo comparativo co seu hospedeiro *Erythrina speciosa* Andr. na Estaçao Chuvosa (São Paulo, SP, Brasil). An Acad Bras Cienc 34:527–544
Bethlenfalvay GJ, Brown MS, Ames RN, Thomas RS (1988) Effects of drought on host and endophyte development in mycorrhizas soybeans in relation to water use and phosphate uptake. Physiol Plant 72:565–571
Bier JE (1961) The relation of bark moisture to the development of canker diseases caused by native, facultative parasites. VI. Pathogenicity studies of *Hypoxylon pruinatum* (Klotzsch) Cke., and *Septoria musiva* PK. on species of *Acer*, *Populus*, and *Salix*. Can J Bot 39:1555–1561
Bier JE, Rowat MH (1962) The relation of bark moisture to the development of canker diseases caused by native, facultative parasites. VII. Some effects of the saprophytes on the bark of poplar and willow on the incidence of *Hypoxylon* cancer. Can J Bot 40:61–69
Biggs AR (1987) Occurrence and location of suberin in wound reaction zones in xylem of 17 tree species. Phytopathology 77:718–725
Boari F, Malone M (1993) Wound-induced hydraulic signals: survey of occurrence in a range of species. J Exp Bot 44:741–746
Boyd RT, Furbank RT, Read DJ (1986) Ectomycorrhiza and the water relations of trees. In: Gianinazzi-Pearson V, Gianinazzi S (eds) Physiological and genetical aspects of mycorrhizae. Proc 1st Europ Symp on Mycorrhizae, INRA, Dijon, pp 689–693
Boyer JS (1995) Biochemical and biophysical aspects of water deficits and the predisposition to disease. Annu Rev Phytopathol 33:251–274
Boyle CD, Hellenbrand KE (1991) Assessment of the effect of mycorrhizal fungi on drought tolerance of conifer seedlings. Can J Bot 69:1764–1771
Briggs SP, Scheffer RP, Haug AR (1984) Osmotic conditions affect sensitivity of oat tissues to toxin from *Helminthosporium victoriae*. Physiol Plant Pathol 25:103–110

Brlansky RH, Lee RF, Collins MH (1985) Structural comparison of xylem occlusions in the trunks of citrus trees with blight and other decline diseases. Phytopathology 75:145–150

Brodbeck BV, Mizell RF, French WJ, Andersen PC, Aldrich JH (1990) Amino acids as determinants of host preference for the xylem feeding leafhopper, *Homalodisca coagulata* (Homoptera: Cicadellidae). Oecologia 83:338–345

Brodbeck BV, Mizell RF, Andersen PC (1993) Physiological and behavioral adaptations of three species of leafhoppers in response to the dilute nutrient content of xylem fluid. J Insect Physiol 39:73–81

Brown MS, Bethlenfalvay GJ (1987) The *Glycine-Glomus-Rhizobium* symbiosis. VI. Photosynthesis in nodulated, mycorrhizal, or N- and P-fertilized soybean plant. Plant Physiol 85:120–123

Buwalda JG, Stribley DP, Tinker PB (1983) Increased uptake of anions by plants with vesicular-arbuscular mycorrhiza. Plant Soil 71:463–467

Cahill DM, Weste GM,Grant BR (1986) Changes in cytokinin concentrations in xylem extrudate following infection of *Eucalyptus marginata* Donn ex Sm with *Phytophthora cinnamomi* Rands. Plant Physiol 81:1103–1109

Calder DM, Bernhardt P (eds) (1983) The biology of mistletoes. Academic Press, Sydney

Chain EB, Mantle PG, Milborrow BV (1971) Further investigations of the toxicity of fusicoccins. Physiol Plant Pathol 1:495–514

Cheung WWK, Marshall AT (1973) Water and ion regulation in cicadas in relation to xylem feeding. J Insect Physiol 19:1801–1816

Christ CR, Schoeneweiss DF (1975) The influence of controlled stresses on susceptibility of European White Birch stems to attack by *Botryosphaeria dothidea*. Phytopathology 65:369–373

Clarke JA, Lisker N, Lamport DTA, Ellingboe AH (1981) Hydroxyproline enhancement as a primary event in the successful development of *Erysiphe graminis* in wheat. Plant Physiol 67:188–189

Clay K (1990) Fungal endophytes of grasses. Annu Rev Ecol Syst 21:275–297

Coetzee J, Fineran BA (1987) The apoplastic continuum, nutrient absorption and plasmatubules in the dwarf mistletoe *Korthalsella lindsayi* (Viscaceae). Protoplasma 136:145–153

Colhoun J (1973) Effects of environmental factors on plant disease. Annu Rev Phytopathol 11:343–364

Cook RJ (1973) Influence of low plant and soil water potentials on diseases caused by soilborne fungi. Phytopathology 63:451–458

Cook RJ, Papendick RI (1972) Influence of water potential of soils and plants on root disease. Annu Rev Phytopathol 10:349–374

Cooper KM (1984) Physiology of VA mycorrhizal associations. In: Powell C, Bagyaraj J (eds) VA mycorrhiza. CRC Press, Boca Raton, pp 35–55

Cruickshank IAM, Rider NE (1961) *Peronospora tabacina* in tobacco: transpiration, growth, and related energy considerations. Aust J Biol Sci 14:45–57

Cull DC, Van Emden HF (1977) The effect on *Aphis fabae* of diel changes in their food quality. Physiol Entomol 2:109–115

Daft MJ, Okusanya BO (1973) Effects of *Endogone* mycorrhiza on plant growth. VI. Influence of infection on the anatomy and reproductive development in four hosts. New Phytol 72:1333–1339

Davidson NJ, True KC, Pate JS (1989) Water relations of the parasite:host relationship between the mistletoe *Amyema linophyllum* (Fenzl) Tieghem and *Casuarina obesa* Miq. Oecologia 80:321–330

Davis MJ, Purcell AH, Thomson SV (1978) Pierce's disease of grapevines: isolation of the causal bacterium. Science 199:75–77

Davis MJ, Whitcomb RF, Gillaspie AG (1981) Fastidious bacteria of plant vascular tissue and invertebrates (including so-called Rickettsia-like bacteria). In: Starr MP, Stolp H, Truper HG, Balows A, Schlegel HG (eds) The procaryotes: a handbook on habits, isolation, and identification of bacteria. Springer, Berlin Heidelberg New York, pp 2172–2188

Dawson P, Weste G (1982) Changes in water relations associated with infection by *Phytophthora cinnamomi*. Aust J Bot 30:393–400

Dawson P, Weste G (1984) Impact of root infection by *Phytophthora cinnamomi* on the water relations of two *Eucalyptus* species that differ in susceptibility. Phytopathology 74:486–490

De Bock F, Fer A (1992) Effects of abscisic acid on the transfer of sucrose from host, *Pelargonium zonale* (L.) Aiton, to a phanerogamic parasite, *Cuscuta reflexa* Roxb. Aust J Plant Physiol 19:679–691

Dimond AE (1955) Pathogenesis in the wilt diseases. Annu Rev Plant Physiol 6:329–350

Dimond AE (1970) Biophysics and biochemistry of the vascular wilt syndrome. Annu Rev Phytopathol 8:301–322

Dix NJ, Webster J (1995) Fungal ecology. Chapman & Hall, London

Dixon RK, Garret HE, Cox GS (1988) Cytokinins in the root pressure exudate of *Citrus jambhiri* Lush. colonized by vesicular-arbuscular mycorrhiza. Tree Physiol 4:9–18

Dörr I, Kollmann R (1976) Strukturelle Grundlagen des Parasitismus bei *Orobanche*. III. Die Differenzierung des Xylemanschlusses bei *O. crenata*. Protoplasma 89:235–249

Drüge U, Schönbeck F (1992) Effect of vesicular-arbuscular mycorrhizal infection on transpiration, photosynthesis and growth of flax (*Linum usitatissimum* L.) in relation to cytokinin levels. J Plant Physiol 141:40–48

Duddridge JA, Malibari A, Read JD (1980) Structure and function of mycorrhizal rhizomorphs with special reference to their role in water transport. Nature 287:834–836

Duniway JM (1971) Water relations of *Fusarium* wilt in tomato. Physiol Plant Pathol 1:537–546

Duniway JM (1973) Pathogen-induced changes in host water relations. Phytopathology 63:458–466

Duniway JM (1977) Changes in resistance to water transport in safflower during the development of *Phytophthora* root rot. Phytopathology 67:331–337

Duniway JM, Durbin RD (1971a) Detrimental effect of rust infection on the water relations of bean. Plant Physiol 48:69–72

Duniway JM, Durbin RD (1971b) Some effects of *Uromyces phaseoli* on the transpiration rate and stomatal response of bean leaves. Phytopathology 61:114–119

Edriss MH, Davis RM, Burger DW (1984) Influence of mycorrhizal fungi on cytokinin production in sour orange. J Am Soc Hortic Sci 109:587–590

Ehleringer JR, Schulze E-D, Ziegler, H, Lange OL, Farquhar GD, Cowan IR (1985) Xylem-tapping mistletoes: water or nutrient parasites? Science 227:1479–1481

Ehleringer JR, Cook CS, Tieszen LL (1986) Comparative water use and nitrogen relationship in a mistletoe and its host. Oecologia 68:279–284

El Mahjoub M, Le Picard D (1985) Scanning electron microscopy of *Fusarium oxysporum* f. sp. *melonis* in xylem vessels of resistant and susceptible melon plants. Cryptogam Mycol 6:119–128

Elmi AA, West CP (1995) Endophyte infection effects on stomatal conductance, osmotic adjustment and drought recovery of tall fescue. New Phytol 131:61–67

Elmi AA, West CP, Turner KE (1989) *Acremonium* endophyte enhances osmotic adjustment in tall fescue. Arkans Farm Res 38:7

El-Sharkawy MA, Cock JH, Del Pilar Henandez A (1986) Differential response of stomata to air humidity in the parasitic mistletoe (*Phthirusa pyrifolia*) and its host, mandarin orange (*Citrus reticulata*). Photosynth Res 9:333–343

English-Loeb GM (1990) Plant drought stress and outbreaks of spider mites: a field test. Ecology 71:1401–1411

Estabrook EM, Yoder JI (1998) Plant–plant communications: rhizosphere signalling between parasitic angiosperms and their hosts. Plant Physiol 116:1–7

Evans K, Trudgill DL, Brown NJ (1977) Effects of potato cyst-nematodes on potato plants. V. Root system development in light and heavy-infected susceptible and resistant varieties, and its importance in nutrient and water uptake. Nematologica 23:153–164

Faber BA, Zasoski RJ, Munns DN, Shakel K (1991) A method for measuring hyphal nutrient and water uptake in mycorrhizal plants. Can J Bot 69:87–94

Fahmy GM (1993) Transpiration and dry matter allocation in the angiosperm root parasite *Cynomorium coccineum* L. and two of its halophytic hosts. Biol Plant 35:603–608

Farrell GM, Preece TF, Wren MJ (1969) Effects of infection by *Phythophthora infestans* (Mont.) de Bary on the stomata of potato leaves. Ann Appl Biol 63:265–275

Fatemy F, Evans K (1986) Growth, water uptake and calcium content of potato cultivars in relation to tolerance of cyst nematodes. Rev Nematol 9:171–179

Fisher JT (1983) Water relations of mistletoes and their hosts. In: Calder DM, Bernhardt P (eds) The biology of mistletoes. Academic Press, Sydney, pp 161–183

Fitter AH (1985) Functioning of vesicular-arbuscular mycorrhizas under field conditions. New Phytol 99:257–265

Foster RC (1981) Mycelial strands of *Pinus radiata* D. Don: ultrastructure and histochemistry. New Phytol 88:705–712

Gauslaa Y (1990) Water relations and mineral nutrients in *Melampyrum pratense* (Scrophulariaceae) in oligo- and mesotrophic boreal forests. Acta Oecol 11:525–537

Glatzel G (1983) Mineral nutrition and water relations of hemiparasitic mistletoes: a question of partitioning. Experiments with *Loranthus europaeus* on *Quercus petraea* and *Quercus robur*. Oecologia 56:193–201

Goldstein G, Rada R, Sternberg L, Burguera JL, Orozco M, Montilla M, Zabala O et al. (1989) Gas exchange and water balance of a mistletoe species and its mangrove hosts. Oecologia 78:176–183

Graham JH, Syvertsen JP (1984) Influence of vesicular-arbuscular mycorrhiza on the hydraulic conductivity of roots of two citrus rootstocks. New Phytol 97:277–284

Graham JH, Syvertsen JP, Smith ML (1987) Water relations of mycorrhizal and phosphorous-fertilized non-mycorrhizal citrus under drought stress. New Phytol 105:411–419

Griffin DM (1977) Water potential and wood-decay fungi. Annu Rev Phytopathol 15:319–329

Guerin V, Trinchant JC, Rigaud J (1990) Nitrogen fixation (C_2H_2 reduction) by broad bean (*Vicia faba* L.) nodules and bacterioids under water-restricted conditions. Plant Physiol 92:595–601

Guerin V, Pladys D, Trinchant JC, Rigaud J (1991) Proteolysis and nitrogen fixation in faba-bean (*Vicia faba*) nodules under water stress. Physiol Plant 82:1–7

Haglund BM (1980) Proline and valine – cues which stimulate grasshopper herbivory during drought stress? Nature 288:697–698

Hancock JG (1981) Osmotic conditions influence estimation of passive permeability changes in diseased tissues. Physiol Plant Pathol 18:117–122

Hardie K (1985) The effect of removal of extraradical hyphae on water uptake by vesicular-arbuscular mycorrhizal plants. New Phytol 101:677–684

Hardie K, Leyton K (1981) The influence of vesicular-arbuscular mycorrhiza on growth and water relations of red clover. I. In phosphate deficient soil. New Phytol 89:599–608

Harrison JAC (1971) Transpiration in potato plants infected with *Verticillium* spp. Ann Appl Biol 68:159–168

Hearon SS, Sherald JL, Kostka SJ (1980) Association of xylem-limited bacteria with elm, sycamore, and oak leaf scorch. Can J Bot 58:1986–1993
Helms JA, Cobb FW, Whitney HS (1971) Effect of infection by *Verticicladiella wagenerii* on the physiology of *Pinus ponderosa*. Phytopathology 61:920–925
Hellmuth EO (1971) Eco-physiological studies on plants in arid and semi-arid regions in Western Australia. IV. Comparison of the field physiology of the host, *Acacia grasbyi*, and its hemi-parasite, *Amyema nestor*, under optimal and stress conditions. J Ecol 59:351–363
Hoffman GD, McEvoy PB (1986) Mechanical limitations on feeding by meadow spittlebugs *Philaenus spumarius* (Homoptera: Cercopidae) on wild and cultivated host plants. Ecol Entomol 11:415–426
Hollinger DY (1983) Photosynthesis and water relations of the mistletoe, *Phoradendron villosum*, and its host, the California valley oak, *Quercus lobata*. Oecologia 60:396–400
Hopkins DL (1989) *Xylella fastidiosa*: xylem-limited bacterial pathogen of plants. Annu Rev Phytopathol 27:271–290
Horsfield D (1977) Relationships between feeding of *Philaenus spumarius* (L.) and the amino acid concentration in the xylem sap. Ecol Entomol 2:259–266
Huang C-Y, Boyer JS, Vanderhoef LN (1975 a) Acetylene reduction (nitrogen fixation) and metabolic activities of soybean having various leaf and nodule water potentials. Plant Physiol 56:222–227
Huang C-Y, Boyer JS, Vanderhoef LN (1975 b) Limitation of acetylene reduction (nitrogen fixation) by photosynthesis in soybean having low water potentials. Plant Physiol 56:228–232
Huang R-S, Smith WK, Yost RS (1985) Influence of vesicular-arbuscular mycorrhiza on growth, water relations, and leaf orientation in *Leucaena leucocephala* (Lam.) De Wit. New Phytol 99:229–243
Hunt S, Layzell DB (1991) The role of oxygen in nitrogen fixation and carbon metabolism in legume nodules. In: Pell EJ, Steffen KL (eds) Active oxygen / oxidative stress and plant metabolism. Am Soc Plant Physiol, Rockville, pp 26–39
Hunt S, Layzell DB (1993) Gas exchange of legume nodules and their regulation of nitrogenase activity. Annu Rev Plant Physiol Plant Mol Biol 44:483–511
Ihl, B, Jacob F, Meyer A, Sembdner G (1987) Investigations on the endogenous levels of abscisic acid in a range of parasitic phanerogams. J Plant Growth Regul 2:77–90
Ikeda T, Susaki T (1984) Influence of pine wood nematodes on hydraulic conductance and water relations in *Pinus thunbergii*. Jpn J For Soc 66:412–420
Ikeda T, Kiyohara T, Kusunoki M (1990) Change in water status of *Pinus thunbergii* Parl. inoculated with species of *Bursaphelenchus*. J Nematol 22:132–135
Irigoyen JJ, Emerich DW, Sánchez-Díaz M (1992 a) Water stress induced changes in concentrations of proline and total soluble sugars in nodulated alfalfa (*Medicago sativa*) plants. Physiol Plant 84:55–60
Irigoyen JJ, Emerich DW, Sánchez-Díaz M (1992 b) Phosphoenolpyruvate carboxylase, malate and alcohol dehydrogenase activities in alfalfa (*Medicago sativa*) nodules under water stress. Physiol Plant 84:61–66
Irigoyen JJ, Sánchez-Díaz M, Emerich DW (1992 c) Transient increase of anaerobically-induced enzymes during short-term drought of alfalfa root nodules. J Plant Physiol 139:397–402
Johnson JM, Choinski JS (1993) Photosynthesis in the *Tapinanthus-Diplorhynchus* mistletoe-host relationship. Ann Bot 72:117–122
Jordan CM, Jordan LS, Endo RM (1988) Association of phenol-containing structures with *Apium graveolens* resistance to *Fusarium oxysporum* f. sp. *apii* race 2. Can J Bot 67:3153–3163
Kao J, Damann KE (1978) Microcolonies of the bacterium associated with ratoon stunting disease found in sugarcane xylem matrix. Phytopathology 68:545–551

Khanna-Chopra R, Koundal KR, Sinha SK (1984) A simple technique of studying water deficit effects on nitrogen fixation in nodules without influencing the whole plant. Plant Physiol 76:254–256

Koide R (1985) The effect of VA mycorrhizal infection and phosphorus status on sunflower hydraulic and stomatal properties. J Exp Bot 36:1087–1098

Kothari SK, Marschner H, George E (1990) Effect of VA mycorrhizal fungi and rhizosphere microorganisms on root and shoot morphology, growth and water relations in maize. New Phytol 116:303–311

Küppers M (1992) Carbon discrimination, water-use efficiency, nitrogen and phosphorus nutrition of the host/mistletoe pair *Eucalyptus behriana* F. Muell and *Amyema miquelii* (Lehm. ex Miq.) Tiegh. at permanently low plant water status in the field. Trees 7:8–11

Küppers M, Küppers BIL, Neales TF, Swan AG (1992) Leaf gas exchange characteristic, daily carbon and water balances of the host/mistletoe pair *Eucalyptus behriana* F. Muell. and *Amyema miquelii* (Lehm. ex Miq.) Tiegh. at permanently low water status in the field. Trees 7:1–7

Kuijt J (1969) The biology of parasitic flowering plants. Univ California Press, Berkeley

Kuijt J (1977) Haustoria of phanerogamic parasites. Annu Rev Phytopathol 17:91–118

Kuroda K (1989) Terpenoids causing tracheid-cavitation in *Pinus thunbergii* infected by the pine wood nematode (*Bursaphelenchus xylophilus*). Ann Phytopathol Soc Jpn 55:170–178

Kuroda K, Yamada T, Minio K, Tamura H (1988) Effects of cavitation on the development of pine wilt disease caused by *Bursaphelenchus xylophilus*. Ann Phytopathol Soc Jpn 54:606–615

Lamont B (1982) Mineral nutrition of mistletoes. In: Calder DM, Bernhardt T (eds), The biology of mistletoes. Academic Press, Sydney, pp 185–204

Lansac AR, Martin A, Roldan A (1995) Mycorrhizal colonization and drought interactions of Mediterranean shrubs under greenhouse conditions. Arid Soil Res Rehabil 9:167–175

Lee-Stadelmann OY, Bushnell WR, Stadelmann EJ (1984) Changes of plasmolysis form in epidermal cells of *Hordeum vulgare* infected by *Erysiphe graminis*: evidence for increased membrane-wall adhesion. Can J Bot 62:1714–1723

Levy Y, Krikum J (1980) Effect of vesicular-arbuscular mycorrhiza on *Citrus jambhiri* water relations. New Phytol 85:25–31

Levy Y, Syvertsen JP, Nemec S (1983) Effect of drought stress and vesicular-arbuscular mycorrhiza on *Citrus* transpiration and hydraulic conductivity of roots. New Phytol 93:61–66

Lösch R, Schmitz U, Cours F (1995) *Cuscuta* am Niederrhein: Verbreitungsfähigkeit und Wasserpotential-gradienten zwischen Wirt und Parasit. Verh Ges Ökol 24:567–570

Lüttge U, Haridasan M, Fernandes GW, de Mattos EA, Trimborn P, Franco AC, Caldas LS, Ziegler H (1998) Photosynthesis of mistletoes in relation to their hosts at various sites in tropical Brazil. Trees 12:167–174

Lynn DG, Chang M (1990) Phenolic signals in cohabitation: implications for plant development. Annu Rev Plant Physiol Plant Mol Biol 41:497–526

Mace ME, Bell AA, Beckman CH (1981) Fungal wilt diseases of plants. Academic Press, New York

Madar Z, Solel Z, Sztejnberg A (1990) The effect of *Diplodia pinea* f. sp. *cupressi* and *Seiridium cardinale* on water flow in cypress branches. Physiol Mol Plant Pathol 37:389–398

Main CE, Walker JC (1971) Physiological responses of susceptible and resistant cucumber to *Erwinia tracheiphila*. Phytopathology 61:518–522

Malone M (1993) Hydraulic signals. Philos Trans Roy Soc Lond B 341:33–39

Malone M, Alarcon J-J, Palumbo L (1994) An hydraulic interpretation of rapid, long-distance wound signalling in the tomato. Planta 193:181–185
Mattson WJ, Haack RA (1987) The role of drought in outbreaks of plant-eating insects. BioScience 37:110–118
Melakeberhan H, Webster JM (1993) The phenology of plant-nematode interaction and yield loss. In: Khan MW (ed) Nematode interactions. Chapman & Hall, London
Melakeberhan H, Toivonen PMA, Vidaver WE, Webster JM, Dube SL (1991) Effect of *Bursaphelenchus xylophilus* on the water potential and water-splitting complex of photosystem II of *Pinus silvestris* seedlings. Physiol Mol Plant Pathol 38:83–91
Musselman LJ (1980) The biology of *Striga*, *Orobanche*, and other root-parasitic weeds. Annu Rev Phytopathol 18:463–489
Musselman LJ (ed) (1987) Parasitic weeds in agriculture. I. *Striga*. CRC Press, Boca Raton
Nelsen CR (1987) The water relations of vesicular-arbuscular mycorrhizal systems. In: Safir GR (ed) Ecophysiology of VA mycorrhizal plants. CRC Press, Boca Raton, pp 71–91
Nelsen CR, Safir GR (1982) Increased drought tolerance of mycorrhizal onion plants caused by improved phosphorus nutrition. Planta 154:407–413
Newcombe G, Robb J (1988) The function and relative importance of the vascular coating response in highly resistant, moderately resistant and susceptible alfalfa infected by *Verticillium albo-atrum*. Physiol Mol Plant Pathol 33:47–58
Newman SE, Davies FT (1988) High root-zone temperatures, mycorrhizal fungi, water relations, and root hydraulic conductivity of container-grown woody plants. J Am Soc Hortic Sci 113:138–146
Nour JJ, Todd P, Yaghmaie P, Panchal G, Stewart GR (1984) The role of mannitol in *Striga hermontica*. In: Parker C, Musselman LJ, Polhill RM, Wilson AK (eds) Proc 3rd Int Symp on parasitic weeds. ICARDA, Damaskus, pp 81–89
Olsen MW, Misaghi IJ, Goldstein D, Hine RB (1983) Water relations in cotton plants infected with *Phymatotrichum*. Phytopathology 73:213–216
Ouellette GB, Baayen RP (2000) Peculiar structures occurring in vessel walls of the susceptible carnation cultivar Early Sam infected with *Fusarium oxysporum* f. sp. *dianthi*. Can J Bot 78:270–277
Ouellette GB, Baayen RP, Simard M, Rioux D (1999) Ultrastructural and cytochemical study of colonization of xylem vessel elements of susceptible and resistant *Dianthus caryophyllus* by *Fusarium oxysporum* f.sp. *dianthi*. Can J Bot 77:644–663
Pankhurst C, Sprent J (1975) Surface features of soybean root nodules. Protoplasma 85:85–98
Panvini AD, Eickmeier WG (1993) Nutrient and water relations of the mistletoe *Phoradendron leucarpum* (Viscaceae): how tightly are they integrated. Am J Bot 80:872–878
Parke JL, Linderman RG, Black CH (1983) The role of ectomycorrhizas in drought tolerance of Douglas-fir seedlings. New Phytol 95:83–95
Pearce RB (1990) Occurrence of decay-associated xylem suberization in a range of woody species. Eur J For Pathol 20:275–289
Pearce RB (1991) Reaction zone relics and the dynamics of fungal spread in the xylem of woody angiosperms. Physiol Mol Plant Pathol 29:41–55
Peña JI, Sánchez-Díaz M, Aguirreolea J, Becana M (1988) Increased stress tolerance of nodule activity in the *Medicago-Rhizobium-Glomus* symbiosis under drought. J Plant Physiol 133:79–83
Press MC, Graves JD (eds) (1995) Parasitic flowering plants. Chapman & Hall, London
Press MC, Whittaker JB (1993) Exploitation of the xylem stream by parasitic organisms. Philos Trans R Soc Lond B 341:101–111
Press MC, Tuohy JM, Stewart GR (1987) Gas exchange characteristic of the sorghum-*Striga* host-parasite association. Plant Physiol 84:814–819

Press MC, Graves JD, Stewart GR (1990) Physiology of the interaction of angiosperm parasites and their higher plant hosts. Plant Cell Environ 13:91–104

Press MC, Parsons AN, Mackay AW, Vincent CA, Cochrane V, Seel WE (1993) Gas exchange characteristics and nitrogen relations of two Mediterranean root hemiparasites: *Bartsia trixago* and *Parentucellia viscosa*. Oecologia 95:145–151

Press MC, Scholes JD, Barker MG (eds) (1999a) Physiological plant ecology, 39th Brit Ecol Soc Symp, York. Blackwell, Oxford, pp 175–197

Press MC, Scholes JD, Watling JR (1999b) Parasitic plants: physiological and ecological interactions with their hosts. In: Press MC, Scholes JD, Barker MG (1999a), pp 175–197

Purcell AH, Hopkins DL (1996) Fastidious xylem-limited bacterial plant pathogens. Annu Rev Phytopathol 34:131–151

Quisenberry SS, Joost RE (eds) (1990) Proceedings of the international symposium on *Acremonium*/grass interactions. Louisiana Agric Exp Station, Baton Rouge

Radomiljac AM, McComb JA, Pate JS (1999) Gas exchange and water relations of the root hemi-parasite *Santalum album* L. in association with legume and non-legume hosts. Ann Bot 83:213–224

Raven JA (1983) Phytophages of xylem and phloem: a comparison of animal and plant sapfeeders. Adv Ecol Res 13:135–234

Reid CPP (1979) Mycorrhiza and water stress. In: Riedacker A, Gagnaire-Michard J (eds) Root physiology and symbiosis. IUFRA Symp Proc, Nancy, pp 392–408

Rhoades DF (1983) Herbivore population dynamics and plant chemistry. In: Denno RF, McClure MS (eds) Variable plants and herbivores in natural and managed systems. Academic Press, New York, pp 155–220

Richardson MD, Chapman GW, Hoveland CS, Bacon CW (1992) Sugar alcohols in endophyte-infected tall fescue under drought. Crop Sci 32:1060–1061

Richardson MD, Hoveland CS, Bacon CW (1993) Photosynthesis and stomatal conductance of symbiotic and nonsymbiotic tall fescue. Crop Sci 33:145–149

Robb J, Smith A, Brisson JD, Busch L (1979) Ultrastructure of wilt syndrome caused by *Verticillium dahliae.* VI. Interpretative problems in the study of vessel coatings and tyloses. Can J Bot 57:795–821

Robb J, Powell D, Street PFS (1987) The chronological development of the coating response in *Verticillium*-infected tomato petioles. Physiol Mol Plant Pathol 31:217–226

Safir GR (1987) Ecophysiology of VA mycorrhizal plants. CRC Press, Boca Raton

Safir GR, Boyer JS, Gerdemann JW (1971) Mycorrhizal enhancement of water transport in soybean. Science 172:581–583

Safir GR, Boyer JS, Gerdemann JW (1972) Nutrient status and mycorrhizal enhancement of water transport in soybean. Plant Physiol 49:700–703

Saikkonen K, Faeth SH, Helander M, Sullivan TJ (1998) Fungal endophytes: a continuum of interactions with host plants. Annu Rev Ecol Syst 29:319–343

Sánchez-Díaz M, Aguirreolea J (1993) Relaciones hídricas. In: Azcon-Bieto J, Talon M (eds) Fisiología y bioquimica vegetal. Interamericana/McGraw-Hill, New York, pp 49–90

Sánchez-Díaz M, Honrubia M (1994) Water relations and alleviation of drought stress in mycorrhizal plants. In: Gianinazzi S, Schüepp H (eds) Impact of arbuscular mycorrhizas on sustainable agriculture and natural ecosystems. Birkhäuser, Basel, pp 167–178

Sánchez-Díaz M, Pardo M, Antolín M, Peña IJ, Aguirreolea JA (1990) Effect of water stress on photosynthetic activity in the *Medicago-Rhizobium-Glomus* symbiosis. Plant Sci 71:215–221

Scheffer RP (1976) Host-specific toxins in relation to pathogenesis and disease resistance. In: Heitefuss R, Williams PH (eds) Physiological plant pathology. Springer, Berlin Heidelberg New York, pp 247–269

Schoeneweiss DF (1975) Predisposition, stress, and plant disease. Annu Rev Phytopathol 13:193–211
Schoeneweiss DF (1986) Water stress predisposition to disease - an overview. In: Ayres PG, Boddy L (eds) Water, fungi, and plants. Cambridge Univ Press, Cambridge, pp 157–174
Schönwitz R, Ziegler H (1989) Interaction of maize roots and rhizosphere organisms. J Plant Nutr Soil Sci 152:217–222
Scholander PF, Hammel HAT, Bradstreet ED, Hemmingsen EA (1965) Sap pressure in vascular plants. Science 148:339–346
Schulze E-D, Ehleringer JR (1984) The effect of nitrogen supply on growth and water-use efficiency of xylem-tapping mistletoes. Planta 162:268–275
Schulze E-D, Turner NC, Glatzel G (1984) Carbon, water and nutrient relations of two mistletoes and their hosts: a hypothesis. Plant Cell Environ 7:293–299
Schulze E-D, Lange OL, Ziegler H, Gebauer G (1991) Carbon and nitrogen isotope ratios of mistletoes growing on nitrogen and non-nitrogen fixing hosts and on CAM plants in the Nambib desert confirm partial heterotrophy. Oecologia 88:457–462
Seel WE, Cooper RE, Press MC (1993a) Growth, gas exchange and water use efficiency of the facultative hemiparasite *Rhinanthus minor* associated with hosts differing in foliar nitrogen concentration. Physiol Plant 89:64–70
Seel WE, Parsons AN, Press MC (1993b) Do inorganic solutes limit growth of the facultative hemiparasite *Rhinanthus minor* L. in the absence of a host? New Phytol 124:283–289
Sempio C, Majerník O, Raggi V (1966) Stomatal and cuticular transpiration of beans (*Phaseolus vulgaris* M.) attacked by rust (*Uromyces appendiculatus* [Pers.] Link). Biol Plant (Praha) 8:316–320
Shah N, Smirnoff N, Stewart GR (1987) Photosynthesis and stomatal characteristics of *Striga hermonthica* in relation to its parasitic habit. Physiol Plant 69:699–703
Sheehy JE, Webb J (1991) Oxygen diffusion pathways and nitrogen fixation in legume root nodules. Ann Bot 67:85–92
Sheehy JE, Minchin FR, Witty JF (1983) Biological control of the resistance to oxygen flux in nodules. Ann Bot 52:565–571
Sheoran IS, Kaur A, Singh R (1988) Nitrogen fixation and carbon metabolism in nodules of pidgeonpea (*Cajanus cajan* L.) under drought stress. J Plant Physiol 132:480–483
Shi J, Mueller WC, Beckman CH (1992) Vessel occlusion and secretory activities of vessel contact cells in resistant or susceptible cotton plants infected with *Fusarium oxysporum* f. sp. *vasinfectum*. Physiol Mol Plant Pathol 40:133–147
Shigo AL (1984) Compartmentalization: a conceptual framework for understanding how trees grow and defend themselves. Annu Rev Phytopathol 22:189–214
Shtienberg D (1992) Effects of foliar diseases on gas exchange processes: a comparative study. Phytopathology 82:760–765
Sieverding E (1981) Influence of soil water regimes on VA mycorrhiza. I. Effect on plant growth, water utilization and development of mycorrhiza. Z Acker- Pflanzenbau 150:400–411
Skinner MF, Bowen GD (1974) The penetration of soil by mycelial strands of ectomycorrhizal fungi. Soil Biol Biochem 6:57–61
Smith S, Stewart GR (1990) Effect of potassium levels on the stomatal behavior of the hemi-parasite *Striga hermonthica*. Plant Physiol 84:1472–1476
Spiller NJ, Koenders L, Tjallingii WF (1990) Xylem ingestion by aphids - a strategy for maintaining water balance. Entomol exp appl 55:101–104
Sprent JI (1971) The effects of water stress on nitrogen-fixing root nodules. I. Effects on the physiology of detached soybean nodules. New Phytol 70:9–17
Sprent JI (1972 a) The effects of water stress on nitrogen-fixing root nodules. II. Effects on the fine structure of detached soybean nodules. New Phytol 71:443–450

Sprent JI (1972 b) The effects of water stress on nitrogen-fixing root nodules. IV. Effects on whole plants of *Vicia faba* and *Glycine max*. New Phytol 71:603–611

Sprent JI, Gallacher A (1976) Anaerobiosis in soybean root nodules under water stress. Soil Biol Biochem 8:317–320

Stewart GR, Press MC (1990) The physiology and biochemistry of parasitic angiosperms. Annu Rev Plant Physiol 41:127–151

Stewart GR, Nour JJ, MacQueen M, Shah N (1984) Aspects of the biochemistry of *Striga*. In: Ayensu ES, Dogget H, Keynes RD, Marton-Lefevre J, Musselman LJ, Parker C, Pickering A (eds) *Striga* biology and control. ICSU/IDRS, Paris, pp 161–178

Street PFS, Cooper RM (1982) The role of endopectin lyase and high molecular weight polysaccharides in occlusion of xylem vessels in tomatoes infected with *Verticillium albo-atrum*. Can J Plant Pathol 4:310

Swaraj K, Topunow AF, Golubeva LI, Kretovich WL (1986) Effect of water stress on enzymatic reduction of leghemoglobin in soybean nodules. Soviet Plant Physiol 33:70–74

Teakle DS, Smith PM, Steindl DRL (1973) Association of a small coryneform bacterium with the ratoon stunting disease of sugarcane. Aust J Agric Res 24:869–874

Ter Borg SJ (eds) (1986) Biology and control of *Orobanche*. LH/VPO, Wageningen

Tessier BJ, Mueller WC, Morgham AT (1990) Histopathology and ultrastructure of vascular responses in peas resistant or susceptible to *Fusarium oxysporum* f. sp. *pisi*. Phytopathology 80:756–764

Timmer LW, Brlansky RH, Graham JH, Sandler HA, Agostini JP (1986) Comparison of water flow and xylem plugging in declining and in apparently healthy citrus trees in Florida and Argentina. Phytopathology 76:707–711

Tissera P, Ayres PG (1988) Hydraulic conductance and anatomy of roots of *Vicia faba* plants infected by *Uromyces viciae-fabae*. Physiol Mol Plant Pathol 32:199–207

Towers B, Stambaugh WJ (1968) The influence of induced soil moisture stress upon *Fomes annosus* root rot of loblolly pine. Phytopathology 58:269–272

Turner NC, Granitti A (1969) Fusicoccin: a fungal toxin that opens stomata. Nature 223:1070–1071

Ullmann I, Lange OL, Ziegler H, Ehleringer J, Schulze E-D, Cowan IR (1985) Diurnal courses of leaf conductance and transpiration of mistletoes and their hosts in Central Australia. Oecologia 67:577–587

Van Doorn WG, Schurer K, De Witte Y (1989) Role of endogenous bacteria in vascular blockage of cut rose flowers. J Plant Physiol 134:375–381

Vannini A, Valentini R (1994) Influence of water relations on *Quercus cerris-Hypoxylon mediterraneum* interaction: a model of drought-induced susceptibility to a weakness parasite. Tree Physiol 14:129–139

Wargo PM (1996) Consequences of environmental stress on oak: predisposition to pathogens. Ann Sci For 53:359–368

Weber HC, Forstreuter W (eds) (1987) Parasitic flowering plants. Proc 4th Int Symp on parasitic flowering plants, Marburg

Webster JM (1975) Aspects of the host-parasite relationship of plant-parasitic nematodes. Adv Parasitol 13:225–250

White J, Strehl CE (1978) Xylem feeding by periodical cycada nymphs on tree roots. Ecol Entomol 3:323–327

White RH, Engelke MC, Norton SJ, Johnson-Cicalease JM, Ruemmele BA (1992) *Acremonium* endophyte effects on tall fescue drought tolerance. Crop Sci 32:1392–1396

White TCR (1976) Weather, food and plagues of locusts. Oecologia 22:119–134

White TCR (1984) The abundance of invertebrate herbivores in relation to the availability of nitrogen in stressed food plants. Oecologia 63:90–105

Whittacker JB (1970) Cercopid spittle as a microhabitat. Oikos 21: 59–64
Wiegert RG (1964) The ingestion of xylem sap by meadow spittlebugs, *Philaenus spumarius* (L.). Am Midl Nat 71: 422–428

Prof. Dr. Rainer Lösch
Dr. Dirk Gansert
Abt. Geobotanik, H. Heine-Universität
Universitätsstraße 1/26.13
40225 Düsseldorf, Germany
e-mail: Loesch@uni-duesseldorf.de
e-mail: Gansert@uni-duesseldorf.de

Pathways and Enzymes of Brassinosteroid Biosynthesis

By Bernd Schneider

Abstract

This review provides an overview of the field of biosynthesis of brassinosteroids, the only known class of plant steroid hormones. The rapidly growing knowledge of biochemical pathways and enzymes involved in the brassinosteroid biosynthetic network is the major subject of this paper. Recent progress in this field is mainly due to mutant studies combined with feeding experiments. Contributions of molecular and, as far as is known, protein biochemical investigations are discussed. The literature in this field up to the end of 2000 has been evaluated.

1 Introduction

The early history of brassinosteroids started in 1941 when Mitchell and Whitehead reported plant growth-promoting activity of pollen extracts from *Zea mays* (Mitchell and Whitehead 1941). It turned out that a special class of polyhydroxylated sterols, collectively named brassinosteroids, are responsible for stimulating plant growth. Grove et al. (1979) isolated the first brassinosteroid from vast amounts of rape pollen (*Brassica napus*), elucidated its structure by spectroscopic methods including x-ray analysis. The compound (22*R*,23*R*,24*S*)-2α,3α,22,23-tetrahydroxy-24-methyl-B-homo-6a-oxa-5α-cholestan-6-one, was given the trivial name brassinolide (BL) (Fig. 1), due to the original Brassicaceae source. More than 40 further brassinosteroids have since been isolated

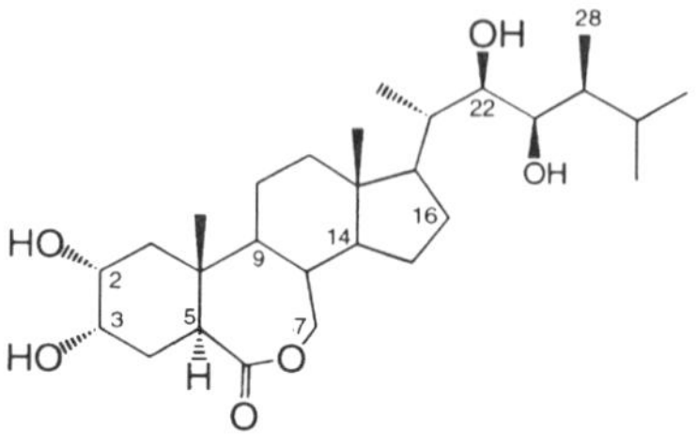

Fig. 1. Structure and numbering of brassinolide

Progress in Botany, Vol. 63

from natural sources, and even more have been obtained by chemical synthesis. Brassinolide is still the most active compound of this type at very low concentrations (for review, see Adam et al. 1999). This fact indicates sophisticated structural optimization during evolution of that class of compounds. Analysis of many plant species has confirmed that the occurrence of brassinosteroids in the plant kingdom is ubiquitous.

Despite numerous efforts to elucidate their biological mode of action, brassinosteroids gained acceptance as plant hormones only when mutants became available in the middle of the last decade. A number of mutants of *Arabidopsis thaliana* and other plants were rescued to the wild-type phenotype by treatment with exogenously applied brassinosteroids, whereas other phytohormones were inactive. These mutants, therefore, were considered to be deficient in one of the brassinosteroid biosynthetic enzymes. Most importantly, the essential role of brassinosteroids in plant growth and development was confirmed by mutant studies (Kauschmann et al. 1996; Li et al. 1996; Szekeres et al. 1996).

Early books about this topic appeared at the beginning of the last decade (Cutler et al. 1991; Khripach et al. 1993). Two books representing the current state of the entire field of brassinosteroid research have appeared recently (Khripach et al. 1999; Sakurai et al. 1999). A number of recent reviews from 1999 and 2000 covering aspects of brassinosteroid research, e.g. molecular physiology (Altmann 1999), biosynthesis inhibitors (Asami and Yoshida 1999), biological effects (Brosa 1999), mode of action (Li and Chory 1999), practical application (Khripach et al. 2000), biosynthesis (Sakurai 1999), as well as biosynthesis and metabolism (Yokota 1999), are recommended for further reading.

2 Brassinosteroid Structure

Brassinosteroids represent polyoxygenated sterols formally derived from the 5α-cholestane skeleton, which is characterized by the trans-fused A/B ring system. Further typical structural features of the most active brassinosteroids are the vicinal diol groups at ring A (C-2α/C-3α) and at C-22-(*R*)OH/C-23-(*R*)OH in the side chain, and 6-oxo or 6a-oxalactone functionality in ring B. Deviation in stereochemistry occurs, for example, in 2β-hydroxy- and 3β-hydroxybrassinosteroids, and usually reduces biological activity drastically. Different numbers of carbon atoms in the side chain enable classification into C-27, C-28- and C-29 brassinosteroids. Biosynthetic intermediates are characterized by a lower degree of functionalization as, for example, missing oxygenation in the side chain, at C-2, or at C-6. In general a brassinosteroid requires at least one hydroxy group at ring A of 5α-cholestane system and another one in the side chain.

3 Methodology Used in Biosynthetic Studies

The common occurrence of structurally similar brassinosteroids and possible precursors first hinted at the existence of biosynthetic relationships. Feeding experiments in early biosynthetic studies on brassinosteroid-specific pathways have been carried out in crown gall cell cultures of *Catharanthus roseus*, because they contained relatively high concentrations of the target compounds (see the chapter by Sakurai et al. in Cutler et al. 1991). Untransformed cell cultures and seedlings of *C. roseus* and other plant species including *Arabidopsis thaliana* were also used. Recently, labeling experiments have been increasingly combined with mutant studies to elucidate individual biosynthetic steps. Interestingly, brassinosteroid-insensitive mutants (*bri*) of *Arabidopsis* accumulated high levels of brassinosteroids (Noguchi et al. 1999b) and have been used successfully in recent feeding experiments.

Chemical and biosynthetic methods have been used to prepare labeled brassinosteroid precursors. Chemical synthesis of tritium- (Yokota et al. 1990; Kolbe et al. 1998a) and deuterium-labeled intermediates (Takatsuto and Ikekawa 1986; Kolbe et al. 1998a) were developed for that purpose. Carbon-labeled early precursors have also been prepared and used in biosynthetic experiments (Suzuki et al. 1995a). Compactin, an inhibitor of mevalonic acid (MVA) biosynthesis, enabled label from MVA to be incorporated into the desired intermediates, such as campesterol (CR) in plant cell cultures growing on nutrient medium containing [^{13}C]- and/or [^{14}C]MVA.

Full-scan gas chromatography – mass spectrometry (GC-MS) and GC-selected ion monitoring (SIM) have been used extensively in brassinosteroid analysis to detect labeled intermediates and to determine the ratio of endogenous and biosynthetically formed brassinosteroids (see the chapter by Takatsuto and Yokota in Sakurai et al. 1999). Methaneboronation of vicinal diol groups and silylation of hydroxyls represent standard derivatization methods in brassinostroid analysis.

The identification of brassinosteroid biosynthesis-deficient mutants of *Arabidopsis*, tomato and garden pea had considerable impact on the elucidation of brassinosteroid biosynthesis. The biosynthetic lesions of these mutants were established by means of rescue experiments with exogenously applied intermediates. The dwarf phenotype of the mutants was restored by brassinosteroids downstream of the deficient enzyme but remained unchanged by upstream intermediates. Molecular cloning, sequencing, comparison with known gene or amino acid sequences and, in part, functional expression in heterologous systems revealed relationships with known proteins, providing most of the current knowledge about the enzymes involved in brassinosteroid biosynthesis. Classical protein biochemical methods were employed for enzymes, which have not been accessible in mutant studies until recently.

4 Early Biosynthetic Steps – Campesterol Biosynthesis

Plant sterols, which are considered precursors of brassinosteroids, are biosynthesized in the cytoplasm via the mevalonate pathway through squalene and cycloartenol (Fig. 2). They are not [or only to a minor degree due to crosstalk between mevalonate and the recently discovered methylerythritol-4-phosphate (MEP) pathway (Eisenreich et al. 1998;

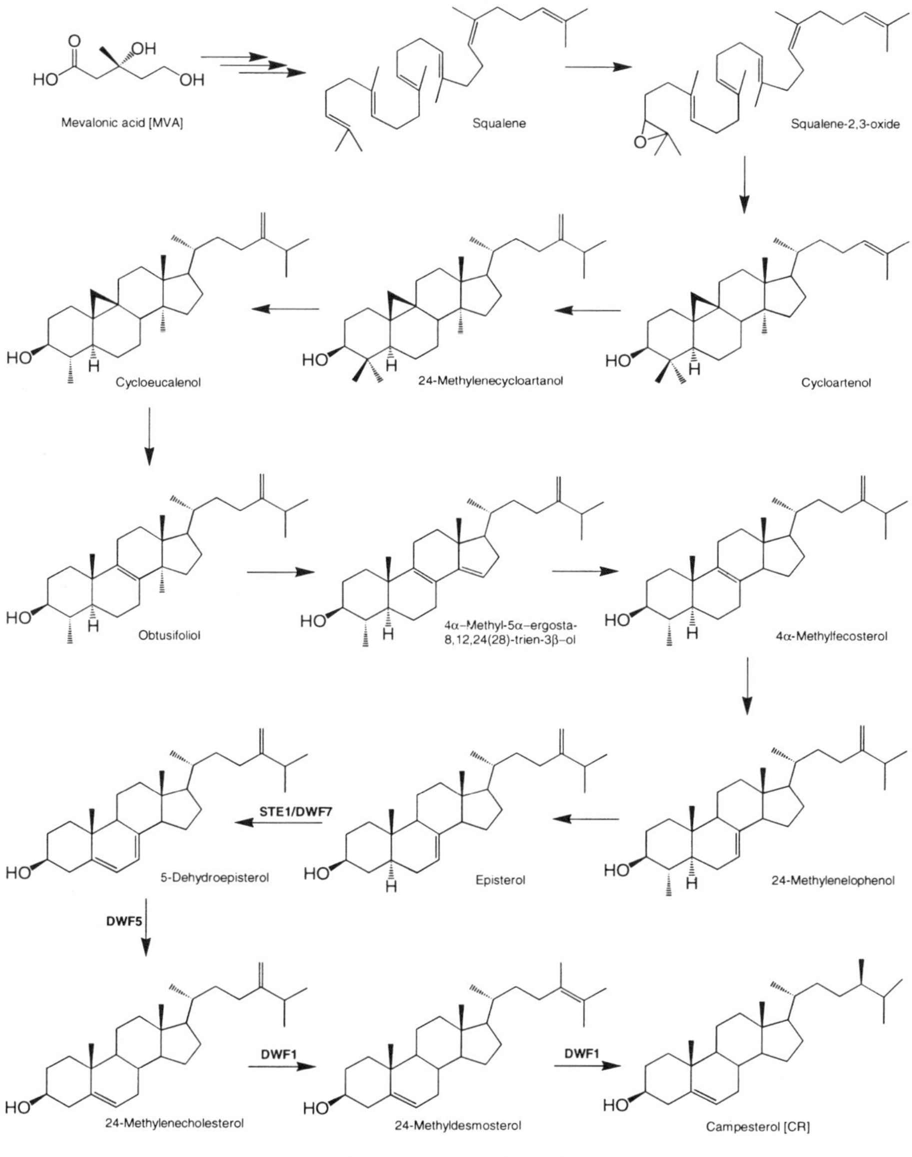

Fig. 2. Biosynthesis of campesterol from mevalonic acid

Rohmer 1999)] synthesized through 1-deoxyxylulose-phosphate (Arigoni et al. 1997; Lichtenthaler et al. 1997). Thus, the mevalonate-independent MEP pathway does not play a significant role in sterol and brassinosteroid biosynthesis.

Cyclisation of squalene-2,3-oxide gives the first tricylic triterpene in the pathway, cycloartenol. This undergoes methylation at C-24, 4β-demethylation followed by opening of the cyclopropane ring, demethylation at C-14, reduction of $\Delta^{14(15)}$ double bond, and $\Delta^{8(9)}$-double bond isomerization to give 24-methylenelophenol, the branching point between the pathways to sitosterol and campesterol. Phytosterols are important not only as precursors of more oxidized steroids, among which brassinosteroids are only one group, but also as membrane components. From this point of view, the steps of early pathways and the enzymes involved are not considered specific for brassinosteroid biosynthesis and, are therefore, not being discussed in more detail here. However, reading the extensive literature in this field is recommended (for review, see Benveniste 1986). As a consequence of overexpression of sterol methyltransferase 2–1 (*SMT 2–1*), which catalyzes C-28-methylation of 24-methylenelophenol, sitosterol biosynthesis in *Arabidopsis* was enhanced, thereby diminishing campesterol. *Arabidopsis* mutants depleted in campesterol exhibited changes in the phenotype, which could be rescued by brassinosteroid treatment (Schaeffer et al. 2001). This suggests campesterol plays a specific role in brassinosteroid supply.

On the biosynthetic branch towards campesterol (Fig. 2), 24-methylenelophenol undergoes 4α-demethylation, subsequent migration of $\Delta^{7(8)}$ double bond of episterol via the $\Delta^{5(6)}\Delta^{7(8)}$ diene intermediate, 5-dehydroepisterol, to $\Delta^{5(6)}$ double bond in 24-methylenecholesterol. *Arabidopsis* dwarf-phenotype mutants *sterol1/dwarf7* (*ste1/dwf7*) have been identified which are defective in sterol C-5 desaturase in the conversion of episterol to 5-dehydroepisterol (Choe et al. 1999a). Molecular characterization of the *STE1* gene indicated some sequence identity with fungi and yeast C-5 desaturases.

Reduction of the $\Delta^{7(8)}$ double bond of 5-dehydroepisterol to give 24-methylenecholesterol is considered to be catalyzed by a sterol C-7 reductase. As in the previous step, *Arabidopsis* mutants have also been identified for reduction of $\Delta^{7(8)}$ double bond (Choe et al. 2000). These mutants are characterized by a dwarfed phenotype, a scarce level of 24-methylenecholesterol, and were rescued to wild-type phenotype upon treatment with brassinolide. Moreover, ^{13}C-MVA was converted to [$^{13}C_5$]episterol, [$^{13}C_5$]methylenecholesterol and [$^{13}C_5$]campesterol in the wild-type. In contrast, in *dwf5*, [$^{13}C_5$]7-dehydrocampesterol and [$^{13}C_5$]7-dehydrocampestanol were found instead. These are supposed to be formed through [$^{13}C_5$]5-dehydroepisterol as a low-concentrated and therefore undetectable intermediate. These findings indicated a biosynthetic lesion before 24-methylenecholesterol. Molecular cloning revealed a sequence identity of less than 40% with sterol reductases only, but many highly conserved domains were found.

Labeling studies using [26,27-$^{13}C_2$]24-methylenecholesterol provided evidence that isomerization of the side chain double bond to the intermediate 24-methyldesmosterol and further reduction to campesterol (CR) represent the next steps (Yamada et al. 1997). *Dwarf* (*dwf1*) mutants which were blocked in this part of the pathway were first isolated by Feldman et al. (1989) without identifying the function of the corresponding enzyme at that time. *Diminuto* (*dim*), *cabbage1* (*cbb1*) (Kauschmann et al. 1996) and *lkb* (Nomura et al. 1997; Nomura et al. 1999), which are alleles of the *dwf1* mutant, have since been identified from *Arabidopsis* and *Pisum sativum*, respectively (for review, see Altmann 1999). Cloning and sequencing of *DIM* (Takahashi et al. 1995) revealed homology only to a domain found in flavine adenine dinucleotide (FAD)-dependent oxidoreductases. Recent studies have indicated that DWF1 is an integral membrane protein probably associated with the endoplasmic reticulum (Klahre et al. 1998; Choe et al. 1999b). The same authors proved by mutant analysis and feeding experiments using deuterium-labeled 24-methylenecholesterol and 24-methylenedesmosterol, that DWF1 is involved both in the isomerization and reduction step. Choe et al. 1999b also speculated that both steps are catalyzed by different domains of the multifunctional DWF1 protein.

The early steps of the sterol/brassinosteroid pathway (shown in Fig. 2), MVA → squalene → squalene-2,3-oxide → cycloartenol → 24-methylenecycloartanol → cycloeucalenol → obtusifoliol → 4α-methyl-5α-ergosta-8,12,24(28)-trien-3β-ol → 4α-methylfecosterol → 24-methylenelophenol → episterol → 5-dehydroepisterol → 24-methylenecholesterol → 24-methyldesmosterol → campesterol, are well characterized. Almost all steps have been confirmed by feeding studies and/or on the gene and enzyme level.

5 Campesterol to Campestanol

Campesterol is generally considered as the starting point of the brassinosteroid-specific biosynthetic pathway. The biosynthetic sequence between CR and campestanol (CN) leads to completion of the carbon skeleton including *trans* stereochemistry of the A/B ring junction. All further steps represent either introduction to, or modifications of, oxygen functions on the scaffold. Formation of CN from CR has been elucidated by labeling experiments employing cell cultures of *Catharanthus roseus*. Biosynthetically prepared ^{13}C- and/or ^{14}C-labeled CR, re-administered to *Catharanthus* cell cultures, was converted to CN as a major metabolite (Suzuki et al. 1995a). Administration of [$^{2}H_6$]campesterol and the complete series of deuterium-labeled intermediates, such as

(24*R*)-[$^2H_6$]24-methylcholest-4-en-3β-ol (4-en-3-ol), (24*R*)-[$^2H_6$]24-methylcholest-4-en-3-one (4-en-3-one), (24*R*)-[$^2H_6$]24-methyl-5α-cholestan-3-one (3-one), resulted in conversion to each of the subsequent compounds in this series, including campestanol (Noguchi et al. 1999a). From these results, the biosynthetic sequence from campesterol to campestanol has been deduced: Isomerization of campesterol to 4-en-3-ol, oxidation of the 3β-hydroxy group to the corresponding 3-oxo intermediate, 4-en-3-one, subsequent stereospecific 5α-reduction of the $\Delta^{4(5)}$ double bond to establish the *trans*-fused A/B ring system of 3-one. Finally, reduction of the 3-oxo functionality results in campestanol.

This pathway was further confirmed by detailed analysis of the *deetiolated2* (*det2*) mutant of *Arabidopsis* (Li et al. 1996). Cloning and sequencing demonstrated that the DET2 protein shares about 40% identity with mammalian steroid 5α-reductases, suggesting involvement of this protein in the 5α-reduction of campesterol. Moreover, when expressed in human kidney cell cultures, specific reduction of the steroid $\Delta^{4(5)}$ double bond, but not of the $\Delta^{5(6)}$ double bond of campesterol was observed (Li et al. 1997). The *Pisum sativum* dwarf mutant *lk* might be allelic with *det2* (Yokota 1997).

Phytochemical analyses demonstrated that the campestanol content of *det2* mutants was significantly below the level in *Arabidopsis* wild-type seedlings, while (24*R*)-24-methylcholest-4-en-3-one accumulated in the mutant above that level (Fujioka et al. 1997). Similar biosynthetic incorporation studies as described for *C. roseus* cell cultures were also carried out with *Arabidopsis* using deuterium-labeled precursors ([$^2H_6$]CR, [$^2H_6$]4-en-3-ol, [$^2H_6$]4-en-3-one, [$^2H_6$]3-one). While incorporation of all precursors into campestanol was detected in the wild-type *Arabidopsis* seedlings, the lesion between (24*R*)-24-methylcholest-4-en-3-one and (24*R*)-24-methyl-5α-cholestan-3-one clearly prevented incor-

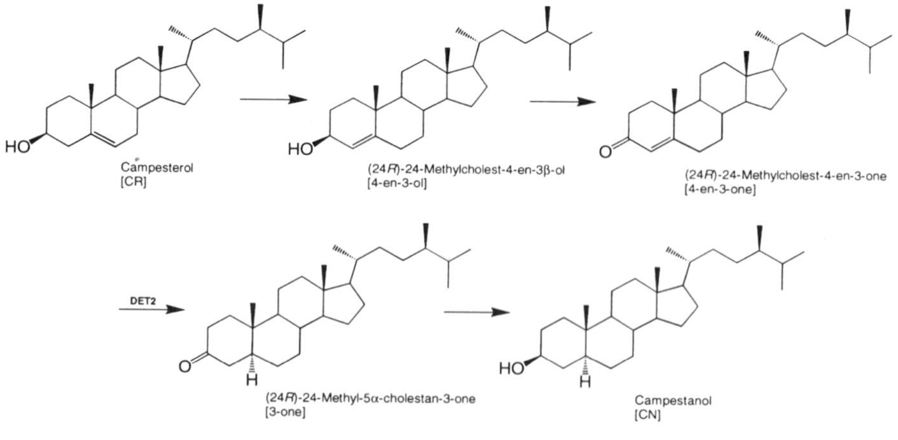

Fig. 3. Conversion of campesterol to campestanol

poration of [$^2H_6$]CR, [$^2H_6$]4-en-3-ol, and [$^2H_6$]4-en-3-one in *det2* mutant seedlings (Noguchi et al. 1999a).

Figure 3 shows the pathway campesterol → (24*R*)-24-methylcholest-4-en-3β-ol → (24*R*)-24-methylcholest-4-en-3-one → (24*R*)-24-methyl-5α-cholestan-3-one → campestanol. While the 5α-reduction step is well characterized on the gene and enzyme level, the other steps have yet to be studied in detail.

6 Campestanol to Castasterone

Campestanol represents an important intermediate in the brassinolide biosynthetic pathway. After 5α-reduction of the $\Delta^{4(5)}$ double bond, the formation of the carbon skeleton is finished. The observation that all naturally occurring brassinosteroids share the same structure of the ring system suggests that all further steps represent either introduction to or modifications of oxygen functions. The oxygenation pattern supports the hypothesis of the pathway from campestanol through teasterone (TE), typhasterol (TY) to castasterone (CS) and finally to brassinolide (Yokota et al. 1991). The sequence TE → TY → CS has been confirmed by feeding experiments using *C. roseus* cultured cells. 6α-Hydroxy-campestanol (6–OHCN) and 6-oxocampestanol (6-oxoCN) were identified by GC-MS upon feeding [^{13}C]- and/or [^{14}C]campesterol. Feeding of [^{13}C]campestanol led to 6α-hydroxycampestanol. [$^2H_6$]6α-Hydroxy-campestanol was converted to 6-oxocampestanol while 6β-hydroxy-campestanol was not, indicating stereospecificity of C-6 hydroxylation of campestanol in this pathway (Suzuki et al. 1995a). It seems likely that a cytochrome P450 monooxygenase is involved in C-6 oxidation of campestanol. Whether or not the tomato dwarf (d) protein, which is involved in C-6 oxidation of 6-deoxocastasterone (Bishop et al. 1999; see below), also catalyzes C-6 oxidation of campestanol has not yet been clarified.

Isolation of cathasterone (CT) from cultured cells of *C. roseus* indicated that 22-hydroxylation is the next biosynthetic step in the sequence (Fujioka et al. 1995). Due to the very small pool size of cathasterone in that species, conversion of precursors to cathasterone remained tentative in *Catharanthus* until recently and therefore was considered the rate-limiting step in brassinosteroid biosynthesis. Conversion of [$^2H_6$]6-oxocampestanol to [$^2H_6$]cathasterone was the last step in the so-called early C-6 oxidation pathway confirmed by precursor administration experiments (Fujioka et al. 2000a). The *Arabidopsis* dwarf mutant *dwarf4* (*dwf4*) has been identified by Azpiroz et al. (1998). The dwarf phenotype of this mutant could be rescued only by application of brassinolide and other 22α-hydroxylated brassinosteroids, suggesting a lesion in the brassinosteroid biosynthesis at the 22-hydroxylation step (Choe et al. 1998). Cloning and sequencing revealed all characteristic signature sequences, typical for microsomal cytochrome P450 monooxygenases, in the DWF4 protein. The protein exhibiting greatest homology (43%) to DWF4 was the CPD protein (CYP90 A) from *Arabidopsis*, which is also involved in brassinosteroid biosynthesis (see below). Therefore, DFW4 was named CYP90B. From these genetic, molecular and biochemical studies, the 22α-hydroxylase has been identified as a cytochrome P450 steroid 22α-hydroxylase. The low level of the *DWF4* transcript supported earlier speculations that the 22α-hydroxylation might be the rate-limiting step in brassinosteroid biosynthesis.

Feeding experiments using deuterium-labeled cathasterone were conducted and teasterone and typhasterol were detected by GC-SIM analysis. The result showed side chain hydroxylation in position 22α followed by 23α-hydroxylation. This was supported by the finding that 23α-hydroxy-6-oxocampestanol, $\Delta^{22(23)}$-6-oxocampestanol and 22,23-epoxy-6-oxocampestanol, which are other possible precursors of teasterone, were not incorporated. The *Arabidopsis constitutive photomorphogenesis* and *dwarfism* (*cpd*) mutant (Szekerez et al. 1996) and the allelic *cbb3* (Kauschmann et al. 1996) were among the first brassinosteroid biosynthetic mutants to be characterized. Feeding studies showed that treatment with C-23 hydroxylated brassinosteroids including teasterone, 3-dehydroteasterone (3-dehydroTE), typhasterol and castasterone rescued the *cpd* dwarf phenotype to wild-type phenotype, whereas brassinosteroids, which do not carry a hydroxyl at C-23, did not alter the *cpd* phenotype (Szekerez et al. 1996). Normalization of *cbb3* by administration of 24-epicastasterone and 24-epibrassinolide both morphologically and on the level of gene expression was reported (Kauschmann et al. 1996). Molecular analysis showed that the *CPD* gene possessed all functionally important domains of cytochrome P450 monooxygenases and, in addition, homology to specific domains of mammalian steroid hydroxylases. Due to its overall identity of less than 40% with other P450 s, the CPD protein was considered the first member of a new P450 family, CYP90 (Szekerez et al. 1996).

Other steroidal biosynthetic pathways, e.g. cardenolides (Kawaguchi et al. 1993) and ecdysteroids (Milner and Rees 1985), report that the inversion of 3β-hydroxy function to 3α-hydroxy proceeds via 3-keto intermediates. $[^2H_6]$3-Dehydroteasterone was converted to typhasterol as a major metabolite and to teasterone as a minor one in cultured cells of *C. roseus*. Although 3-dehydroteasterone was not detectable after feeding both $[^2H_6]$teasterone and $[^2H_6]$typhasterol in *Catharanthus roseus*, this finding suggested involvement of the 3,6-diketo intermediate in reversible interconversion between both epimers (Suzuki et al. 1994a). As demonstrated in tomato cell cultures, conversion of tritium-labeled 3-dehydro-24-epiteasterone to the 3β- and 3α-epimers is strongly influenced by regio- and stereoselective 3β-glycosidation of 24-epiteasterone

(Kolbe et al. 1998b). Isolation of a 3,6-diketo intermediate in the metabolism of tritium-labeled 24-epicastasterone in tomato cell cultures supported the proposed operation of a redox mechanism in the inversion of configuration at C-3 of brassinosteroids (Hai et al. 1996). Very recent feeding experiments using *Arabidopsis* seedlings have provided evidence for conversion of [$^2H_6$]3-dehydroteasterone to typhasterol and teasterone at a ratio of 33:1 (Noguchi et al. 2000). 24-Epiteasterone, 3-dehydro-24-epiteasterone and 24-epityphasterol have been incubated with enzyme preparations from cell suspension cultures of *Lycopersicon esculentum*. When 3-dehydro-24-epiteasterone was incubated with a microsomal fraction, 24-epiteasterone was formed in an NADPH-dependent reaction. Formation of 24-epityphasterol from the same substrate was achieved by means of a cytosolic fraction preferring NADH as a cofactor (Winter et al. 1999). The same cytosolic fraction was able to dehydrogenate 24-epityphasterol to 3-dehydro-24-epiteasterone in the presence of NAD. These results suggest that at least two enzymes are involved in epimerization at C-3 in *L. esculentum*. Conversion of 24-epiteasterone to 3-dehydro-24-epiteasterone in the presence of NAD was demonstrated by a cytosolic fraction of *Arabidopsis* (Stuendl and Schneider, submitted).

Using a cytosolic fraction of cultured cells of the liverwort *Marchantia polymorpha*, Park et al. (1999) demonstrated conversion of teasterone to typhasterol. In contrast to studies in *Arabidopsis* (Stuendl and Schneider, submitted), teasterone was converted in vitro to 3-dehydroteasterone without the presence of any cofactor. Incubation of the latter product under identical conditions yielded typhasterol.

Upon feeding of [$^2H_6$]teasterone, tritium-labeled typhasterol and castasterone were detected as metabolites in *Arabidopsis* (Noguchi et al. 2000). Clearly, after C-3-epimerization, another hydroxylation at the brassinosteroid skeleton takes place at C-2α of typhasterol yielding castasterone. This conversion was found earlier in cultured cells of *C. roseus* (Suzuki et al. 1994b), and seedlings of *C. roseus, Oryza sativa* and *Nicotiana tabacum* (Suzuki et al. 1995b). [$^2H_6$]Typhasterol was also converted to castasterone in *bri1-5* mutants of *Arabidopsis* but not in wild-type seedlings (Noguchi et al. 2000). The *dwf8* mutant of *Arabidopsis* seems to be a candidate for being blocked in the hydroxylation of typhasterol because this compound accumulated in *dwf8* tissue (see Clouse and Feldman, in: Sakurai et al. 1999). However, more investigation is necessary.

In summary, as shown in Fig. 4, the route from campestanol $\rightarrow$ 6α-hydroxycampestanol $\rightarrow$ 6-oxocampestanol $\rightarrow$ cathasterone $\rightarrow$ teasterone $\rightarrow$ 3-dehydroteasterone $\rightarrow$ typhasterol $\rightarrow$ castasterone is now generally accepted as the early C-6-oxidation pathway. Although suggestions about which enzymes catalyze that part of the brassinosteroid bio-

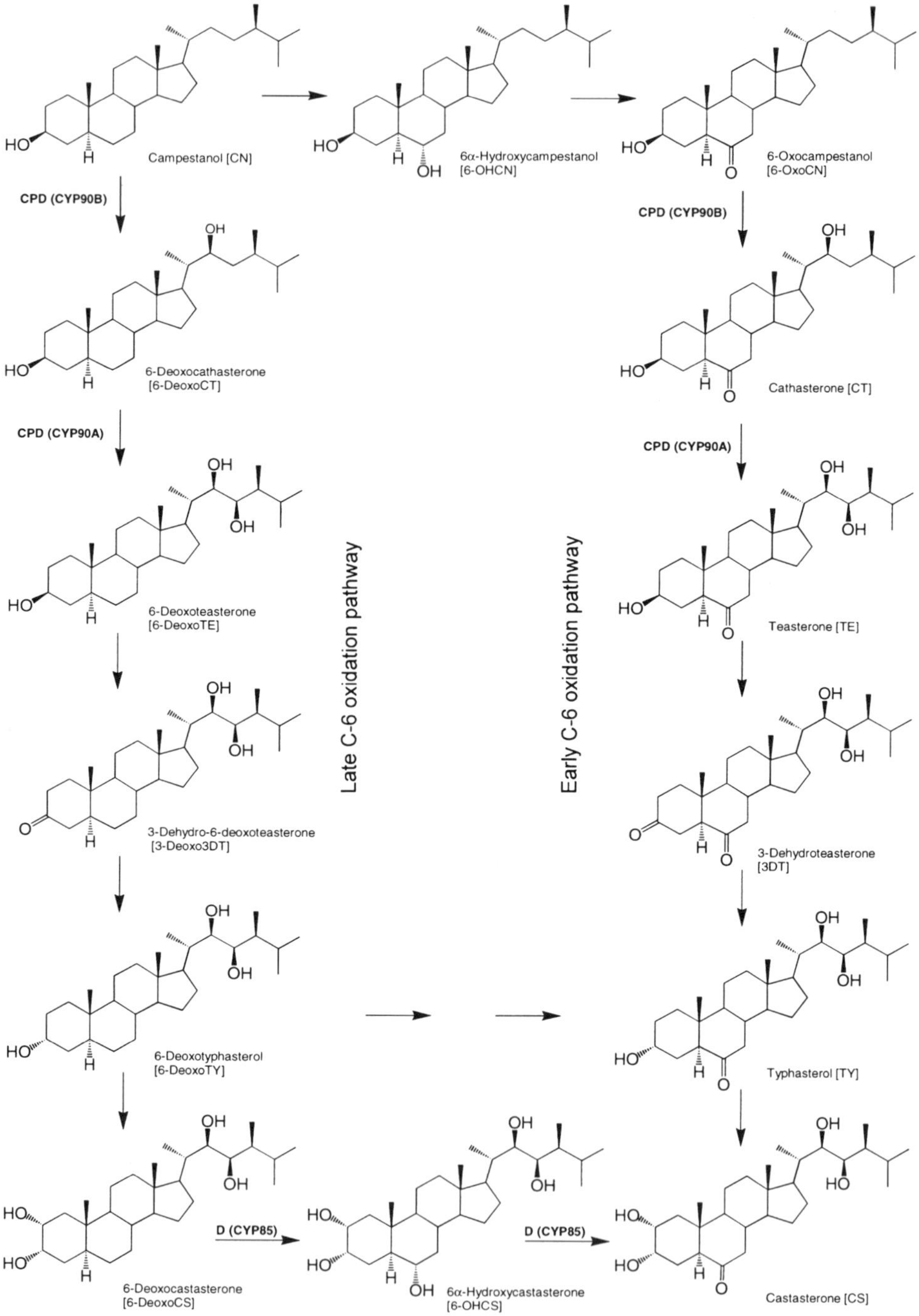
Campestanol [CN]
6α-Hydroxycampestanol [6-OHCN]
6-Oxocampestanol [6-OxoCN]
CPD (CYP90B)
6-Deoxocathasterone [6-DeoxoCT]
CPD (CYP90B)
Cathasterone [CT]
CPD (CYP90A)
6-Deoxoteasterone [6-DeoxoTE]
CPD (CYP90A)
Teasterone [TE]
Late C-6 oxidation pathway
Early C-6 oxidation pathway
3-Dehydro-6-deoxoteasterone [3-Deoxo3DT]
3-Dehydroteasterone [3DT]
6-Deoxotyphasterol [6-DeoxoTY]
Typhasterol [TY]
6-Deoxocastasterone [6-DeoxoCS]
D (CYP85)
6α-Hydroxycastasterone [6-OHCS]
D (CYP85)
Castasterone [CS]

Fig. 4. Conversion of campestanol to castasterone through the early and late C-6 oxidation pathway

synthetic pathway exist for most of these steps, until recently only two P450 steroid monooxygenases (CYP90A and CYP90B) had been clearly identified and a number of other enzymes partially characterized.

The occurrence of 6-deoxobrassinosteroids such as 6-deoxocastasterone (6-deoxoCS), 6-deoxoteasterone (6-deoxoTE), and 6-deoxotyphasterol (6-deoxoTY) in several plants (for review see: Sakurai et al. 1999) suggested a parallel or alternative brassinosteroid biosynthetic route (Griffiths et al. 1995). This late C-6 oxidation pathway was worked out mainly in *Catharanthus roseus*, *Arabidopsis thaliana* and *Lycopersicon esculentum*.

From their chemical structures, 6-deoxocathasterone (6-deoxoCT) and 6-deoxoteasterone (Fig. 4), which have been identified as endogenous brassinosteroids in tomato (Bishop et al. 1999) and *C. roseus* (Fujioka et al. 2000a), were expected to be formed from campesterol. This has been confirmed recently by administration of [$^2H_6$]campestanol to *Arabidopsis* (Noguchi et al. 2000). 6-Deoxocathasterone and 6-deoxoteasterone were found to rescue the *dwf4* mutant even more efficiently than do corresponding intermediates belonging to the early C-6 oxidation pathway. It was concluded that the DWF4 cytochrome P450 monooxygenase protein (CYP90B, see above) catalyzes both 22α-hydroxylation in early and late C-6 oxidation.

The severe phenotypic effects of the *cpd* mutation imply that the *cpd* plants do not contain significant concentrations of biologically active brassinosteroids. Hence, C-23 hydroxylation of the side chain seems to be blocked both in early and late C-6 oxidation pathways. The CPD protein (CYP90A, see above) therefore was considered to have dual function in both branches of brassinolide biosynthesis. In tomato, which seem to be depleted in the intermediates of the early C-6 oxidation route (Bishop et al. 1999), the *dumpy* (*dpy*) mutant was identified (Koka et al. 2000). Application of brassinolide and castasterone rescued the *dpy* phenotype, as did C-23 hydroxylated 6-deoxo intermediates. Campesterol, campestanol, and 6-deoxocathasterone failed to rescue, suggesting that *dpy* may be a tomato analogue of the *cpd* mutant of *Arabidopsis*.

Administration of deuterium-labeled 6-deoxoteasterone and 3-dehydro-6-deoxoteasterone (3-deoxo3DT) revealed conversion to 3-deoxotyphasterol in cultured cells of *C. roseus* (Choi et al. 1997). Another series of feeding experiments with 6-deoxoteasterone, 3-dehydro-6-deoxoteasterone, and 6-deoxotyphasterol in *Arabidopsis* indicated, analogously with the early C-6 oxidation pathway, reversible epimerization of the 3-hydroxy group (Noguchi et al. 2000). 6-Deoxotyphasterol was converted to 6-deoxocastasterone both in *C. roseus* (Choi et al. 1997) and *Arabidopis* (Noguchi et al. 2000).

Historically, the biosynthesis of deuterium-labeled brassinolide from exogenously administered 6-deoxocastasterone via castasterone in cultured cells and seedlings of *Catharanthus roseus* and conversion of 6-

deoxocastasterone to castasterone in *Oryza sativa* and *Nicotiana tabacum* was the first experimental evidence for the existence of the late C-6 oxidation pathway (Choi et al. 1996). Conversion of 6-deoxocastasterone to castasterone via 6α-hydroxycastasterone (6-OHCS) has been found recently in brassinosteroid-accumulating *bri1-5* mutants of *Arabidopsis*. In these experiments, from both the feeding of deuterium-labeled 6-deoxocastasterone and 6α-hydroxycastasterone, castasterone was detected as a metabolite (Noguchi et al. 2000). Conversion of 6α-hydroxycastasterone to castasterone was also found in *C. roseus* (Fujioka et al. 2000a).

The tomato *dwarf* (*d*) mutation has been known since the middle of the nineteenth century. The *Dwarf* (*D*) gene was isolated by heterologous transposon tagging and determined by means of the amino acid sequence as a cytochrome P450 monooxygenase (CYP85). Involvement in brassinosteroid biosynthesis has been proposed as the biological function of this protein (Bishop et al. 1996). Quantitative analysis of endogenous brassinosteroid levels in tomato plants showed mainly the presence of 6-deoxobrassinosteroids. Moreover, the *extreme dwarf* (d^x) allele lacks castasterone but does accumulate 6-deoxocastasterone. Exogenous application of C-6 oxobrassinosteroids to d^x plants enhanced hypocotyl elongation while 6-deoxocastasterone had little effect, indicating that the Dwarf/CYP85 protein catalyzed C-6 oxidation. Functional expression in yeast confirmed this suggestion. In these experiments, 6α-hydroxycastasterone and castasterone were identified, confirming the involvement of Dwarf/CYP85 enzyme both in C-6α hydroxylation and oxidation of hydroxyl in the conversion of 6-deoxocastasterone to castasterone (Bishop et al. 1999). No conversion of campestanol was observed when it was incubated with the heterologous enzyme.

In addition to the early C-6 oxidation route (see above) and the late C-6-oxidation sequence campestanol → 6-deoxocathasterone → 6-deoxoteasterone → 3-dehydro-6-deoxoteasterone → 6-deoxotyphasterol → 6-deoxocastasterone → 6α-hydroxycastasterone → castasterone (Fig. 4), cross-links between both branches seem to exist. This was demonstrated by feeding experiments using deuterium-labeled 6-deoxoteasterone, 3-dehydro-6-deoxoteasterone, and 6-deoxotyphasterol, in which typhasterol was identified as a metabolite (Noguchi et al. 2000; Fig. 4).

7 Final Biosynthetic Steps – Castasterone to Brassinolide

Biological Baeyer-Villiger oxidation of the six-membered B-ring (containing an oxo group at position 6) of castasterone to give the seven-membered 6a-oxalactone structure is considered the final step in brassinolide biosynthesis. Castasterone has been suggested as the immediate

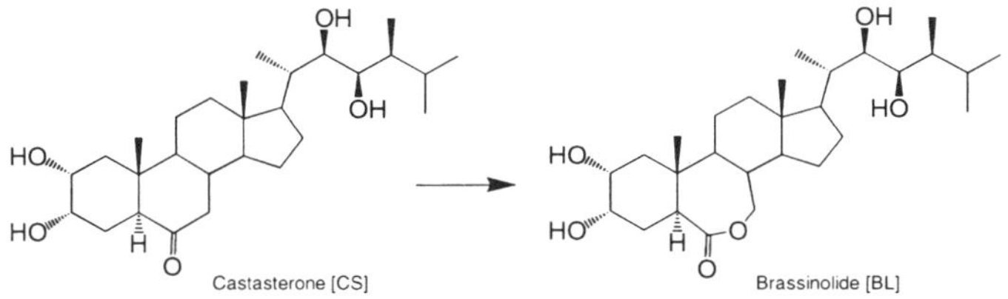

Castasterone [CS] → Brassinolide [BL]

Fig. 5. Conversion of castasterone to brassinolide

precursor of brassinolide, since it has been isolated as the second brassinosteroid (Yokota et al. 1982). This hypothesis was proven by feeding tritium-labeled castasterone to crown gall cells of *Catharanthus roseus* (Yokota et al. 1990) and further confirmed in non-transformed *C. roseus* cell cultures and seedlings using deuterated castasterone as a precursor (Suzuki et al. 1993; Suzuki et al. 1995b; Fig. 5).

However, attempts to demonstrate conversion of castasterone to brassinolide in other species, e.g. seedlings of *Nicotiana tabacum* and *Oryza sativa* (Suzuki et al. 1995b) and wild-type seedlings of *Arabidopsis* (Noguchi et al. 2000) failed. Some cases found epimerization of the 3α-hydroxyl group instead. Cell suspension cultures of *Ornithopus sativus*, for example, did not form 24-epibrassinolide when fed 24-epicastasterone but, after epimerization of 3α-hydroxyl, formed mainly fatty acid esters and pregnane-type metabolites (Kolbe et al. 1996). However, a cytosolic fraction of cultured cells of *Lycopersicon esculentum* was successfully used in vitro to demonstrate the Baeyer-Villiger reaction of a brassinosteroid, 24-epicastasterone (Winter et al. 1999). In vivo feeding experiments using the same species and the same substrate failed to demonstrate this reaction, but did result in deactivation reactions such as glycoside formation (Hai et al. 1996). Brassinosteroid-insensitive (*bri1*) mutants, which accumulated very high concentrations of endogenous brassinosteroids (Noguchi et al. 1999b) and which were probably upregulated in brassinosteroid biosynthesis, were successfully used to demonstrate conversion of castasterone to brassinolide (Noguchi et al. 2000). Mutants exhibiting a lesion in the Baeyer-Villiger oxidation step hitherto are not known. The only information on the enzyme involved in the Baeyer-Villiger reaction comes from the above mentioned in vitro-transformation studies (Winter et al. 1999) of 24-epicastasterone in *L. esculentum* cell cultures, indicating an NADPH-dependent cytosolic protein.

8 Hypothetical Subpathways of Brassinolide Biosynthesis

The fact that 22-hydroxycampesterol (22-OHCR) was able to rescue *dwf4* plants suggests that the metabolism of that compound may represent a new subpathway in brassinosteroid biosynthesis (Choe et al. 1998). The *dwf4* mutant was also rescued by (22*S*,24*R*)-22-hydroxyergost-4-en-3-one (22-OH-4-en-3-one) and (22*S*,24*R*)-22-hydroxy-5α-ergostan-3-one (22-OH-3-one), suggesting a subroute from 22α-hydroxycampesterol through these intermediates and a merger with the late C-6 oxidation pathway at the 6-deoxocathasterone stage (Ephritikhine et al. 1999b; Fig. 6). In that series of experiments, root growth and hypocotyl elongation of *det2* was rescued by administration of 22-OH-3-one but did not respond to 22-OHCR and 22-OH-4-en-3-one. In contrast, the recently described *sax1* dwarf mutant of *Arabidopsis* (Ephritikhine et al. 1999a) was rescued by 22-OH-4-en-3-one, 22-OH-3-one and downstream intermediates of brassinosteroid biosynthesis, such as 6-deoxocathasterone and cathasterone. Thus, the SAX1 protein catalyzes the oxidation and isomerization of 22α-hydroxycampesterol to the 22-OH-4-en-3-one (and probably other 3β-hydroxyl-$\Delta^{5(6)}$-precursors to 3-oxo-$\Delta^{4(5)}$-steroids). However, confirmation of this finding and the hypothetical subpathway 22α-hydroxycampesterol → 22-OH-4-en-3-one → 22-OH-3-one → 6-deoxocathasterone would require the natural occurrence of these intermediates, which have hitherto not been found in plants.

Accumulation of the $\Delta^{24(28)}$ intermediates, 24-methylenecholest-4-ene-3-one and 24-methylene-5α-cholestane-3β-ol, in the *dim* mutant, an allele of *dwf1* which is blocked in isomerization and reduction of 24-methylenecholesterol (see above), revealed the induction of an alterna-

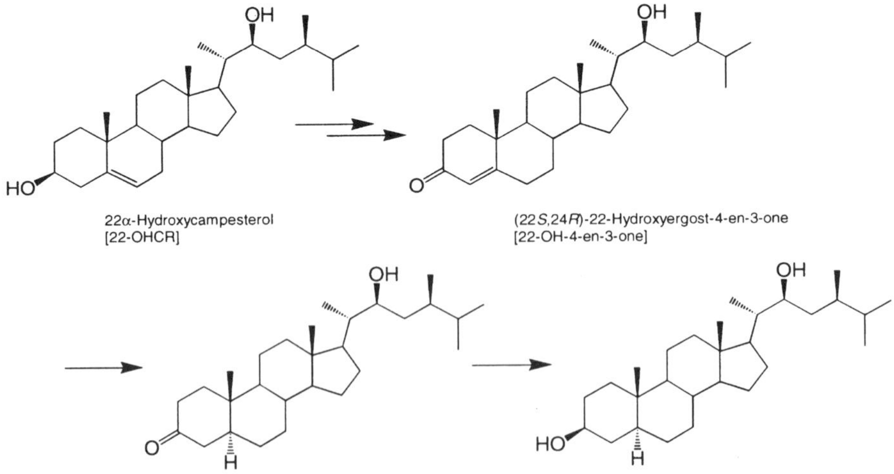

22α-Hydroxycampesterol [22-OHCR]

(22*S*,24*R*)-22-Hydroxyergost-4-en-3-one [22-OH-4-en-3-one]

(22*S*,24*R*)-22-Hydroxy-5α-ergostan-3-one [22-OH-3-one]

6-Deoxocathasterone [6-DeoxoCT]

Fig. 6. Hypothetical conversion of 22α-hydroxycampesterol to 6-deoxocathasterone

tive route (Klahre et al. 1998). Whether or not 24-methylene-5α-cholestane-3β-ol can be converted to campestanol in wild-type plants is an interesting question; if so, it would add another subroute to the network of brassinosteroid biosynthesis.

How endogenous brassinosteroids, which do not fit into the established pathways, are formed remains to be explained. A number of suggestions for hypothetical biosynthetic relationships with established pathways exist. For example, 3-epi-6-deoxocathasterone, which was recently found in *C. roseus* might be a precursor of 6-deoxotyphasterol (Fujioka et al. 2000a). Otherwise, 3-epi-6-deoxocathasterone could be considered a metabolite of 6-deoxocathasterone, which is formed by the late C-6 oxidation pathway.

9 Biosynthesis of Brassinosteroids Possessing Different Side Chain Structures

The biosynthetic pathways described above deal mainly with the C-28 brassinosteroids. Brassinolide, the most biologically active member of the C-28 brassinosteroids, is formed from C-28 precursors. Parallel pathways are assumed to be operating for brassinosteroids having different side chain skeletons, e.g. different numbers of carbon atoms (Takatsuto et al. 1999) and/or different stereochemistry of 24*R*-or 24*S* alkyl groups as, for example, 24-epibrassinosteroids (Winter et al. 1999). Based on common occurrence, the route sitosterol → 24-ethylcholest-4-en-3β-ol (not detected) → 24-ethylcholest-4-en-3-one → 24-ethyl-5α-cholestan-3-one → sitostanol was suggested as a part of the early biosynthesis of C-29 brassinosteroids in seeds of *Triticum aestivum* and *Setaria italica* (Takatsuto et al. 1999). Brassinosteroids therefore are suggested to be biosynthesized from corresponding sterols without modification of the side chain skeleton. However, 28-norcastasterone, which was assumed (but not experimentally demonstrated) to be derived from cholesterol (Yokota et al. 1997), was found very recently as a metabolite of castasterone in seedlings of *Arabidopsis thaliana*, *Oryza sativa* and *Lycopersicon esculentum*, and in cultured cells of *C. roseus* (Fujioka et al. 2000b; Fig. 7).

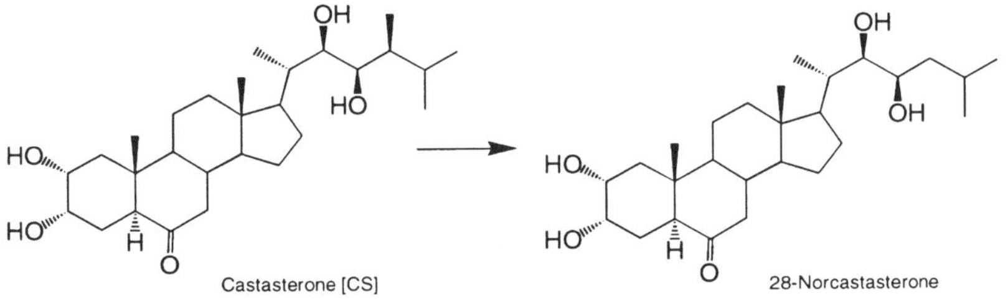

Castasterone [CS] → 28-Norcastasterone

Fig. 7. Conversion of castasterone to 28-norcastasterone

This unexpected result is the only known case of the side chain skeleton of a brassinosteroid changing late in biosynthesis. Furthermore, typhasterol was predicted as a precursor of 28-nortyphasterol, a new brassinosteroid, in *Arabidopsis* by analogy to 28-norcastasterone biosynthesis from castasterone. Moreover, co-occurrence of 28-norcastasterone and 28-nortyphasterol gave rise to speculations that there might be another parallel subpathway to 28-norcastasterone from 28-nortyphasterol (Fujioka et al. 2000b).

10 Conclusions

Among the six classes of plant hormones, brassinosteroids were the last to be discovered. Only recently have they gained acceptance as phytohormones. However, tremendous progress has been made in the elucidation of biosynthetic pathways and genetic characterization of mutants. Individual biosynthetic steps are still being uncovered which combine to form subpathways resulting in a network of brassinosteroid biosynthesis. A number of enzymes have been characterized, based on sequence homology with known proteins, or by means of biochemical methods. However, more detailed knowledge is required to understand the enzyme properties and, for example, their substrate specificity with respect to involvement in analogous steps of parallel subpathways. Complete knowledge of the *Arabidopsis* genome will greatly aid further research in this field. Future studies could address the role of early and late C-6 oxidation (and possible further) pathways in plants. Light dependence and species specificity are two possible explanations for how such different routes might operate. Detailed knowledge of biosynthetic pathways is a prerequisite for future progress in the elucidation of molecular mechanisms of brassinosteroid action. The identification of brassinosteroids as plant hormones, active in cell elongation, morphogenesis and stress response, also leads to questions about the evolution of steroid hormones in general.

Acknowledgements. Emily Wheeler, Jena, is gratefully acknowledged for editorial and linguistic support in the preparation of the manuscript. The Deutsche Forschungsgemeinschaft, Bonn, and the Fonds der Chemischen Industrie, Frankfurt, financially supported our own work in this field.

References

Adam G, Schmidt J, Schneider B (1999) Brassinosteroids. In: Herz W, Falk H, Kirby GW, Moore RE, Tamm Ch (eds) Progress in the chemistry of organic natural products, vol 78. Springer, Berlin Heidelberg New York, pp 1–46

Altmann T (1999) Molecular physiology of brassinosteroids revealed by the analysis of mutants. Planta 208:1–11

Arigoni D, Sagner S, Latzel C, Eisenreich W, Bacher A, Zenk MH (1997) Terpenoid biosynthesis from 1-deoxyxylulose in higher plants by intramolecular skeletal rearrangement. Proc Natl Acad Sci USA 94:10600–10605

Asami T, Yoshida S (1999) Brassinosteroid biosynthesis inhibitors, TIPS 4:348–353

Azpiroz R, Wu YW, LoCascio JC, Feldmann KA (1998) An *Arabidopsis* brassinosteroid-dependent mutant is blocked in cell elongation. Plant Cell 10:219–230

Benveniste P (1986) Sterol biosynthesis. Ann Rev Plant Physiol 37:275–307

Bishop GJ, Harrison K, Jones JDG (1996) The tomato *Dwarf* gene isolated by heterologous transposon tagging encodes the first member of a new cytochrome P450 family. Plant Cell 8:959–969

Bishop GJ, Nomura T, Yokota T, Harrison K, Noguchi T, Fujioka S, Takatsuto S, Jones JDG, Kamiya Y (1999) The tomato dwarf enzyme catalyses C-6 oxidation in brassinosteroid biosynthesis. Proc Natl Acad Sci USA 96:1761–1766

Brosa C (1999) Biological effects of brassinosteroids. Crit Rev Biochem Mol Biol 34:339–358

Choe S, Dilkes BD, Fujioka S, Takatsuto S, Sakurai A, Feldman KA (1998) The *DWF4* gene of *Arabidopsis* encodes a cytochrome P450 that mediates multiple 22α-hydroxylation steps in brassinosteroid biosynthesis. Plant Cell 10:231–243

Choe S, Noguchi T, Fujioka S, Takatsuto S, Tissier CP, Gregory BD, Ross AS, Tanaka A, Yoshida S, Tax FE, Feldmann KA (1999a) The *Arabidopsis* dwf7/ste1 mutant is defective in the Delta(7) sterol C-5 desaturation step leading to brassinosteroid biosynthesis. Plant Cell 11:207–221

Choe S, Dilkes BP, Gregory BD, Ross AS, Yuan H, Noguchi T, Fujioka S, Takatsuto S, Tanaka A, Yoshida S, Tax FE, Feldmann KA (1999b) The Arabidopsis *dwarf1* mutant is defective in the conversion of 24-methylenecholesterol to campesterol in brassinosteroid biosynthesis. Plant Physiol 119:897–907

Choe S, Tanaka A, Noguchi T, Fujioka S, Takatsuto S, Ross AS, Tax FE, Yoshida S, Feldman KA (2000) Lesions in the sterol Delta(7) reductase gene of *Arabidopsis* cause dwarfism due to a block in brassinosteroid biosynthesis. Plant J 21:431–443

Choi YH, Fujioka S, Harada A, Yokota T, Takatsuto S, Sakurai A (1996) A brassinolide biosynthetic pathway via 6-deoxocastasterone. Phytochemistry 43:593–596

Choi YH, Fujioka S, Nomura T, Harada A, Yokota T, Takatsuto S, Sakurai A (1997) An alternative brassinolide biosynthetic pathway via late C-6 oxidation. Phytochemistry 44:609–613

Cutler HC, Yokota T, Adam G (eds.) (1991) Brassinosteroids: chemistry, bioactivity and applications. ACS Symp Ser 474, Am Chem Soc, Washington, DC

Eisenreich W, Schwarz M, Cartayrade A, Arigoni D, Zenk MH, Bacher A (1998) The deoxyxylulose phosphate pathway of terpenoid biosynthesis in plants and microorganisms. Chem Biol 5:R221-R233

Ephritikhine G, Fellner M, Vannini C, Lapous D, Barbier-Brygoo H (1999a) The *sax1* dwarf mutant of *Arabidopsis thaliana* shows altered sensitivity of growth responses to abscisic acid, auxin, gibberellins and ethylene and is partially rescued by exogenous brassinosteroid. Plant J 18:303–314

Ephritikhine G, Pagant S, Fujioka S, Takatsuto S, Lapous D, Caboche M, Kendrick RE, Barbier-Brygoo H (1999b) The *sax1* mutation defines a new locus involved in the brassinosteroid biosynthesis pathway in *Arabidopsis thaliana*. Plant J 18:315–320

Feldman KA, Marks DM, Christianson ML, Quatrano RS (1989) A dwarf mutant of *Arabidopsis* generated by T-DNA insertion mutagenesis. Science 243:1351–1354

Fujioka S, Inoue T, Takatsuto S, Yanagisawa T, Yokota T, Sakurai A (1995) Identification of a new brassinosteroid, cathasterone, in cultured cells of *Catharanthus roseus* as a biosynthetic precursor of teasterone. Biosci Biotech Biochem 59:1543–1547

Fujioka S, Li JM, Choi YH, Seto H, Takatsuto S, Noguchi T, Watanabe T, Kuriyama H, Yokota T, Chory J, Sakurai A (1997) The *Arabidopsis deetiolated2* mutant is blocked in early brassinosteroid biosynthesis. Plant Cell 11:1951–1962

Fujioka S, Noguchi T, Watanabe T, Takatsuto S, Yoshida S (2000a) Biosynthesis of brassinosteroids in cultured cells of *Catharanthus roseus*. Phytochemistry 53:549–553

Fujioka S, Noguchi T, Sekimoto M, Takatsuto S, Yoshida S (2000) 28-norcastasterone is biosynthesized from castasterone. Phytochemistry 55:97–101

Griffiths PG, Sasse JM, Yokota T, Cameron DW (1995) 6-Deoxotyphasterol and 3-dehydro-6-deoxoteasterone, possible precursors of brassinosteroids in the pollen of *Cupressus arizonica*. Biosci Biotech Biochem 59:956–959

Grove MD, Spencer GF, Rowedder WK, Mandava N, Worley JF, Warte JD, Steffens GL, Flippen-Anderson JL, Cook, JK (1979) Brassinolide, a plant growth-promoting steroid isolated from *Brassica napus* pollen. Nature 281:216–217

Hai T, Schneider B, Porzel A, Adam G (1996) Metabolism of 24-epicastasterone in cell suspension cultures of *Lycopersicon esculentum*. Phytochemistry 41:197–201

Kauschmann A, Jessop A, Koncz C, Szekeres M, Willmitzer L Altmann T (1996) Genetic evidence for an essential role of brassinosteroids in plant development. Plant J 9:701–713

Kawaguchi K, Hirotani M, Furuya T (1993) *Strophanthus* species (members of the dogbane family): in vitro culture and the production of cardenolides. In: Bajaj YPS (ed) Biotechnology in agriculture and forestry, vol 21. Medicinal and aromatic plants. Springer, Berlin Heidelberg New York, pp 371–386

Khripach VA, Lakhvich FA, Zhabinskii VN (1993) Brassinosteroids. Science and Technique, Minsk

Khripach VA, Zhabinskii VN, de Groot Æ (1999) Brassinosteroids, Steroidal plant hormones. Academic Press, San Diego

Khripach V, Zhabinskii V, de Groot Æ (2000) Twenty years of brassinosteroids: steroidal plant hormones warrant better crops for the XXI century. Ann Bot 86:441–447

Klahre U, Noguchi T, Fujioka S, Takatsuto S, Yokota T, Nomura T, Yoshida S, Chua NH (1998) The Arabidopsis *DIMINUTO/DWARF1* gene encodes a protein involved in steroid synthesis. Plant Cell 10:1677–1690

Koka CV, Cerny RE, Gardner RG, Noguchi T, Fujioka S, Takatsuto S, Yoshida S, Clouse SD (2000) A putative role for the tomato genes *DUMPY* and *CURL-3* in brassinosteroid biosynthesis and response. Plant Physiol 122:85–98

Kolbe A, Schneider B, Porzel A, Adam G (1996) Metabolism of 24-epicastasterone and 24-epibrassinolide in cell suspension cultures of *Ornithopus sativus*. Phytochemistry 41:163–167

Kolbe A, Schneider B, Voigt B, Adam G (1998a) Labelling of biogenetic brassinosteroid precursors. J Lab Comp Radiopharm 41:131–137

Kolbe A, Schneider B, Porzel A, Adam G (1998b) Metabolic inversion of the 3-hydroxy function in brassinosteroids. Phytochemistry 48:467–470.

Li JM, Nagpal P, Vitart V, McMorris TC, Chory J (1996) A role for brassinosteroids in light-dependent development of *Arabidopsis*. Science 272:398–401

Li JM, Biswas MG, Chao A, Russell DW, Chory J (1997) Conservation of function between mammalian and plant steroid 5α-reductases. Proc Natl Acad Sci USA 94:3554–3559

Li JM, Chory J (1999) Brassinosteroid actions in plants, J Exp Bot 50:275–282

Lichtenthaler HK, Schwender J, Disch A, Rohmer (1997) Biosynthesis of isoprenoids in higher plants proceeds via a mevalonate-independent pathway. FEBS Lett 400:271–274

Milner NP, Rees HH (1985) Involvement of 3-dehydroecdysone in the 3-epimerization of ecdysone. Biochem J 231:369–374

Mitchell JW, Whitehead MR (1941) Responses of vegetative parts of plants following application of extract of pollen from *Zea mays* Bot Gaz 102:770–791

Noguchi T, Fujioka S, Takatsuto S, Sakurai A, Yoshida S, Li J, Chory J (1999a) *Arabidopsis det2* is defective in the conversion of (24*R*)-24-methylcholest-4-en-3-one to (24*R*)-24-methyl-5α-cholestan-3-one in brassinosteroid biosynthesis. Plant Physiol 120:833–840

Noguchi T, Fujioka S, Choe S, Takatsuto S, Yoshida S, Yuan H, Feldman KA, Tax FE (1999b) Brassinosteroid-insensitive dwarf mutants of *Arabidopsis* accumulate brassinosteroids. Plant Physiol 121:743–752

Noguchi T, Fujioka S, Choe S, Takatsuto S, Tax FE, Yoshida S, Feldman KA (2000) Biosynthetic pathway of brassinolide in *Arabidopsis*. Plant Physiol 124:201–209

Nomura T, Nakayama M, Reid JB, Takeuchi Y, Yokota T (1997) Blockage of brassinosteroid biosynthesis and sensitivity causes dwarfism in *Pisum sativum*. Plant Physiol 113:31–37

Nomura T, Kitasaka Y, Takatsuto S, Reid JB, Fukami M, Yokota T (1999) Brassinosteroid/sterol synthesis and plant growth as affected by *lka* and *lkb* mutations of pea. Plant Physiol 119:1517–1526

Park S-H, Han, K-S, Kim T-W, Shim J-K, Takatsuto S, Yokota T, Kim S-K (1999) In vivo and in vitro conversion of teasterone to typhasterol in cultured cells of *Marchantia polymorpha*. Plant Cell Physiol 40:955–960

Rohmer M (1999) The discovery of a mevalonate-independent pathway for isoprenoid biosynthesis in bacteria, algae and higher plants. Nat Prod Rep 16:565–574

Sakurai A (1999) Brassinosteroid biosynthesis. Plant Physiol Biochem 37:351–361

Sakurai A, Yokota T, Clouse SD (eds) (1999) Brassinosteroids: steroidal plant hormones. Springer, Berlin Heidelberg New York

Schaeffer A, Bronner R, Benveniste P, Schaller H (2001) The ratio of campesterol to sitosterol that modulates growth in *Arabidopsis* is controlled by sterol methyltransferase 2-1. Plant J 25:605–615

Stündl U, Schneider B 3β-Brassinosteroid dehydrogenase activity in *Arabidopsis* and tomato (submitted)

Suzuki H, Fujioka S, Takatsuto S, Yokota T, Murofushi N, Sakurai A (1993) Biosynthesis of brassinolide from castasterone in cultured cells of *Catharanthus roseus*. J Plant Growth Regul 12:101–106

Suzuki H, Inoue T, Fujioka S, Takatsuto S, Yanagisawa T, Yokota T, Murofushi N, Sakurai A (1994a) Possible involvement of 3-dehydroteasterone in the conversion of teasterone to typhasterol in cultured cells of *Catharanthus roseus*. Biosci Biotech Biochem 58:1186–1188

Suzuki H, Fujioka S, Takatsuto S, Yokota T, Murofushi N, Sakurai A (1994b) Biosynthesis of brassinolide from teasterone via typhasterol and castasterone in cultured cells of *Catharanthus roseus*. J. Plant Regul 13:21–26

Suzuki H, Inoue T, Fujioka S, Saito T, Takatsuto S, Yokota T, Murofushi N, Yanagisawa T, Sakurai A (1995a) Conversion of 24-methylcholesterol to 6-oxo-24-methylcholestanol, a putative intermediate of the biosynthesis of brassinosteroids, in cultured cells of *Catharanthus roseus*. Phytochemistry 40:1391–1397

Suzuki H, Fujioka S, Takatsuto S, Yokota T, Murofushi N, Sakurai A (1995b) Biosynthesis of brassinolide in seedlings of *Catharanthus roseus, Nicotiana tabacum* and *Oryza sativa*. Biosci Biotech Biochem 59:168–172.

Szekeres M, Koncz C (1998) Biochemical and genetic analysis of brassinosteroid metabolism and function in *Arabidopsis*. Plant Physiol Biochem 36:145–155

Szekeres M, Nemeth K, Koncz-Kalman Z, Mathur J, Kauschmann A, Altmann T, Redei GP, Nagy F, Schell J, Koncz C (1996) Brassinosteroids rescue the deficiency of CYP90, a cytochrome P450, controlling cell elongation and de-etiolation in *Arabidopsis*. Cell 85:171–182

Takahashi T, Gasch, A, Nishizawa N, Chua N-H (1995) The *DIMINUTO* gene of *Arabidopsis* is involved in regulating cell elongation. Genes Dev 9:97–107

Takatsuto S, Ikekawa N (1986) Synthesis of deuterio-labelled brassinosteroids, [26,28-$^{2}H_{6}$]brassinolide, [26,28-$^{2}H_{6}$]castasterone, [26,28-$^{2}H_{6}$]typhasterol and [26,28-$^{2}H_{6}$]teasterone. Chem Pharm Bull 34:4045–4049

Takatsuto S, Kosuga N, Abe B, Noguchi T, Fujioka S, Yokota T (1999) Occurrence of potential brassinosteroid precursor steroids in seeds of wheat and foxtail millet. J Plant Res 112:27–33

Winter J, Schneider B, Meyenburg S, Strack D, Adam G (1999) Monitoring brassinosteroid biosynthetic enzymes by fluorescent tagging and HPLC analysis of their substrates and products. Phytochemistry 51:237–242

Yamada J, Morisaki M, Iwai K, Hamada H, Sato N, Fujimoto Y (1997) 24-methyl and 24-ethyl-$\Delta^{24(25)}$-cholesterols as immediate biosynthetic precursors of 24-alkylsterols in higher plants. Tetrahedron 53:877–884

Yokota T (1997) The structure, biosynthesis and function of brassinosteroids. TIBS 2:137–143

Yokota T (1999) Brassinosteroids. In: Hooykaas PJJ, Hall MA, Libbenga KR (eds) Biochemistry and molecular biology of plant hormones. pp 277–293

Yokota T, Arima M, Takahashi N (1982) Castasterone, a new phytosterol with plant hormone potency, from chestnut insect gall. Tetrahedron Lett 23:1275–1278

Yokota T, Ogino Y, Takahashi N, Saimoto H, Fujioka S, Sakurai A (1990) Brassinolide is biosynthesized from castasterone in *Catharanthus roseus* crown gall cells. Agric Bio Chem 54:1107–1108

Yokota T, Ogino Y, Suzuki H, Takahashi N, Saimoto H, Fujioka S, Sakurai A (1991) Metabolism and biosynthesis of brassinosteroids. In: Cutler HC, Yokota T, Adam G (eds) Brassinosteroids: chemistry, bioactivity and applications. ACS Symp Ser 474, Am Chem Soc, Washington, DC, pp 86–96,

Yokota T, Nomura T, Nakayama M (1997) Identification of brassinosteroids that appear to be derived from campesterol and cholesterol in tomato shoots. Plant Cell Physiol 38:1291–1294

Priv.-Doz. Dr. Bernd Schneider
Max-Planck-Institute for Chemical Ecology
Carl-Zeiss-Promenade 10
07745 Jena, Germany
e-mail: schneider@ice.mpg.de

Ecology

Stomatal Water Relations and the Control of Hydraulic Supply and Demand

By Thomas N. Buckley and Keith A. Mott

1 Introduction

Stomata have fascinated plant biologists for well over 100 years. It is difficult to think of another plant system that responds to so many factors or displays such complexity at so many levels. Indeed, when one considers the number of feedback loops involving stomatal conductance and all of the potential interactions among these feedbacks, it is really quite remarkable that stomata work at all.

There is now general agreement about the environmental factors to which stomata respond, and the advent of relatively low-cost, easy to use gas-exchange systems and porometers has resulted in a surfeit of data describing stomatal conductance responses in natural and laboratory situations. Despite this, the mechanisms by which stomata respond to environmental factors are largely unknown. It has been clear for some time that active regulation of guard cell osmotic pressure plays a large role in most stomatal responses, and there has been some recent progress in elucidating the signal transduction chains and ion transport processes responsible for this aspect of stomatal function. It seems likely that many of these subcellular processes will be worked out in the near future. However, in order to be useful in predicting and interpreting gas exchange, these metabolic processes must be translated into dynamics of stomatal conductance at larger scales. This in turn requires a detailed understanding of the hydraulic factors that link stomatal aperture to guard cell osmotic content and to the water relations of the epidermis. Short-term stomatal responses to perturbations in hydraulic supply or demand are an appropriate context in which to develop this understanding, not only because of the ecological significance of these stomatal responses and their consequent need for explanation, but also because they provide a window on epidermal hydraulics: the system is prodded from different angles, one by one, producing apparently disparate responses that lend themselves to logical synthesis. Our goal in this review is a parsimonious, synthetic, mechanistic explanation of these responses.

Progress in Botany, Vol. 63

2 Hydraulics of Stomatal Responses to Environmental Factors

Although most stomatal responses are driven ultimately by changes in guard cell solute concentration, stomata are fundamentally hydraulic entities, and it is impossible to understand conductance responses to environmental perturbations without taking hydraulics into account. Stomatal dynamics emerge from a highly connected hydraulic medium: water evaporates from a mesh of mesophyll and epidermal cells to which guard cells are hydraulically slaved. Two factors ensure the dynamic and spatial significance of hydraulic interactions among stomata. First, individual stomatal apertures are controlled by both guard cell turgor and epidermal cell turgor, which affect aperture in opposite ways but are hydraulically governed by the same sources and sinks. Second, recent data suggest that in some species the proximity of adjacent stomata causes neighboring guard cell pairs to be, in effect, partially slaved to one another for access to water. It is now becoming apparent that these factors interact in complex and often counterintuitive ways that cannot be understood or predicted without extending the spatial context of stomatal physiology to at least the leaf scale.

An important facet of stomatal hydraulics is that the effect of epidermal turgor on stomatal aperture is greater than that of guard cell turgor by a factor of 1.5 or more (Sharpe et al. 1987; Franks et al. 1998). Thus, for equal increases in guard cell and epidermal cell turgor, stomatal aperture will decrease. In most stomatal movements, this 'mechanical advantage' enjoyed by the epidermal cells is more than offset by the much larger, osmotically induced changes in turgor pressure of the guard cells. Nevertheless, the effect of epidermal turgor on aperture is an important component of most stomatal responses. In light of the importance of both guard and epidermal cell turgor in determining stomatal aperture, four questions are of interest for understanding how aperture responds to environmental perturbations. First, how do guard and epidermal turgor pressures determine aperture? Second, what controls guard cell turgor? Third, what controls epidermal cell turgor? And fourth, how do the factors controlling these two parameters interact? These four questions will be discussed in order, below.

Although some models of stomatal functioning have assumed linear effects of guard and epidermal cell turgor (Cowan 1972; Delwiche and Cooke 1977; Haefner et al. 1997), recent work shows that these effects are, in fact, strongly nonlinear (Franks et al. 1998). In the three species examined by Franks et al. (1998), aperture increased in a saturating fashion with guard cell turgor when epidermal turgor was zero. Nonzero epidermal turgor depressed stomatal apertures for any value of guard cell turgor, and the overall relationship between aperture and guard cell turgor became sigmoidal (Fig. 1, redrawn from Franks et al. 1998, with

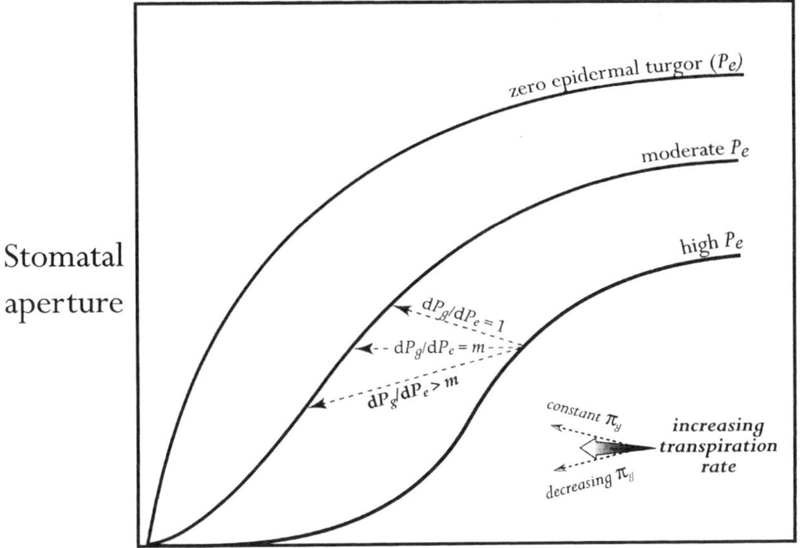

Fig 1. Diagram illustrating the effects of epidermal cell and guard cell turgor pressures (P_e and P_g) on stomatal aperture, based on the data of Franks et al (1998). Superposed on these relationships are three heuristic trajectories for increasing transpiration rate, for a steady-state system governed by the properties of stomatal mechanics implied by these data, and by the assumption that guard cells and epidermal cells share the same water potential at equilibrium. These trajectories show that in such a system, for stomatal aperture to decline as transpiration rate increases, guard cells must actively control osmotic pressure (π_g) in order to overcome the epidermal mechanical advantage, *m*, defined here as the ratio of the sensitivities of aperture to P_e and P_g. Were π_g allowed to remain constant by passive equilibrium, P_e and P_g would change by equal amounts, causing aperture to increase with transpiration rate (*upper trajectory*); and with some osmotic control, but not enough, aperture would remain constant as P_e and P_g declined (*middle trajectory*). The *lower trajectory*, annotated in **bold**, shows the correct response

annotations). These relationships must be taken into account as the turgor pressures of guard and epidermal cells vary, as discussed below.

Guard cell turgor is a function of both osmotic pressure and water potential. The former is controlled by the cellular processes that affect the osmotic concentration of the guard cells. These processes have been the subject of several recent reviews (Assmann 1999; Assmann and Shimazaki 1999; Blatt 2000) and will not be discussed here. Guard cell water potential is completely controlled by hydraulics, at least in the short term, and is therefore influenced by transpiration rate, soil water potential, and the conductance for water flow from the soil to leaf. All of these have important implications for stomatal responses, and will be discussed in detail in a later section. Transpiration rate, however, is interesting in the immediate discussion because of its interaction with stomatal aperture.

As guard cells accumulate solutes, their turgor pressure and volume increase. This opens the stomatal pore and allows more transpiration, which in turn lowers the water potential and turgor pressure of the surrounding cells (Shackel and Brinckmann 1985; Nonami and Schulze 1989; Nonami et al. 1990; Mott and Franks, in press), which are presumably in hydraulic equilibrium with the guard cells. This decline in guard cell water potential tempers the effect of solute accumulation on turgor, and the magnitude of this decline is controlled by evaporative demand (Δw, the gradient for water vapor between the leaf and air): larger values of Δw will result in larger effects of aperture on transpiration and therefore on guard cell water potential. Therefore, a given increase in guard cell osmotic content will translate into a smaller increase in guard cell turgor at high values of Δw than at low values. The effect of solute accumulation on turgor is also reduced by dilution as the guard cell increases in volume. The dilution of solutes as volume increases occurs in all plant cells, but most plant cells undergo relatively small (~15%) changes in volume as they move from maximum turgor to zero turgor (e.g., Steudle et al. 1977). However, because of the exceptionally large changes in guard cell osmotic pressure and therefore turgor pressure (Franks et al. 1995; Franks et al., in press), volume changes for guard cells will have a substantial effect on osmotic pressure. As turgor pressure increases from 10 to 45 bars in the guard cells of *Vicia faba*, volume increases by approximately 50% (Franks et al, in press). Thus, solute concentration must actually increase by 500% over this range of pressure, instead of by 350%, as would be the case if volume were relatively constant. These considerations make it impossible to define a simple, unique rule that allows one to predict changes in guard turgor from changes in guard osmotic pressure.

The factors controlling epidermal turgor have been studied in much less depth than those controlling guard cell turgor, probably because in most of the species studied, epidermal turgor appears to be a passive function of epidermal water potential. In a notable exception to this generalization, Klein et al. (1996) observed that epidermal cells of *Vicia faba* lost a substantial amount of turgor (estimated by the difference between bulk epidermal water potential and osmotic pressure of detached epidermis) early in the morning before stomata opened, and they concluded that epidermal turgor did not play an important role in controlling stomatal aperture during the day. However, as noted above, pressure probe studies have shown that epidermal turgor and mesophyll turgor respond rapidly to changes in transpiration rate that are caused by perturbations in ambient humidity. Thus, epidermal and mesophyll turgor appear to depend on the prevailing balance between water supply and demand: transpiration removes water from the epidermis, and flow from the soil to the leaf replaces it. This leads to some interesting and well-documented kinetics. For example, when humidity is suddenly lowered, transpiration increases, causing a rapid decline in epidermal turgor (and presumably guard cell turgor). Because of the mechanical advantage of the epidermal cells, the pore actually opens in response to this decrease in turgor, which further increases transpiration. This sequence of events would constitute a 'runaway' reaction if it were not for the fact that the indirect effect of epidermal turgor on transpiration (via conductance) is much smaller than the direct effect of transpiration on epidermal turgor. The transient opening response persists until active processes in the guard cell cause the osmotic pressure, and hence the turgor pressure, of the guard cell to increase sufficiently to cause a reduction in steady-state aperture. A similar sequence of events, often called the Iwanoff effect (Iwanoff 1928), occurs when leaf turgor is suddenly reduced by cutting off the water supply to the leaf (Raschke 1970), and the opposite sequence of events occurs when leaf water potential is increased by pressurizing the roots (Comstock and Mencuccini 1998). It has also been suggested that the guard cells experience the lion's share of water loss, and the resulting imbalance in water potential drawdown between the epidermal and guard cells is sufficient to overcome the epidermal mechanical advantage; in that case, stomatal closure in low humidity would not require active efflux of osmotica (see Cowan 1994 for a discussion). This hypothesis is discussed in a later section (*Stomata as hydraulic integrators*), but here we note only that it is difficult to explain the transient responses described above with such a hypothesis.

Perhaps more important than these transient effects, but less obvious, is that continuous feedback exists between epidermal turgor pressure and stomatal aperture, even for stomatal movements that are initiated by active changes in guard cell osmotic pressure, rather than by hydraulic perturbations to the epidermis or guard cells. For example, when stomata are induced to open by light, guard cells initiate the response by the active uptake of ions. This results in an increase in guard cell turgor, which overcomes the backpressure of the epidermis and begins to open the pore. As transpiration increases, both epidermal and guard cell turgor decline, with the result (because of the mechanical advantage of the epidermis) that the pore opens even more. Thus, the effect of guard cell turgor on aperture is amplified by the effect of transpiration on epidermal and guard cell turgor. Since the effect of aperture on transpiration is larger at lower atmospheric humidity, this amplification effect is also larger at low atmospheric humidity. Therefore, for a given pumping rate, stomata respond more rapidly to environmental perturbations in dry air, because hydraulic interactions with epidermal cells reinforce the opening or closing responses. This effect has been noted several times in the literature (e.g., Assmann and Grantz 1990; Mott et al. 1999), and may have adaptive value because rapid responses to environmental perturbations will be more important for carbon water balance when humidity is low and water loss rates are high.

It is interesting to note that rapid changes in epidermal osmotic pressure have been observed for some monocot species and could contribute to stomatal responses in these species (Raschke and Fellows 1971). Little information is available concerning the mechanisms for these changes in osmotic pressure or their effects on stomatal aperture, and this area seems worthy of further research.

The interactions between all of these factors are complicated further by the fact that they occur in a hydraulically connected spatial context. The epidermal cells within an areole do not have independent water supply pathways. This is understood most easily by considering the epidermal cells in the center of an areole, which are distal to all other cells in the water supply pathway and thus necessarily share the same water supply as many, if not all, other cells in the areole. Additionally, localized changes in evaporative water loss from a group of epidermal cells will lower the water potential in those cells, drawing water from neighboring cells despite constant evaporative demand from the latter cells. Thus, any changes in the aperture of one stoma will necessarily influence the water status and therefore the aperture of surrounding stomata. Experimental evidence for these interactions has been provided by several studies (Mott et al. 1997, 1999; Mott and Franks, in press). Other evidence suggests that changes in hydraulic demand in one region of a leaf can influence stomata in a distant region of the same leaf. When stomatal closure is induced in half of a wheat leaf by decreasing the photon flux density (PFD), stomatal conductance increases in the other half of the same leaf (Buckley and Mott 2000). It has been suggested that such long-distance interactions may help to coordinate whole-leaf gas exchange in the event of localized perturbations in hydraulic supply (by cavitation of minor leaf veins) or heterogeneity in the initial kinetics of stomatal responses to any environmental change (Mott and Buckley 1998, 2000).

3 Stomatal Responses to Hydraulic Environmental Factors

Years of research have substantially clarified the nature of what has historically been termed the 'stomatal response to humidity.' It has been experimentally demonstrated that stomatal conductance does not respond directly to atmospheric humidity, but instead responds to the rate of water loss from the leaf (Mott and Parkhurst 1991). These studies have been supported by a subsequent study (Monteith 1995) showing that stomatal responses to humidity from the literature are consistent with a linear response of conductance to transpiration rate up to some critical transpiration rate above which the response becomes nonlinear. Despite this progress, no consensus has emerged concerning the mechanism behind this response. It is also clear that stomata can respond to the rate of water supply from the roots (Sperry 2000). Numerous studies have shown that stomatal conductance is proportional to the hydraulic conductivity between the soil and the leaf (e.g., Meinzer and Grantz 1990; Meinzer et al. 1995; Saliendra et al. 1995; Comstock 2000). Also, stomata show rapid, reversible responses to perturbations in the supply of water to the leaf. These perturbations have been effected experimentally by pressurizing roots or by changing the hydraulic conductance between the soil and the leaf (usually by xylem cavitation). When roots are pressurized, stomata open over a period of many minutes, and often display an initial transient closing response similar to that observed when Δw is decreased (Saliendra et al. 1995; Comstock and Mencuccini 1998). Subsequent depressurization causes the opposite response. Furthermore, the stomatal response to humidity can be essentially completely reversed by pressurizing the roots (Saliendra et al. 1995; Comstock and Mencuccini 1998). Reductions in xylem conductivity also cause stomata to close relatively rapidly (Hubbard et al. 2001). The similarities between the stomatal responses to humidity and water supply raise the intriguing possibility that these two responses may be caused by the same mechanism (Saliendra et al. 1995), and this idea is explored in the paragraphs below.

Two possible mechanisms for the stomatal response to water supply are immediately obvious. First, the response could be due to a signal (chemical or other) that is generated by the roots or xylem tissue in response to changes in water availability or water transport rate. Although the effects of chemical signals from the roots that are produced in response to soil drought are well-documented (e.g., Lösch and Schulze 1994), these signals must be carried from the roots to the leaves in the transpiration stream and therefore show a rather slow response time. It seems unlikely that they could be responsible for rapid reversible movements that stomata display in response to root pressurization and xylem cavitation (Sperry 2000). The second possible mechanism is that stomata are responding to the water potential (or turgor pressure) of the leaf, which is controlled by the transpiration rate, the soil water potential, and the hydraulic conductivity between the soil and the leaf. Mesophyll or epidermal cells could be involved in this response, but it is unlikely that guard cells themselves could serve as the sensing site, since they undergo such large changes in turgor pressure independent of those caused by changes in water supply. Therefore, it is necessary to also postulate the existence of some signal, generated by cells in the leaf other than the guard cells, that causes changes in guard cell osmotic concentration. There is some evidence for such an effect (Grantz and Schwartz 1988), and it would explain stomatal responses to several different experimental perturbations in water supply, and, as we shall see below, it could also explain stomatal responses to humidity.

There are several possible mechanisms by which stomata could respond to the rate of transpiration from the leaf. One possibility is that the stomata could sense transpiration rate via some molecule in the transpiration stream that accumulates at the guard cells in proportion to the transpiration rate. Abscisic acid (ABA) is one possible candidate for such a role (Lösch and Schulze 1994), but it has recently been shown that ABA-insensitive and ABA-deficient mutants of *Arabidopsis* have stomatal responses to humidity that are similar to that of the wild type (Assmann et al. 2000). Ewert et al (2000) have proposed that sucrose could fulfill this role, and they demonstrated that mannitol, applied through the transpiration stream, could accumulate in guard cell walls of a transpiring leaf to a high enough level to substantially lower the turgor pressure in the guard cells. Although attractive in principle, this hypothesis is not consistent with data showing that stomata are capable of responding to humidity in darkness, when sucrose concentrations in the apoplast should be quite low (Kappen and Haeger 1991). It also cannot serve as the mechanism for stomatal responses to cavitation and root pressurization since neither of these stimuli will have a direct effect on transpiration rate and therefore will have no effect on the delivery of the signal molecule to the guard cell. Indeed, as stomata open in response to root pressurization, the transpiration rate is actually increased, which should increase the delivery of the signal molecule to the guard cells and cause stomatal closure.

Another possibility is that stomatal responses to humidity are caused by direct transpiration-induced reductions in the turgor pressure of the guard cells. This hypothesis requires that the water potential drawdown is substantially larger for the guard cells than for the surrounding epidermal cells (to overcome the mechanical advantage of the epidermis), and this assumption in turn demands either that more water evaporates from the guard cells than from the surrounding epidermal cells, or that there is a very low hydraulic conductivity between the guard cells and the surrounding epidermal cells, or both. The evidence concerning these postulates is mixed and contradictory (see Cowan 1994). However, in some species stomatal responses to humidity can take upwards of an hour before they are complete, and it seems unlikely that the hydraulic conductivity between guard and epidermal cells could be low enough to cause this slow a response. If guard cells were the dominant site of evaporation but had such difficulty obtaining water, it seems likely that they would plasmolyze following any substantial increase in the rate of water loss, causing complete and immediate stomatal closure followed by a very slow recovery of aperture. It is also important to note that this mechanism, like the chemical-signal hypothesis discussed above, can not explain the stomatal responses that are observed when leaf water supply is perturbed. Pressurization of the roots would increase the turgor pressure of both the epidermal cells and the guard cells by roughly equal amounts, and (because of the mechanical advantage of the epidermal cells) this would lead to stomatal closure, rather than stomatal opening. Nevertheless, these counter-arguments are circumstantial and theoretical, and there is no direct and unequivocal evidence to disprove the hypothesis that guard cells are the primary and initial sensor of changes in transpiration rate.

The last possibility that will be considered here is that the stomatal response to humidity is mediated by water potential or turgor pressure somewhere in the leaf other than the guard cells. In this scenario, the stomatal response to humidity would actually be a response to water potential or turgor pressure in specific cells in the leaf. This hypothesis requires that these physical variables be transduced into a signal that is perceptible by guard cells, and although there is no direct evidence for such a signal, there is good circumstantial evidence for such a mechanism. First, although bulk leaf water potential is often relatively constant as Δw is varied, pressure probe studies show that epidermal and mesophyll cells experience rapid changes in turgor pressure in response to changes in transpiration rate (Shackel and Brinckmann 1985; Nonami and Schulze 1989; Nonami et al. 1990; Mott and Franks, in press). Thus, these cells could generate the necessary signal. Second, there is good evidence that most of the stomatal response to humidity is caused by metabolic events in the guard cells rather than purely by hydraulics (see Grantz 1990 for a discussion). Third, this mechanism can accurately account for both the kinetics and steady-state responses of stomata to humidity (Haefner et al. 1997; Jarvis et al. 1999). And fourth, it is possible to explain stomatal responses to perturbations in the supply of water using the same mechanism (Fig. 2), and this provides an attractive unifying theme for how stomata respond over short time periods to balance the supply and loss of water from the leaf.

It has been pointed out that such a mechanism cannot account for stomatal responses that reduce transpiration as Δw is increased (Farquhar 1978). This type of response occurs occasionally at very high values of Δw, and has been termed 'feedforward' (Cowan and Farquhar 1977). Water loss directly from the outside of the guard cells has been proposed as a possible mechanism for this response (see Cowan 1994 for a discussion), but as discussed above, mechanisms based on direct evaporation from the guard cells have severe theoretical limitations. It has also been proposed that these feedforward responses of stomata to humidity are associated with patchy stomatal closure (Mott and Parkhurst 1991). This idea has been challenged by Bunce (1997), who showed that feedforward responses could occur in the absence of patchy stomatal conductance. However, in that study deviations in the relationship between CO_2 assimilation rate (A) vs. intercellular CO_2 (c_i) were used as

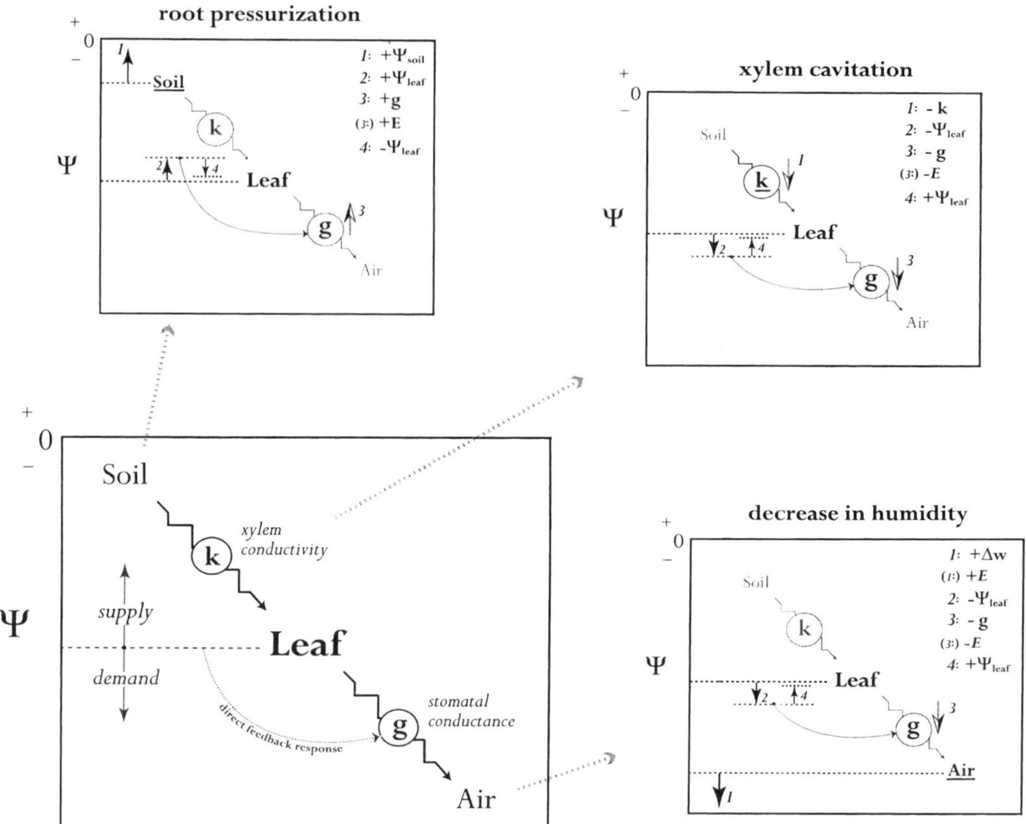

Fig. 2. Diagram showing how the short-term stomatal responses to several hydraulic perturbations (root pressurization, increasing evaporative demand (D), and xylem cavitation) may be synthesized under the auspices of a single mechanism wherein stomata respond by direct feedback to leaf water potential, Ψ (or a surrogate thereof). Leaf water potential is determined by the balance between hydraulic supply (through the xylem, and ultimately from the soil) and demand (by the atmosphere, via transpirational water loss). The sequence of events following each of these three hydraulic perturbations is shown at the *bottom*; in each case, the variable that changes first is *underlined*, shifts in leaf water potential are shown by *solid arrows*, and responses in stomatal conductance (g) or xylem conductance (k) are shown with *half-shaded arrows* to clarify that those responses do not relate to the vertical axis of the diagram, which represents water potential. In real leaves, these dynamics may involve several cycles of the feedback loop before equilibrium is achieved, but for clarity, only one cycle is shown in each case

an indicator of patchiness, and subsequent studies have shown that most patchy conductance distributions will not cause substantial effects on this relationship (Buckley et al. 1997).

We propose that the 'feedforward' phenomenon may be explained by a feedback response of stomatal conductance to water potential if it is recognized that (1) not only hydraulic demand (transpiration), but also hydraulic supply through the xylem, can influence water potential, Ψ, (2) changes in xylem hydraulic conductance, K, are generally irreversible in the short term (Sperry 2000), and (3) the response of K to Ψ is sigmoidal (Fig. 3, top right), with no insignificant loss of conductivity above a certain threshold Ψ, but dramatic declines in K for lower values of Ψ. As illustrated heuristically in Fig. 3, there is very little loss of K as water potential declines with increasing Δw for most of the range of Δw. At higher values of Δw, however, xylem cavitation hysteretically changes the relationship between transpiration rate (E) and water potential, and as a result, a plot of E vs. D appears inconsistent with a feedback mechanism (hence the term 'feedforward'). The difficulty is resolved when the movement of the system through a third dimension (K, shown in contour form in the lower two plots of Fig. 3) is considered. This hypothesis is consistent with experimental evidence showing that feedforward responses are irreversible in the short term, and tend to occur only at high values of Δw (Franks et al. 1997; Cowan and Farquhar 1977; Farquhar 1978).

4 Stomata as Integrators of Hydraulic Supply and Demand

Most of this review has focused on identifying a single mechanism that can explain short-term stomatal responses to a variety of hydraulic factors. In this section, we take an entirely different tact and ask *why*, rather than *how*, stomata respond to these factors. We begin with the assumption that the evolution of stomatal function has been driven by the need to satisfy certain ecological 'goals' by precise control of gas exchange. It is important to review our analysis from this perspective for at least two reasons. First, these ecological goals if appropriately framed, are universal, and are thus likely to unify stomatal behavior even when mechanisms differ (as, for example, may be the case for stomatal responses to humidity and gradual soil drought). Second, there are two seemingly disparate ecological 'goals' that may drive short-term stomatal responses to hydraulic perturbations, and it is necessary to determine whether the single mechanism that can reproduce the phenomenology of these responses can also satisfy these distinct, and presumably underlying goals.

Stomatal behavior impacts fitness in several obvious ways. By controlling leaf gas exchange, stomata determine leaf energy balance, mitigate the contingency of xylem cavitation, regulate competition for water, and control the balance between short-term carbon gain and water loss. Traditionally, optimality theory has considered carbon gain to be a bottom line, primarily because carbon can be invested to mitigate these other fitness impacts (for instance, in roots to acquire soil water at the

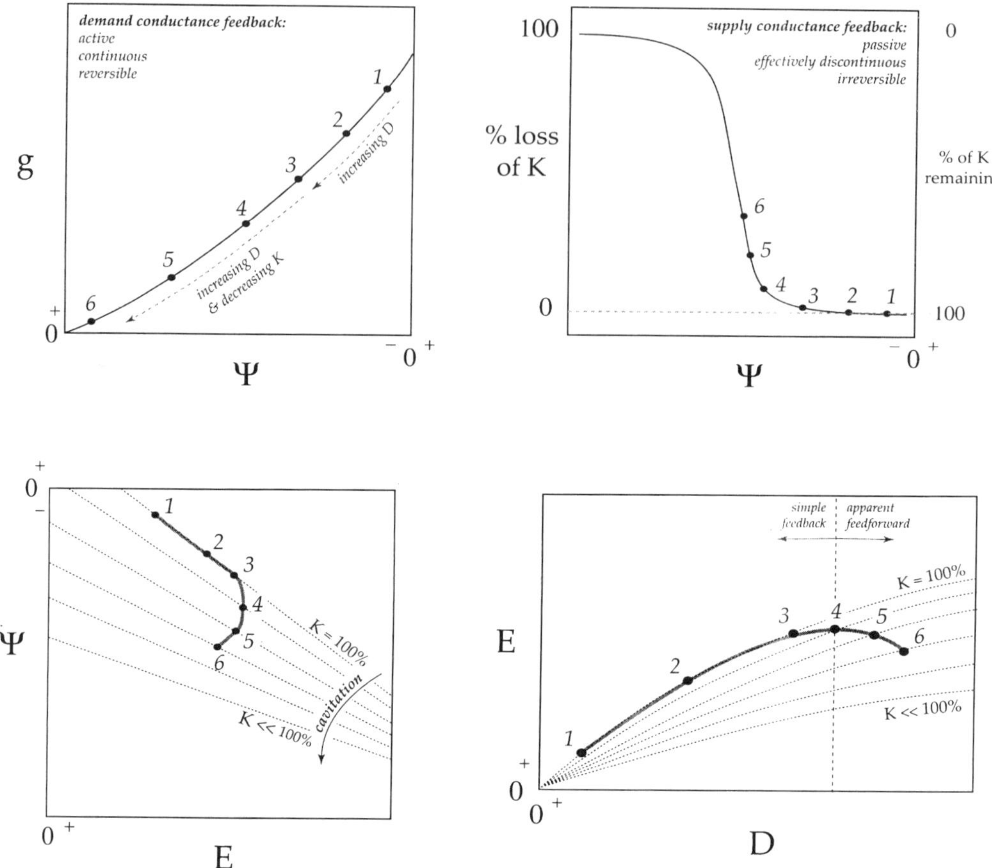

Fig. 3. Qualitative relationships between stomatal conductance (*g*), water potential at the evaporating site (*Y*), and xylem hydraulic conductance (*K*), transpiration rate (*E*), and evaporative demand (*D*), annotated with a hypothetical sequence of six steady-state points along a trajectory of increasing D. The intent of this figure is to illustrate how the 'feedforward response of stomata to humidity,' in which E declines with increasing D at high values of D (as shown in the plot at *lower right*) may be explained by the hypothesis that stomata respond by direct feedback to Y (plot at *upper left*), in conjunction with the observation that K remains nearly constant above a certain threshold Y (plot at *upper right*). Increasing evaporative demand draws down water potential via transpiration (points *1*, *2* and *3* in each plot), but as Y declines further, decreases in K become very substantial, effectively throwing the system onto a series of new trajectories corresponding to progressively lower values of K (points *4*, *5* and *6*). The resulting plot relationship between E vs. D is non-unique, which is inconsistent with a feedback mechanism (hence the term 'feedforward'); however, the true feedback is neither of E to D, nor of E to Y, but of g to Y. The distinction is clarified by showing that the trajectory of E moves through a third dimension (K, shown in contour form in the lower two plots) when plotted against either D or Y

expense of a competitor, in sapwood to increase xylem flow capacity and limit cavitation, and in leaf pubescence to increase reflectivity to reflect excess radiation). Fixed carbon can also, of course, be used to create viable and dispersible seeds, which are the ultimate bottom line for fitness. In this section, we discuss stomatal hydraulics in the context of carbon/water balance and cavitation prevention, assuming for the moment that these are separable ecological goals in their own right. Recent work (see Sperry 2000) has suggested that stomata act to maintain leaf water potential above some critical value to prevent runaway xylem cavitation. The goal in this case is apparently to keep the rate of water loss below a specified discrete maximum value. A reasonable question is then, 'How close to that threshold can stomata go?' A different line of thought preempts that question by suggesting that stomata modulate transpiration rate continuously to keep gas exchange near a mathematically identifiable optimum location – a fixed ratio of the incremental increases in carbon gain and water loss resulting from an increment in conductance. We proceed by identifying precisely what is required, both empirically and mechanistically, for stomatal dynamics to fulfill each role, and then determine whether a single mechanism can meet these requirements.

If the rate of water loss were able to increase without bounds as evaporative demand increased, then water potential throughout the plant's hydraulic continuum could decrease indefinitely. Therefore, in order to prevent runaway xylem cavitation (which would result if xylem water potential were allowed to drop below some critical threshold; Tyree and Sperry (1989)), the transpiration rate must either (1) reach a maximum and subsequently decline with further increases in evaporative demand, (2) reach a maximum and stay there, not responding at all to further increases in Δw, or (3) asymptotically approach a predetermined maximum value as Δw is increased. What underlying mechanistic responses can produce each of these three gas exchange patterns? None of these three options can depend entirely on feedback from xylem conductance (K) or any variables that may be influenced by K, which does not change significantly with increasing Δw until Δw reaches fairly large values (Sperry 2000). Furthermore, cavitation prevention also requires that stomata respond to changes in hydraulic supply (via soil water potential or K) as well as demand. Because this requires stomatal conductance to decline before any change in transpiration rate (E) and without any change in evaporative demand, the mechanism cannot depend entirely on a stomatal response to either E or Δw. Therefore, all three options for the response of transpiration to Δw that satisfy the cavitation-prevention role also require that stomatal conductance be controlled by feedback from hydraulic demand (via E or Δw) under certain conditions, and by feedback from hydraulic supply (K) under other conditions.

Mathematical analysis can reveal patterns of stomatal behavior that maximize carbon gain for a given water supply. These patterns sometimes demand that stomata actually close in the middle of the day, under conditions of high evaporative demand (Cowan and Farquhar 1977). This cannot be achieved with a direct feedback response of stomata to evaporative demand or transpiration rate *per se*, because it requires the 'feedforward' pattern – transpiration must eventually decline and approach zero continuously as Δw increases. However, as discussed above, this response cannot depend entirely on stomatal sensitivity to K either, because K only begins to decline significantly at high values of Δw. Therefore, the continuous-optimization role also demands that stomata respond by direct feedback both to hydraulic supply and demand.

The most parsimonious synthesis of these theoretical and empirical considerations would involve regulation of stomatal conductance by direct, reversible feedback from a single variable that is influenced by both hydraulic supply and demand. Water potential at the evaporating site (or a variable that is directly and reliably linked to that water potential, such as epidermal turgor pressure) is the most obvious candidate for this sensor. By postulating a direct feedback response of stomatal conductance to water potential, the short-term stomatal responses to humidity, root pressurization, and xylem cavitation are unified by a single role for stomata as integrators of hydraulic supply and demand.

5 Concluding Remarks

Stomata are the nexus of hydraulic supply and demand. They bear the heavy ecological burden of integrating all immediate and contingent threats to the continual supply of water, because that supply is critical for carbon acquisition and thus, ultimately, for reproductive success. In the last decade, we have developed a more thorough understanding of the hydraulic aspects of stomatal function at several scales, from single guard and epidermal cells to whole leaves, and this information is critical for interpreting cellular process of guard cells in terms of whole leaf stomatal conductance. Furthermore, we have recognized and substantially characterized the interactions between stomatal behavior and xylem conductance. Analysis of this new knowledge suggests a synthetic theory for the mechanism of stomatal responses to hydraulic perturbations. This theory postulates water potential at the evaporating site as a primary sensor, and we have put forward a version of this mechanism for debate. Although consensus on the fine details of this mechanism remains elusive, the basic theory is compelling for several reasons. First, when placed in the context of recent discoveries about the response of xylem conductance to water potential, this theory may explain the 'feedforward' response of stomata to humidity. Second, the mechanism appears to explain the short-term stomatal responses to three different factors: humidity, root pressurization, and xylem cavitation. Finally, this mechanism appears to be consistent with seemingly disparate ecological goals postulated for stomatal behavior: optimization of carbon/water balance by continuous response to immediately perceptible environmental conditions, and prevention of the contingent threat, not immediately perceptible by stomata, of runaway xylem cavitation.

References

Assmann SM (1999) The cellular basis of guard cell sensing of rising CO_2. Plant Cell Environ 22:629–637

Assmann SM, Grantz DA (1990) Stomatal response to humidity in sugarcane and soybean: effect of vapour pressure difference in the kinetics of the blue light response. Plant Cell Environ 13:163–169

Assmann SM, Shimazaki K (1999) The multisensory guard cell. Stomatal responses to blue light and abscisic acid. Plant Physiol 119:809–815

Assmann SM, Snyder JA, Lee YJ (2000) ABA-deficient (*aba1*) and ABA-insensitive (*abi1-1, abi2-1*) mutants of *Arabidopsis* have a wild-type stomatal response to humidity. Plant Cell Environ 23:387–395

Blatt MR (2000) Cellular signaling and volume control in stomatal movements in plants. Annu Rev Cell Dev Biol 16:221–241

Buckley TN, Mott KA (2000) Stomatal responses to non-local changes in PFD: evidence for long-distance hydraulic interactions. Plant Cell Environ 23: 301–309

Buckley TN, Farquhar GD, Mott KA (1997) Qualitative effects of patchy stomatal conductance distribution features on gas exchange calculations. Plant Cell Environ 20:867–880

Bunce JA (1997) Does transpiration control stomatal responses to water vapour pressure deficit? Plant Cell Environ 20:131–135

Comstock J, Mencuccini M (1998) Control of stomatal conductance by leaf water potential in *Hymenoclea salsola* (T.&G.), a desert shrub. Plant Cell Environ 21:1029–1038

Comstock JP (2000) Variation in hydraulic architecture and gas-exchange in two desert sub-shrubs, *Hymenoclea salsola* (T.&G.) and *Ambrosia dumosa* (Payne). Oec 125:1–10

Cowan IR (1972) Oscillations in stomatal conductance and plant functioning associated with stomatal conductance: observations and a model. Planta 106:185–219

Cowan IR (1994) As to the mode of action of the guard cells in dry air. In: Schulze E-D, Caldwell MM (eds) Ecophysiology of photosynthesis. Springer, Berlin Heidelberg New York, pp 205–299

Cowan IR, Farquhar GD (1977) Stomatal function in relation to leaf metabolism and environment. Symp Soc Exp Biol 31:471–505

Delwiche MJ, Cooke JR (1977) An analytical model of the hydraulic aspects of stomatal dynamics. J Theor Biol 69:113–141

Ewert MS, Outlaw WH, Zhang S, Aghoram K, Riddle KA (2000) Accumulation of an apoplastic solute in the guard-cell wall is sufficient to exert a significant effect on transpiration in *Vicia faba* leaflets. Plant Cell Environ 23:195–203

Farquhar GD (1978) Feedforward responses of stomata to humidity. Aust J Plant Physiol 5:787–800

Franks PJ, Cowan IR, Tyerman SD, Cleary AL, Lloyd J, Farquhar GD (1995) Guard cell pressure/aperture characteristics measured with the pressure probe. Plant Cell Environ 18:795–800

Franks PJ, Cowan IR, Farquhar GD (1997) The apparent feedforward response of stomata to air vapour pressure deficit: information revealed by different experimental procedures with two rainforest trees. Plant Cell Environ 20:142–145

Franks PJ, Cowan IR, Farquhar GD (1998) A study of stomatal mechanics using the cell pressure probe. Plant Cell Environ 21:94–100

Franks PJ, Buckley TN, Shope JC, Mott KA (2001) Guard cell pressure and volume measured concurrently by confocal microscopy and the cell pressure probe. Plant Physiol (in press)

Grantz DA (1990) Plant responses to atmospheric humidity. Plant Cell Environ 13:667–679

Grantz DA, Schwartz A (1988) Guard cells of *Commelina communis* L. do not respond metabolically to osmotic stress in isolated epidermis: implications for stomatal responses to drought and humidity. Planta 174:166–173

Haefner JW, Buckley TN, Mott KA (1997) A spatially explicit model of patchy stomatal responses to humidity. Plant Cell Environ 20:1087–1097

Hubbard RM, Ryan MG, Stiller V, Sperry JS (2001) Stomatal conductance and photosynthesis vary linearly with plant hydraulic conductance in ponderosa pine. Plant Cell Environ 24:113–121

Iwanoff L (1928) Zur Methodik der Transpirations-bestimmung am Standort. Ber Dtsh Bot Ges 46: 306–310

Jarvis AJ, Young PC, Taylor CJ, Davies WJ (1999) An analysis of the dynamic response of stomatal conductance to a reduction in humidity over leaves of *Cedrella oderata*. Plant Cell Environ 22:913–924

Kappen L, Haeger S (1991) Stomatal responses of *Tradescantia albiflora* to changing air humidity in light and in darkness. J Exp Bot 42:979–986

Klein M, Cheng G, Chung M, Tallman G (1996) Effects of turgor potentials of epidermal cells neighbouring guard cells on stomatal opening in detached leaf epidermis and intact leaflets of *Vicia Faba* L (faba bean). Plant Cell Environ 19:1399–1407

Lösch R, Schulze E-D (1994) Internal coordination of plant responses to drought and evaporational demand. In: Schulze E-D, Caldwell MM (eds) Ecophysiology of photosynthesis. Springer, Berlin Heidelberg New York, pp 185–204

Meinzer FC, Grantz DA (1990) Stomatal and hydraulic conductance in growing sugarcane: stomatal adjustment to water transport capacity. Plant Cell Environ 13:383–388

Meinzer FC, Goldstein G, Jackson P, Holbrook NM, Butierrez MV, Cavelier J (1995) Environmental and physiological regulation of transpiration in tropical forest gap species: the influence of boundary layer and hydraulic conductance properties. Oec 101:514–522

Monteith JL (1995) A reinterpretation of the stomatal response to humidity. Plant Cell Environ 18:357–364

Mott KA, Buckley TN (1998) Stomatal heterogeneity. J Exp Bot 49:407–417

Mott KA, Buckley TN (2000) Patchy stomatal conductance: emergent collective behaviour of stomata. Trends Plant Sci 5:258–262

Mott KA, Franks PJ (2001) The role of epidermal turgor in stomatal interactions following a perturbation in humidity. Plant Cell Environ (in press)

Mott KA, Parkhurst DF (1991) Stomatal responses to humidity in air and helox. Plant Cell Environ 14:509–515

Mott KA, Denne F, Powell J (1997) Interactions among stomata in response to perturbations in humidity. Plant Cell Environ 20:1098–1107

Mott KA, Shope JC, Buckley TN (1999) Effects of humidity on light-induced stomatal opening: evidence for hydraulic coupling among stomata. J Exp Bot 50:1207–1213

Nonami H, Schulze E-D (1989) Cell water potential, osmotic potential, and turgor in the epidermis and mesophyll of transpiring leaves. Planta 177:35–46

Nonami H, Schulze E-D, Ziegler H (1990) Mechanisms of stomatal movement in response to air humidity, irradiance and xylem water potential. Planta 183:57–64

Raschke K (1970) Stomatal responses to pressure changes and interruptions in the water supply of detached leaves of *Zea mays* L. Plant Physiol 45:415–423
Raschke K, Fellows MP (1971) Stomatal movement in *Zea mays*: shuttle of potassium and chloride between guard cells and subsidiary cells. Planta 110:296–316
Saliendra NZ, Sperry JS, Comstock JP (1995) Influence of leaf water status on stomatal response to humidity, hydraulic conductance, and soil drought in *Betula occidentalis*. Planta 196:357–366
Shackel KA, Brinckmann E (1985) In situ measurement of epidermal cell turgor, leaf water potential, and gas exchange in *Tradescantia virginiana* L. Plant Physiol 78:66–70
Sharpe PJH, Wu H, Spence RD (1987) Stomata mechanics. In: Zeiger E, Farquhar GD (eds) Stomatal function. Stanford University Press, Stanford, pp 91–114
Sperry JS (2000) Hydraulic constraints on plant gas exchange. Ag For Meteor 104:13–23
Steudle E, Zimmermann U, Luttge U (1977) Effect of turgor pressure and cell size on the wall elasticity of plant cells. Plant Physiol 59:285–289
Tyree MT, Sperry JS (1989) Vulnerability of xylem to cavitation and embolism. Annu Rev Plant Physiol Molec Biol 40:19–38

Thomas N. Buckley
Department of Forest Resources
Utah State University
Logan, Utah 84322-5215,
e-mail: Tom.Buckley@alumni.jmu.edu

Keith A. Mott
Biology Department
Utah State University
Logan, Utah 84322-5305, USA
Tel.: +01-435-797-3563
Fax: +01-435-797-1575
e-mail: kmott@biology.usu.edu

Spatially Explicit Vegetation Models: What Have We Learned?

By F. Jeltsch and K.A. Moloney

1 Why Worry About Space? – General Philosophy and Historical Motivation

Although ecology is a relatively young science, it has produced numerous insights into the natural world through a variety of approaches, ranging from direct observation to pure theory. Interestingly enough, most modern ecological research has been conducted without an explicit consideration of spatial relationships. In fact, in many cases space has been willfully excluded from ecological studies. Why is this so? "Space complicates" or "space confounds" would be the common, but generally unstated, reason. What this really means is that most ecological studies are designed to eliminate environmental or ecological variability in space, so that "pure and uncontaminated" ecological processes and relationships can be examined.

Space can be a problem for ecological experiments conducted under field conditions. The response to a particular treatment can depend upon the local environmental conditions, as well as upon the treatment of interest. If care is not exercised, the results may incorporate an unintended spatial effect, since environmental conditions are highly correlated in space (Burrough 1983). Because of this, a great deal of effort has been expended to develop experimental designs that efficiently minimize the potential for contamination of experimental outcomes from uncontrolled, "correlated" factors, such as those associated with space. As a consequence, "*randomization, block, split-plot*" has become the mantra of statistical design for field ecologists. This reflects a legacy that is hard to overturn in the interest of understanding the importance of spatial relationships in natural ecological systems. However, experimental designs and statistical techniques for exploring spatial relationships in ecological systems are now being developed, although it is difficult (e.g., Dale 1999).

On another front, ecological theory has followed the lead of theoretical physics by focusing on an analytical approach that primarily utilizes abstract mathematical models, most of which ignore space, viewing it as a complicating factor that is of little interest. Most of these models treat

Progress in Botany, Vol. 63

space as a zero-dimensional point, where all individuals experience the same environment at the same time. Additionally, these models assume that populations are "well mixed", with all individuals having an equal probability of interacting with all other individuals. While this may be an appropriate starting point for understanding the general principles of ecological dynamics, it clearly leaves much to be desired in attempting to understand natural ecological phenomena; ecological systems are embedded in heterogeneous environments and their dynamics derive, in part, from the action of organisms with limited mobility, especially plants. Simply put, space is not homogeneous and organisms are not well mixed. If we want to understand ecology within this more complex, natural framework, then new theoretical approaches must be developed, just as new experimental protocols are required.

A number of studies have shown that if we relax the assumptions of simple analytical models by adding a spatial component, very interesting, and sometimes counterintuitive things occur. For example, Alan Hastings (1993) made an extremely simple, spatial extension to the logistic model. Instead of modeling a single, nonspatial population, he considered two subpopulations, each characterized by logistic growth and connected by a low rate of dispersal (D). Although each subpopulation had the same intrinsic rate of increase (r) and carrying capacity (K), he found that if the two were initialized at different densities the dynamics were quite different from a single population characterized by the same r and K. For instance, in Hasting's primary example with r=3.8 and K=1, the nonspatial model exhibited complex, chaotic dynamics, but the spatially subdivided model (D=0.15) exhibited two potential outcomes: a cycle between two distinct population sizes (a 2-cycle) or a constant "equilibrium" population size (Fig. 1). The outcome observed for the spatial population model depended critically upon initial population densities, with an infinitesimally small change in initial densities potentially shifting the outcome from a two cycle to an equilibrium situation or vice versa.

There are several interesting points to be taken from Hasting's study. The addition of space caused a stabilization of the dynamics of the system, shifting it from a chaotic system to either a 2-cycle or an "equilibrium" system. This contradicts the general conclusion that spatial relationships will complicate the ecological setting. Hasting's results also show us that, if we ignore space, we run the risk of not being able to accurately predict the dynamics of even a very simple ecological system, no matter how accurately we measure the underlying demographic rates. Another, perhaps more technical, point to be taken from Hasting's example is that we cannot easily extend the traditional analytical approach to spatial systems. The dynamics of the nonspatial model could be easily determined analytically, but a determination of the dynamics for the spatial model required numerical solutions provided by computer simulation models. There is a very forceful message here: *even an extremely simple analytical model may become intractable analytically once it includes a consideration of space.* In fact, it can generally be stated that analytical models become intractable once spatial variability is introduced, necessitating numerical solution through simulation (Renshaw 1991). This truism is rarely acknowledged by ecological theorists, but it is indeed a fact that must be reckoned with.

Then how do we proceed to develop a better understanding of space in ecological systems? One possibility would be to use spatially explicit simulation models. Computer simulations can easily explore a broader

Non-spatial model **Simple spatial model**

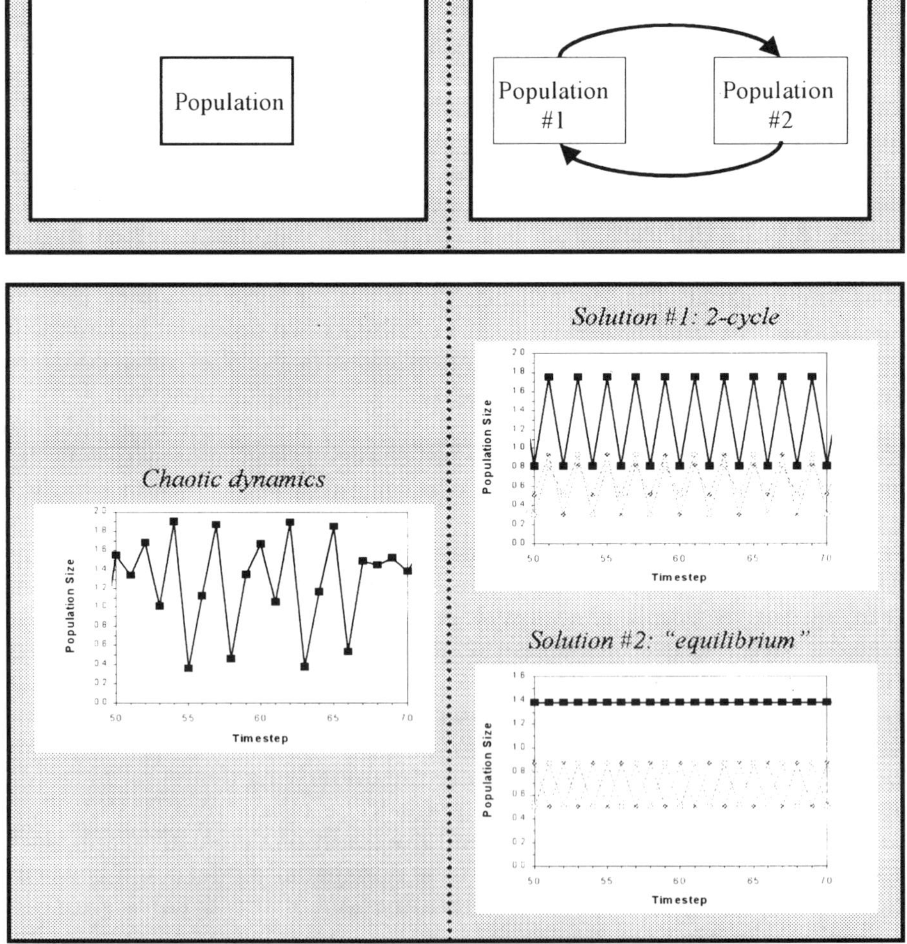

range of conditions, spatially and temporally, than is possible in an experimental system. Simulation models can also avoid some of the problems faced by analytical models in dealing with space. How is this? Principally, a simulation model is a hybrid between an experimental and an analytical system. Simulation models are not solved analytically, but they can be studied as if they were an experimental system, even though they are developed by implementing "formulae" or "rules" in a computer. The formulae or rules are generally developed from basic ecological knowledge combined with hypotheses about how a system functions. The model can then be used to test the hypotheses by comparing model output to the dynamics of the system being studied. Another approach is to use the models in a more theoretical framework to examine "what if" questions.

Clearly, a case can be made that spatially explicit models will provide a very useful tool in the study of ecological systems. But have these types of models been implemented? The answer is a resounding "Yes!". Spatially explicit computer simulation models have become increasingly popular among ecologists over the last decade or so because of several reasons. First, spatially explicit simulation models have allowed ecologists to explore more complex situations than could be dealt with experimentally. Second, these models are, or at least superficially appear to be, easier to implement than analytical models. Perhaps even more important has been the rapid deployment of powerful desktop computers and the development of relatively easy-to-use programming languages. There is also the undeniable attraction of the two-dimensional spatial output from these models. This can be colorfully displayed, allowing for a quick, but perhaps superficial, visual assessment of the development of spatial pattern as a response to the ecological processes coded into the model structure. This connection between pattern and process has a strong appeal to the ecologist as it has been, historically, the paradigm driving much ecological research.

Fig. 1. Results of Hasting's simple logistic model for population dynamics either with (*right side*) or without (*left side*) space considered as a factor. Change in population *i*'s density $x_i(t)$ for time step *t* was determined using a density dependent difference equation model, $\tilde{x}_i(t) = r * x_i(t) * (1 - x_i(t))$, where $\tilde{x}_i(t)$ is population density after local demographic dynamics are complete at time step *t*, but prior to dispersal. For the non-spatial model this means that $x_i(t+1) = \tilde{x}_i(t)$. In the spatial model, the dynamics are determined by two equations coupled through dispersal D: $x_i(t+1) = \tilde{x}_i(t) + D * [\tilde{x}_j(t) - \tilde{x}_i(t)]$ for subpopulation *i* and $x_j(t+1) = x_j(t) + D * [x_i(t) - x_j(t)]$ for subpopulation *j*. The results shown above are produced from a model with r=3.8 and D=0.15 (spatial model). The magnitude of the population density for the non-spatial model was doubled to produce a comparable scale across models. In the spatial model, the total density for the two subpopulations is indicated by the *black line connecting solid black squares* (this also represents total population density in the non-spatial model) and the densities of the two subpopulations are indicated by the *gray lines*

Given that there is now a relatively easy-to-use tool for exploring the connections between pattern and process, that this tool has been employed across a range of systems and that it has been used to study a range of theoretical questions, how much have we really learned? Are there any new insights to be gained? Or are we barking up the wrong tree? Generally, we believe that there is much to be learned through the application of well-crafted, spatially explicit models. But, as with any tool, spatially explicit models must be well designed and appropriately applied.

We will not give a detailed overview about techniques of spatially explicit modeling here. Excellent methodological reviews are, for example, given in Durrett and Levin (1994), Bascompte and Solé (1995) and Czárán (1998). What follows instead is a discussion of the state-of-the-art and, eventually, a look to what future insights might be gained by using this relatively new tool.

2 What Have We Learned?

In principle, ecological models can be built to accomplish many different goals. One commonly stated goal is to predict the future state of a system being modeled (e.g., Jesse 1999). Other possible goals include the testing of hypotheses derived from field observations and experiments (e.g., Ratz 1995), the development of new hypotheses on the basis of theoretical assumptions (e.g., Jeltsch et al. 1998), and the codification of existing knowledge as a mechanism for testing the current conceptual understanding of a system (e.g., Wiegand et al. 1995). In all of these cases, the final aim is either to improve our understanding of a specific system or to elucidate more general ecological principles. These two foci – case-specific modeling and conceptual, theory-oriented modeling – are the two major directions in spatial vegetation modeling, just as they are in other modeling approaches. We will provide a brief overview of these two approaches in the sections below.

a) Theoretical Studies

A majority of the spatial vegetation models that have been developed recently deal with more general, theoretical questions. As stated above, space has always been identified as a crucial factor in plant ecology (Clements 1916; Gleason 1920; Watt 1947; Kershaw 1964; Greig-Smith and Chadwick 1965), even though methodological problems have, until recently, prohibited the explicit consideration of space as a factor. Now that we have suitable computational tools, ecological processes incorporating a spatial component are being seriously examined. These new, spatially motivated studies are tackling a range of issues, some of the more important of which include an examination of the role of spatial interactions at small spatial scales (e.g., Winkler and Klotz 1997; Grist 1999; Berger and Hildenbrandt 2000), theories regarding the impact of spatial interactions on the coexistence of species (e.g., Wiegand et al. 1995; Jeltsch et al. 1996; Moloney and Levin 1996) and the relationship between spatio-temporal pattern formation and ecological processes (e.g., Green 1989; Silvertown et al. 1992; Ratz 1995).

α) Development of the Spatial Modeling Approach

Some of the earliest vegetation models developed with an explicit spatial component were those of Pacala and Silander (e.g., Pacala 1986; Pacala and Silander 1990). They employed fairly simple, yet elegant, mathematical models to examine the importance of intra- and interspecific neighborhood competition for individual plant performance and population dynamics. The Pacala and Silander models were tested using experimental studies in the field. Although they found that incorporating space into their models did not substantially improve their predictive power, they did not interpret these results as an indication that space was unimportant (Pacala and Silander 1990). Instead, they viewed the lack of a significant spatial component to be indicative of the fact that they were examining a fairly simple system involving two annual weed species grown at high density in a fairly homogeneous environment. This was perhaps not the best system to use in examining spatial interactions, but it was a start.

Almost in parallel with Pacala and Silander, several other research groups were developing a new approach to modeling vegetation dynamics in a spatial context (e.g., Crawley and May 1987; Hobbs and Hobbs 1987; Czaran 1989; also see other studies mentioned in Silvertown et al. 1992). They were adapting a technique being used by physicists, known as cellular automata (CA), to address questions about plant competition (see Wolfram 1984 for a good introduction to CA models). Cellular automata are relatively simple simulation models that are run on a grid of cells. Each cell is characterized by its state. For example, in a CA developed by Crawley and May, the state of each cell in the model was characterized by whether it was occupied by an annual plant, a perennial plant, or was empty (Crawley and May 1987). A second trait of the typical CA model is that changes in the state of the cells in the model are determined by simple rules that depend only upon the states of neighboring cells and nothing else. Even with fairly simple rules complex spatial patterns can arise and new insights into ecological processes can be gained (see below).

The introduction of the CA approach to modeling vegetation by Crawley and May, and others, in many ways revolutionized spatial modeling in ecology. More and more models began to appear that utilized this simple formulation for studying ecological dynamics in space (e.g., Green 1989; Iwasa et al. 1991; Silvertown 1992; Jeltsch and Wissel 1993). As the models became more sophisticated, new elements were introduced to examine more realistic scenarios. For example, Colasanti and Grime (1993) used rules derived from plant strategy theory (Grime 1977) to explore the extent to which fundamental differences in the resource dynamics of component populations could generate familiar vegetation patterns and processes. This was accomplished by developing a model on a spatial grid containing orthogonal gradients in resource availabilities and rates of disturbance. They showed that simple differences in resource utilization patterns by founder populations were sufficient enough to generate a realistic pattern of succession, with phases of dominance and decline in individual species over time. The predicted patterns were consistent with earlier theories of succession (i.e., Egler's (1954) initial floristic composition hypothesis, Odum's (1969) theory of ecosystem maturation and Connell and Slayter's (1977) inhibition model of secondary succession). Perhaps even more important was their finding that, with the addition of a disturbance gradient, differences in resource utilization patterns among species may also permit the stable coexistence of populations that would otherwise become extinct (Colasanti and Grime 1993).

β) Spatial Coexistence

Subsequent to the study of Colasanti and Grime (1993), many plants ecologists and modelers have developed spatially explicit models that examine the conditions under which different plant species or types can coexist in a spatially explicit context (e.g., Lavorel and Chesson 1995; Wiegand et al. 1995; Jeltsch et al. 1996, 1998, 2000; Schwinning and Parsons 1996; Grist 1999; Hovestadt et al. 2000; Oborny et al. 2000) and in many of these studies disturbance plays a very important role (e.g., Hobbs and Hobbs 1987; Moloney and Levin 1996; Jeltsch et al. 1998). In another development, while most of the early spatial vegetation models treated space as being environmentally homogenous, a number of studies began abandoning this approach and began to consider models within which space was treated as being environmentally heterogeneous. In one of the first examples, Palmer (1992) developed a model that clearly demonstrated the importance of considering spatial heterogeneity. His model was designed so that two critical factors in the spatial distribution of resources could be varied independently of one another. First, he varied the overall degree of heterogeneity in resource availabilities by controlling the total spatial variance in the distribution of resources. Second, he varied the spatial pattern so that the distribution of resources ranged from a smooth gradient at one extreme to a completely, spatially random distribution of resources at the other. What he clearly demonstrated was that spatial pattern mattered as much as total variance in determining overall species richness. This was true for average species richness within individual cells (α-diversity) and for species richness at the scale of the entire landscape (γ-diversity). Palmer's (1992) findings were very important and yet they still remain to be tested experimentally.

The inclusion of environmental variability within spatially explicit vegetation models represents a very important trend in spatial modeling, which increases the ability to examine more realistic ecological settings. This trend can be seen in an increasing number of models being produced that incorporate a more realistic, environmentally heterogeneous landscape in their design (compare Holt et al. 1995; Kareiva and Wennergren 1995; Liu et al. 1995; Tappeiner et al. 1998; Hiebeler 2000; Shugart 2000; Wiegand et al. 2000b). Although this expands our ability to examine more realistic scenarios, there is a cost in terms of complexity of design and analysis that must be considered.

γ) Pattern and Process

The direct study of the interaction between spatial pattern and ecological process (sensu Levin 1992) represents another important topic being addressed today by means of conceptual spatial models (e.g., see Silvertown et al. 1992; Molofsky 1994; Grimm et al. 1996). Indeed, spatial modeling offers the opportunity to investigate two aspects of this interaction that are extremely difficult to explore in an empirical context: the impact of pattern on the outcome of process (e.g., Silvertown et al. 1992) and the impact of different processes on the formation of vegetation pattern (e.g., Iwasa et al. 1991; Jeltsch and Wissel 1994; Ratz 1995; Thiéry et al. 1995).

In spatially explicit vegetation models, pattern formation primarily occurs as a consequence of local dynamics and spatial interactions. Neighborhood interactions between plant parts, individual plants or plant assemblages generally determine the local rules that dictate how neighboring elements (e.g., adjacent grid cells) influence each other's state and dynamics. A number of models have shown that even a few simple rules of local interaction can lead to rather complex spatio-temporal dynamics at a broader scale (e.g., Iwasa et al. 1991; Halley et al. 1994; Jeltsch and Wissel 1994; Thiéry et al. 1995; Jeltsch et al. 1997a; Gassmann et al. 2000; for a more general discussion of this topic see

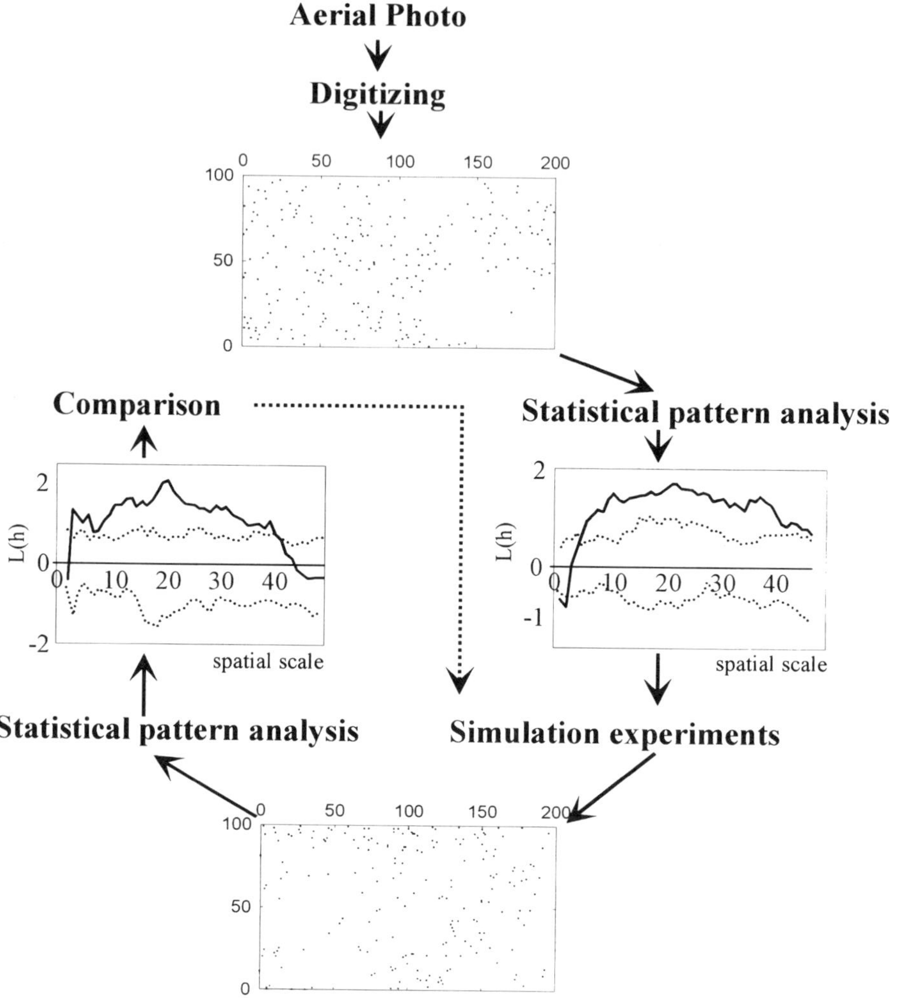

Fig. 2. Schematic diagram of a proposed research cycle coupling empirical data to spatial modeling, using as an example tree distributions from the southern Kalahari (Jeltsch et al. 1999). First, real vegetation patterns (e.g., from aerial photographs, field-mapped data or high resolution satellite images) are analyzed using advanced geostatistical methods (e.g., Ripley's K- or L-function in the Kalahari example; see Haase 1995) or other suitable techniques. Next, spatially explicit simulation models are designed, based on hypotheses generated from the pattern analysis and other sources of ecological information. Simulations are run and the spatial output is analyzed by the same methods used in analyzing the field data. A systematic comparison of the pattern in the field with the output of the simulation experiments may then be used to identify crucial processes involved in pattern formation (e.g., Jeltsch et al. 1999) and may point out sources of conceptual misunderstanding about the system that need to be modified. This information may lead to modification of the model, followed by further comparison of model output with field data. In addition to the steps indicated in the figure, there is also a role for experimentation to play in the process (see the text for further discussion of this point)

Durrett and Levin 1994 or Bascompte and Solé 1995). This process of self-organization is further complicated if interactions operate over a wider area than nearest neighbors (e.g., Jeltsch and Wissel 1994) or if influences occur over even longer distances through processes such as long distance seed dispersal (e.g., Wiegand et al. 1995; Moloney and Levin 1996). In addition to the more direct biotic effects discussed above, other exogenous (biotic or abiotic) processes, such as disturbances (Moloney and Levin 1996) or climatic effects (Jeltsch et al. 1996), may also produce a pattern at either the local (e.g., patch formation by local disturbances) or regional (e.g., all plants are affected by a drought) scale.

It is likely that, in real landscapes, spatial patterns in ecological systems are the result of a combination of factors, including heterogeneous edaphic conditions and the aforementioned processes of pattern formation. These factors will operate across a range of scales and will produce quite complex, overlapping patterns in space. Because of the complexity, disentangling and identifying the various influences that are responsible for the production of pattern in ecological systems remains as one of the grand challenges in spatial ecology. As a consequence, there are an increasing number of studies that focus on developing techniques for detecting underlying ecological processes from spatial patterns (e.g., Ratz 1995; Lobo et al. 1998; Wiegand et al. 1998, 2000a; Jeltsch et. al. 1999; Klaas et al. 2000). This approach is still in its infancy. However, the combined use of spatial simulation models, sophisticated spatial statistical methods (e.g., Haase 1995) and pattern recognition techniques for the analysis of aerial photographs or high resolution remote sensing data represents a promising direction to take in intensifying the study of this topic (Fig. 2).

b) Specific Case Studies

In addition to the more theoretically oriented studies discussed above, there are also a growing number of papers focusing on the utilization of spatial models as tools for exploring the dynamics of specific systems. The types of questions that have been studied cover a wide range of topics that are also of general interest to plant ecologists. For example, a number of papers have examined pattern formation in different vegetation systems (e.g., Jeltsch and Wissel 1993; Thiéry et al. 1995; Puigdefábregas at al. 1999), the impact of disturbance (e.g., Coffin and Lauenroth 1989; Wu and Levin 1994; Moloney and Levin 1996), as well as the importance of colonization processes (e.g., Chiarello and Barrat-Segretain 1997) and the mechanisms leading to long-term coexistence of species (e.g., Wiegand et al. 1995; Winkler and Klotz 1997; Köhler and Huth 1998). These examples include both single species studies in a dynamic landscape context (e.g., Valverde and Silvertown 1997; Wiegand et al. 1999) and the study of the spatial distribution pattern of multiple species (e.g., Gao et al. 1996).

It is remarkable that, in principle, spatial modeling has been applied to most of the major research areas generally explored in empirical, ecological research. This indicates that modeling should not be seen as a separate discipline in plant ecology but, more appropriately, as an important tool to be used in addition to or supplementing empirical studies. This is further emphasized by the use of spatial modeling in more applied research (e.g., Turner et al. 1995). For example, spatial models have recently been employed to explore the impact of land use in different ecological systems, in an attempt to develop better management guidelines or to critically examine current approaches in management (e.g., Jeltsch et al. 1997a,b; Bradstock et al. 1998; Weber et al. 1998, 2000; Gustafson et al. 2000; Weber and Jeltsch 2000). Spatial models offer the possibility of testing different scenarios of land use or management that can hardly be tested in reality, either because of the long time spans required to observe a response (i.e., system responses to human impact often take decades before trends can be separated from natural variation), or the inability to obtain sufficient spatial replication in the system being studied, or because of the possible destructive effects of manipulating the species or communities of interest. In these situations computer simulations can act as an experimental alternative to the direct manipulation of the system. However, models in general, and spatial models in particular, should not be seen as a mirror image of reality (Starfield 1997). At best, models can include the factors and processes that are viewed as being crucial for addressing specific questions and can help us obtain a better understanding of the possible implications of certain management or land use actions. Typically, (spatial) modeling studies can be used to give a relative evaluation of alternative management scenarios rather than making precise quantitative predictions.

A representative example of the use of spatial models in the arena of management can be seen in a series of studies exploring the effects of cattle grazing on the process of encroachment by less palatable or unpalatable shrubs in the southern Kalahari of South Africa (Jeltsch et al. 1997a,b; Weber et al. 1998, 2000; Weber and Jeltsch 2000). These models include the factors and processes thought to be relevant for adequately characterizing the long-term vegetation dynamics of a semi-arid savannah, i.e., rainfall, competition for space and soil moisture among annuals, perennial grasses and shrubs, spatially explicit (patchy) reduction of grass biomass by cattle grazing and (in some of the models) fire. The simulations show that the long-term (i.e., decades long) dynamics of shrub encroachment exhibit a threshold effect in response to grazing intensity. This result suggests that actual stocking strategies are probably based on an underestimate of the actual risks of overstocking (Jeltsch et al. 1997b). Further investigations have shown that the level of this threshold, i.e., the stocking rate at which shrub encroachment is almost unavoidable for larger time scales, strongly depends on the spatial heterogeneity or patchiness of cattle grazing used in managing the system (Weber et al. 1998). Thus, the more homogenous the grazing regime is (e.g., high intensity stocking for short periods of time leads to homogeneous grazing at a small scale), the lower the risk of shrub encroachment is predicted to be.

The types of results discussed above, being oriented more towards a general understanding of management practices and land use patterns than towards precise quantitative predictions, are perhaps typical for what should be expected from general, applied spatial vegetation models. Exact quantitative predictions are usually not possible because of the natural stochasticity and complexity of the modeled systems, the general lack of precise data for parameterizing and testing the models, and the necessary restriction of the models to a consideration of only the most crucial factors and processes (for critical discussions of this topic see Ruckelshaus et al. 1996; Smith and Bull 1997; Hartway et al. 1998; Ruxton and Saravia 1998). Nonetheless, spatial models can provide very valuable insights that can be used effectively in making critical management decisions, if they are used wisely and prudently in combination with more empirical approaches.

3 Where Are We Going? Where Should We Head?

Reviewing a representative cross-section of the existing literature on spatially explicit vegetation models demonstrates an impressive array of models incorporating a broad spectrum of spatial and temporal scales and levels of ecological organization in their design. The questions asked in these studies also range from specific case-oriented questions (e.g., Wiegand et al. 1995) to questions of a more general theoretical nature (e.g., Lavorel and Chesson 1995). A variety of approaches also exist in the methods used in developing the models. These range from the more classical reaction-diffusion approaches (e.g., Jesse 1999) to pure rule-based computer simulations developed without the incorporation of mathematical equations (e.g., Jeltsch et al. 1997a).

Although there is a wide range of variability in the types of models that have been employed in modeling spatial interactions, a general statement can be made in examining their results: *space does indeed matter*. Numerous examples show that the explicit inclusion of space in a modeling framework has a significant influence on the predicted outcomes of all aspects of plant ecology, including individual plant performance, population dynamics, coexistence of species, and community, landscape, and ecosystem scale dynamics. Thus, spatially explicit models have become an integral and important part of modern plant ecology, particularly as the field has broadened to incorporate an explicit consideration of spatial processes in all of its aspects. However, even though we have made some strides in understanding spatial dynamics in ecological systems, in many respects we are still at the beginning of our exploration of this area of research. We still need to learn much more about the consequences of interactions of spatial processes at different spatial scales. For example, we have learned quite a lot about possible mechanisms of coexistence with the help of spatial models (e.g., Lavorel and Chesson 1995; Wiegand et al. 1995; Jeltsch et al. 1996; Grist 1999; Oborny et al. 2000), but most modeling studies have been restricted to a consideration of only a few species or strategic plant types at a time. We are still far from understanding long-term coexistence in natural, species-rich plant communities.

Also, in another central area of spatial plant ecology – the interrelation of pattern and process – we are still just beginning. Spatial models have provided us with some insights into the principles of pattern formation (Durrett and Levin 1994; Jeltsch and Wissel 1994; Molofsky 1994; Bascompte and Solé 1995) and have helped explain existing patterns in a few specific cases (e.g., Sato and Iwasa 1993; Thiéry et al. 1995; Jeltsch et al. 1997a; Puigdefábregas 1999). However, there is still nothing approaching what could be called a general theory of pattern formation. We need to develop a better understanding of how to predict the formation of spatial patterns in plant communities from a knowledge of correlation and synchronization effects produced by local dynamics and spatial interactions. Further issues that could be addressed through the development of appropriate spatial models might be: How do we use an understanding of spatial interactions to improve our ability to detect the underlying processes driving an ecological system from existing spatial patterns? How can we differentiate between biotic self-organization and external forces in a (dynamic) heterogeneous environment when we are interested in identifying the source of spatial pattern in plant communities? In the future, when considering broader temporal scales, we will also have to ask: What role does evolution play in determining the spatio-temporal dynamics of an ecological system? In asking this last question, it is important to point out that evolution as a process has been neglected in most, if not all, spatially explicit models developed so far. This is not a particular failing of the spatial models, however, as evolutionary processes are rarely included in ecological studies.

All of the questions posed above are also relevant in the development of applied ecological research. Indeed, the success of applied plant sciences and nature conservation, as well as the success of more basic research, may be enhanced by incorporating at least some component of spatially explicit modeling in the research program. For example, modeling the spatial spread and control of a plant disease or pathogen (e.g.,

Filipe and Gibson 1998) or evaluating the impact of land use measures, such as grazing or fire, on plant community dynamics (e.g., Jeltsch et al. 1997b; van Oene et al. 1999) are just two examples of where spatially explicit modeling can help in management and conservation.

In both case-specific and more basic plant ecological research, there is clearly a need for an improved interaction between modelers and empirical researchers. In general, there still exists a mismatch between experimental and field collected data and the data required for the development of spatially explicit models. For example, most spatial models have shown us the crucial importance of neighborhood interactions in determining the overall spatio-temporal dynamics of the system, yet there are very little empirical (especially ecophysiological) data available regarding plant interactions and autocorrelation patterns in ecological systems (although see, for example, Schlesinger et al. 1996 and Dale 1999). It is critical that more experiments be designed and more data be collected for direct use in the development and analysis of spatially explicit models. Vice versa, modeling studies should make a more concerted effort to provide suggestions regarding how the results and hypotheses arising out of the modeling effort might be tested in reality. Only through a productive interaction between empirical and theoretical studies will significant progress in (spatial) plant ecology result.

Acknowledgements. We are grateful to J. Groeneveld and V. Grimm for valuable comments on earlier versions of this manuscript. Kirk Moloney was supported, in part, by a grant from the US Environmental Protection Agency (NAGW-3124).

References

Bascompte J, Solé RV (1995) Rethinking complexity: modelling spatiotemporal dynamics in ecology. Tree 10:361–366

Berger U, Hildenbrandt H (2000) A new approach to spatially explicit modelling of forest dynamics: spacing aging and neighborhood competition of mangrove trees. Ecol Model 132:287–302

Bradstock RA, Bedward M, Kenny BJ, Scott J (1998) Spatially-explicit simulation of the effect of prescribed burning on fire regimes and plant extinctions in shrubland typical of south-eastern Australia. Biol Conserv 86:83–95

Burrough PA (1983) Multiscale sources of spatial variation in soil. I. The application of fractal concepts to nested levels of soil variation. J Soil Sci 34:577–597

Chiarello E, Barrat-Segretain M-H (1997) Recolonization of cleared patches by macrophytes: modelling with point processes and random mosaics. Ecol Model 96:61–73

Clements FE (1916) Plant succession. An analysis of the development of vegetation. Carnegie Institution, Washington

Coffin DP, Lauenroth WK (1989) Disturbances and gap dynamics in a semiarid grassland: a landscape-level approach. Landscape Ecol 3:19–27

Colasanti RL, Grime JP (1993) Resource dynamics and vegetation processes: a deterministic model using two-dimensional cellular automata. Funct Ecol 7:169–176

Connell JH, Slayter OR (1977) Mechanisms of succession in natural communities and their role in community stability and organization. Am Nat 111:1119–1144

Crawly MJ, May RM 1987 Population dynamics and plant community structure: competition between annuals and perennials. J Theor Biol 125:475–489
Czárán T (1989) Coexistence of competing populations along an environmental gradient: a simulation study. Coenoses 4:113–120
Czárán T (1998) Spatiotemporal models of population and community dynamics. Chapman and Hall, London
Dale MRT (1999) Spatial pattern analysis in plant ecology. Cambridge studies in ecology. Cambridge University Press, Cambridge
Durrett R, Levin SA (1994) Stochastic spatial models: a user's guide to ecological applications. Phil Trans R Soc Lond B 343:329–350
Egler FE (1954) Vegetation science concepts. I Initial floristic composition, a factor in old-field vegetation development. Vegetatio 4: 412–417
Filipe JAN, Gibson GJ (1998) Studying and approximating spatio-temporal models for epidemic spread and control. Philos Trans R Soc Lond B 353:2153–2162
Gao Q, Li JD, Zheng HY (1996) A dynamic landscape simulation model for the alkaline Grasslands on Songnen Plain in northeast China. Landscape Ecol 11:339–349
Gassmann F, Klötzli F, Walther G-R (2000) Simulation of observed types of dynamics of plants and plant communities. J Veg Sci 11:397–408
Gleason HA (1920) Some applications of the quadrat method. Bull Torrey Bot Club 47:21–33
Green DG (1989) Simulated effects of fire, dispersal and spatial pattern on competition within forest mosaics. Vegetatio 82:139–153
Greig-Smith P, Chadwick MJ (1965) Data on pattern within plant communities. III. Acacia - Capparis semi desert shrub in Sudan. J Ecol 53: 465–474
Grime JP (1977) Evidence for the existence of three primary strategies in plants and its relevance to ecological and evolutionary theory. Am Nat 111:1169–1194
Grimm V, Frank K, Jeltsch F, Brandl R, Uchmanski J, Wissel C (1996) Pattern oriented modelling in population ecology. Sci Total Environ 183: 151–166
Grist EPM (1999) The significance of spatio-temporal neighbourhood on plant competition for light and space. Ecol Model 121:63–78
Gustafson EJ, Shifley SR, Mladenoff DJ, Nimerfro KK, He HS (2000) Spatial simulation of forest succession and timber harvesting using Landis. Can J For Res 30:32–43
Haase P (1995) Spatial pattern analysis in ecology based on Ripley's K-function: introduction and methods of edge correction. J Veg Sci 6:575–582
Halley JM, Comins HN, Lawton JH, Hassel MP (1994) Competition, succession and pattern in fungal communities: towards a cellular automaton model. OIKOS 70:435–442
Hartway C, Ruckelshaus M, Kareiva P (1998) The challenge of applying spatially-explicit models to a world of sparse and messy data. In: Bascompte J, Solé RV (eds) Modelling spatio-temporal dynamics in ecology. Springer, Berlin Heidelberg New York, pp 215–223
Hastings A (1993) Complex interactions between dispersal and dynamics: lessons from coupled logistic equations. Ecology 74:1362–1372
Hiebeler D (2000) Populations on fragmented landscapes with spatially structured heterogeneities: landscape generation and local dispersal. Ecology 81:1629–1641
Hobbs RJ, Hobbs VJ (1987) Gophers and grassland - a model of vegetation response to patchy soil disturbance. Vegetatio 69:141–146
Holt RD, Pacala SW, Smith TW, Liu J (1995) Linking contemporary vegetation models with spatially explicit animal population models. Ecol Appl 5[1]:20–27
Hovestadt T, Poethke HJ, Messner S (2000) Variability in dispersal distances generates typical successional patterns: a simple simulation model. OIKOS 90:612–619
Iwasa Y, Kazunori S, Nakshima S (1991) Dynamic modelling of wave regeneration (Shimagare) in subalpine *Abies* forests. J Theor Biol 152:143–158

Jeltsch F, Wissel C (1993) Modelling factors which may cause stand-level dieback in forest. In: Huettl RF, Mueller-Dombois D (eds) Forest decline in the Atlantic and Pacific region. Springer, Berlin Heidelberg New York, pp 251–260
Jeltsch F, Wissel C (1994) Modelling dieback phenomena in natural forests. Ecol Model 75/76:111–121
Jeltsch F, Milton SJ, Dean WRJ, van Rooyen N (1996) Tree spacing and coexistence in semiarid savannas. J Ecol 84:583–595
Jeltsch F, Milton SJ, Dean WRJ, van Rooyen N (1997a) Simulated pattern formation around artificial waterholes in the semi-arid Kalahari. J Veg Sci 8:177–188
Jeltsch F, Milton SJ, Dean WRJ, van Rooyen N (1997b) Analysing shrub encroachment in the southern Kalahari: a grid-based modelling approach. J Appl Ecol 34:1497–1509
Jeltsch F, Milton SJ, Dean WRJ, van Rooyen N, Moloney KA (1998) Modelling the impact of small-scale heterogeneities on tree-grass coexistence in semi-arid savannas. J Ecol 86:780–794
Jeltsch F, Moloney KA, Milton SJ (1999) Detecting process from snap-shot pattern: lessons from tree spacing in the southern Kalahari. Oikos 85:451–467
Jeltsch F, Weber GE, Grimm V (2000) Ecological buffering mechanisms in savannas: a unifying theory of long-term tree-grass coexistence. Plant Ecol 150 (1/2): 161–171
Jesse KJ (1999) Modelling of a diffuse Lotka-Volterra-System: the climate-induced shifting of tundra and forest realms in North-America. Ecol Model 123:53–64
Kareiva P, Wennergren U (1995) Connecting landscape patterns to ecosystem and population processes. Nature 373:299–302
Kershaw KA (1964) Quantitative and dynamic ecology. Edward Arnold, London
Klaas BA, Moloney KA, Danielson BJ 2000 The tempo and mode of gopher mound production in a tallgrass prairie remnant. Ecography 23:246–256
Köhler P, Huth A (1998) The effects of tree species grouping in tropical rainforest modelling: simulations with the individual-based model FORMIND. Ecol Model 109: 301–321
Lavorel S, Chesson P (1995) How species with different regeneration niches coexist in patchy habitats with local disturbances. OIKOS 74:103–114
Levin SA (1992) The problem of pattern and scale in ecology – The Robert H. MacArthur Award Lecture. Ecology 73:1943–1967
Liu J, Dunning JB Jr, Pulliam HR (1995) Potential effects of a forest management plan on Bachman's Sparrows (*Aimophila aestivalis*): linking a spatially explicit model with GIS. Conserv Biol 9:62–75
Lobo A, Moloney KA, Chic O, Chiariello N (1998) Analysis of fine-scale spatial pattern of a grassland from remotely-sensed imagery and field collected data. Landscape Ecol 13:111–131
Molofsky J (1994) Population dynamics and pattern formation in theoretical population. Ecology 75:30–39
Moloney KA, Levin SA (1996) The effects of disturbance architecture on Landscape-Level Population dynamics. Ecology 77:375–394
Oborny B, Kun Á, Czárán T, Bokros S (2000) The effect of clonal integration on plant competition for mosaic habitat space. Ecology 81: 3291–3304
Odum EP (1969) The strategy of ecosystem development. Science 164:262–270
Pacala SW (1986) Neighborhood models of plant population dynamics. 2. Multi- species models of annuals. Theor Popul Biol 29:262–292
Pacala SW, Silander JA (1990) Field tests of neighborhood population dynamic models of two annual weed species. Ecol Monogr 60:113–134
Palmer M (1992) The coexistence of species in fractal landscapes. Am Nat 139:375–397
Puigdefábregas J, Gallart F, Biaciotto O, Allogia M, del Barrio G (1999) Banded vegetation patterning in a subantarctic forest of Tierra del Fuego, as an outcome of the interaction between wind and tree growth. Acta Oecol 20:135–146

Ratz A (1995) Long-term spatial patterns created by fire: a model oriented towards boreal forests. Int J Wildl Fire 5:25–34
Renshaw E (1991) Modelling biological populations in space and time. Cambridge studies in mathematical biology. Cambridge University Press, Cambridge
Ruckelshaus M, Hartway C, Kareiva P (1996) Assessing the data requirements of spatially explicit dispersal models. Conserv Biol 11:1298–1306
Ruxton GD, Saravia LA (1998) The need for biological realism in the updating of cellular automata models. Ecol Model 107:105–112
Sato K, Iwasa Y (1993) Modeling of wave regeneration in subalpine *Abies* forests: population dynamics with spatial structure. Ecology 74:1538–1550
Schlesinger WH, Raikes JA, Hartley AE, Cross AF (1996). On the spatial pattern of soil nutrients in desert ecosystems. Ecology 77:364–374
Schwinning S, Parsons AJ (1996) A spatially explicit population model of stoloniferous N-fixing legumes in mixed pasture with grass. J Ecol 84: 815–826
Shugart HH (2000) Importance of structure in the longer-term dynamics of landscape. J Geophys Res-Atmos 105:20065–20075
Silvertown J, Holtier S, Johnson J, Dale P (1992) Cellular automata models of interspecific competition for space – the effect of pattern on process. J Ecol 80:527–534
Smith GC, Bull DS (1997) Spatial and temporal ordering of events in discrete time cellular automata – an overview. Ecol Model 96:305–307
Starfield AM (1997) A pragmatic approach to modeling for wildlife management. J Wild Manage 61:261–270
Tappeiner U, Tasser E, Tappeiner G (1998) Modelling vegetation patterns using natural and anthropogenic influence factors: preliminary experience with a GIS model based applied to an Alpine area. Ecol Model 113[1–3]:225–237
Thiéry JM, D'Herbès J-M, Valentin C (1995) A model simulating the genesis of banded vegetation patterns in Niger. J Ecol 83:497–507
Turner MG, Arthaud GJ, Engstrom RT, Hejl SJ, Lui J, Loeb S, McKelvey K (1995) Usefulness of spatially explicit population models in land management. Ecol Appl 5:12–16
Valverde T, Silvertown J (1997) An integrated model of demography, patch dynamics and seed dispersal in a woodland herb, *Primula vulgaris.* OIKOS 80:67–77
Van Oene H, van Deursen EJM, Berendse F (1999) Plant-Herbivore Interaction and Its consequences for succession in wetland ecosystems: a modeling approach. Ecosystems 2:122–138
Watt AS (1947) Pattern and process in the plant community. J Ecol 35:1–22
Weber GE, Jeltsch F (2000) Long-term impacts of livestock herbivory on herbaceous and woody vegetation in semiarid savannas. Basic Appl Ecol 1:13–23
Weber GE, Jeltsch F, van Rooyen N, Milton SJ (1998) Simulated long-term vegetation response to spatial grazing heterogeneity in semiarid rangelands. J Appl Ecol 35:687–699
Weber G, Moloney KA, Jeltsch F (2000) Simulated long-term vegetation response to alternative stocking strategies in savanna rangelands. Plant Ecol 150 (1/2):77–96
Wiegand T, Milton SJ, Wissel C (1995) A simulation model for a shrub ecosystem in the semiarid Karoo, South Africa. Ecology 76:2205–2221
Wiegand T, Moloney K, Milton SJ (1998) Population dynamics, disturbance, and pattern evolution identify the fundamental scales of organization in a model ecosystem. Am Nat 152:321–337
Wiegand K, Jeltsch F, Ward D (1999) Analysis of the population dynamics of desert-dwelling Acacia trees with a spatially-explicit computer simulation model. Ecol Model 117:203–224
Wiegand K, Ward D, Thulke H, Jeltsch F (2000a) From snap-shot information to long-term population dynamics of *Acacias* by a simulation model. Plant Ecol 150 (1/2):97–114

Wiegand K, Schmidt H, Jeltsch F, Ward D (2000b) Reflections on linking a spatially-explicit model of Acacias to GIS and remotely-sensed data. Folia Geobot 35: 211–230
Winkler E, Klotz S (1997) Long-term control species abundances in a dry grassland: a spatially explicit model. J Veg Sci 8: 189–198
Wolfram S (1984) Cellular automata as models of complexity. Nature 311: 419–424
Wu JG, Levin SA (1994) A spatial patch dynamic modeling approach to pattern and process in an annual grassland. Ecol Monogr 64: 447–464

F. Jeltsch
Plant Ecology and Nature Conservation
Institute for Biochemistry and Biology
University of Potsdam
Maulbeerallee 2
14469 Potsdam, Germany
e-mail: jeltsch@rz.uni-potsdam.de

K.A. Moloney
Department of Botany
Iowa State University
143 Bessey Hall
Ames, Iowa 50011–1020, USA
e-mail: kmoloney@iastate.edu

The Role of Mycorrhizal Fungi in the Composition and Dynamics of Plant Communities: A Scaling Issue

By Michael F. Allen, Jennifer Lansing, and Edith B. Allen

1 Introduction

Mycorrhizae are by now a well-known feature of plant communities in terrestrial ecosystems. These symbioses, by definition, are mutualistic and always between plants and fungi localized in the root or rhizoid, and are found in every terrestrial ecosystem except the Dry Valleys of Antarctica. For over a century, the importance of mycorrhizae in facilitating (and occasionally inhibiting) nutrient uptake by plants has been documented for hundreds of plant species and in many different soils and environmental conditions.

We have a good understanding of the physiological interactions at the individual plant level, and a solid and rapidly improving understanding of the molecular mechanisms regulating those interactions. However, the critical limiting factors for mycorrhizal functioning may well change as complexity increases by adding many different fungi and plants to the mix. As a plant grows into its substrate, the root encounters a mycorrhizal fungus. As it continues to grow, the fungus may continue to colonize that root, or it may outrun that fungus and encounter a second fungus. As that root sends out a second root, that root may encounter the initial or may encounter a completely new different mycorrhizal fungus.

The same basic function occurs for fungi. They extract energy from a plant root. This allows them to grow out from that root segment and look for additional roots. The root they encounter may be the same root, the same plant, or an entirely different plant species. This process gives rise to the complex belowground structure that community ecologists are just beginning to discover.

It is critical to note that the roots and hyphae may be interconnected across potentially large areas. However, we know that plants still compete, deplete resources, and pursue life as individuals. How then, do mycorrhizae change the outcomes of competition, resource extraction, and reproduction? This becomes a challenging and critical question to understanding the dynamics of plant communities.

Progress in Botany, Vol. 63

2 Diversity of Mycorrhizal Types

There are several types of mycorrhizae demonstrating multiple evolutionary events. Arbuscular mycorrhizae (AM) evolved early, likely prior to, or just as, the invasion of the land occurred. Vesicles, arbuscules and hyphal structures found in fossils from the Devonian epoch are indistinguishable from modern structures. The fungi forming AM are a deeply divergent group, the Glomales, which appears to go back to an ancestral fungus forming an association with algae in the Ordovician or Silurian epoch. Early divergence between groups such as the Glomaceae and Geosiphon (a related fungus forming a symbiosis with *Nostoc*) occurred a very long time ago, likely as early as the Silurian or Devonian epochs (Redecker et al. 2000). A further divergence between the *Gigaspora* and the *Glomus* lineages occurred sometime between 350 and 450 million years ago. Today, the systematic relationships are changing. With recent additions, eight AM genera and over 200 species are currently recognized, *Glomus, Acaulospora, Sclerocystis, Entrophospora, Archaeospora, Paraglomus, Gigaspora* and *Scutellospora* (Morton and Redecker 2001). Further, *Geosiphon* is taxonomically within the Glomales, although it does not form AM.

Most early land plants formed AM. These include some of the non-vascular plants (mosses) and those forming primitive vascular structures (whisk ferns). Interestingly, many recently evolved plant groups, such as grasses, also form arbuscular mycorrhizae. Arbuscular plants are, in fact, so ubiquitous that it is simpler to list all other types. There is almost no terrestrial habitat in which AM cannot be found. The only exception appears to be the Antarctic Dry Valleys, where no multicellular plants or fungi occur.

Ectomycorrhizae (EM) evolved considerably later, although there is considerable debate as to the exact times. Clearly and importantly, EM evolved many times and in many groups of fungi! One group of Zygomycetes, the Endogonales, forms EM. The remaining EM fungi are Ascomycetes and Basidiomycetes. These independent evolutionary events make generalizations for all EM, in terms of functional relationships, very difficult. However, the internal root structuring is surprisingly similar which led to studying mycorrhizae as a functional unit in the first place. This type of mycorrhiza is largely found in trees and shrubs, long-lived woody plants. The mycorrhizae are largely in temperate extending into tropical forests. In drier soils of the arctic and alpine, EM predominate as well. Arbutoid mycorrhizae are a variant of EM found in ericaceous plants.

Structural differences follow the phylogenic differences. In AM, the fungi proliferate through infectable portions of the roots and penetrate individual cortical cell walls with an extensive hyphal/plant membrane interface (arbuscules or coils) where most of the nutrient transfer occurs. In EM, the fungi encase the root with an extensive inter-cellular interface (hartig net) for transfer. In both of these cases, there is an extensive external mycelial network that encompasses multiple plants and a large surface area.

There are two additional mycorrhizal types, orchid mycorrhizae and ericoid mycorrhizae. These co-exist with specific plant groups and consist of specialized fungal symbionts. Interestingly, both groups of fungi are imperfects, orchid mycorrhizae are basidiomycete imperfects in the genus *Rhizoctonia*, and ericoid mycorrhizae are ascomycetes in the genus *Oidiodendron*. Both also promote plant growth by directly mineralizing complex organic materials, often in low pH conditions, and taking up the nutrients directly from that organic material. Both also have minimal external hyphae but release complex enzymes into the substrate.

What we have learned is that "a mycorrhiza is not a mycorrhiza is not a mycorrhiza". Mycorrhizal associations have evolved independently many times over the past 450 million years. What this means is that we should not expect a uniform "mycorrhizal" response. Variations, such as conditions and possibly even species, wherein a mycorrhiza becomes more of a parasite than a mutualist (e.g., Johnson et al. 1997), should be expected, and not be a surprise. This means that many differing types of physiological and morphological structures depending on the particular combination of plant and fungus is also to be expected. Mycorrhizae rapidly enter the realm of complex interactions, the surface of which has only been scratched.

3 Interactions Between Multiple Mycorrhizal Fungi and Individual Plants

a) Structure/Function Relationships

The basic functioning of all mycorrhizae is tightly coupled to both its internal (within the host) and external structure (within the soil). External hyphae ramify through the soil or along a root searching out susceptible root segments. In most cases, this means an uninfected segment located just behind the region of elongation. This hypha penetrates a susceptible root and forms an appressorium. The internal hyphae then proliferate between cells. For AM, these form either an Arum-type mycorrhiza, in which intracellular arbuscules emerge from intercellular hyphae, or a Paris-type mycorrhiza in which sequential coils and infrequent arbuscules emerge from inter- or intracellular hyphae. In all cases, there remain intact membranes separating plant and fungus cytoplasm. For EM, a massive intercellular network, the hartig net is formed between the cortical cells.

Carbon is transported back from the intercellular into the external hypha. The external hyphae appear to form two distinct functions. The first is a mycelial network that continues to spread looking for new root segments. Associated with the internal network is also a hyphal network that spreads outward searching for soil resources. We do not know the limits to this external expansion. For EM, sporocarps can be found many meters away from the host. For AM, the extension is probably only a few centimeters.

This means that there are three distinct functions of the fungal mycelium. The first element is the internal hyphae. Carbon is taken up by the fungal hyphae within the plant. Evidence to date shows that much, if not all, of the internal hyphae can take up sugars (Bago et al. 2000). In AM, ATPases are concentrated in the arbuscules, in EM in the hartig net. Importantly, fungi generally convert the simple sugars to trehalose or other complex sugars to prevent or reduce re-absorption. Within the root cortex, hormonal composition also changes. The hormonal compositional changes are probably primarily associated with the loading and unloading of carbon and nutrients. However, they also lead to other shifts in rooting morphology such as increased lateral root formation. These changes can affect the entire dynamics of root form and chemistry.

The second element is the external absorbing network. In AM, this is a dichotomous branching mycelium that starts as a single thick hypha extending out beyond the root hairs, then dividing into ever-finer hyphae. This network appears to consist of 80–120 cm of hyphae per root penetration in up to an eighth order branching structure (Friese and Allen 1991). At the base, this hypha is 10–12 μm in diameter. Towards the hyphal tip, it may be as fine as 2 μm. The outward limit appears to range from 2 to 8 cm, corresponding to a depletion zone surrounding each individual mycorrhizal root. The basic functioning of AM to an individual plant is based on this group of hyphae and has been rather well described. The external hyphae pick up soil resources, and transport those resources to the host in exchange for carbon. Most soil resources are transported to the plant via this network, including both macro- and micronutrients and water.

For EM, we have no data on the extent of the absorbing network. Read and colleagues (e.g., Taylor et al. 2000) have clearly demonstrated that these fungi have a wide array of strategies (not surprisingly, given the wide diversity in fungi) for obtaining soil resources. Some retain rhizomorphs extending for long distances, then unraveling when a patch of resources is encountered. In other instances, (especially for ascomycetes), a finely branched network probably occurs, resembling the intense utilization of AM.

The third element is the runner hyphal network. These hyphae clearly are involved in searching out new root tips for invasion. They may or may not be functional in the transport of resources between plants or over longer distances. In AM, they consist of single large (10–15 μm in diameter) hyphae, or can even exist as multiple hyphae wrapped around each other in a primitive, rhizomorph-like structure. In EM, they can exist as individual hyphae or even as rhizomorphs consisting of a wrapped network of hyphae.

No similar data set exists for EM, possibly because of the much larger variation among EM fungi. Another factor could be that because EM evolved many times, there is no similarity within an "EM".

In AM, these external hyphae do not appear to have the enzymatic capacity to take up sugars (Bago et al. 2000). In EM, they probably vary greatly. However, they do appear to move carbon and water internally. In some EM, these even form primitive "vessel-like" elements that rapidly transport water (Brownlee et al. 1983). These hyphae probably also move nutrients depending on the flow rates, direction, and sink strength. Importantly, it is this mycelium network that connects multiple roots of a single plant and multiple plants in a complex network of mycorrhizal partners, the common mycelial network. We do not know the full extent of any single mycelium, or the relative transport and uptake characteristics of this network for any single mycorrhiza, much less if they are interacting.

b) Physiological Relationships

Physiologically, mycorrhizal plants are very different from nonmycorrhizal plants. Photosynthesis increases with or without growth enhancements (e.g., Allen et al. 1981). This may occur due to the increased sink strength (Smith and Read 1997) because of the support for the mycelium or just for the increased nutrient exchange. Nutrient uptake is increased, due initially to the increasing absorbing surface area created by the mycorrhizal fungi (Smith and Read 1997). While we have an improving understanding of transport patterns between plant and fungus for resources (e.g., Chalot and Brun 1998; Podila and Douds 2000), the factors regulating the amounts of materials transported *between* symbionts are poorly understood. In part, this comes from the plant-centric view of mycorrhizal functioning. Virtually all studies have concentrated on demonstration of the presence of elemental transport from soil to plant, via the fungal hyphae and the forms of those elements. However, there is little quantitative work on the function of the total amount present or offshoots of mechanisms regulating the transport.

Some of the change is a direct consequence of the increased resource uptake and carbon sinks, but some may be indirect caused by the mechanisms regulating resource transfer. An example is the hormonal balance of the host. Studies have demonstrated changes in the host hormonal balance beyond any direct production and transport from fungus to plant (e.g., Allen et al. 1980, 1982; Duan et al. 1996; Green et al. 1998). This is likely due to the roles of phytohormones in C and nutrient loading and unloading. However, the resulting shifts can affect everything from stomatal opening to root branching. Stomata remain open longer into a drying cycle without the increased nutrient concentrations (e.g., Allen and Allen 1986) associated with both changes in hormonal regulation of stomata and altered osmotic pressure associated with the presence of the mycorrhizal fungus (e.g., Allen and Boosalis 1983; Duan et al. 1996; Green et al. 1998). Generally, the shifting balances appear to be beneficial to both plant and fungus.

Nutrient transport from soil to plant via the mycorrhizal fungus has been the hallmark of mycorrhizal studies for over a century (see Allen 1991; Smith and Read 1997). However, it is important to recall that the fungus itself requires considerable nitrogen (N) and phosphorus (P). As an example, N concentrations in fungi can be up to 10% of the dry mass, compared with 1–3% in most root tissue. This means that in order for the fungus to transport N to a host, it must have an excess of N (compared with C). The same situation exists for P and probably all other nutrients.

This becomes rather crucial in interpreting the results from ecological studies. In the case of both AM and EM, transport of P and N from soil to plant has been repeatedly demonstrated. However, how much and under what conditions does this occur? In *Cenococcum geophilum*, Lussenhop and Fogel (1999) found that the fungus took up P during the winter/spring but released it to the host after photosynthesis (and fungal C gain) was initiated in late spring. In recent studies Yoshida (1999) and McFarland and colleagues (in prep.) evaluated the recovery of injected ^{15}N. Importantly, in both cases (AM and EM), the vast majority of ^{15}N was found retained in the soil microbial or organic pools. Only a small fraction was actually transported to the host. These pools include the mycorrhizal hyphae themselves.

Many evaluations have estimated that the mycorrhizal fungal mass within a root system ranges from 20 to 40% fungus for EM tips. We have looked at the branching orders of tree roots (Pregitzer et al. 2001) and the mycorrhizal infection (Lansing and Allen, unpublished data). Using a conservative estimate of 20% being EM fungal tissue by mass, for first order roots, then, of 100 g of tissue, 2 g of N would reside in the fungal tissue with only 0.5–1 g of N in the root tissue. In heavily EM plants like balsam poplar and pinyon pine, 70–80% of the first order tip N is likely to be fungal. In the case of AM, less of the fungal tissue is located in the root tips and more in the region of elongation. In these areas, approximately 10–15% of the "root" biomass is fungal. For AM plants such as maple and juniper, approximately 55–75% of the "root" N could actually be fungal N. Again, as with EM plants, a large fraction of the "root N" localized in the fine roots is fungal. For the fungus to exchange N for C, the N uptake by the mycelium would have to exceed the direct fungal demand.

At this stage, we have a rather good and rapidly improving understanding of mycorrhizal dynamics and mechanisms at the individual plant/fungus scale. Largely, these results come from pot, chamber and petri dish experimental systems. However, subtle mechanisms that are associated with fungal growth and maintenance, infection, and plant and fungal genetic differences can add up to quite different roles when viewed on the scale of a field, community or landscape.

c) Community Structure

In the field, each plant connects with as many mycorrhizal fungi as it encounters and is compatible with. Plants also have variable responses to differing mycorrhizal types and species. Surprisingly, analyses of the mycorrhizal fungal communities have found remarkable congruity in the community structure of both EM and AM fungi. Allen et al. (1995) hypothesized that the richness may peak in temperate, low-diversity forests. More recent surveys are finding that the richness of mycorrhizal fungi is certainly not correlated with plant diversity, but remains rather stable at the scale of an individual plant to a stand. Is this a real pattern and, if so, what is the cause?

In EM fungal communities, if the plants are isolated from one another, the numbers tend to be rather low. Gehring et al. (1998) found that isolated individual pinyon pines in northern Arizona had 1–5 taxa (based on restriction fragment length polymorphism – RFLP patterns). Across the stand, where plants were largely isolated, the fungal species richness was 15–19, with only a few fungi in common among the plants. In New Mexico, we found that as pinyon pines merge into smaller stands, the number of symbionts per tree increases from an average of 7 species per plant to 20 species per 900 m^2 stand consisting of 6–8 trees (Lansing and Allen, unpublished data). In individual seedlings of coast live oaks in California, the number of RFLP types average 12 per seedling, with up to 20 across the stand. The richness or morphotypes per seedling in an Alaskan bioassay of Balsam Poplar and White Spruce ranged from 5 to 7, and 3 to 7, respectively (Helm et al. 1999). That number increased to 17 and 11, respectively, across the chronosequence (Helm et al. 1996). In 900-m^2 plots, the EM fungal species richness values were surprisingly similar, ranging from 28 Balsam Poplar in central Alaska to 25 in Red Pine in Michigan, and 20 in Oak in Georgia and Pinyon Pine in New Mexico (Lansing and Allen, unpubl. data).

Across landscapes the diversity increases. Taylor and Bruns (1999) reported 37 RFLP types across their Bishop Pine stands on Point Reyes,

California. In individual stands of live oak scattered across Camp Pendleton, California, we found between 20 to 30 species of EM fungi per 900-m^2 plot with a total across the landscape of 53 species of EM fungi (Allen et al. 2001).

On the regional/vegetation type scale, the numbers increased rather dramatically. Trappe (1977) estimated that there are as many as 2,000 species of EM fungi associated with Douglas fir. In coast live oak, we have identified 74 species of EM fungi fruiting across southern California over a 10-year period.

For AM, the situation may, in fact, not be dramatically different. There are only around 150 species of AM fungi described and few characters on which to base morphological taxonomy. However, for any individual plant in the field, the richness appears to average 5–20 species. In tallgrass prairie communities, Eom et al. (2000) reported 16 species of AM fungi associated with 5 species of plants with an average species richness of 10 per plant. Bever et al. (1996) found 23 species of AM fungi in a multispecies old field in North Carolina. In a harsher shrub steppe environment, looking at a single plant subspecies, *Artemisia tridentata* ssp. *tridentata*, across 99 stands, we averaged 5 species per stand with 48 species across western North America, in a one-time survey. Upon examining one stand of *A. tridentata* over 8 years, the number increased from 5 (single year) to 12 species (total) (Allen et al. 1995).

Even in the tropics, the richness of the mycorrhizal fungi at the stand level does not change dramatically. In a high plant diversity tropical forest, (the Chamela Biological Field Station, Jalisco, Mexico, with over 2,000 species of AM plants), we found 15 species of AM fungi (Allen et al. 1998) by looking at the roots and rhizosphere of 12 tree species over a 3-year period. At the El Eden Ecological Reserve (Quintana Roo), we have found 42 species of AM fungi by looking along the roots of six plant species across the Reserve in an area with over 400 species of plants over a 5-year sampling period.

These fungi separate spatially and temporally even when associated with the same plants. We have found species of both *Acaulospora* and *Scutellospora* associated with A. *tridentata* in the same stand. However, during years with winter rains, the predominant sporulating fungus is *S calospora*. But, in years with summer rains, *A. elegans* and *Archaeospora gerdemannii* is found. In stands of *Adenostoma fasciculatum*, during drought years, AM predominated but during wet years, EM including *Cenococcum sp.*, *Rhizopogon mengei* and *Pisolithus tinctorius* was found (Allen et al. 1998). Klironomos et al. (1999) found that *Acaulospora spp.*, *Glomus spp* and *Scutellospora spp* were spatially segregated in the under-canopy, edge of canopy and interspace areas (Fig 1).

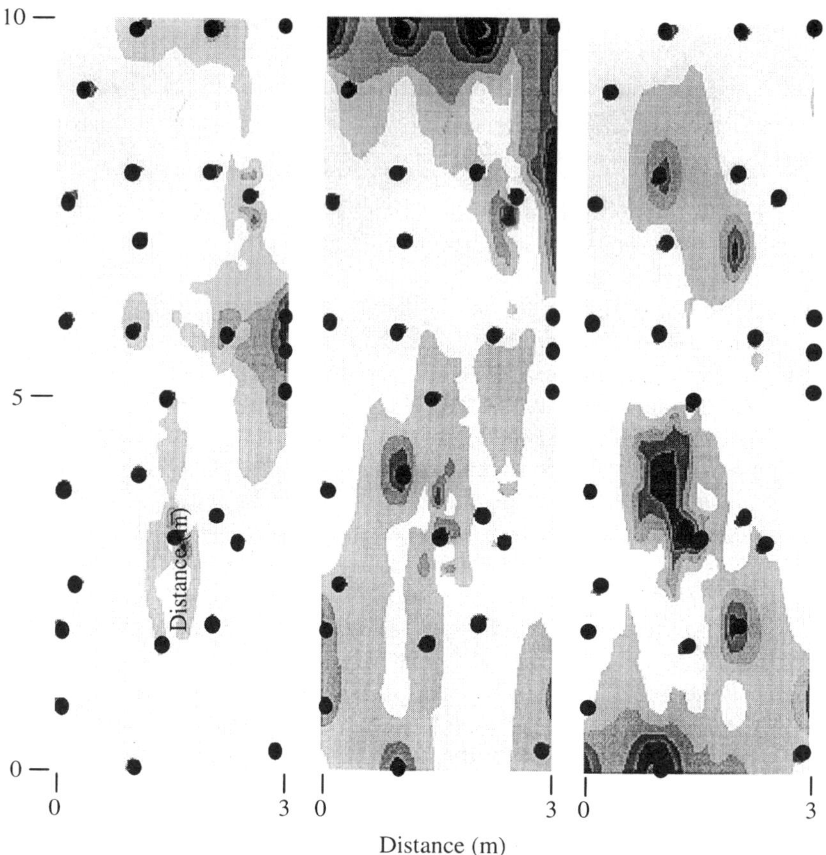

Fig. 1. Population isopleths for (*left*) *Acaulospora*, (*center*) *Glomus*, and (*right*) *Scutellospora* spore levels in the 3×10-m plot. *Black filled circles* represent locations for *Adenostoma fasciculatum* shrubs. Spore number intervals for *Acaulospora* = >0, >4, >8, >12, >16, >20 spores g^{-1} soil; *Glomus* = >0, >10, >20, >30, >40, >50 spores g^{-1} soil; *Scutellospora* = >0, >6, >12, >18, >24, >30 spores g^{-1} soil. (Klironomos et al. 1999)

4 Diversity Patterns of Mycorrhizal Fungi

While the potential global taxon richness is extremely high for both plants and fungi-forming mycorrhizae, there are clearly limits at both the individual plant and stand level. In one sense, the diversity could approach infinity even on an individual stand scale. After all, individual hyphae range from 2 to 10 µm thus they could easily pass one another by.

However, these surveys suggest that α diversity may be relatively similar within many habitats (within 1 ha) for both EM and AM, with a richness approaching 5 to 20. This suggests a limit to the community functional properties of mycorrhizal fungi. However, the turnover in species (β diversity) is high (sensu Whittaker 1970). This is because the fungal species change gradually as one moves from one individual tree to the next, and so on, across the landscape (e.g., Helm et al. 1996, 1999; Gehring et al. 1998; Lansing and Allen, unpubl. data). The γ diversity varies, as to be expected, but also appears to reach an upper limit of somewhere in the range of 30–50 species (per landscape), again for both AM and EM. A major question arises, why does this limit exist (see Sect. 7)?

It is only at the regional level that the diversity level of EM and AM really diverges. Across North America, for *Artemisia tridentata*, we have found only 48 species and only 150 have been described worldwide. For Douglas fir across western North America, there are well over 2,000 associated species and worldwide, the numbers of described species alone is in the tens of thousands. This difference likely lies in the evolutionary history. Evidence still suggests that AM evolved a single time and AM fungi are still considered a monophyletic group (Redecker et al. 2000). However, clearly EM evolved multiple times, from many distinct groups of fungi (Molina et al. 1992). These independent evolutionary events would naturally lead to very different physiological and growth characteristics.

The γ level diversity, however, can have important implications for mycorrhizal functioning at the level of plant communities. Van der Heijden et al. (1998) reported that an increasing richness of AM fungi was associated with increased diversity in plants. Perry et al. (1989) reported that individual Douglas fir plants grew better when supported by multiple mycorrhizal fungi, especially in competition with ponderosa pine. Different fungi, like plants, have different niches and thus physiological characteristics. In EM, fungi range from those such as *Boletus spp.*, which appear to take up mostly organic N, to *Pisolithus* and *Laccaria* that appear to primarily use inorganic N (Read 1997). Indeed, we have found that different EM sporocarps associated with *Q. agrifolia* found in the same stand had highly variable $\delta^{15}N$ values ranging from 2 to 11 (Cario and Allen, unpubl. data). These data indicate that the different species of fungi acquired different N sources.

The different fungi can undertake different ecological roles across space and time associated with a single plant. For example, *Cenococcum* forms a highly branched fine mycelium that appears to intensively exploit the local soil for resources whereas *Pisolithus* forms rhizomorphs that extend several meters into the surrounding soils running from one resource patch to another. *Rhizopogon subcaerulescens* lived deep in the soil surviving fire and re-inoculating bishop pine, whereas *Thelephora* had to re-invade from the surrounding forest (Taylor and Bruns 1999).

Even in AM, where the fungi are of a single group (Glomales), the differing fungi matter to the host plant. Plants of the same species have

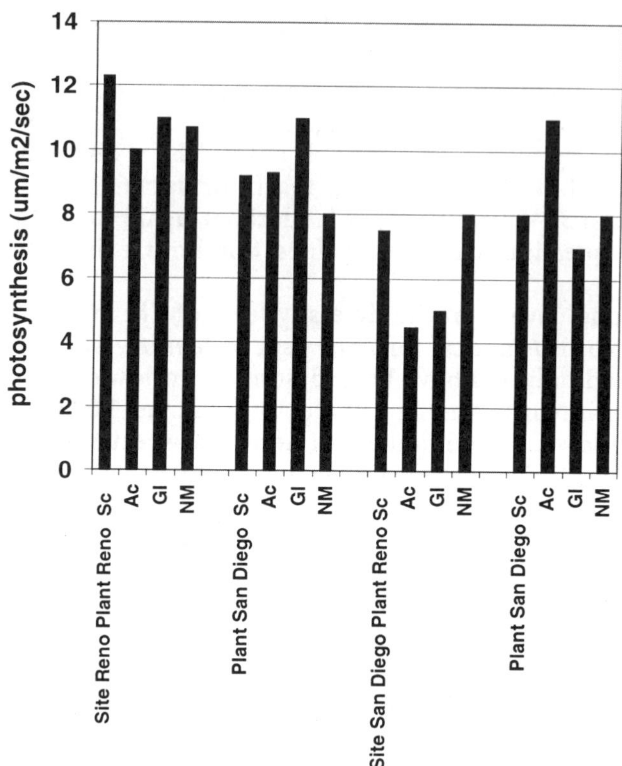

Fig. 2. Comparative photosynthetic rates in the field at a site near Reno, Nevada, and a site near San Diego, California, of *Artemisia tridentata* ssp. *tridentata* seedlings from plants collected at the Reno and San Diego sites. They were inoculated with *Scutellospora calospora* from Reno (*Sc*), *Acauspora elegans* from San Diego (*Ac*), a mix of *Glomus* spp. from Reno at the Reno site and San Diego at the San Diego site (*Gl*) or no mycorrhizae (*NM*) (from Hickson 1993). In this case, the Reno plants photosynthesize best growing at its own site with *S. calospora* collected from its original rhizosphere and worst growing with the *A elegans* from San Diego in San Diego. The similar but slightly different pattern emerges in that the San Diego plants photosynthesize best at the Reno site with *Glomus* spores, or at the San Diego site with its *A. elegans*. It does the worst with *Glomus* collected from the San Diego site

been shown to have different growth rates and physiological characteristics depending on the species and ecotype of fungus (Stahl and Smith 1984). We found that different seeds from the same plant of *A. tridentata* had significantly different rates of photosynthesis and patterns of growth depending on which fungus it was associated with (Fig. 2).

The associations can also change under different conditions. With elevated CO_2, we found that AM shifted from *Glomus* spp. to the *G. tenue* (the fine-endophytic fungus) (Rillig et al. 1999). We have also documented this shift in areas of high N deposition. These data agree with earlier reports that the fine endophyte fungus is adapted to areas of long-term fertilization including agricultural fields (Wang et al. 1993).

Many plants can form multiple types of mycorrhizae, even simultaneously. Australian eucalyptus and North American poplar shifted with age from AM to EM (Chilvers et al. 1987; Lodge and Wentworth 1990). In an oak savannah, oaks form both AM and EM but individual root tips shifted between them based on age and location (Egerton-Warburton and Allen 2000).

Clearly, having many mycorrhizal fungi can hedge the bets for a plant. Differing fungi respond to different conditions in the environment and allow for the acquisition of different resources.

5 Mycorrhizae and Multiple Plant Species

In the field, each mycorrhizal fungus connects with as many plants as it encounters and is compatible with. This pattern exists for all mycorrhizal types in which the plants are relatively nearby. These can coexist or form temporal or spatial separation mechanisms. For example, Massicotte et al. (1999) reported considerable overlap of EM fungi with Ponderosa pine, Douglas fir, tanoak, Grand fir, and madrone. AM fungi are notoriously promiscuous and almost any AM fungus can form a mycorrhiza with virtually any AM plant in pot culture (Smith and Read 1997). We found that AM fungi could survive both in new habitats and associated with differing plant genotypes and species (e.g., Friese and Allen 1991; Weinbaum et al. 1996).

However, the more important feature here may be the ability of a fungus to maximize its own C gain by utilizing multiple plants temporally, spatially, or over successional time. Allen et al. (1984) found that a *Glomus fasciculatum* was found in the rhizosphere of interspersed *Agropyron smithii* (a C_3 grass) and *Bouteloua gracilis* (a C_4 grass). In this case, the arbuscules were formed on the different plants at different seasons depending on the photosynthetic activity of the host. Eom et al. (2000) reported that several species of *Glomus* were found with all plants surveyed. A suite of Glomus species and *Scutellospora calospora* was found with *Artemisia tridentata* completely across its range from Canada to Mexico. M.F. Allen and E.B. Allen (1990) also showed using $\delta^{13}C$ values of *Glomus* spp. spores, that, upon planting a C_4 plant into a matrix of C_3 grasses, that the $\delta^{13}C$ signature shifted from the C_3 to a mixed C_3/C_4 signature. At this stage, we cannot differentiate whether new mycelial networks formed on each plant or if the mycelium integrated the mixed C sources. In gaps formed in tropical forest, Allen et al. (1998) found no changes in AM fungi compared with the neighboring mature forest. Across a larger spatial scale, Bidartondo et al. (2000) found that *Rhizopogon ellenae* was distributed across an *Abies magnifica* stand.

This fungus associated with both the *A. magnifica* and an ericaceous plant, *Sarcodes sanguinea*, and actually increased its density where these two plants were bunched. Away from this combination, the ectomycorrhizal fungal community diversified to other species.

Across larger scales, Grogan et al. (2000) found that some species, for example, *Rhizopogon subcaerulescens* survived fire colonizing plants both before, in the mature forest, and after, in the seedlings. This allows for long-term persistence of the fungus. Horton et al. (1999) found that several fungi persisted as Douglas fir invaded manzanita stands. Thus, the fungi persisted through successional time.

Together, these data indicate that mycorrhizal fungi also hedge their survival bets by forming a symbiosis with many individuals and, often, with many species of plants. There is evidence for some specialist EM fungi but most are probably much more general (Molina et al. 1992). In the case of AM fungi, specialists remain unknown, although persistence does appear to be associated with preferential plant associations (e.g., Weinbaum et al. 1996). For the more specialized types of mycorrhizae, this remains a very open field.

6 Mycorrhizae and Plant Community Dynamics

These observations have led to a wide array of speculation that mycorrhizae change the very nature of plant community dynamics. On one hand, plants are known to compete for resources (e.g., Grace and Tilman 1990). Alternatively, because common mycelial networks interconnect multiple plants, speculation abounds that helper plants overcome traditional competition (see Wilkinson 1999). A large number of datasets now clearly show that mycelial networks connect plants and that both C and nutrients move from one plant to another via mycorrhizal fungal mycelia (e.g., Reid and Woods 1969; Chiarello et al. 1982; Francis et al. 1986; Finlay and Read 1986a,b; Read 1997; Simard et al. 1997a). The issue, then, is not whether it occurs, but do these data sets mean that traditional ideas of plant community dynamics are no longer valid? There is no question, from a mycocentric view, that fungi gain from obtaining C from multiple sources (M.F. Allen and E.B. Allen 1990; Fitter et al. 1998). However, do the plants? Further and critically, do plants gain differentially such that the community structure or composition changes? I believe that the question of relationship to the overall plant community is addressed best using three questions. First, do the resources moving from a plant indirectly alter the surrounding plants by maintaining a viable hyphal network? Second, are there conditions under which resources move from a plant directly to the surrounding plants and alter the composition of that community? And third, does the presence of multiple species of fungi radiating from a plant alter the composition of the surrounding community?

1. Do the resources moving a plant maintain a viable mycelial network of mycorrhizal fungi supporting additional plants?

In several cases, data from the literature clearly support such a hypothesis. For example, Francis and Read (1984) found that seedlings tapped directly into an existing EM hyphal network-facilitated infection, compared to infection when the mycelium was disconnected from surrounding plants. Finlay and Read (1986a) found a similar pattern for AM. Ek et al. (1996) found that 4% of the total allocated C from a birch seedling ended up in the EM fungal mycelium, of which, 1.5% ended up in the neighboring spruce root/mycorrhiza. Graves et al. (1997) also reported that 36% of the C fixed was transported to the AM fungal mycelium. Most of these studies represent this C loss as a C drain. However, as Allen et al. (1981, 1984) and Wang et al. (1989) noted, there was an increase in photosynthetic rates of up to 60% with mycorrhizae: there remained a net C gain for the plant.

The outcome is that plants with high photosynthetic rates and allocation of C to mycorrhizae can preferentially support the formation and maintenance of the common mycelial network. Seedlings clearly rapidly tap this resource (mycorrhizal fungi) earlier and do not have to support the development of a massive mycelium. However, tapping an existing mycelium may or may not alter competitive outcomes among plants.

2. Are there conditions under which resources move from a plant to the surrounding plants via the common mycelium and alter the composition of that community?

Clearly, there are many studies demonstrating that mycorrhizae can affect the outcomes of competition (E.B. Allen and M.F. Allen 1990). Competition can be altered between mycorrhizal and nonmycorrhizal plants (Allen and Allen 1984) and among mycorrhizal plants (e.g., Fitter 1977; Marler et al. 1999a,b).

Moreover, there are clear studies that demonstrate that resources can be transported from a plant to neighboring plants by the mycorrhizal mycelium. The studies listed above, all demonstrated that C, N and P were transported into a neighboring host. Importantly, the resources largely remained in the roots, with only a small fraction being found in the leaves of receiver plants. This led Fitter and colleagues (e.g., Fitter et al. 1998) to conclude that the receiver C was largely just fungal tissue within the roots of the neighbor. However, these were pot studies in which a mycelial network first must be constructed. Building this network may result in different dynamics than in a field with an existing network maintained by mature plants.

In the field, Simard et al. (1997b) reported a net transfer of C among EM plants with no transfer to an AM plant. They estimated an up to 6% net C gain by Douglas fir seedlings, from the donor neighbor. Further, shading had a significant effect on that value by creasing a greater C demand by the shaded plant. In the lab, bi-directional transport was apparent, with the greater (slightly) gain by the plant with lower photosynthetic rates (Simard et al. 1997a). We found that seedlings of *Salsola kali*, a nonmycotrophic plant, actually had reduced establishment and growth in an existing grass stand with AM. The fungus tapped the *S. kali* plant for C without providing benefits to the host. In this case, resources were extracted but not reciprocated (Allen and Allen 1988; Allen et al. 1989, M.F. Allen and E.B. Allen 1990).

Resources acquired by the fungus do move through the mycelium and, in some cases, appear to be dispersed among plants (Smith and Read 1997). However, source/sink dynamics make interpretation of the reallocation very difficult. When P or N was added to the soil within a common mycelial network, it was taken up by both plants with little indication of re-translocation (e.g., Finlay and Read 1986b, Ek et al. 1996).

The second part of the question remains, though. Is the community composition altered? Here the data remain equivocal. Grime et al. (1987) found that the addition of mycorrhizae increased the plant diversity in mesocosms by improving the growth of mycorrhizal, less abundant forbs. Van der Heijden et al. (1998) and Klironomos et al. (2000) reported an additive effect on plants, as one added mycorrhizal fungal taxa. In field studies, Caldwell et al. (1985) found that mycorrhizae facilitated uptake of P preferentially for some species over others. However, there was no indication that this was due to a re-allocation. Allen and Allen (1988) found that mycorrhizae altered the composition of the resulting community, but largely through the elimination of a nonmycotrophic competitor. However, in neither instance, could the competitive outcomes be attributed to resource exchange between plants. The changes in competitive outcomes could be accounted for assuming mycorrhizal fungi are simply another competitive variable. The evidence to date indicates that mycorrhizal fungi simply mediate the direction of flow of resources based on their own C gain.

3. Does the presence of multiple species of fungi radiating from a plant alter the composition of the surrounding plants?

Clearly, if the prior question cannot be answered, this question only complicates the answer. Certainly there is evidence that, there is not clear host specificity between most AM fungi and likely most EM fungi and that among host species within a mycorrhizal type, there are preferences. Stahl and Smith (1984) reported ecotypic preferences among AM fungi. Bever (1994) reported preferential "feedbacks" between plants and "soil communities" and also reported host-dependent preferences in plant growth rates (Bever et al. 1996). Similar types of responses have been reported for the EM forests for almost a century (Allen 1991). However, the mechanisms of such preferences are not yet apparent. Continued work on specificity and molecular interactions is needed.

7 Biodiversity of Mycorrhizae: A Synthesis

Mycorrhizae have been shown through numerous experiments to be important components in communities, but not silver bullets for agricultural and forest production. The reason is likely that mycorrhizal fungi have their own needs, not necessarily related to grain and board feet yield. Further, mycorrhizal fungi are clearly not a homogenous group of organisms that act in concert for plant production. They are a heterogeneous group of fungi that co-evolved independently many times from many independent lineages. The question is, can we make any generalizations as to mycorrhizal functioning in plant communities, or is this such an unwieldy high diversity and a hodgepodge of interactions that interactions approach being random?

At one level, mycorrhizae have a distinct importance for many rare species. Many of these, for example orchids, have distinctive mycorrhizae (e.g., Shefferson et al. 2001) associated only with them. These individuals must establish and persist through time. This creates a unique set of characters that must be incorporated into any conservation plan.

On a broader scale, we postulate that there are, in fact, distinct limits to symbiotic diversity using a feedback analysis between plants and fungi. Further, these feedbacks lead to the diversity of plants found in a stand. Finally, the community-level interactions between mycorrhizal fungi and plants are critical for regulating higher-level ecosystem processes of soil stabilization, production, and likely ecosystem resilience. To illustrate this, we will synthesize the dynamics that comprise a mycorrhiza at the stand level, and then we will describe results from some recent papers that illustrate these interactive dynamics.

Three key processes probably underlie mycorrhizal dynamics within communities. First, specific plants preferentially - but not exclusively - associate with specific fungi (e.g., Molina et al. 1992; Bever et al. 1996). This means that in condition X where fungus 1 predominates, it preferentially supports plant A and reduces the growth of plant B. But as one moves to condition Y, fungus 2 preferentially supports plant B over plant A. Plant A still is able to survive, however, with both fungi 1 and 2 and remains competitive with plant B with either fungus. Plant C outcompetes plant B in the absence of mycorrhizae but cannot compete with plant A if plant A has fungus 1. These conditions can be either spatial or temporal. Based on growth responses, one can thus create assembly rules based on the spatial characteristics of the community.

Mycorrhizal fungi are not always crucial to plant growth. Seastedt and Knapp (1993) noted that there was no single limiting factor in most plant communities, but as resources shifted temporally, limiting resources also changed. Allen and Allen (1986) found that AM influenced plant growth and survival by increasing water throughput when drought was initiated. This allowed plants to extend their growing season a couple of weeks into the drought period. They labeled this time the "ecological crunch". We see similar patterns in nutrient uptake and release (e.g., Lussenhop and Fogel 1999). This is clearly illustrated in alpine conditions where N and P become available at different times, and N is generally in the form of NO_3^-, a form preferentially utilized by the plants, not the AM fungi. In this case, NO_3^- became available as soon as the snow melted, before AM fungi had colonized the plant. Later, as the soil dried, AM predominated and P uptake, largely via AM, became important (Mullen and Schmidt 1993; Mullen et al. 1998). Thus, mycorrhizae would have times when they are critical, and times when they are not. Since different fungi preferentially utilize different resources (e.g., N) then the timing of their effectiveness would be limiting. Plants clearly also partition the habitat and resource extraction, and even their timing of connected mycorrhiza (Allen et al. 1984).

Mycorrhizae do not act independently of the rest of the community. Mycorrhizal activity can be reduced by herbivory, including both insects (Gehring and Whitham 1995; Gehring et al. 1997) and large ungulates (Rossow et al. 1997). Reduction in numbers or changes in composition could open gaps, temporarily changing both importance and composition of mycorrhizae.

Mycorrhizae are extremely dynamic. Ectomycorrhizal root tips and rhizomorphs generally last for a few months. In pinyon pine, we found that EM occupied virtually all the root tips but these appear to be discarded annually (Pregitzer et al. 2001). EM root tips and fungal rhizomorphs appear to last only a few months as measured by both minirhizotron observations and ^{14}C analyses (Treseder and Allen, unpubl. data). Where the mycorrhizae tend to extend to the second order roots (red pine in Michigan), some of these mycorrhizae appear to survive several years (Treseder, Lansing, Allen and Pregitzer, unpubl. observ.). In AM, a mycorrhizal infection unit forms immediately behind the region of elongation and develops an external absorbing network. These are short-lived and are produced and disappear depending on the inoculum present and rates of root growth (Allen 2001).

Finally disturbance is relative. Clearly there is large-scale disturbance that initiates succession. With severe disturbance, the addition of mycorrhizae in patches allows for the establishment of mycorrhizal plants in a matrix of nonmycorrhizal plants or bare areas (E.B Allen and M.F

Allen 1990). But even within a plant community, there are small-scale disturbances that provide low inoculum patches for individual plants. Pocket gophers and badgers dig soil that either focuses or reduces inoculum depending on the soil layering and history (e.g., Allen et al. 1999). Ants concentrate inoculum in arid regions by weaving infected roots into their seed-caching areas of the mound and increasing local nutrients. Roots from neighboring plants will send new roots into these areas continually to increase nutrient extraction (Friese and Allen 1993). During succession, small-scale disturbances create patches in a soil matrix without mycorrhizae in which mycorrhizae are either added by animal or wind vectors, or provide safe sites to where mycorrhizae preferentially establish (Allen 1988).

Thus, in the illustration, the "outcomes" of interactions means that different plants emerge "victors" at different instances across space and time. Mycorrhizae become dynamic partners in creating these communities. An example comes from the studies of Hartnett and Wilson (1999). They found that when they reduced AM using benomyl, the dominant C_4 grasses were dramatically reduced in growth releasing C_3 grasses and forbs from that intense competition. This actually resulted in an overall increase in community diversity but a shift to many early-seral, weedy species. When the mycorrhizal plants were subordinate, the addition of mycorrhizae increased diversity (Grime et al. 1987; Gange et al. 1993).

Are there limits? As we discussed above, the α diversity appears to be limited in both AM and EM stands. Klironomos et al. (2000) noted that plant production increased with species richness. At low plant species richness, mycorrhizae improved that production but became asymptotic very rapidly (4–6 species) when mycorrhizal. Van der Heijden and colleagues (1998) found that increasing the mycorrhizal fungal species richness could also increase plant species diversity and production but again became asymptotic (6–14 species) by extracting more soil resources. These studies were on small herbs in small mesocosms. By extrapolative asymptotes to the scale of other studies, we find that these levels of richness per unit area per size plant for both plants and fungi may very well correlate with the field levels of diversity found for many other studies for both EM and AM.

As human populations expand, we are changing and, in some cases, losing mycorrhizal fungi at least at the local level. For example, high NH_4^+ deposition in the Netherlands eliminated many of the EM fungi within stands (Karen and Nylund 1997; Wallander et al. 1997). Egerton-Warburton and Allen (2000) found the members of the Gigasporaceae and larger-spored *Glomus* spp. were mostly eliminated with high levels of N (largely NO_x) deposition. Intensive P fertilization eliminated most of the AM fungi from long-term agricultural plots (Wang et al. 1993).

Naeem et al. (2000) reported that co-dependency among members of trophic webs led to increasing diversity overall, but that loss of one member could have dramatic consequences across the community as a whole. Lundberg et al. (2000) found that often, loss of a species does not necessarily result in replacement of that species or its function, or lead to cascading extinction. At this stage, it is impossible to determine if loss of AM fungi is due to invasion of alien grasses and the reduction in local native species, or loss of native species is due to a reduction in diversity of AM fungi due to soil nutrient enrichment. Most likely it is due to a complex feedback and cascading change in members of the community.

To this end, it becomes of critical importance to understand mycorrhizae from the scale of the entire community, not just at the individual root segment. Mycorrhizae are clearly important in structuring plant communities. Mycorrhizal fungi do many things beyond taking up P or N for individual plants at the scale of a plant community (Allen 1991). They take C from hosts, grow at different times, preferentially use one elemental form over another, and die or segregate at different time scales from plants. This includes a large fraction of the soil resources on which a plant depends and they serve as a large C sink. Mycorrhizal fungi are diverse and connect multiple plants and the reverse is true. However, the divergence between extremely high diversity for EM and low for AM fungi appears to reside at the γ-diversity level where the longer evolutionary history probably plays a large role. At the α- and likely β-diversity level, the richness is limited. However, the mechanism(s) for that limitation are not clear and constitute a crucial research question. It likely depends on a range of mechanisms operating across a multi-resource limitation in space and time. This notion squares well with the "plurality" hypothesis of Wilkinson (1999). If one could tip a community over, the other side might not be so different - a group of interacting organisms extracting resources, feeding on each other, and affecting each other's abundance.

Acknowledgements. Work on this project was supported by the Ecosystem Studies (DEB-9996211) and Biocomplexity (DEB-9981548) programs of the U.S. National Science Foundation and the University of California Agricultural Experiment Station. We also acknowledge the contributions of the Sevilleta LTER and the Riverside County Multiple Species Reserve.

References

Allen EB, Allen MF (1984) Competition between plants of different successional stages: mycorrhizae as regulators. Can J Bot 62:2625–2629

Allen EB, Allen MF (1986) Water relations of xeric grasses in the field: interactions of mycorrhizas and competition. New Phytol 104:559–572

Allen EB, Allen MF (1988) Facilitation of succession by the non-mycotrophic colonizer *Salsola kali* (Chenopodiaceae) on a harsh site: effects of mycorrhizal fungi. Am J Bot 75:257–266

Allen EB, Allen MF (1990) The mediation of competition by mycorrhizae in successional and patchy environments. In: Grace JB, Tilman GD (eds) Perspectives on plant competition. Academic Press, New York, pp 367–389

Allen EB, Rincon E, Allen MF, Perez-Jimenez A, Huante P (1998) Disturbance and seasonal dynamics of mycorrhizae in a tropical deciduous forest in Mexico. Biotropica 30:261–274

Allen EB, Brown JS, Allen MF (2001) Restoration and Biodiversity. In: Levin S (ed) Encyclopedia of biodiversity. Academic Press, San Diego (in press)

Allen MF (1988) Re-establishment VA of mycorrhizae following severe disturbance: comparative patch dynamics of a shrub desert and a subalpine volcano. Proc R Soc Edinb 94:63–71

Allen MF (1991) The ecology of mycorrhizae. Cambridge University Press, New York

Allen MF (2001) Modeling arbuscular mycorrhizal infection: is % infection an appropriate variable? Mycorrhiza 10:255–258

Allen MF, Allen EB (1990) Carbon source of VA mycorrhizae fungi associated with Chenopodiaceae from a semi-arid steppe. Ecology 71:2019–2021

Allen MF, Boosalis MG (1983) Effects of two species of VA mycorrhizal fungi on drought tolerance of winter wheat. New Phytol 93:67–76

Allen MF, Moore TS Jr, Christensen M (1980) Phytohormone changes in *Bouteloua gracilis* infected by vesicular-arbuscular mycorrhizae: I. cytokinin increases in the host plant. Can J Bot 58:371–374

Allen MF, Smith WK, Moore TS Jr, Christensen M (1981) Comparative water relations and photosynthesis of mycorrhizal and non-mycorrhizal *Bouteloua gracilis* (J.B.K.) Lag ex Steud. New Phytol 88:683–693

Allen MF, Moore TS Jr, Christensen M (1982) Phytohormone changes in *Bouteloua gracilis* infected by vesicular-arbuscular mycorrhizae. II. Altered levels of gibberellin-like substances and abscisic acid in the host plant. Can J Bot 60:468–471

Allen MF, Allen EB, Stahl PD (1984) Differential niche response of *Bouteloua gracilis* and *Pascopyrum smithii* to VA mycorrhizae. Bull Torrey Bot Club 111:361–365

Allen MF, Allen EB, Friese CF (1989) Responses of the non-mycotrophic plant *Salsola kali* to invasion by vesicular-arbuscular mycorrhizal fungi. New Phytol 111:45–50

Allen MF, Morris SJ, Edwards F, Allen EB (1995) Microbe-plant interactions in mediterranean-type habitats: shifts in fungal symbiotic and saprophytic functioning in response to global change. In: Morene JM (ed) Global change and mediterranean-type ecosystems. Ecological studies vol 117. Springer, Berlin Heidelberg New York, pp 287–305

Allen MF, Allen EB, Zink TA, Harney S, Yoshida LC, Siguenza C, Edwards F, Hinkson C, Rillig M, Bainbridge D, Doljanin C, MacAller R (1999) Soil microorganisms. In: Walker LR (ed) Ecosystems of the world; ecosystems of disturbed ground. Elsevier, New York, pp 521–544

Bago B, Shachar-Hill Y, Pfeffer PE (2000) Dissecting carbon pathways in arbuscular mycorrhizas with NMR spectroscopy. In: Podila GK, Douds DD Jr (eds) Current advances in mycorrhizae research. The American Phytopathological Society, St. Paul, Minnesota, p 193

Bever JD (1994) Feedback between plants and their soil communities in an old field community. Ecology 75:1965–1977

Bever JD, Morton JB, Antonovics J, Schultz PA (1996) Host-dependent sporulation and species diversity of arbuscular mycorrhizal fungi in a mown grassland. J Ecol 84:71–82

Bidartondo MI, Kretzer AM, Pine EM, Bruns TD (2000) High root concentration and uneven ectomycorrhizal diversity near *Sarcodes sanguinea* (Ericaceae): a cheater that stimulates its victims? Am J Bot 87:1783–1788

Brownlee C, Duddridge JA, Malibari A, Read DJ (1983) The structure and function of mycelial systems of ectomycorrhizal roots with special reference to their role in forming inter-plant connections and providing pathways for assimilate and water transport. Plant Soil 71:433–443

Caldwell MM, Eissenstat DM, Richards JH, Allen MF (1985) Competition for phosphorus: differential uptake from dual-isotope-labeled soil interspaces between shrub and grass. Science 229:384–386

Chalot M, Brun A (1998) Physiology of organic nitrogen acquisition by ectomycorrhizal fungi and ectomycorrhizas. FEMS Microbiol Rev 22:21–44

Chiarello N, Hickman JC, Mooney HA (1982) Endomycorrhizal role for interspecific transfer of phosphorus in a community of annual plants. Science 217:941-943

Chilvers GA, Lapeyrie FF, Horan DP (1987) Ectomycorrhizal vs. endomycorrhizal fungi within the same root system. New Phytol 107:441-448

Duan XG, Neuman DS, Reiber JM, Green CD, Saxton AM, Auge RM (1996) Mycorrhizal influence on hydraulic and hormonal factors implicated in the control of stomatal conductance during drought. J Exp Bot 47:1541–1550

Egerton-Warburton LM, Allen EB (2000) Shifts in arbuscular mycorrhizal communities along an anthropogenic nitrogen deposition gradient. Ecol Appl 10:484–496

Ek H, Andersson S, Soderstrom B (1996) Carbon and nitrogen flow in Silver birch and Norway spruce connected by a common mycorrhizal mycelium. Mycorrhiza 6:465–467

Eom AH, Harnett DC, Wilson GWT (2000) Host plant species effects on arbuscular mycorrhizal fungal communities in tallgrass prairie. Oecologia 122:435–444

Finlay RD, Read DJ (1986a) The structure and function of the vegetative mycelium of ectomycorrhizal plants: I. translocation of carbon-14 labeled carbon between plants interconnected by a common mycelium. New Phytol 103:143–156

Finlay RD, Read DJ (1986b) The structure and function of the vegetative mycelium of ectomycorrhizal plants: II. the uptake and distribution of phosphorus by mycelial strands interconnecting host plants. New Phytol 103:157–166

Fitter AH (1977) Influence of mycorrhizal infection on competition for phosphorus and potassium by two grasses. New Phytol 79:19-125

Fitter AH, Graves JD, Watkins NK, Robinson D, Scrimgeour C (1998) Carbon transfer between plants and its control in networks of arbuscular mycorrhizas. Funct Ecol 12:406–412

Francis R, Read DJ (1984) Direct transfer of carbon between plants connected by vesicular-arbuscular mycorrhizal mycelium. Nature 307:53–56

Francis R, Finlay RD, Read DJ (1986) Vesicular-arbuscular mycorrhiza in natural vegetation systems: IV. Transfer of nutrients in inter-specific and intra-specific combinations of host plants. New Phytol 102:103–111

Friese CF, Allen MF (1991) The spread of VA mycorrhizal fungal hyphae in the soil: inoculum types and external hyphal architecture. Mycologia 83:409–418

Friese CF, Allen MF (1993) The interaction of harvester ants and VA mycorrhizal fungi in a patchy semi-arid environment: the effects of mound structure on fungal dispersion and establishment. Funct Ecol 7:13–20

Gange AC, Brown VK, Sinclair GS (1993) Vesicular-arbuscular mycorrhizal fungi: a determinant of plant community structure in early succession. Funct Ecol 7:616–622

Gehring CA, Whitham TG (1995) Duration of herbivore removal and environmental stress affect the ectomycorrhizae of pinyon pines. Ecology 76:2118–2123

Gehring CA, Cobb NS, Whitham TG (1997) Three-way interactions among ectomycorrhizal mutualists, scale insects, and resistant and susceptible pinyon pines. Am Nat 149:824–841

Gehring CA, Theimer TC, Whitham TG, Kiem P (1998) Ectomycorrhizal fungal community structure of pinyon pine growing in two environmental extremes. Ecology 79:1562–1572

Grace JB, Tilman GD (1990) Perspectives on plant competition. Academic Press, New York

Graves JD, Watkins NK, Fitter AH, Robinson D, Scrimgeour C (1997) Intraspecific transfer of carbon between plants linked by a common mycorrhizal networks. Plant Soil 192:153–159

Green CD, Stodola A, Auge RM (1998) Transpiration of detached leaves from mycorrhizal and nonmycorrhizal cowpea and rose plants given varying abscisic acid, pH, calcium and phosphorous. Mycorrhiza 8:93–99

Grime JP, Mackey JML, Hillier SH, Read DJ (1987) Floristic diversity in a model system using experimental microcosms. Nature 328:420–422

Grogan P, Baar J, Bruns TD (2000) Below-ground ectomycorrhizal community structure in a recently burned Bishop pine forest. J Ecol 88:1051–1062

Hartnett DC, Wilson WT (1999) Mycorrhizae influence plant community structure and diversity in tallgrass prairie. Ecology 80:1187–1195

Helm DJ, Allen EB, Trappe JM (1996) Mycorrhizal chronosequence near exit glacier, Alaska. Can J Bot 74:1496–1506

Helm DJ, Allen EB, Trappe JM (1999) Plant growth and ectomycorrhiza formation by transplants on deglaciated land near exit glacier, Alaska. Mycorrhiza 8:297–304

Hickson LD (1993) The effects of vesicular-arbuscular mycorrhizal fungi on the light harvesting, gas exchange, and architecture of sagebrush *(Artemisia tridentata* ssp. *tridentata)*. MS Thesis, San Diego State University, San Diego, 95 pp

Horton TR, Bruns TD, Parker VT (1999) Ectomycorrhizal fungi associated with *Arctostaphylos* contribute to *Pseudotsuga menziesii* establishment. Can J Bot 77:93–102

Johnson NC, Edgerton-Warburton LM, Allen EB (1997) Mycorrhizal responses to nitrogen eutrophication at six mesic to semiarid sites. Bull Ecol Soc Am 78:118

Karen O, Nylund JE (1997) Effects of ammonium sulphate on the community structure and biomass of ectomycorrhizal fungi in a Norway spruce stand in southwestern Sweden. Can J Bot 75:1628–1642

Klironomos JN, Rillig MC, Allen MF (1999) Designing belowground field experiments with the help of semi-variance and power analyses. Appl Soil Ecol 12:227–238

Klironomos JN, McCune J, Hart M, Neville J (2000) The influence of arbuscular mycorrhizae on the relationship between plant diversity and productivity. Ecol Lett 3:137–141

Lodge DJ, Wentworth TR (1990) Negative associations among VA-mycorrhizal fungi and some ectomycorrhizal fungi inhabiting the same root systems. Oikos 57:347–356

Lundberg P, Ranta E, Kaitala V (2000) Species loss leads to community closure. Ecol Lett 3:465–468

Lussenhop J, Fogel R (1999) Seasonal change in phosphorus content of *Pinus strobus-Cenococcum geophilum* ectomycorrhizae. Mycologia 91:742–746

Marler MJ, Zabinski CA, Callaway RM (1999a) Mycorrhizae indirectly enhance competitive effects of an invasive forb on a native bunchgrass. Ecology 80:1180–1186

Marler MJ, Zabinski CA, Wojtowicz T, Callaway RM (1999b) Mycorrhizae and fine root dynamics of *Centaurea maculosa* and native bunchgrasses in western Montana. Northwest Sci 73:217–224

Massicotte HB, Molina R, Tackaberry LE, Smith JE, Amaranthus MP (1999) Diversity and host specificity of ectomycorrhizal fungi retrieved from three adjacent forest sites by five host species. Can J Bot 77:1053–1076

Molina R, Massicotte H, Trappe JM (1992) Specificity phenomena in mycorrhizal symbiosis: community-ecological consequences and practical implications. In: Allen MF (ed) Mycorrhizal functioning. Chapman and Hall, New York, pp 357–423

Morton JB, Redecker D (2001) Two new families of Glomales, Archaeosporaceae and Paraglomaceae, with two new genera *Archaeospora* and *Paraglomus*, based on concordant molecular and morphological characters. Mycologia 93:181–195

Mullen R, Schmidt S (1993) Mycorrhizal infection, phosphorus uptake, and phenology in *Ranunculus adoneus*: implications for the functioning of mycorrhizae in alpine systems. Oecologia 94:229–234

Mullen RB, Schmidt SK, Jaeger CH (1998) Nitrogen uptake during snowmelt by the snow buttercup, *Ranunculus adoneus*. Arctic Alpine Res 30:121–125

Naeem S, Daniel R, Schuurman G (2000) Producer-decomposer co-dependency influences biodiversity effects. Nature 403:762–764

Perry DA, Margolis H, Choquette C, Molina R, Trappe JM (1989) Ectomycorrhizal mediation of competition between coniferous tree species. New Phytol 112:501–512

Podila GK, Douds DD Jr (2000) Current advances in mycorrhizae research. American Phytophathological Society, St. Paul, Minnesota

Pregitzer KS, DeForest JL, Burton AJ, Allen MF, Ruess RW, Hendrick RL (2001) Fine root length, diameter, specific root length and nitrogen concentration of nine tree species across four North American biomes. Ecology (in press)

Read DJ (1997) Mycorrhizal fungi: the ties that bind. Nature 388:517–518

Redecker D, Morton JB, Bruns TD (2000) Ancestral lineages of arbuscular mycorrhizal fungi (glomales). Mol Phylogen Evol 14:276–284

Reid CPP, Woods FW (1969) Translocation of C^{14}-labeled compounds in mycorrhizae and its implications in interplant nutrient cycling. Ecology 50:179–181

Rillig MC, Wright SF, Allen MF, Field CB (1999) Rise in carbon dioxide changes in soil structure. Nature 400:628

Rossow LJ, Bryant JP, Kielland K (1997) Effects of above-ground browsing by mammals on mycorrhizal infection in an early successional taiga ecosystem. Oecologia 110:94–98

Seastedt TR, Knapp AK (1993) Consequences of nonequilibrium resource availability across multiple time scales: the transient maxima hypothesis. Am Nat 141:621–633

Shefferson RP, Sandercock BK, Proper J, Beissinger SR (2001) Estimating dormancy and survival of a rare herbaceous perennial using mark-recapture models. Ecology 82:145–156

Simard SW, Jones MD, Durall DM, Perry DA, Myrold DD, Molina R (1997a) Reciprocal transfer of carbon isotopes between ectomycorrhizal *Betula papyrifera* and *Pseudotsuga menziesii*. New Phytol 137:529–542

Simard, SW, Perry DA, Jones MD, Myrold DD, Durall DM, Molina R (1997b) Net transfer of carbon between ectomycorrhizal tree species in the field. Nature 388:579–582

Smith SE, Read DJ (1997) Mycorrhizal symbiosis. Academic Press, San Diego

Stahl PO, Smith WK (1984) Effects of different geographic isolates of *Glomus* on the water relations of *Agropyron smithii*. Mycologia 76:261-267

Taylor DL, Bruns TD (1999) Community structure of ectomycorrhizal fungi in a *Pinus muricata* forest: minimal overlap between the mature forest and resistant propagule communities. Mol Ecol 8:1837–1850

Taylor AFS, Martin F, Read DJ (2000) Fungal diversity in ectomycorrhizal communities of Norway spruce (*Picea abies* (L.) Karst.) and beech (*Fagus sylvatica* L.) along north-south transects. In: Schulze ED (ed) Carbon and nitrogen cycling in European forest ecosystems. Ecological studies, vol 142. Springer, Berlin Heidelberg New York, pp 343–365

Trappe JM (1977) Selection of fungi for ectomycorrhizal inoculation in nurseries. Annu Rev Phytopathol 15:203–222

Van der Heijden MGA, Klironomos JN, Ursic M, Moutoglis P, Streitwolf-Engel R, Boller T, Wiemken A, Sanders IR (1998) Mycorrhizal fungal diversity determines plant biodiversity, ecosystem variability and productivity. Nature 396:69–72

Wallander H, Arnebrant K, Ostrand F, Karen O (1997) Uptake 15 N-labelled alanine, ammonium and nitrate in *Pinus sylvestris* L. ectomycorrhiza growing in forest soil treated with nitrogen, sulphur or lime. Plant Soil 195:329–338

Wang GM, Coleman DC, Freckman DW, Dyer MI, McNaughton SJ, Acra MA, Goeschl JD (1989) Carbon partitioning patterns of mycorrhizal versus non-mycorrhizal plants: real-time dynamic measurements using $^{11}CO_2$. New Phytol 112:489–493

Wang, GM, Stribley DP, Tinker PB, Walker C (1993) Effects of pH on arbuscular mycorrhiza. I. Field observations on the long-term liming experiments at Rothamsted and Woburn. New Phytol 124:465–472

Weinbaum BS, Allen MF, Allen EB (1996) Survival of arbuscular mycorrhizal fungi following reciprocal transplanting across the Great Basin, USA. Ecol Appl 6:1365–1372

Whittaker RH (1970) Communities and ecosystems. MacMillan, New York

Wilkinson DM (1999) Mycorrhizal networks are best explained by a plurality of mechanisms: a comment on *Fitter et al.*, 1998. Funct Ecol 13:435–436

Yoshida LC (1999) Ammonium and nitrate additions to a mycorrhizal annual invasive grass, *Bromus madritensis*, and *Artemisia californica* in coastal sage scrub: greenhouse and field experiments. PhD, University of California, Riverside

Prof. Dr. Michael F. Allen
Jennifer Lansing
Dr. Edith B. Allen
Center for Conservation Biology
University of California
Riverside, California 92521–0334, USA
Tel.: +1-909-787-5494
Fax: +1-909-787-4625
e-mail: mallen@citrus.ucr.edu

History of Flora and Vegetation During the Quaternary

By Burkhard Frenzel

This report is a continuation of the discussions on the Quaternary paleoclimatology and paleoecology of the Tibetan Plateau given in Progress in Botany, vol. 59, 1998. It was made possible by the translation of several new relevant Chinese papers by Prof. Dr. Liu Shijian, Institute of Mountain Disasters and Environment, Chinese Academy of Sciences, Chengdu, when he worked as a fellow of the Max Planck Gesellschaft zur Förderung der Wissenschaften from the month of June 2000 to January 2001 at the Institute of Botany of the Hohenheim University. I am strongly indebted to the Max-Planck-Gesellschaft for this financial and scientific help.

1 Modern Sedimentation of Sporomorphs on the Tibetan Plateau

A prerequisite for reliable interpretations of fossil pollen samples in terms of the former vegetation is a good knowledge of the interrelations existing between modern vegetation and recent sedimentation of sporomorphs at the sites studied. Valuable information is given meanwhile by Xu (1999) on the Mt. Qomolangma region, by Schlütz (1999) on the Nianbaoyeze Shan, northeastern Tibet, and by van Campo et al. (1996) on the surroundings of Lake Bangong, western Tibet. Observations of this type were used by Tang and Shen (1999) for the description of main regional trends in the Late-Glacial to Holocene vegetation history of the Tibetan Plateau. The background for these studies is a good knowledge of the present-day distribution pattern of the most important vegetation-types. They are given, e.g. in the Atlas of the Tibetan Plateau (1990), yet the data shown there are not always correct. It must be stressed, for instance, that in the very vicinity of the Nianbaoyeze Shan, vast and dense forests of various fir, spruce, birch and alder species do thrive, interrupted by juniper/*Sabina* stands, on sites which formerly were cleared of forests, though this vegetation is not mapped in the atlas just mentioned. The same holds for several other sites, most of all in eastern and eastern-central Tibet in the vicinity of the former natural boundary of forests against the steppe-areas (see Frenzel 1998, Fig. 8; 2000). Thus, in this respect much more research-work has to be done.

In western Tibet, i.e. the Bangong Co area, van Campo et al. (1996) stress the importance of the *Artemisia* to Chenopodiaceae pollen-ratio for describing former paleoecological conditions. High values are held to indicate somewhat moister conditions than low values do. This may hold for the region studied, but the author feels that this conclusion should not uncritically be extrapolated to other regions of the Tibetan Plateau since (1) according to van Campo et al. (1996), the modern pollen curves of *Artemisia*, Chenopodiaceae, Poaceae and Cyperaceae in the region studied show a very striking *Gleichläufigkeit* (parallelism) in time (September 1989 to the end of August 1990); this points to governing meteorological factors for all of these taxa studied, influencing their pollen production in more or less the same way; (2) the ecological requirements of the various *Artemisia*-taxa of the Tibetan Plateau are very different from one another; and (3) according to the author's own experiences in central and eastern Tibet, Chenopodiaceae are relatively rare there, and mostly concentrated on local haline sites only. Thus at least in central and eastern Tibet the *Artemisia/Chenopodiaceae* pollen-ratio does not seem to unequivocally inform about past major vegetation types.

2 Old and Middle Quaternary Vegetation History

Thanks to deep geological borings, which have recently been made in various parts of the Tibetan Plateau and in the surrounding mountain systems, some information on vegetation history of various parts of the Old and Middle Quaternary could meanwhile be obtained. These investigations concentrated on the Mt. Qomolangma region (Cheng 1999; Xu 1999), the area of Tianshuihai, western Kunlun Mts. (Liu et al. 1998), the Qingshuihe region, northern Tibet at the Qinghai-Xizang highway (Tang and Wang 1999) and on the Zoigê-basin, northwestern Sichuan (Liu et al. 1994). Within the Mt. Qomolangma region the following stratigraphical sequence seems in general to be accepted now.

Upper Quaternary	Rongpude stadial = Little Ice Age Holocene Climatic Optimum Qomolangma Glaciation with Yali Interstadial
Middle Quaternary	Jiabula Interglacial Nyanyaxungla Glaciation
Lower Quaternary	Pali Interglacial Xixabangma Glaciation

Whether this stratigraphical scheme can directly be compared to that of other parts of the globe is uncertain. According to Cheng (1999) and Xu (1999) in several geological profiles of the Mt. Qomolangma region, pollen and even macrofossil floras (Xu 1999) could be detected. In general, the sum of sporomorphs analyzed per sample is small (between 24 and 275 sporomorphs, mean value about 96 sporomorphs). This impairs a reliable reconstruction of vegetation history.

According to Xu (1999), the Pliocene pollen flora dating from shortly before the onset of the Quaternary on the southern flank of Mt. Nyanyaxungla, to the northeast of Mt. Xixabangma, reflects a *Quercus* forest with some *Abies*, *Picea*, *Pinus*, *Tsuga*, *Betula* and *Lithocarpus*, though the site is at an elevation of 4850 m a.s.l., some 1800 to 2300 m higher up than their modern equivalents. A comparable situation is reported by Cheng (1999) concerning the Pali and the Jiabula interglacials. In sediments of the Jiabula Interglacial imprints of *Sabina recurva*, *Picea spinulosa*, *Populus* sp. and *Lespedeza* sp. were found at an elevation of 5000 to 5200 m a.s.l. The question is how old the fossils are in reality. The author thinks that the correlation of this Jiabula Interglacial with the Last Interglacial cannot be correct since this would mean an uplift of the region studied, by about 2500 m within ca. 120,000 years. To conclude from the very scarce pollen flora, the old-Pleistocene glacial vegetation, on the other hand, seems to have been an alpine dwarf-shrub and meadow community (Cheng 1999).

Liu et al. (1998) report on the late Middle and Upper Pleistocene vegetation history of the Tianshuihai region, western Kunlun Mts. The age data given in this paper are based on ^{14}C and U/Th data. It seems that the last 240,000 years are reflected in the 56.32-m-deep boring. Yet pollen analytical work suffers from a very low pollen density. In general ca. 20 sporomorphs per sample could be analyzed only; occasionally up to ca. 300 were counted. It seems that independently of climate history, the vegetation there at an elevation of about 4900 m a.s.l. was always governed by *Chenopodiaceae* and some *Artemisia* or *Ephedra* only. Yet, between about 59,000 to 17,000 B.P., the share of *Artemisia* had increased considerably, indicating somewhat moister conditions. This is supported by other comparable observations (Progress in Botany, vol. 59, pp. 612, 613). On the other hand, Tang and Wang (1999) report on strong changes in the Upper-Quaternary vegetation in the northern part of the Tibetan Plateau, i.e. within the catchment area of Qingshuihe, 4400 m a.s.l. Here, a 203-m-deep geological boring was analyzed palynologically. It is said that two phases rich in arboreal pollen (*Picea, Pinus, Betula*, in the second phase even *Quercus* and *Carpinus*) were divided from one another by a forest-steppe and desert phase. However, I feel that the very small number of pollen grains counted does not allow these conclusions to be drawn.

Liu et al. (1994) studied palynologically a 120.4-m-long boring made in the Zoigê-basin, northwestern Sichuan, at an elevation of 3396 m a.s.l. Here, too, the content of sporomorphs in peaty, silty and sandy layers is very low (in general some 10,000 grains per 100 g sediment). According to paleomagnetic, conventional ^{14}C and AMS datations, it is considered that the base dates from about 826,000 B.P. The profile ends at ca. 150,000 years B.P. At present, the vast tectonic Zoigê-basin is surrounded by forest-clad mountain systems. If the datations given should approach reality, there have repeatedly been forest phases (*Abies, Picea, Quercus,* some *Tsuga* and *Betula*) at about 826,000–813,000, 805,000–793,000, 480,000–450 000, and 250,000–234,000 B.P. At about 730,000–710,000, 650,000–630,000, 592,000–505,000, 410,000–330,000, 330,000–298,000, and 211,000–150,000 years B.P., on the other hand, a forest steppe with *Abies* and *Picea* at various times, seems to have surrounded the basin. All these phases were divided from one another by periods in which, either no sporomorphs were sedimented, or they were destroyed. Though from a general point of view, the "forest-phases" mentioned might have existed, the small number of sporomorphs counted and the very simple pollen diagram seem to discourage a more detailed and critically done analysis of what had happened in reality.

The various "forest-phases" discussed just now are repeatedly used to reconstruct the history of climate (Liu et al. 1994; Cheng 1999; Tang and Wang 1999; Xu 1999). If this is based on sufficient material (sporomorphs and/or macrofossils), the results will be quite reliable, yet if a very small number of pollen grains per sample is counted, the results may be strongly misleading. Another difficulty is caused by an inaccurate dating of sediment-layers. The Jiabula Interglacial just mentioned may serve as an example. It is characterized, at the sites studied, by an exacting mesic forest vegetation although the site investigated is situated now at an elevation of 4900 to 5100 m under extremely cold conditions. It is stressed that modern equivalents of the fossil flora and vegetation are thriving some 1600–1800 m deeper in moister valleys (Cheng 1999). No doubt the sediments studied belong to the uppermost interglacial layers at the site investigated. But have these sediments been formed during the Last Interglacial? If so, the rate of uplift of the region studied should have amounted to ca. 16 mm/year. This should have influenced the climate and thus the development of paleoecological conditions strongly. But is the correlation of the Jiabula Interglacial with the Last Interglacial proven? We have to discuss these problems again when the Holocene is dealt with.

3 The Last Glaciation

Most of the relevant geological and botanical problems of the Last Glaciation on the Tibetan Plateau have already been dealt with in Progress in Botany, vol. 59, 1998. Again, Kuhle (1998, 2000) discusses the problem of the hypothetical existence of a huge inland ice during the Last Glaciation on the Tibetan Plateau. For reasons of space and time, this will not be elucidated here. Only the following may be stated: In the eastern part of the West-Tibetan lake district, middle-paleolithic stone artifacts could be found repeatedly in situ (the Siling Co, Dagce Co area) at an elevation of about 4600 to 4730 m a.s.l., still lying in the context of their former production (Frenzel et al. 2001). The area is situated near the center of the hypothetical inland ice region. This contradicts the existence of this ice sheet. Yet, it is another problem why sediments of the last glaciation are so rare on the Tibetan Plateau. As already stated (Progress in Botany, vol. 59, 1998, p. 609 ff.) glacigene sediments and landforms dating from early and full glacial times of the last glaciation are repeatedly found there in mountain systems, higher than ca. 4900 m, and in their immediate vicinity, but not on most regions of the plateau itself. Lehmkuhl et al. (2001) stressed that, according to the results of the joint 1996 Chinese-German expedition to central and western Tibet, repeatedly loesses and loess-like sediments could be found on the plateau. In general, they date from Late-Glacial times only. On the other hand, the results of the author's own expeditions to eastern, central, southeastern and western Tibet (1989, 1992, 1996) enabled the conclusion to be drawn that on the Tibetan Plateau and within the mountain systems of western Sichuan, even thick full glacial loesses of various glaciations do occur (Progress in Botany, 59, p. 608, 1998), however, only at levels below about 4000 m a.s.l., as could be shown by a new analysis of the observations contained in the diaries of these expeditions. Evidently the climate on the plateau had favored periglacial, aeolian denudation processes most of all, so that now full glacial sediments are lacking in vast regions. In consequence, palynological data on the full glacial vegetation of the Tibetan Plateau are very scarce and even from the late glacial stadials, only very scanty observations on a desert-steppe vegetation, poor in botanical taxa, are available (e.g. Tang and Shen 1999). On the other hand Schlütz (1998, 1999) reported very convincingly and generally based on good pollen samples, on the vegetation history of the last glaciation in the Qinling Shan, central China. Though at present no physical datations are available, Schlütz convincingly demonstrates that in all probability, the vegetation of this mountain system had changed repeatedly and intensively during the last glaciation. Evidently during the equivalents of deep-sea stages (MIS) 5c and 5a, a forest vegetation had existed there at about 2500 m a.s.l., which strongly resembled that of the present day natural vegetation, whereas during the stadials, an alpine meadow vegetation had developed into which loesses or other silty material was transported, together with several long distance transported forest-sporomorphs.

Hydrological and paleoecological conditions prevailing during the last glaciation in Mongolia and the Qinghai Hu (KuKuNor) basin are extensively dealt with by Walther (1998), Wünnemann et al. (1998), and Lister et al. (1991).

4 The Holocene

Several papers deal with problems of vegetation history and paleoecology of the Late-Glacial and Holocene on the Tibetan Plateau and in surrounding regions: northwestern Tibet, the Karakorum and the northwestern Himalayas: Avouac et al. (1986); van Campo et al. (1996); Fan et al. (1996); Fontes et al. (1996); Gasse et al. (1996); Schlütz (1998, 1999); central and eastern Himalaya: Beug and Miehe (1998); Schlütz (1999); Xu (1999); Nianbaoyeze Shan, northeastern Tibet: Schlütz (1999); a summary of some important papers on the Tibetan Holocene vegetation history: Tang and Shen (1999); lake depressions in Mongolia and the Qinghai Hu area: Lister et al. (1991); Wünnemann et al. (1998); Naumann and Walther (2000). These investigations are paralleled by paleopedological studies in the western Tien Shan and in Nepal by Bäumler and Zech (2000); in Kirgizia and Uzbekistan by Zech et al. (2000); dendroclimatology in eastern Tibet: Bräuning (1994a,b, 1999, 2000); Bräuning and Lehmkuhl (1996); Zimmermann et al. (1997); and in northwestern Himalaya: Esper et al. (1996). These very interesting details should be studied in the papers mentioned. Here, another general problem will be dealt with instead. This is the major trend in the paleoecological evolution of the Tibetan Plateau.

It is a well-known fact now that the water budget in vast regions of the Tibetan Plateau and in the lower lying modern desert regions to the north had remarkably improved by about 10,000 to 9600 years B.P. (conventional ^{14}C ages). This is stressed in nearly all papers mentioned, as far as the Late-Glacial to Holocene transition has been analyzed. It could already be shown that this important climatological and edaphic change had happened in various regions of the study area not exactly at the same time, but it was felt in southern regions about 1300 years earlier than in most other regions, whereas farther to the north it seems to have happened somewhat later (Frenzel 1994). On the other hand, the phase of optimal hydrological conditions had lasted up to the end of Mid-Holocene times (e.g. see Gasse et al. 1996; van Campo et al. 1996; Wünnemann et al. 1998; Naumann and Walther 2000). Within this phase of optimal climatic conditions there had repeatedly been minor oscillations to drier conditions. Yet the termination of the general phase of optimal climate has evidently not been a sudden change, but the consequences were felt from about 6000 or 6200 B.P. to about 3500 to 2500 B.P., with remarkable regional differences in its timing. At approximately the same time, forest seems to have retreated in various regions of the (at present) moister parts of the Tibetan Plateau and in western Sichuan. In northwestern Sichuan, in the Zoigê-basin, this retreat of the forests had begun at about 5200 to 4600 B.P., it was strongly intensified at ca. 2400 to about 2200 B.P. Thelaus (1992) and Frenzel (1994) had ascribed these two phases of forest retreat to the activities of early herdsmen. But this view is not held by other authors. Instead, it is thought that climate change had caused remarkable changes in vegetation there. It is thought that human impact there dates from the last 800 to about 2500 years or so (e.g. Beug and Miehe 1998; Schlütz 1999; Tang and Sheng 1999).

No doubt, the climate had changed during the Holocene repeatedly. Yet the question is how strong these climate changes might have been. Xu (1999) describes holocene imprints of *Lonicera* cf. *hispida*, *Viburnum* cf. *erubescens*, *Rhododendron* cf. *hypenanthum*, *Rosa* sp., *Spiraea* sp., *Rhamnus* sp., *Lonicera* cf. *tomentella* and *Salix* sp. in the Mt. Qomolangma region. They were found in holocene travertine layers at an elevation of 4300 m, 500 to 900 m higher up than these plants are said to occur in nearby valleys today. From this, it is concluded that the area studied should have been tectonically uplifted by about 500 m or so. Man's influence in depressing the vertical distribution boundaries of these taxa is not taken into consideration, though it is said that, synchronously with the plants mentioned, man had been there, as is evidenced by "fine artifacts".

At present, several archaeological observations of man's activities on the Tibetan Plateau since paleolithic times do already exist: Hou (1991); Wu (2000); literature in Frenzel et al. (2001). Yet these data cannot help prove or reject the hypothesis of an important human impact on the vegetation of the Tibetan Plateau since about 5000 B.P. or so. Instead, another hypothesis is put forward here: If climate change should have triggered the retreat of forests and the spread of grasslands on the Tibetan Plateau, these changes should have influenced other ecological processes too. For the investigation of this problem, Fig. 1 shows the age data (in conventional ^{14}C years) for the beginning of peat-formation, of now fossil humus-layers, of the beginning (and if possible, the end) of glacier advances on the Tibetan Plateau and of lake level changes of Qinghai Hu. These data are given by Lister et al. (1991), Thelaus (1992), Frenzel (1994), Bräuning and Lehmkuhl (1996), Fontes et al. (1996), van Campo et al. (1996), Beug and Miehe (1998), Schlütz (1998, 1999), Bäumler and Zech (2000), Zech et al. (2000), Lehmkuhl et al. (2001) and Frenzel (unpubl.).

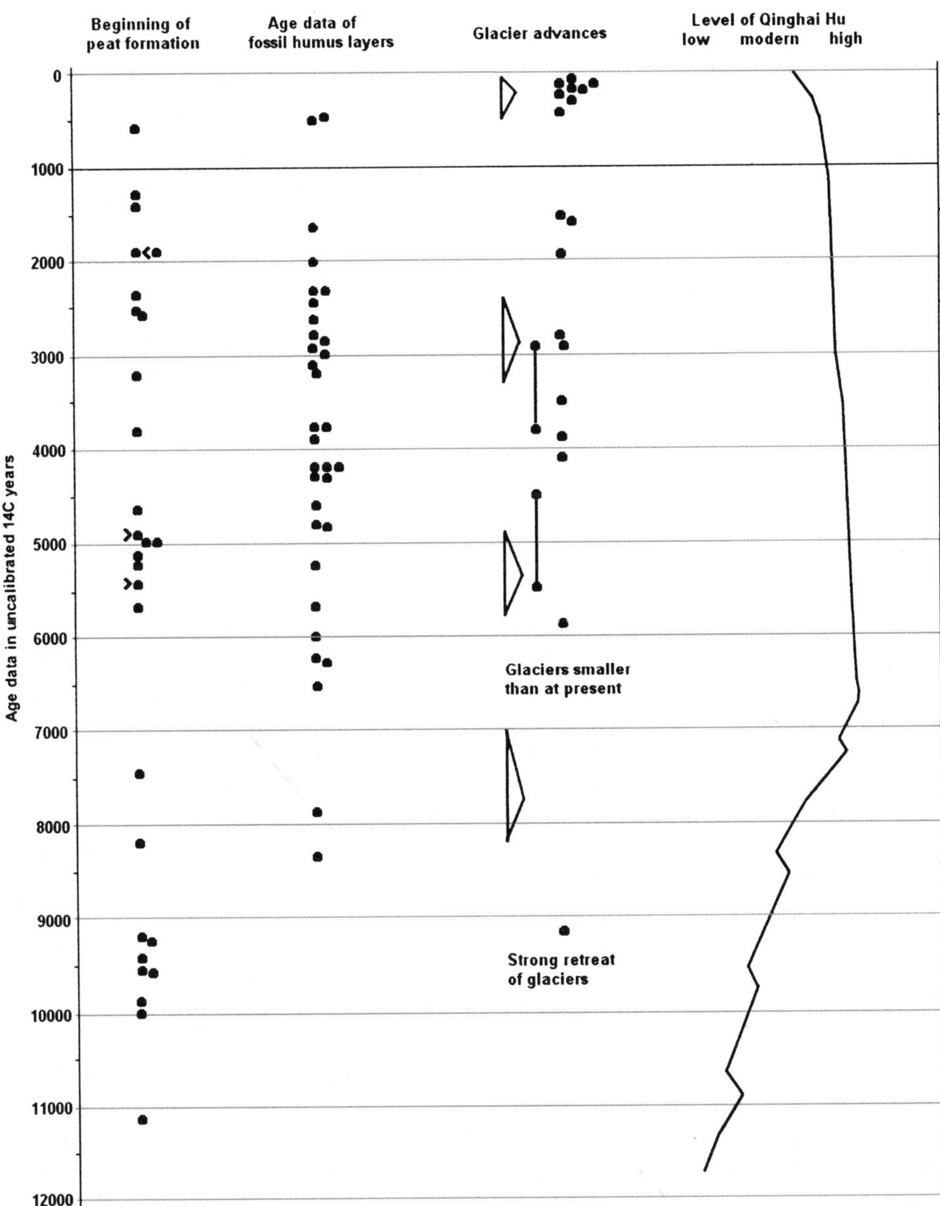

Fig. 1. Age data for the beginning of peat growth, the formation of fossil humus horizons and glacier advances together with a curve of lake-level changes of Qinghai Hu (Kuku Nor). As to the literature used, see text. - In column "*glacier advances*", the point of the *triangles* shown to the *right* indicates the time of most extensive glacier advance

The beginning of peat-formation indicates a surplus of moisture (locally or regionally). Within the relatively dry to arid regions studied, the formation of humus-layers which are now fossil, reflects first of all, phases of slope stability and sufficient moisture for at least a steppe vegetation to thrive in these mountainous regions. The termination of these pedogenetic processes indicates on the other hand an increase in instability of the slopes, either by periglacial processes, or by destruction of the vegetation cover higher up. In view of the generally very cold climates of the Tibetan Plateau and its surrounding mountain systems (permafrost is widely spread there, see Map of Snow, Ice and Frozen Ground in China 1988) glacier advances will have been caused during the Holocene mostly by an increase in moisture, at least during summer times. As seen in Fig. 1, there was such a phase of beginning of peat formation between about 10,000 and 9,000 B.P. This is in parallel with the observations mentioned that at about 10,000 to 9,500 B.P. a remarkable increase in moisture available was felt on the Tibetan Plateau. Another phase of intensified peat formation had happened from about 5700 to 4600 B.P. This is not paralleled by observations in the Bangong Co area, western Tibet (van Campo et al. 1996; Gasse et al. 1996). Yet, this region is much more influenced by the westerlies during winter time, than are most parts of central and eastern Tibet. Fan et al. (1996) and van Campo et al. (1996) state that in western Tibet between about 8600 and 7800 B.P., some short-term dry spells had occurred. As can be seen in Fig. 1, in eastern-central Tibet, at about 8300 and 7850 B.P., phases of peat-formation in a fossil lake basin had happened. Yet this lake seems to have been dammed up (in the Bien Ba-region) by mountain slope disasters (observations of the 1992 expedition). So these layers must not be interpreted as indicating phases of a drier climate. A third phase of peat formation seems to have existed from about 2500 to 1250 B.P. The level of Qinghai Hu was lowering at that time. This seems to be in contradiction to a phase of peat-formation beginning in nearby regions (Fig. 2).

It might be argued that though a beginning of peat-formation reflects moister local or regional conditions, peat-growth may have been interrupted repeatedly later on by the climate becoming drier. Figures 3–6 exemplify that there is no general tendency in the growth of various peat bogs studied, which might point to climatically induced retardations or a stop in peat-growth. Moreover, the age data of fossil humus layers or of glacier advances (Fig. 1) cannot evidently be interpreted as reflecting a climate becoming adverse or dangerous for vegetation. These data may only show that the climate from about 6000 or 6500 B.P. had become more variable than it had been before. Yet this is a well-known fact (Matthews et al. 1993; Nesje 1997). Thus the impression is that even on the Tibetan Plateau, there has not, since about 6000 B.P. (or so) been a general tendency of the climate becoming dangerous for forest growth,

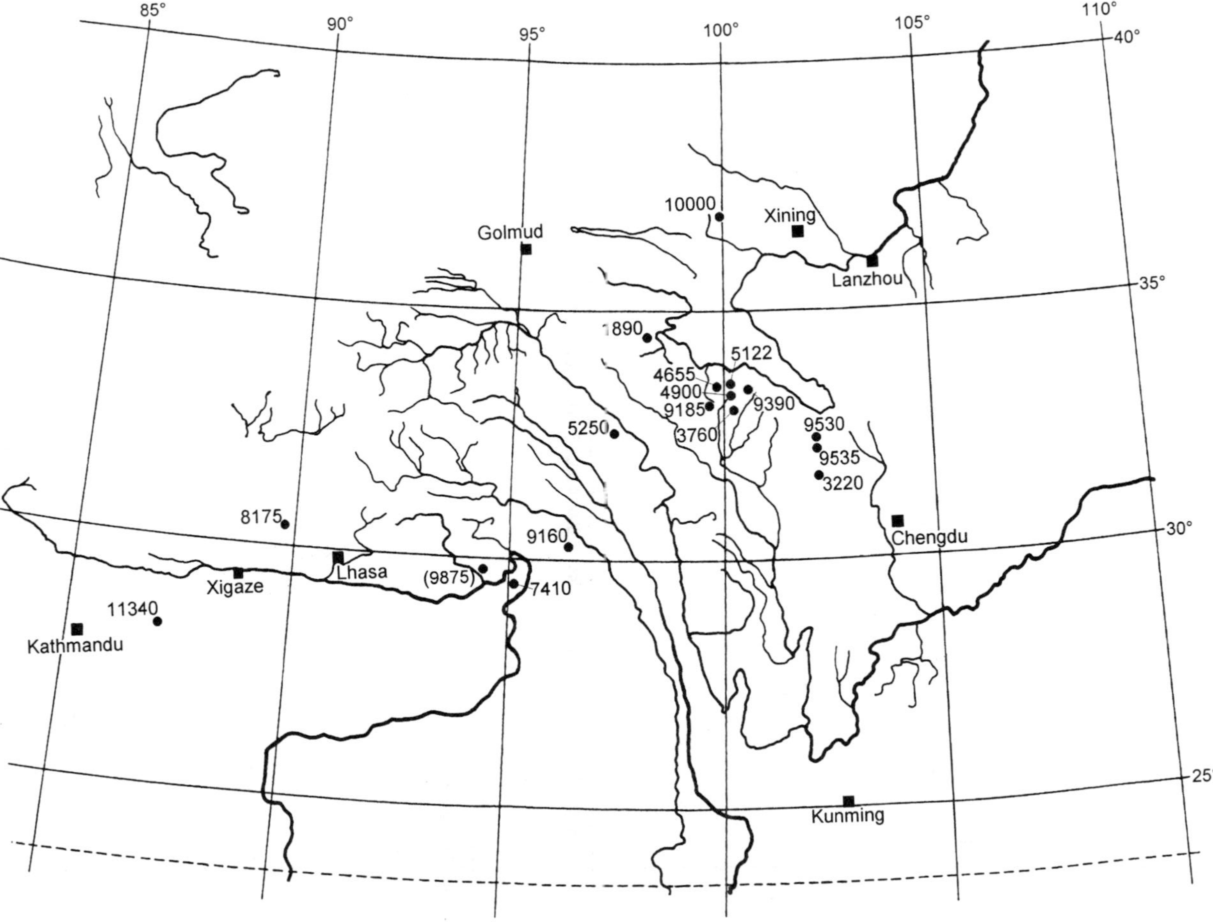

Fig. 2. Geographical distribution pattern of the beginning of peat growth in Tibet and Sichuan. Literature used, see text

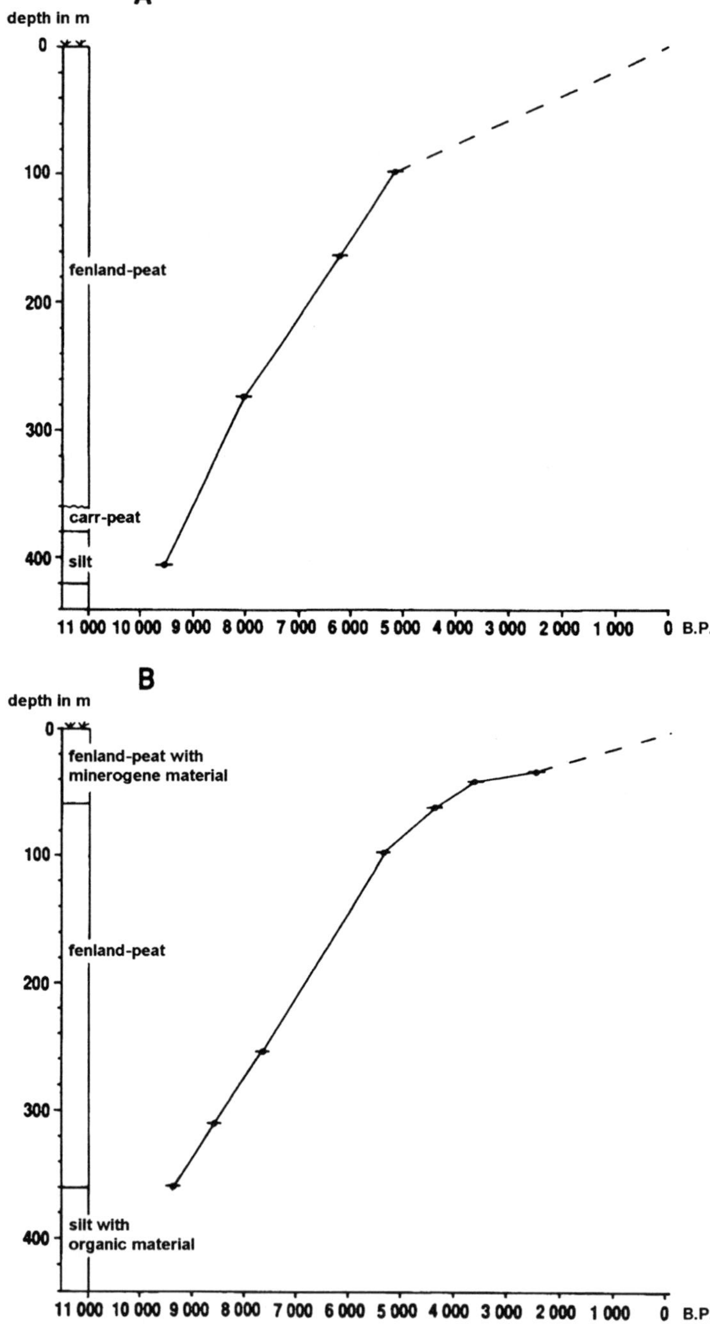

Fig. 3. Growth curves of the Hung Yuan peat bog, northwestern Sichuan, according to Thelaus (1992) [A] and Frenzel (1994) [B]

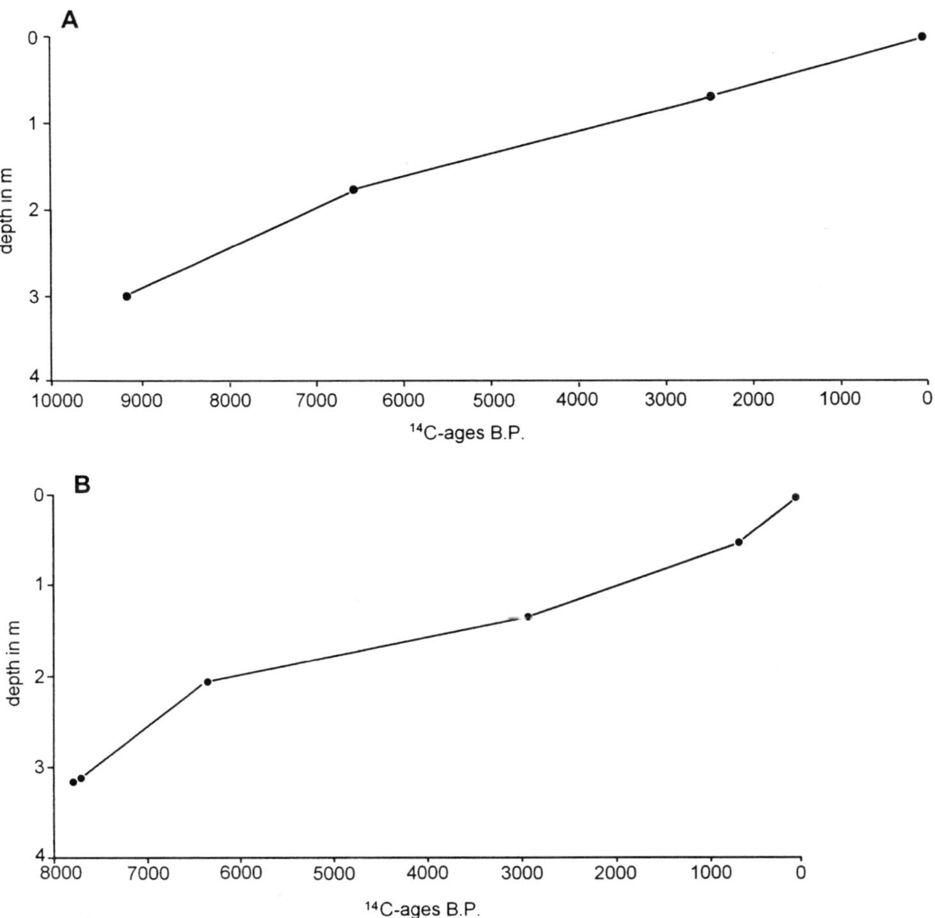

Fig. 4. Growth curves of two peat bogs, investigated by Schlütz (1999). A Peat bog of the Nianbaoyeze Shan, northeastern Tibet. B Peat bog of the Nanga Parbat region, "Märchenwiese", northwestern Himalaya

though on a local scale, the repeatedly occurring changes of climate may sometimes have become adverse or – in other places or at other times – favorable for tree growth. In this respect it must be stressed that, according to Schlütz (1999), there had been strong regional differences in the evolution of vegetation during the younger part of the Holocene.

All this taken together favors the view that the general retreat of forests in eastern and southern Tibet during the last 5000 years has not primarily been caused by climate. If so, human impact seems to have been much more important. These anthropogenic changes in the vegetation of various parts of the Himalayas as well as in southern, central, eastern Tibet and western Sichuan have been repeatedly studied or discussed: Beug and Miehe (1998); Damm (1998); Miehe et al. (1998, 2000);

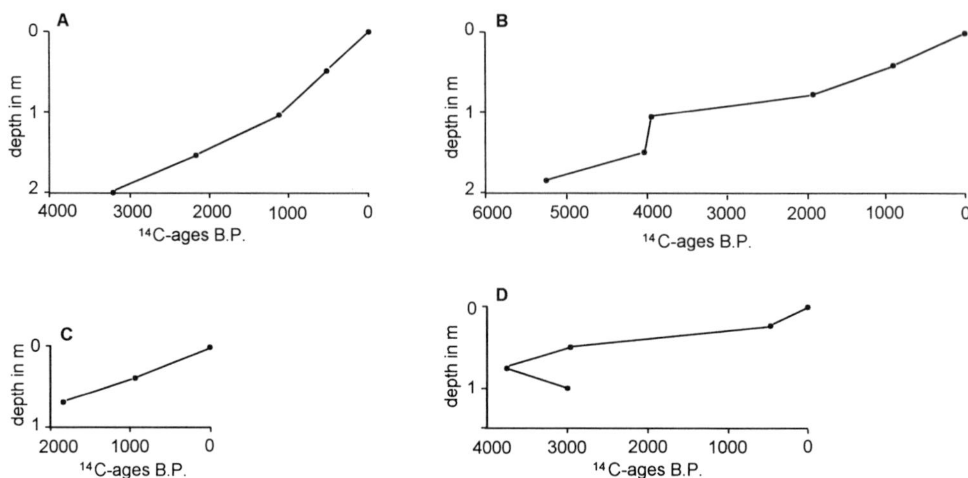

Fig. 5. Growth curves of four peat bogs in western Sichuan and eastern Tibet (Frenzel, unpubl.). A "Yak" between Hung Yuan and Aba; B Shi qü to the west of Yü shu; C Yen ma-pass, Bayan Kara Shan; D Ta xi lung ma, northwest of Ban ma, immediately to the west of Nianbaoyeze Shan

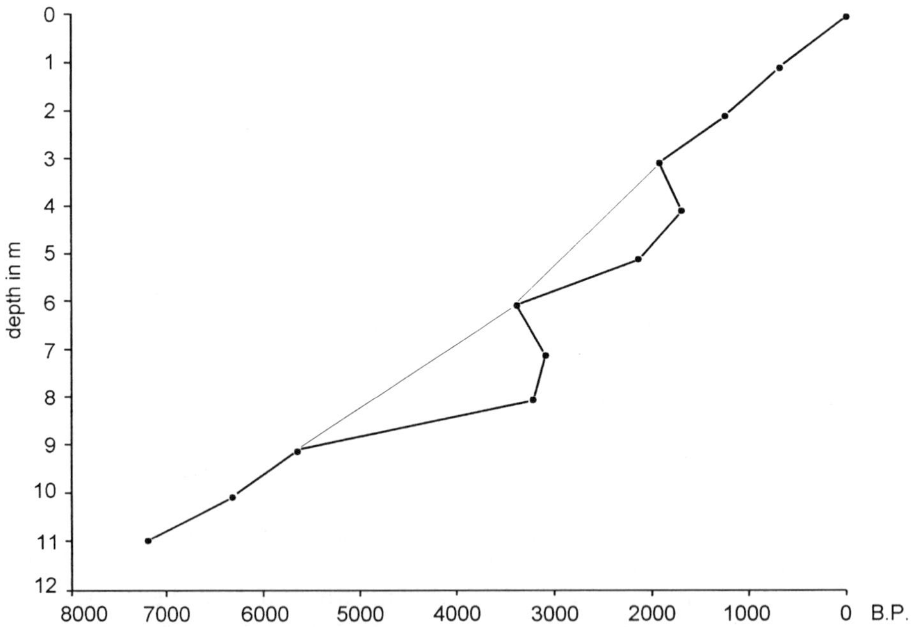

Fig. 6. Growth curve of the swimming peat bog in a glacial lake of Hai ze Shan, to the northwest of Garzê, central-eastern Tibet (Frenzel, unpubl.). The deviations from the straight line at a depth of 3–5 and 7–8 m were probably caused by yaks, which sank into the swimming bog (report of herdsmen)

Baade and Mäusbacher (2000); Miehe and Miehe (2000); Peer (2000); Schickhoff (2000); Wu (2000) (as to the Wutai Shan, Shanxi, China: Rost and Sun 2000).

In this respect another problem is very interesting: It is a well-known fact that, at present, forests or tree stands in western Sichuan and eastern Tibet seem to be confined to those mountain slopes which are exposed to the northerly quadrant. V. Wissmann (1959) explained this situation by the extremely intensive solar radiation within the high mountain systems there. Yet Winkler (1998, 2000) states that this distribution pattern of present-day forests is man-made and Frenzel (2000) mapped forests and isolated groves of various tree-taxa occurring within the regions studied at this exposed site according to observations made in his 1989, 1992 and 1996 expeditions to western Sichuan, eastern, southern, central and western Tibet. Doing so, Frenzel (2000) arrives at the same conclusions as Winkler (1998, 2000). This is proven by the fact that after stronger human impact has ceased, a re-immigration and re-colonization of formerly bare southern slopes by various tree-taxa can be observed nearly everywhere in eastern Tibet and western Sichuan. On the other hand, it is extremely difficult to prove, by means of pollen analyses, when human impact had begun there and when it began to be important. In Europe, the history of man-made forest clearances is studied very intensively by means of pollen analysis and botanical macrofossils. Here a wealth of synanthropic plant taxa enables this to be done quite reliably. In western Sichuan and eastern and central Tibet, as well, the situation is much less favorable. Certainly, there do exist synanthropic plant-taxa too. Yet they (1) originate in the spontaneous vegetation and (2) in general they are insect-pollinated. Thus their pollen production is generally small and wind does not transport the relevant pollen grains intensively. Fortunately this anthropo-zoogene influence on the Tibetan vegetation has been studied intensively, even using plant remains in the excrements of most of the Tibetan ungulates meanwhile: Ram (1992); Rikhari et al. (1992, 1993); Fu et al. (1993); Negri et al. (1993); Harris and Miller (1995); Schaller (2000). Moreover, the impact of vole grazing on the vegetation has been analyzed provisionally (Schaller 2000; Samjaa et al. 2000). This is important since the activity of voles can become very deleterious to the steppe vegetation there, as has already been repeatedly stressed by Hedin (1909). Thus some clues can be obtained now, as to which plant-taxa are favored by grazing and browsing of the Tibetan herbivores. Yet since nearly all of these taxa are insect-pollinated, this needs very comprehensive pollen analyses to be done, in which several hundred pollen grains per sample have to be determined. Several of the pollen analyses mentioned in this report seem to show that human influence on the Tibetan vegetation has been a long-lasting one (e.g. occurrence of *Saussurea* and *Aster* types, of *Pedicularis*, *Leontopodium*, *Caragana*, *Polygonum aviculare*, *Polygonum bistorta* types, *Salvia* type, etc.). However, relevant research is only at its very beginning in these fascinating regions of the globe. This holds for sound paleoecological investigations, too.

References

Atlas of the Tibetan Plateau (1990) Beijing, Institute of Geography, Chinese Academy of Sciences (in Chinese)

Avouac J-P, Dobremez J-F, Bourjot L (1996) Palaeoclimatic interpretation of a topographic profile across middle Holocene regressive shorelines of Longmu Co (Western Tibet). Palaeogeogr Palaeoclimatol Palaeoecol 120:93–104

Baade J, Mäusbacher R (2000) Environmental change and settlement history – preliminary results from the Muktinath Valley, Inner Himalayas, Nepal. Marburger Geogr Schr 135:40–52

Bäumler R, Zech W (2000) Soil development as an indicator of the pleistocene and holocene landscape history in Western Tien Shan and Nepal. Marburger Geogr Schr 135:69–82

Beug HJ, Miehe G (1998) Vegetationsgeschichtliche Untersuchungen in Hochasien. 1. Anthropogene Vegetationsveränderungen im Langtang-Tal, Himalaya, Nepal. Petermanns Geogr Mitt 142:141–148

Bräuning A (1994a) Dendrochronology for the last 1400 years in eastern Tibet. GeoJournal 34:75–95

Bräuning A (1994b) Dendrochronologische Untersuchungen an osttibetischen Waldgrenzstandorten. Göttinger Geogr Abh 95:185–192

Bräuning A (1999) Zur Dendroklimatologie Hochtibets während des letzten Jahrtausends. Dissertationes Botanicae, vol 312. J Cramer, Berlin, 164 pp

Bräuning A (2000) Ecological division of forest regions of eastern Tibet by use of dendroecological analyses. Marburger Geogr Schr 135:111–127

Bräuning A, Lehmkuhl F (1996) Glazialmorphologische und dendrochronologische Untersuchungen neuzeitlicher Eisrandlagen Ost- und Südtibets. Erdkunde 50:341–359

Cheng S (1999) Discussion on the Quaternary Palaeogeography from the spore-pollen analyses in the Mt. Qomolangma region. In: Zhen B, Shen Y (eds) Formation, evolution and environmental changes of cryosphere in Qinghai-Tibetan Plateau. Contributions collections of research on Quaternary glaciation and environments in the Qinghai-Tibetan Plateau, vol 2. Lanzhou, Inst of Glaciology and Geocryology, Chinese Acad of Sciences, pp 528–535 (in Chinese)

Damm B (1998) Landschaftsdegradation in Hochweideregionen Tibets als Folge natürlicher und anthropogener Einflüsse. Petermanns Geogr Mitt 142:175–179

Esper J, Bosshard A, Schweingruber FH, Winiger M (1996) Tree-rings from the upper timberline in the Karakorum as climatic indicators for the last 1000 years. Dendrochronologia 13:79–88

Fan H, Gasse F, Huc A, Li Y, Siffedine A, Soulié-Märsche J (1996) Holocene environmental changes in Bangong Co basin (Western Tibet). Part3: Biogenic remains. Palaeogeogr Palaeoclimatol Palaeoecol 120:65–78

Fontes J-C, Gasse F, Gibert E (1996) Holocene environmental changes in Lake Bangong basin (Western Tibet). Part 1: Chronology and stable isotopes of carbonates of a Holocene lacustrine core. Palaeogeogr Palaeoclimatol Palaeoecol 120:25–47

Frenzel B (1994) Über Probleme der holozänen Vegetationsgeschichte Osttibets. Göttinger Geogr Abh 95:143–166

Frenzel B (1998) History of flora and vegetation during the quaternary. Prog Bot 59:599–633

Frenzel B (2000) Nacheiszeitliche Veränderungen des Waldlandes in der Osthälfte des Tibetischen Plateaus. Akad-J 2:2–7

Frenzel B, Huang W, Liu S (2001) Stone artefacts from south-central Tibet, China. Quartär 51/52:1–21

Fu Y, Li E, Jia D (1993) The vegetation succession of alpine *Kobresia capillifolia* to *Carex sp*. The symposium of the expeditions to the Tibetan Plateau. Abstr p 61, Beijing (in Chinese)

Gasse F, Fontes, JC, Campo E v, Wei K (1996) Holocene environmental changes in Bangong Co basin (Western Tibet). Part 4: Discussion and conclusions. Palaeogeogr Palaeoclimatol Palaeoecol 120:79–92

Harris, RB, Miller DJ (1995) Overlap in summer habitats and diets of Tibetan Plateau ungulates. Mammalia 59:197–212

Hedin S (1909) Transhimalaja 2. Leipzig

Hou S (1991) An outline of Tibetan archaeology. Lhasa: People's Publishing House of Tibet. 244 pp (in Chinese, Tibetan and English summaries)

Kuhle M (1998) Neue Ergebnisse zur Eiszeitforschung Hochasiens in Zusammenhang mit den Untersuchungen der letzten 20 Jahre. Petermanns Geogr Mitt 142:219–226

Kuhle M (2000) Pleistocene glaciations in the Himalayas and Tibet: New findings from the Himalaya north slopes to central and western Tibet. Marburger Geogr Schr 135:1–14

Lehmkuhl F, Klinge M, Rees-Jones J, Rhodes, EJ (2001) Late Quaternary aeolian sedimentation in central and south-eastern Tibet. (Submitted)

Lister GS, Kelts K, Zao Ch, Yu JQ, Niessen F (1991) Lake Qinghai, China: Closed-basin lake levels and the oxygen isotope record for ostracoda since the latest pleistocene. Palaeogeogr Palaeoclimatol Palaeoecol 84:141–162

Liu G, Shen Y, Zhang P, Wang S (1994) Pollen record and its palaeoclimatic significance between 800–150 ka from RH-core in Zoige Basin in Qinghai-Xizang Plateau. Acta Sedimentol Sin 12:101–109 (in Chinese, English summary)

Liu G, Wang R, Li Sh, Li B, Zhu Zh (1998) Palynological evidence of ecological environment change since 240 ka BP for the Tianshuihai Lake, West Kunlun Mountain. J Glaciol Geocryol 20:21–24 (in Chinese, English summary)

Map of Snow, Ice and Frozen Ground in China (1988) Academia Sinica, Institute of Glaciology and Geocryology, Lanzhou

Mathews JA, Ballantyne CK, Harris Ch, McCarroll D (1993) Solifluction and climatic variation in the Holocene: discussion and synthesis. In: Frenzel B, Matthews JA, Gläser B (eds) Solifluction and climatic variation in the Holocene. Paläoklimaforschung – Palaeoclimate Research, vol 11. G Fischer, Stuttgart, pp 339–361

Miehe G, Miehe S (2000) Environmental changes in the pastures of Xizang (Contributions to ecology, phytogeography and environmental history of High Asia, 3). Marburger Geogr Schr 135:282–311

Miehe G, Miehe S, Huang J, Otsu T (1998) Forschungsdefizite und -perspektiven zur Frage der potentiellen natürlichen Bewaldung in Tibet. Petermanns Geogr Mitt 142:155–164

Miehe S, Miehe G, Huang J, Otsu, T, Tuntsu T, Tu Y (2000) Sacred forests of south-central Xizang and their importance for the restoration of forest resources. Contributions to ecology, phytogeography and environmental history of High Asia, 2. Marburger Geogr Schr 135:228–249

Naumann S, Walther M (2000) Mid-Holocene lake-level fluctuations of Bayan Nuur (North-West Mongolia). Marburger Geogr. Schr 135:15–27

Negri GCS, Rikhari HC, Singh SP (1993) Plant regrowth following selective horse and sheep grazing and clipping in an Indian Central Himalayan alpine meadow. Arctic Alpine Res 25:211–215

Nesje A (1997) Holocene glacier and climate variations in the Jostedalsbreen and Hardangerjøkulen regions, southern Norway. In: Frenzel B, Boulton GS, Gläser B,

Huckriede U (eds.). Glacier fluctuations during the Holocene. Paläoklimaforschung – Palaeoclimate Research, vol 24. G Fischer Stuttgart, pp 105–114

Peer T (2000) The highland steppe of the Hindukush Range as indicators of centuries old pasture farming. Marburger Geogr Schr 135:312–325

Ram J (1992) Effects of clipping on aboveground biomass and total herbage yields in a grassland above treeline in central Himalaya, India. Arctic Alpine Res 24:78–81

Rikhari HC, Negri GCS, Pant GB, Rana BS, Singh SP (1992) Phytomass and primary productivity in several communities of Central Himalayan alpine meadow, India. Arctic Alpine Res 24:344–351

Rikhari HC, Negri GCS, Ram J, Singh SP (1993) Human induced secondary succession in an alpine meadow of Central Himalaya, India. Arctic Alpine Res 25:8–14

Rost KT, Sun J (2000) Traces of Late-Holocene deforestation and climate changes in the upper Wutai Shan (Shanxi, China) derived from loess-like sediments. Marburger Geogr Schr 135:94–110

Samjaa R, Zöphel U, Peterson J (2000) The impact of the vole *Microtus brandti* on Mongolian steppe ecosystems. Marburger Geogr Schr 135:346–360

Schaller GB (2000) Wildlife of the Tibetan steppe. University Press, Chicago, 373 pp

Schickhoff U (2000) Persistence and dynamics of long-lived forest stands in the Karakorum under the influence of climate and man. Marburger Geogr Schr 135:250–264

Schlütz F (1998) Vegetationsgeschichtliche Untersuchungen in Hochasien. 2. Zum Holozän des Karakorum und zum letzten Glazial im Qinling Shan. Petermanns Geogr Mitt 142:149–154

Schlütz F (1999) Palynologische Untersuchungen über die holozäne Vegetations-, Klima- und Siedlungsgeschichte in Hochasien (Nanga Parbat, Karakorum, Nianbaoyeze, Lhasa) und des Pleistozän in China (Qinling-Gebirge, Gaxun Nur). Dissertationes Botanicae, vol 315. J Cramer, Berlin, 182 pp

Tang L, Shen C (1999) Holocene vegetation and climate in the Qinghai-Xizang Plateau. Formation, evolution and environmental changes of cryosphere in Qinghai-Tibetan Plateau. In: Zheng B, Shen Y (eds) Contributions collections of research on quaternary glaciation and environments in the Qinghai-Tibetan Plateau 2. Inst of Glaciology and Geocryology, Chinese Acad of Sciences, Lanzhou, pp 545–551 (in Chinese)

Tang L, Wang Y (1999) Pollen composition and significance of a 203 m drilling core in Qingshuihe, Qinghai-Xizang Highway. Formation, evolution and environmental changes of cryosphere in Qinghai-Tibetan Plateau. In: Zheng B, Shen Y (eds) Contributions collections of research on quaternary glaciation and environments in the Qinghai-Tibetan Plateau 2. Inst of Glaciology and Geocryology, Chinese Acad of Sciences, Lanzhou, pp 497–505

Thelaus M (1992) Some characteristics of the mire development in Hongyuan County, eastern Tibet Plateau. Proc 9th Int Peat Congr 1992, Uppsala, vol 1, pp 334–351

Van Campo E, Cour P, Hang S (1996) Holocene environmental changes in Bangong Co basin (Western Tibet). Part 2: The pollen record. Palaeogeogr Palaeoclimatol Palaeoecol 120:49–63

Walther M (1998) Paläoklimatische Untersuchungen zur jungpleistozänen Landschaftsentwicklung im Changai-Bergland und der nördlichen Gobi (Mongolei). Petermanns Geogr Mitt 142:207–217

Winkler D (1998) Die waldfreien Sonnhänge Osttibets am Beispiel Jiuzhaigous (Zitsa Degu) und der Einfluß von Feuer und Weidewirtschaft. Petermanns Geogr Mitt 142:165–174

Winkler D (2000) Patterns of forest distribution and the impact of fire and pastoralism in the forest region of Tibet. Marburger Geogr Schr 135:201–227

Wissmann, H v (1959) Die heutige Vergletscherung und Schneegrenze in Hochasien. Abhandl der math-nat Klasse 14. Akademie der Wissenschaften und der Literatur, Mainz

Wu N (2000) Vegetation pattern in Western Sichuan, China and human kind's impact on its dynamics. Marburger Geogr Schr 135:188–200

Wünnemann B, Pachur H-J, Li J, Zhang H (1998) Chronologie der pleistozänen und holozänen Seespiegelschwankungen des Gaxun Nur/Sogo Nur and Baijian Hu, Innere Mongolei, Nordwestchina. Petermanns Geogr Mitt 142:191–206

Xu R (1999) Study on the quaternary paleobotany in the Mt. Qomolangma region. In: Zheng B and Shen Y (eds) Formation, evolution and environmental changes of cryosphere in Qinghai-Tibetan Plateau. In: Zheng B, Shen Y (eds) Contributions collections of research on quaternary glaciation and environments in the Qinghai-Tibetan Plateau 2. Inst of Glaciology and Geocryology, Chinese Acad of Sciences, Lanzhou, pp 518–527 (in Chinese)

Zech W, Bäumler R, Guggenberger G, Petrov M, Ni A, Lemzin J (2000) Pleistocene and holocene landscape development in the Kichik Alay and Hissar Ranges (Khyrgyzia and Uzbekistan) as deduced from soil morphology. Marburger Geogr Schr 135:53–68

Zimmermann B, Schleser GH, Bräuning A (1997) Preliminary results of a Tibetan stable C-isotope chronology dating from 1200 to 1994. Isotopes Environ Health Stud 33:157–165

Prof. Dr. Burkhard Frenzel
Institut für Botanik
und Botanischer Garten
Universität Hohenheim (210)
70593 Stuttgart, Germany

Diversity and Ecology of Biological Crusts

By Burkhard Büdel

1 Introduction

Biological crusts are composed of algae, cyanobacteria, bacteria, microfungi and lichens, sometimes also mosses. Only in rare cases are all components involved in crust formation, quite often there is only one or two of these groups involved. Biological crusts occur on rock and soil surfaces and on the bark of trees, on leaves, but also on man-made substrates. This review focuses on the two main terrestrial habitats – soil and rock – where biological crusts are predominantly found in semiarid and arid regions or under arid microclimatic conditions in all climatic ecoregions on Earth.

There is recent evidence from a carbonaceous ancient soil of eastern Transvaal (South Africa), showing that biological crusts in the form of microbial mats with an age of 2.6 billion years might belong to the oldest terrestrial community of organisms (Watanabe et al. 2000). Nevertheless, in terms of biodiversity and ecology, biological crusts, together with the remaining top meter of the soil belong to the most poorly researched habitats on Earth (Moore 1998; Copley 2000).

Biological soil crusts are reviewed in a recent publication of Evans and Johansen (1999) and there is a special volume of the Ecological Studies series (Springer-Verlag) in print, completely dedicated to biological soil crusts. Cyanobacterial crusts on rock or soil are treated in the chapters of Stal (2000), Vincent (2000) and Wynn-Williams (2000) in the recent book of Whitton and Potts (2000), *The Ecology of Cyanobacteria.*

2 Crust Types and Distribution

Biological crusts exist in many different expressions, ranging from a thin film, less than 1 mm thick, consisting of one or two cyanobacterial or algal species only, to highly complex crusts more than 1 cm thick, composed of several groups of organisms (e.g. cyanobacteria, fungi, lichens and mosses).

Progress in Botany, Vol. 63

Table 1. Definition of crust types associated with soil (edaphic)

Author	Term	Description
Friedmann et al. (1967)	Epedaphic	Cyanobacteria, algae, lichens; surface of the soil
	Endedaphic	Cyanobacteria, algae; beneath the surface of the soil
Komáromy (1976)	Dispers	Mainly green algae, but also cyanobacteria, inside the soil
	Stratose	Filamentous green algae like *Zygogonium* or *Klebsormidium*; soil surface
	Microcoleus type	Filamentous non-branched cyanobacteria, e.g. *Microcoleus* spp.; inside the soil
	Ramose	Filamentous branched cyanobacteria; inside the soil
	Mucose	Unicellular and filamentous cyanobacteria and algae with strong production of mucilage; soil surface
	Glutinose	Mainly mucilagenous diatoms; soil surface
Eldridge and Green (1994)	Hypermorph	Aboveground
	Perimorph	At ground level
	Cryptomorph	Hidden belowground
Belnap (2001)	Smooth	Cyanobacteria, algae; inside the soil; surface roughness 0–1 cm
	Rugose	Lichens and mosses above the soil surface (in addition to organisms inside the soil); roughness 1–3 cm; no frost heave
	Rolling	Lichens and mosses above the soil surface (in addition to organisms inside the soil); roughness 1–5 cm, frost heave
	Pinnacled	Cyanobacteria, algae, etc. inside the soil, with or without lichens and mosses above ground; roughness 3–10 cm; frost heave

a) Crust Types

Biological crusts on soil result from an intimate association between soil particles and cyanobacteria, algae, microfungi, lichens, and bryophytes (in different proportions) which live within, or immediately on top of, the uppermost millimeters of soil. By definition (Belnap et al. 2001), biological soil crusts form a coherent layer on the surface of the ground, due to the aggregation of soil particles by the presence and activity of the biota. Several types of biological soil crusts can be distinguished and differing terminologies exist (Table 1). Since it facilitates a worldwide comparison of soil crusts, the classification of Belnap et al. (2001) is suggested to follow here.

On rock substrates biological crusts consist mainly of cyanobacteria, algae, microfungi and lichens. Mosses are usually not as tightly attached to the rock surface as for example cyanobacteria or lichens and are therefore not a common constituent of rock crusts. Thin layers of cyanobacteria and/or algae alone are also sometimes called "films" or "biofilms" (e.g. Büdel 1999; Castenholz and Garcia-Pichel 2000). Films and crusts can be found at the surface (epilithic) and/or below (endolithic) the rock surface. The different types of rock crusts are summarized in Fig. 1 and Table 2.

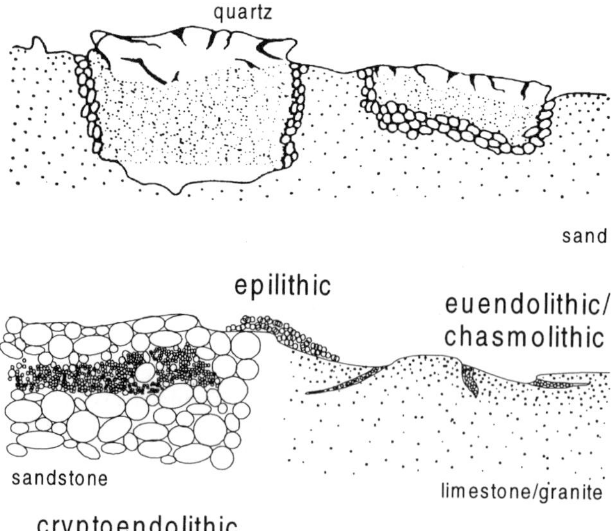

Fig. 1. Growth types of cyanobacteria and algae on rock (hypolithic type taken from Cameron and Blank 1966)

Table 2. Definition of crust types associated with rock (lithophytic, adapted from Golubic et al. 1981)

Term	Organisms involved, microhabitat
Epilithic (to a certain degree also edaphic)	Cyanobacteria, algae and lichens; rock surface
Hypolithic	Cyanobacteria, algae; at the bottom of translucent stones of desert pavements
Endolithic types	Living inside the rock
a) Euendolithic	Cyanobacteria, algae; actively boring (so far only known from calcareous rocks)
b) Chasmolithic	Cyanobacteria, algae; within cracks and fissures of rocks (e.g. granite)
c) Cryptoendolithic	Cyanobacteria, algae; within natural cavities of porous rocks (e.g. sandstone)

It should also be mentioned here that biofilms and crusts exist, which are composed of cyanobacteria, lichens and mosses, in the phyllosphere of the tropical rainforest of Central America and West Africa (Lücking 1992; Freiberg 1998a; Lücking et al. 1998). However, this specific type of an epiphytic crust will not be treated further in this review.

b) Distribution

Biological crusts most often occur under harsh conditions that include at least a periodic lack of water and extremes in temperature and light, decreasing competition from phanerogameous vegetation, e.g. in hot and cold deserts of the world, savannas, steppe formations, tundras and all types of rock formations.

α) Soil Crusts

Soil crusts occur in all biomes of the world containing an arid element that decreases competition. Only the tropical evergreen rain forest appear to lack biological soil crusts. In Fig. 2, our present knowledge of the worldwide distribution of soil-crust communities is depicted. This map is based upon the information summarized in Belnap and Rosentreter (2001) for North America, Büdel (2001a) for South America, Hansen (2001) for Arctic Greenland, Türk and Gärtner (2001) for the Alps, Büdel

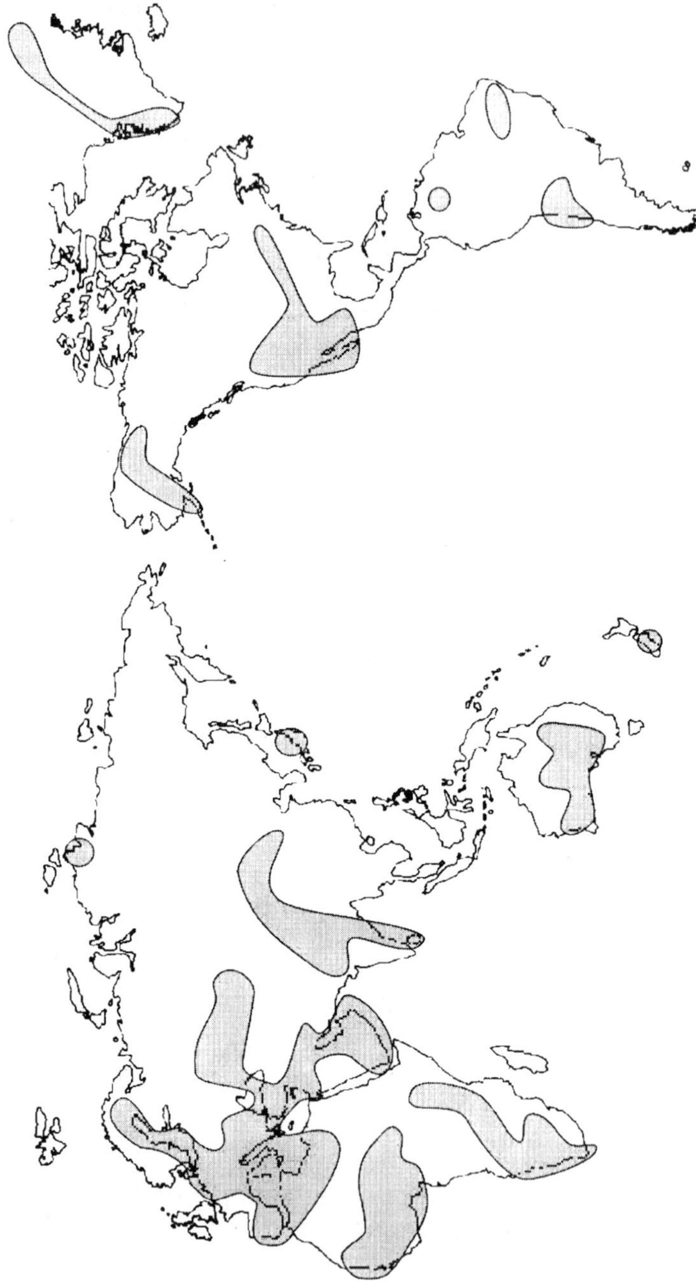

Fig. 2. World distribution of soil crusts

(2001b) for the European and Mediterranean regions, Büdel (2001c) for Asia, Galun and Garty (2001) for the Middle East, Ullmann and Büdel (2001a) for Africa, Eldridge (2001) for Australia and Green and Broady (2001) for Antarctica, upon personal experience, and that of colleagues. However, their distribution may be much more extensive. Detailed studies of the diversity and phytogeography of soil crust formation by cyanobacteria together with mainly green algae from the Sahara-Sind, the Irano-Turan and the Gobi-Pamir subregions are summarized in Novichkova-Ivanova (1980). Recent papers dealing with the structure and composition of soil crusts are that of Reynaud and Lumpkin (1988) for the Lanzhou region in China, Kidron (1995) for the Negev Desert in Israel, Dor and Danin (1996) for the Dead Sea Valley in Israel, Malam Issa (1998), Malam Issa et al. (1999) and Hahn and Kusserow (1998) for the Sahel region in Africa, Eldridge and Koen (1998) for Australia, De Winder (1990) for coastal dunes in northern Europe and Paus (1997) for northern Germany.

β) Rock Crusts

Seemingly naked rock surfaces that are fully exposed to sunlight and thus can be considered as edaphically dry, occur in all biomes on earth. Closer examination revealed that almost all rock surfaces are covered by a dense film or crust of cyanobacteria together with algae, lichens and microfungi. These films and crusts grow on all types of rocks, including man-made rock substrates in tropical and subtropical biomes (Garty 1989). Such mostly cyanobacterial dominated crusts occur epilithic on the surface of the table mountains in the Guyana region in South America (Büdel 1999; Büdel et al. 2000b), endolithic along exposed rock surfaces (Bell et al. 1986; Büdel and Wessels 1991; Wessels and Büdel 1995; Weber et al. 1996), epilithic and partly endolithic together with cyanobacterial lichens and rarely green algae on inselbergs (isolated rock outcrops) in dry and humid savannas (Welwitsch 1868; Golubic 1967b; Wessels and Büdel 1989; Sarthou et al. 1995; Büdel et al. 1994, 1997a, 2000a; Büdel 1999). The characteristic color of the rock surface of inselbergs is due to the composition of the organismic cover of the rock. While the black color of inselbergs and other exposed rock surfaces in humid savannas and rainforest biomes is caused by a dense film or crust of cyanobacteria, often intermingled with cyanobacterial microlichens, the characteristic ochre color of rock surfaces and inselbergs in dry savannas, semi-deserts and deserts is created by a dense crust dominated by cyanobacterial lichens, mainly of the genus *Peltula*, in combination with surface oxidation processes of the rock itself (Hambler 1964; Büdel et al. 2000a). The structure of cyanobacterial/cyanolichen crusts on inselbergs in humid savannas closely resembles that of a micro forest and

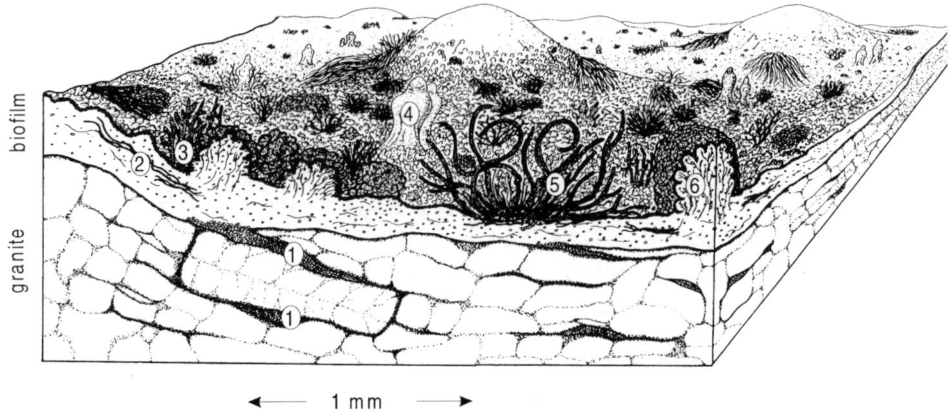

Fig. 3. "Micro-forest" on the rock surface of Brazilian inselbergs; cyanobacterial rock crust with cyanolichens. *1* Chasmo- and cryptoendolithic cyanobacteria; *2* film of the unicellular cyanobacterium *Gloeocapsa* spp. and the filamentous *Schizothrix* spp.; *3 Stigonema* spp.; *4* young specimens of the cyanolichen *Peltula tortuosa*; *5 Scytonema* spp.; *6* thalli of the cyanolichen *Phylliscum* spp. and/or *Paulia* spp. (Weber 1997)

can be divided into an understorey and a canopy (Weber 1997; Fig. 3). Hypolithic crusts, also dominated by cyanobacteria, were reported by Vogel (1955) for southern Africa, by Friedmann et al. (1967) and Friedmann and Galun (1974) for the Negev and the Sonoran desert and by Büdel and Wessels (1991) for the Namib and the Sonoran desert. Rummrich et al. (1989) reported diatoms as hypolithic algae from the Namib desert. The biological nature of the black rock varnishes, typical for rocks in all deserts of the world, was studied and explained by Krumbein and Jens (1981). A recent and detailed review on cyanobacteria in deserts of the world can be found in Wynn-Williams (2000).

In temperate, boreal and arctic regions exposed rock surfaces are often coated by bluish or black stripes. These so-called Tintenstriche (German for "ink stripes") are composed of a typical set of unicellular and filamentous cyanobacteria and rarely of a few cyanolichens (Diels 1914; Jaag 1945; Lüttge 1997). In these regions calcareous rocks are often covered by characteristic communities of cyanobacteria and algae (e.g. Golubic 1967a; Miszalski et al. 1995). Many components of the former-mentioned cyanobacterial communities can be found again as epi- and endolithic crusts on ancient monuments, constructed of the same rock type, often combined with a more or less intensive deterioration activity of that crust (Anagnostidis et al. 1983).

Besides epilithic and chasmolithic cyanobacterial crusts (Broady 1996), exposed rocks of Antarctica bear a special type of crust: an endolithic (cryptoendolithic) community living in the upper few millimeters of the Beacon sandstone. These crusts form an olive-green and often a second, bluish-black layer parallel to the rock surface. They are composed of cyanobacteria, algae, microfungi and lichens (Friedmann 1980; Hale 1987; Friedmann et al. 1988)

3 Diversity of Organisms and Communities

Despite their widespread occurrence, comparing the biogeography and ecology of crust biota and communities on a global basis is almost impossible. This is because many of the data were collected using different methodologies, different taxonomic systems, and are often focused on a single group of organisms only (i.e. lichens, algae, cyanobacteria or mosses) to the exclusion of other groups. In addition, studies have been unevenly distributed over the earth, and only a very small percentage of studies have taken the ecologically-important non-lichenized fungi and heterotrophic bacteria into consideration.

a) Soil Crusts

Since many floristic studies of biological soil crusts do not distinguish between species that substantially contribute to crust formation and those that live in or on the top of crusts, but do not contribute to the formation of the crust, our knowledge of soil crust biota from different geographical regions is rather dissimilar. In addition, most floristic studies conducted on biological soil crusts are restricted to one or two groups of organisms and thus are nearly always incomplete (Evans and Johansen 1999). This might be one of the reasons for the high species richness of different algal groups reported for soils of arid regions in North Africa, Middle East and Asia (Novichkova-Ivanova 1980).

In total, 35 cyanobacteria genera, 68 eukaryotic algae genera, 13 cyanolichen genera, 69 phycolichen genera, and 62 moss and liverwort genera are reported from soil crusts of the earth (Büdel 2001d). The highest number of genera for algae (51, including 17 cyanobacteria), lichens (44, including 8 cyanolichens) are reported for North-America (26 mosses/liverworts), followed by Europe (44 algae including 18 cyanobacteria; 34 lichens including 5 cyanolichens; 14 mosses and liverworts), Africa (30 algae including 18 cyanobacteria; 19 lichens including 4 cyanolichens; 7 mosses and liverworts), Middle East (27 algae including 14 cyanobacteria; 16 lichens including 8 cyanolichens; 8 mosses and liverworts), Asia (26 algae including 17 cyanobacteria; 24 lichens including 7 cyanolichens; 2 [?]mosses), South America (18 algae including 15 cyanobacteria; lichens unknown; 5 mosses and liverworts), and Greenland (42 lichens including 4 cyanolichens, with the number of algae and mosses unknown). With 42 genera, the highest number of mosses and liverworts are reported from Australia (15 algae including 10 cyanobacteria; 26

lichens including 4 cyanolichens). For Antarctica, extensive cyanobacterial crusts are reported, and even *Microceoleus vaginatus*, a common species in soil crusts of deserts and semi-deserts, was found in crusts from the Dry Valleys (Broady 1996).

The most common lichens of soil crusts in arid environments of the earth are the members of the "community of colored lichens" (e.g. *Fulgensia fulgens, Psora decipiens, Squamarina* spp., *Toninia sedifolia, Catapyrenium* spp., *Diploschistes* spp., *Endocarpon* spp., *Collema* spp.). The "Bunte Erdflechtengesellschaft" (community of colored lichens) occurs world-wide in comparable habitats like the Central European steppe formations, the mediterranean region, and the North American, Australian and African deserts and semi-deserts (Büdel 2001d).

Cyanobacterial genera, in many cases also species, are generally more widely distributed than eukaryotic algae. The two most common genera are *Microcoleus* (*M. vaginatus, M. paludosus, M. chtonoplastes, M. sociatus*) and *Nostoc* spp. Other common genera are *Calothrix, Lyngbya, Oscillatoria, Phormidium, Scytonema* and *Tolypothrix*. Common green algae in soil crusts are *Chlorella, Chlorococcum, Coccomyxa* and *Klebsormidium*. Typical mosses of soil crusts are *Bryum* species, *Crossidium* spp. and *Tortula* spp., and among the liverworts, the genus *Riccia* is the most widespread one, followed by *Fossombronia*.

Poor sandy soils are generally dominated by cyanobacterial crusts; the proportion of lichens increases with carbonate, gypsum and silt contents of the substrate. If the salt influence of the soil increases, the filamentous cyanobacterium *Microcoleus chtonoplastes* becomes the main constituent of the crust (Ullmann and Büdel 2001b). Reliable records of moss-dominated soil crusts are very rare and only three regions in the Middle East, Brazil and Australia are known where mosses dominate (Frey and Kürschner 1991; Eldridge and Tozer 1996; Bastos et al. 1998). Soil crusts dominated by green algae and/or other eukaryotic algal groups appear restricted to soils in temperate regions, where they appear as a successional phase, or where soils are slightly acidic. However, our knowledge of this crust type is far too limited to draw any further conclusions (Büdel 2001d).

b) Rock Crusts

Biological crusts on rock are even more widespread than soil crusts, but the data available do not allow us to even estimate their total coverage on earth yet. Their diversity however, is somehow better investigated in terms of species composition and communities, compared to that of soil crusts (e.g. Welwitsch 1868; Diels 1914; Jaag 1945; Friedmann et al. 1967, 1988; Golubic 1967a,b; Büdel et al. 1994, 1997a, 2000a,b; Sarthou et al.

1995; Wessels and Büdel 1995; Broady 1996; Weber et al. 1996; Büdel 1999).

Obviously all types of rock crusts are formed or at least dominated by cyanobacteria, often accompanied by green algae (Zygnematales, Oedogoniales, Charophyceae) and diatoms. Extensive lists of species and communities for the "Tintenstriche" and other rock crusts and films of temperate regions can be found in Jaag (1945) and Golubic (1967a). For the tropics, such lists are given in Sarthou et al. (1995), Büdel (1999) and Büdel et al. (2000a) for epi- and endolithic crusts in semi-arid and humid tropical regions; desert regions of the world are treated in Novichkova-Ivanova (1980) and, more recent, Wynn-Williams (2000).

The typical structure of crusts and films on rock outcrops in the tropics closely resemble that of a micro-forest (Fig. 3) with a "lower story", composed of unicellular cyanobacteria (e.g. *Chroococcidiopsis* sp., *Chroococcus* sp., *Gloeocapsa* spp., *Gloeothece* spp.) and a few filamentous ones (e.g. *Schizothrix* spp.), with the genus *Gloeocapsa* being the most common one. The "canopy" is formed by the filamentous cyanobacteria *Scytonema* spp. and *Stigonema* spp., together with cyanolichens of the genera *Paulia, Peltula, Phylliscum* and in rare cases, a few others (Büdel et al. 2000a; Weber 1997). The unicellular cyanobacterium *Xenococcus* sp. often occurs as an "epiphyte" on the filaments of *Scytonema* spp. and *Stigonema* spp. In many cases of biofilms and crusts on rock, their species composition exposed a typical rarity of members of the non-heterocyte forming order Oscillatoriales. This might be due to high light irradiance, characteristic of such open rock habitats. The presence of a light-protecting, intensely colored sheath, usually not found in the Oscillatoriales, might be a prerequisite for colonizing such an extreme environment (Castenholz and Garcia-Pichel 2000). One of the few members of the Oscillatoriales possessing a colored sheath is the genus *Schizothrix*, which can thus be found in highly insolated habitats, often in connection with early soil formation, e.g. ephemeral rock pools (Büdel 1999).

A comparison of the epilithic cyanobacterial flora of the tropics and lithophytic floras from temperate regions reveals an astonishing similarity in species. Only the cyanolichens involved are different. While in tropical crusts mainly the genera *Paulia, Peltula* and *Phylliscum* occur, this role is taken over by the cyanolichen genera *Collema, Leptogium, Synalissa* and rarely *Anema* in crusts of temperate regions. However, a modern and thorough revision of the species concept in cyanobacteria, involving molecular methods, would probably reveal a larger difference in species composition between tropical and temperate regions (Büdel 1999).

4 Ecology and Ecophysiology

Most ecosystem balances do not take biological crusts into consideration, neither at the level of CO_2 sinks, nor as important primary colonizers or initial steps of succession. Until recently, a "missing sink" of 24.6% of anthrogenic CO_2 was hypothesized (Schlesinger 1997). However, on the basis of an O_2/N_2 gradient, Keeling et al. (1996) postulated, that the global oceans and the land biota removed the equivalent of approximately 30% of the fossil-fuel CO_2 emissions, while the remaining CO_2 is responsible for the atmospheric increase. They conclude the land biota as being the "missing sink". In the light of this new results, it would be extremely interesting to learn more about the role of biological crusts on rock and soil, towards their possible role as a part of the land biota sink.

Although biological soil crusts occur in many biomes of the world and are a widespread phenomenon (Fig. 2), our knowledge of their ecological function, their role in ecosystems, as well as their floristic and phytogeography is still rather incomplete or even unknown in many regions. For future floristic and phytogeographical research in biological soil crusts, the application of a standardized and commonly used protocol would contribute greatly towards a reliable international basis of comparison.

a) Biomass

The biomass found in the different crust types (rock or soil) was between 17 and 1000 mg chlorophyll a or a + b/m^2, with the highest values occurring in soil crusts (soil: mean 224 mg/m^2; rock: mean 108 mg/m^2; Table 3). Compared to the average chlorophyll content per square meter of desert or savanna biomes of the world (100–1500 mg chlorophyll/m^2; Larcher 1994), crusts thus can contribute considerably to the biomass of semi-arid and arid biomes. Nevertheless, biological crusts are not normally taken into consideration when dealing with ecosystem balances.

b) Carbon and Nitrogen

Carbon and nitrogen fixation and dynamics of biological soil crusts are comprehensively reviewed in Evans and Johansen (1999) and, especially for their cyanobacterial part, in Stal (2000). In all environments, water availability is the factor controlling both activities, CO_2 and nitrogen fixation (either as N_2 or combined N). Almost all crust types are dominated by cyanobacteria, whose metabolism can only be activated by water in the liquid phase, not by water vapor alone, as is the case for green

Table 3. Biomass of biological crusts (expressed as chlorophyll content per m^2)

Crust type	Chlorophyll content (mg/m^2) a+b	a	Source of information
Rock crust; Africa, Ivory Coast, inselbergs		151.7±42.2	Büdel (1999)
Rock crust; Africa, South Africa, rock plateau in savanna		29.3±20.4	Büdel (1999)
Rock crust; Africa, South Africa; rock cliffs		63±19.4	Büdel (1999)
Rock crusts, South Africa, inselbergs		218.6±68.8	Büdel (1999)
Rock crust; South America, Venezuela, table mountain		122±14.6–211±50	Büdel (1999)
Rock crust; South America, Venezuela, inselbergs		63±19.7–133±40.6	Büdel (1999)
Rock crusts, North America, Arizona, Colorado Plateau		87.3±22.3	Bell and Sommerfeld (1987)
Rock crust; Australia, granite rock outcrop		141±68	Büdel (1999)
Rock crust; Australia, inselbergs		38.5±20.3–48.8±16.3	Büdel (1999)
Soil crust; Europe, Czech Republic, open pine forest		498–986	Büdel and Lukesova (unpubl.)
Soil crust; Europe, Germany, local steppe formation	130–981		Lange (2001)
Soil crust; Europe, The Netherlands, sand dunes		32–319	De Winder (1990)
Soil crust; Africa, Namibia, desert soil	508		Lange et al. (1994b)
Soil crust; Africa, Tunisia, desert soil		102 - 180	Ullmann and Büdel (2001a)
Soil crust; Middle East, Israel, Negev desert		17 - 51	Kidron (1995)
Soil crust; North America, Utah, desert soil	133–381	58.9 - 155	Lange et al. (1997); Lange et al. (1998)
Soil crust; North America, Arizona, Juniper-pinyon woodland		64	Beymer and Klopatek (1991)
Soil crust; North America, Utah, desert soil		6.3–31.6	Garcia-Pichel and Belnap (1996)
Soil crust; North America, Utah, desert soil		53	Belnap et al. (1994)
Soil crust; Australia, Western Australia, desert soil		50–82	Ullmann and Büdel (2001b)

Table 4. Maximal net photosynthesis at natural ambient CO_2 concentration and optimal conditions

Crust type	Maximal rates (μmol CO_2 m^2 s^{-1})	Source of information
Mixed soil crust with dense population of *Klebsormidium* spp., open pine forest, Czech Republic	11.5	Büdel and Lukesova, (unpubl.)
Soil crust dominated by the phycolichen *Lecidella crystallina*, Namib Desert, Namibia	5.9	Lange et al. (1994b)
Mixed soil crust with cyanobacteria dominating and few green algae, sand dunes Negev Desert, Israel	1.12	Lange et al. (1992)
Mixed soil crust, cyanobacteria and green algae, Orinoco Llanos, Venezuela	4.7	San José and Bravo (1991)
Mixed soil crust with *Microcoleus vaginatus* dominating, southeastern Utah, USA	1.17	Garcia-Pichel and Belnap (1996)
Rock crust with the cyanobacteria *Gloeocapsa sanguinea* and dominating *Stigonema panniforme*, Auyan Tepui, Venezuela	2.2	Büdel (1999)

algae (Büdel and Lange 1991). Consequently, if water is available as rain or dew, the crust organisms must respond fast and should have a wide range of thallus water content, where positive net photosynthesis is possible (e.g. Lange et al. 1992, 1994a,b, 1997, 1998). The only cyanobacterium that could be activated by water vapor alone was the filamentous *Microcoleus sociatus* from soil crusts of the Negev Desert. However, since a relative humidity above 95% was necessary for a period of at least 9 h, this way of activation of net photosynthesis seems to be ecologically irrelevant (Lange et al. 1994a).

Maximal rates of net photosynthesis are reached when water content, temperature and light are just at the optimal point for CO_2 diffusion and the metabolism of the crust organisms. Such conditions are only rarely met with under natural conditions (e.g. Green et al. 1995). Nevertheless, maximal rates can be used to characterize different crust types and to draw conclusions on their possible efficiency (Table 4).

Compared with the average maximum net photosynthesis of leaves of C_3 crop plants (20–40 μmol CO_2 m^{-2} s^{-1}) or needles of evergreen coniferous tress (4–8 μmol CO_2 m^{-2} s^{-1}; Larcher 1994), CO_2 fixation of biological crusts certainly plays an important role in ecosystems and should no longer be neglected. Concerning the land biotic sink for anthropogenic annual CO_2 fixation by biological crusts might potentially contribute to that sink.

Cyanobacteria use a variety ofnitrogen sources. Ammonia can be taken up, either by diffusion or in the protonated form by a specific uptake system and nitrate and nitrite are also important sources of nitrogen. In addition, many cyanobacteria are capable of using dinitrogen (N_2) as their source of nitrogen. The nitrogen fixation and cycle in cyanobacteria-dominated mats and crusts is described in detail by Stal (2000). The role of crust- and biofilm-cyanobacteria for the nitrogen content of soils was demonstrated by several authors for different regions. Input of nitrogen into savanna soils via cyanobacterial soil crusts has been shown by Isichei (1980) for a Nigerian savanna ecosystem. Medina (in Büdel 1999) and Büdel (1999) could demonstrate that the soil surrounding inselbergs in the humid savanna along the Orinoco in Venezuela and the dry savanna of Northern Transvaal in South Africa, exposed a considerably higher nitrogen content compared to the soil further away from the inselbergs. It was concluded that this enrichment in nitrogen is due to the abrasion by desquamation, water, and biodeterioration of lichens and cyanobacteria. At least in the humid savanna, the formation of gallery forests surrounding the inselbergs might be due to the nitrogen input by crusts. Nitrogen fixation by the cyanobacterial biofilm from leaves was shown by Freiberg (1998b).

Nitrogen is not only necessary for the amount of phycobiliproteins and, consequently, the efficiency of light harvesting for photosynthesis. It might also play a major role in the supply of UV-protecting compounds (Castenholz and Garcia-Pichel 2000) that are a common feature and prerequisite of rock and soil-inhabiting cyanobacteria and lichens that often experience full solar radiation in their open habitat (Büdel et al. 1997b).

c) Influence on Biodiversity

There is not very much yet that can be stated on the influence of biological crusts on the biodiversity of their environment and even recent reports are conflicting. A certainly positive influence of rock crusts on inselbergs towards the surrounding soil in humid savannas was shown by Büdel (1999). There in the humid savannas of Venezuela and Ivory Coast, where the availability of water is surely not limiting tree growth, a luxuriant growth of trees can be observed, resulting in the formation of gallery-like forests along the base of inselbergs. To a certain degree, the positive influence of cyanobacterial soil crusts via nitrogen input and the positively influenced soil water household is postulated by Isichei (1980) and later on by Malam Issa et al. (1999) for the tiger bush sequence of the Sahel region of Niger. However, Prasse (1999) clearly showed the negative effect of cyanobacterial soil crusts for the establishment of phanerogamous vegetation in dune areas of the Negev Desert in Israel.

Except for a few reports, showing soil lichens in the Namib Desert as a source of food for beetles (Wessels et al. 1979; Joubert et al. 1982), practically nothing is known about the role of biological crusts as a source of food. Much more research towards the interaction of biological crusts with phanerogamic vegetation and animals is necessary before the real influence of biofilms and crusts on the biodiversity of their environment can be conclusively answered.

Acknowledgements. I would like to express my sincere thanks to all authors providing me with their manuscripts in print for the special volume of Ecological Studies, *Biological Soil Crusts: Structure, Function and Management* (Jayne Belnap and Otto L. Lange eds.).

References

Anagnostidis K, Economou-Amili A, Roussomoustakaki M (1983) Epilithic and chasmolithic microflora (Cyanophyta, Bacillariophyta) from marbles of the Parthenon (Acropolis-Athens, Greece). Nova Hedwigia 38:227–287

Bastos CJP, Albertos B, Bôas SBV (1998) Bryophytes from some Caatinga areas in the state of Bhia (Brazil). Trop Bryol 14:69–75

Bell RA, Sommerfeld MR (1987) Algal biomass and primary production within a temperate zone sandstone. Am J Bot 74:294–297

Bell RA, Athey PV, Sommerfeld MR (1986) Cryptoendolithic algal communities of the Colorado plateau. J Phycol 22:429–435

Belnap, J (2001) Comparative structure of physical and biological soil crusts. In: Lange OL, Belnap J (eds) Biological soil crusts. Ecological studies. Springer, Berlin Heidelberg New York, pp 177–191

Belnap J, Harper KT, Warren SD (1994) Surface disturbance of cryptiobiotic soil crusts: nitrogenase activity, chlorophyll content, and chlorophyll degradation. Arid Soil Res Rehabil 8:1–8

Belnap J, Büdel B, Lange OL (2001) Biological soil crusts: Characteristics and distribution. In: Lange OL, Belnap J (eds) Biological soil crusts. Ecological studies. Springer, Berlin Heidelberg New York, pp 3–300

Beymer RJ, Klopatek JM (1991) Potential contribution of carbon by microphytic crusts in pinyon-juniper woodlands. Arid Soil Res Rehabil 5:187–198

Broady P (1996) Diversity, distribution and dispersal of Antarctic terrestrial algae. Biodiv Conserv 5: 1307–1335

Büdel B (1999) Ecology and diversity of rock inhabiting cyanobacteria in tropical regions. European J Phycol 34:361–370

Büdel B (2001a) Biological Soil crusts of South America. In: Lange OL, Belnap J (eds) Biological soil crusts. Ecological studies. Springer, Berlin Heidelberg New York, pp 51–55

Büdel B (2001b) Biological soil crusts of European temperate and mediterranean regions. In: Lange OL, Belnap J (eds) Biological soil crusts. Ecological studies. Springer, Berlin Heidelberg New York, pp 75–86

Büdel B (2001c) Biological soil crusts of Asia including the Don and Volga region. In: Lange OL, Belnap J (eds) Biological soil crusts. Ecological studies. Springer, Berlin Heidelberg New York, pp 87–94

Büdel B (2001d) Synopsis: comparative biogeography and ecology of soil crust biota and communities. In: Lange OL, Belnap J (eds) Biological soil crusts. Ecological studies. Springer, Berlin Heidelberg New York, pp 141–152

Büdel B, Lange OL (1991) Water status of green and blue-green phycobionts in lichen thalli after hydration by water vapor uptake: do they become turgid? Bot Acta 104:361–366

Büdel B, Wessels DCJ (1991) Rock inhabiting blue-green algae/cyanobacteria from hot arid regions. Arch Hydrobiol Suppl 92 (Algol Stud 64): 385–398

Büdel B, Lüttge U, Stelzer, R, Huber O, Medina E (1994) Cyanobacteria of rocks and soils in the Orinoco region and in the Guyana highlands, Venezuela. Bot Acta 107:422–431

Büdel B, Becker U, Porembski S, Barthlott W (1997a) Cyanobacteria and cyanobacterial lichens from inselbergs of the Ivory Coast, Africa. Bot Acta 110:458–465

Büdel B, Karsten U, Garcia-Pichel F (1997b) Ultraviolet-absorbing scytonemin and mycosporine-like amino acid derivates in exposed, rock inhabiting cyanobacterial lichens. Oecologia 112:165–172

Büdel B, Becker U, Follmann G, Sterflinger K (2000a) Algae, fungi, and lichens on Inselbergs. In: Porembski S, Barthlott W (eds) Inselbergs – biotic diversity of isolated outcrops in tropical and temperate regions. Ecological studies 146. Springer, Berlin Heidelberg New York, pp 69–90

Büdel B, Schultz M, Lakatos M, Woitke M (2000b) Ökologie lithophytischer Cyanobakterien und Cyanobakterien-Flechten des Guyana Hochlands und des Orinoco Tieflands (Venezuela). In: Walter H, Breckle S-W, Schweizer B, Arndt U (eds) Ergebnisse weltweiter ökologischer Forschungen. Beiträge des 1. Symposiums der AFW Schimper-Stiftung von H und E Walter, pp 209–217

Cameron RE, Blank GB (1966) Desert algae: soil crusts and diaphanous substrata as algal habitats. JPL Tech Rep Jet Propulsion Lab, California Inst Tech Pasadena N 32–971

Castenholz RW, Garcia-Pichel F (2000) Cyanobacterial responses to UV-radiation. In: Whitton BA, Potts W (eds) The ecology of cyanobacteria. Kluwer, Dordrecht, pp 591–611

Copley J (2000) Ecology goes underground. Nature 406:452–454

De Winder (1990) Ecophysiological strategies of drought-tolerant phototrophic microorganisms in dune soils. Academisch Proefschrift, University of Amsterdam, Amsterdam

Diels L (1914) Die Algen-Vegetation der Südtiroler Dolomitriffe. Ein Beitrag zur Ökologie der Lithophyten. Ber Dtsch Bot Gesellschaft 32:502–526

Dor I, Danin A (1996) Cyanobacterial desert crusts in the Dead Sea Valley, Israel. Algol Stud 83:197–206

Eldridge DJ (2001) Biological soil crusts of Australia. In: Lange OL, Belnap J (eds) Biological soil crusts. Ecological studies. Springer, Berlin Heidelberg New York, pp 119–131

Eldridge DJ, Green RSB (1994) Microbiotic soil crusts: a review of their roles in soil and ecological processes in the rangelands of Australia. Aust J Soil Res 32:389–415

Eldridge DJ, Koen TB (1998) Cover and floristics of microphytic soil crusts in relation to indices of landscape health. Plant Ecol 137:101–114

Eldridge DJ, Tozer ME (1996) Distribution and floristics of bryophytes in soil crusts in semi-arid and arid eastern Australia. Aust J Bot 44:223–247

Evans RD, Johansen JR (1999) Microbiotic crusts and ecosystem processes. Crit Rev Plant Sci 18:183–225

Freiberg E (1998a) Influence of microclimate on the occurrence of cyanobacteria in the phyllosphere in a premontane rain forest of Costa Rica. Plant Biol 1:244–252

Freiberg E (1998b) Microclimatic parameters influencing nitrogen fixation in the phyllosphere in a Costa Rican premontane rain forest. Oecologia 17:9–18

Frey E, Kürschner H (1991) Lebenstrategien von terrestrischen Bryophyten in der Judäischen Wüste. Bot Acta 104:172–182

Friedmann EI (1980) Endolithic microbial life in hot and cold deserts. Origins Life 10:223–235

Friedmann EI, Galun M (1974) Desert algae, lichens and fungi.. In: Brown GW (ed) Desert biology, vol II. Academic Press, London, pp 165–212

Friedmann EI, Lipkin Y, Ocampo-Paus R (1967) Desert algae of the Negev (Israel). Phycologia 6:185–200

Friedmann EI, Hua M, Ocampo-Friedmann R (1988) Cryptoendolithic lichen and cyanobacterial communities of the Ross Desert, Antarctica. Polarforschung 58:251–259

Galun M, Garty J (2001) Biological soil crusts of the Middle East. In: Lange OL, Belnap J (eds) Biological soil crusts. Ecological studies. Springer, Berlin Heidelberg New York, pp 95–106

Garcia-Pichel F, Belnap J (1996) Microenvironments and microscale productivity of cyanobacterial desert crusts. J Phycol 32: 774–782

Garty J (1989) Influence of epilithic microorganisms on the surface temperature of building walls. Can J Bot 68:1349–1353

Golubic S (1967a) Algenvegetation der Felsen. Eine ökologische Algenstudie im dinarischen Karstgebiet. In: Elster HJ, Ohle W (eds) Die Binnengewässer, vol 23:1–183

Golubic S (1967b) Die Algenvegetation an Sandsteinfelsen Ost-Venezuelas (Cumaná). Int Rev Hydrobiol 52:693–699

Golubic S, Friedmann EI, Schneider J (1981) The lithobiontic ecological niche, with special reference to microorganisms. J Sediment Petrol 51:475–478

Green TGA, Broady P (2001) Biological soil crusts of Antarctica. In: Lange OL, Belnap J (eds) Biological soil crusts. Ecological studies. Springer, Berlin Heidelberg New York, pp 142–139

Green TGA, Meyer A, Büdel B, Zellner H, Lange OL (1995) Diel patterns of CO_2 exchange for six lichens from a temperate rain forest in New Zealand. Symbiosis 18:251–273

Hahn A, Kusserow H (1998) Spatial and temporal distribution of algae in soil crusts in the Sahel of W Africa: preliminary results. Wildenowia 28:227–238

Hale ME (1987) Epilithic lichens in the beacon sandstone formation, Victorialand, Antarctica. Lichenologist 19:269–287

Hambler DJ (1964) The vegetation of granitic outcrops in western Nigeria. J Ecol 52:573–594

Hansen ES (2001) Lichen-rich soil crusts of Arctic Greenland. In: Lange OL, Belnap J (eds) Biological soil crusts. Ecological studies. Springer, Berlin Heidelberg New York, pp 57–65

Isichei AO (1980) Nitrogen fixation by blue-green algal soil crusts in Nigerian savanna In: Rosswall T (ed) Nitrogen cycling in West Africa ecosystems. Royal Swedish Acad Sci Stockholm, pp 191–198

Jaag O (1945) Untersuchungen über die Vegetation und Biologie der Algen des nackten Gesteins in den Alpen, im Jura und im schweizerischen Mittelland. Beitr Kryptogamenflora Schweiz 9:1–560

Joubert JJ, Steyn PL, Britz TJ, Wessles DCJ (1982) Chemical composition of some lichen species occurring in the Namib Desert, South West Africa. Dinteria 16:33–43

Keeling RF, Piper SC, Heimann M (1996) Global and hemispheric CO_2 sinks deduced from changes in atmospheric O_2 concentration. Nature 381:218–221

Kidron G (1995) The impact of microbial crust upon rainfall-runoff-sediment yield relationships on longitudinal dune slopes, Nizzana, western Negev desert, Israel. PhD Thesis, The Hebrew University of Jerusalem (English summary)

Komáromy ZP (1976) Soil algal growth types as edaphic adaptations in Hungarian forest and grass steppe ecosystems. Acta Bot Acad Sci Hung 22:373–379

Krumbein WE, Jens K (1981) Biogenic rock varnishes of the Negev Desert (Israel), an ecological study of iron and manganese transformation by cyanobacteria and fungi. Oecologia 50:25–38

Lange OL (2001) Photosynthesis of soil-crust biota as dependent on environmental factors. In: Lange OL, Belnap J (eds) Biological soil crusts. Ecological studies. Springer, Berlin Heidelberg New York, pp 217–240

Lange OL, Kidron GJ, Büdel B, Meyer A, Kilian E, Abeliovich A (1992) Taxonomic composition and photosynthetic characteristics of the 'biological soil crusts' covering sand dunes in the western Negev Desert. Funct Ecol 6:519–527

Lange OL, Meyer A, Büdel B (1994a) Net-photosynthesis of a desiccated cyanobacterium without liquid water in high air humidity alone. Experiments with *Microcoelus sociatus* isolated from a desert soil crust. Funct Ecol 8:52–57

Lange OL, Meyer A, Zellner H, Heber U (1994b) Photosynthesis and water relations of lichen soil crusts: field measurements in the coastal fog zone of the Namib Desert. Funct Ecol 8:253–264

Lange OL, Belnap J, Reichenberger H, Meyer A (1997) Photosynthesis of green algal soil crust lichens from arid lands in southern Utah, USA: role of water content on light and temperature responses of CO_2 exchange. Flora 192:1–15

Lange OL, Belnap J, Reichenberger H (1998) Photosynthesis of the cyanobacterial soil-crust lichen *Collema tenax* from arid lands in southern Utah, USA: role of water content on light and temperature responses of CO_2 exchange. Funct Ecol 12:195–202

Lange OL, Kidron GJ, Büdel B, Meyer A, Kilian E, Abeliovich A (1992) Taxonomic composition and photosynthetic characteristics of the 'biological soil crusts' covering sand dunes in the western Negev Desert. Funct Ecol 6:519–527

Larcher W (1994) Ökophysiologie der Pflanzen. Verlag Eugen Ulmer, Stuttgart

Lücking R (1992) Zur Verbreitungsökologie foliikoler Flechten in Costa Rica, Zentralamerika. Nova Hedwigia 54:309–353

Lücking R, Becker U, Follmann G (1998) Foliikole Flechten aus dem Tai-Nationalpark, Elfenbeinküste (Tropisches Afrika). II. Ökologie und Biogeografie. Herzogia 13:207–228

Lüttge U (1997) Cyanobacterial Tintenstrich communities and their ecology. Naturwissenschaften 84:526–534

Malam Issa O (1998) Role of microbiotic soil crusts in two sahelian ecosystems (fallow lands and tiger bush) of Niger. Micromorphology, physical and biogeochemical properties. Diss, CNRS-Université d'Orleans

Malam Issa O, Trichet J, Défarge C, Couté A, Valentin C (1999) Morphology and microstructure of microbiotic soil crusts on a tiger bush sequence (Niger, Sahel). Catena 37:175–196

Miszalski Z, Büdel B, Lüttge U (1995) Sensitivity of terrestrial cyanobacteria to light and sulphite stress. Polish J Environ Stud 4:55–59

Moore PD (1998) Life in the upper crust. Nature 393:419–420

Novichkova-Ivanova LN (1980) Soil algae of the phytocoenoses of the Saharan-Gobi desert region. Nauka, Leningrad (in Russian)

Paus SM (1997) Die Erdflechtenvegetation Nordwestdeutschlands und einiger Randgebiete. Vegetationsökologische Untersuchungen unter besonderer Berücksichtigung des Chemismus ausgewählter Arten. Biblio Lichenol 66:1–222

Prasse R (1999) Experimentelle Untersuchungen an Gefässpflanzenpopulationen auf verschiedenen Geländeoberflächen in einem Sandwüstengebiet. Universitätsverlag Rasch, Osnabrück

Reynaud PA, Lumpkin TA (1988) Mikroalgae of the Lanzhou (China) cryptogamic crust. Arid Soil Res Rehab 2:145–155

Rosentreter R, Belnap J (2001) Biological soil crusts of North America. In: Lange OL, Belnap J (eds) Biological soil crusts. Ecological studies. Springer, Berlin Heidelberg New York, pp 31–50

Rummrich U, Rummrich M, Lange-Bertalot H (1989) Diatomeen als "Fensteralgen" in der Namib-Wüste und anderen ariden Gebieten von SWA/Namibia. Dinteria 20:23–29

San Jose JJ, Bravo CR (1991) CO_2 exchange in soil algal crusts occurring in the trachypogon savannas of the Orinoco Llanos, Venezuela. Plant Soil 135:233–244
Sarthou C, Thérézien Y, Couté A (1995) Cyanophycées de l'inselberg des Nourages (Guyane francaise). Nova Hedwigia 61:85–109
Schlesinger WH (1997) Biogeochemistry, 2nd edn. Academic Press, San Diego
Stal LJ (2000) Cyanobacterial mats and stromatolites. In: Whitton BA, Potts W (eds) The ecology of cyanobacteria. Kluwer, Dordrecht, pp 61–120
Türk R, Gärtner G (2001) Biological soil crusts of the subalpine, alpine and nival areas in the Alps. In: Lange OL, Belnap J (eds) Biological soil crusts. Ecological studies. Springer, Berlin Heidelberg New York, pp 67–73
Ullmann I, Büdel B (2001a) Biological soil crusts of Africa. In: Lange OL, Belnap J (eds) Biological soil crusts. Ecological studies. Springer, Berlin Heidelberg New York, pp 107–118
Ullmann I, Büdel B (2001b) Ecological determinants of species composition of biological soil crusts on a landscape scale. In: Lange OL, Belnap J (eds) Biological soil crusts. Ecological studies. Springer, Berlin Heidelberg New York, pp 203–213
Vincent WF (2000) Cyanobacterial dominance in the polar regions. In: Whitton BA, Potts W (eds) The ecology of cyanobacteria. Kluwer, Dordrecht, pp 321–340
Vogel S (1955) Niedere "Fensterpflanzen"in der südafrikanischen Wüste. Eine ökologische Schilderung. Beitr Biol Pflanz 31:45–135
Watanabe Y, Martini JEJ, Ohmoto H (2000) Geochemical evidence for terrestrial ecosystems 2.6 billion years ago. Nature 408: 574–578
Weber B, Wessels DCJ, Büdel B (1996) Biology and ecology of cryptoendolithic cyanobacteria of a sandstone plateau in North-Transvaal, South Africa. Algol Stud 83: 565–579
Weber HM (1997) Ein Biofilm auf freiem Fels: Cyanobakterien der Inselberge Brasiliens. Diploma Thesis, Universities of Bonn and Rostock
Welwitsch F (1868) The Pedras Negras of Pungo Andongo in Angola. J Travel Nat Hist 1:22–36
Wessels DCJ, Büdel B (1989) A rockpool lichen community in Northern Transvaal, South Africa: composition and distribution patterns. Lichenologist 21:259–277
Wessels DCJ, Büdel B (1995) Epilithic and cryptoendolithic cyanobacteria of Clarens sandstone cliffs in the Golden Gate Highlands National Park, South Africa. Bot Acta 108:220–226
Wessels DCJ, Wessels LA, Holzapfel WH (1979) Preliminary report on lichen-feeding Coleoptera occurring on *Teloschistes capensis* in the Namib desert, South West Africa. Bryologist 82:270–273
Whitton BA, Potts W (eds) (2000) The ecology of cyanobacteria, Kluwer, Dordrecht
Wynn-Williams DD (2000) Cyanobacteria in deserts – life at the limit? In: Whitton BA, Potts W (eds) The ecology of cyanobacteria. Kluwer, Dordrecht, pp 341–366

Prof. Dr. Burkhard Büdel
Lehrstuhl für Allgemeine Botanik
Universität Kaiserslautern
FB Biologie
Postfach 3049
67653 Kaiserslautern, Germany
Tel.: +49-0631-205-2360
Fax: +49-0631-205-2998
e-mail: buedel@rhrk.uni-kl.de

Subject Index

Printing: Saladruck, Berlin
Binding: H. Stürtz AG, Würzburg